Partial Differential Equations and Boundary Value Problems with Maple

Second Edition

Partial Differential Equations and Boundary Value Problems with Maple

Second Edition

George A. Articolo

AMSTERDAM • BOSTON • HEIDELBERG • LONDON
NEW YORK • OXFORD • PARIS • SAN DIEGO
SAN FRANCISCO • SINGAPORE • SYDNEY • TOKYO

Academic Press is an imprint of Elsevier

Academic Press is an imprint of Elsevier
30 Corporate Drive, Suite 400, Burlington, MA 01803, USA
525 B Street, Suite 1900, San Diego, California 92101-4495, USA
84 Theobald's Road, London WC1X 8RR, UK

Library of Congress Cataloging-in-Publication Data
Articolo, George A.
   Partial differential equations and boundary value problems with Maple/George A. Articolo. – 2nd ed.
      p. cm.
   Includes bibliographical references and index.
   ISBN 978-0-12-374732-7 (pbk. : alk. paper) 1. Differential equations, Partial—Data processing.
2. Boundary value problems—Data processing. 3. Maple (Computer file) I. Title.
   QA377.A82 2009
   515'.3530285–dc22

                                                                                    2009010098

British Library Cataloguing-in-Publication Data
A catalogue record for this book is available from the British Library.

ISBN: 978-0-12-374732-7

For information on all Academic Press publications
visit our Web site at *www.elsevierdirect.com*

Typeset by: diacriTech, India.

Printed in the United States
09 10 11   9 8 7 6 5 4 3 2 1

Contents

Preface

This is the second edition of the text *Partial Differential Equations and Boundary Value Problems with Maple*, Academic Press, 1998. The text has been updated from Maple release 4 to release 12. In addition, based on recommendations and suggestions of the many helpful reviewers of the first edition, the text incorporates more of the macro commands in Maple to be used as a means of checking solutions. Similar to what was done in the first edition, I continued the presentation of the solutions to problems using the traditional, fundamental, mathematical approach so that the student gets a firm understanding of the mathematical basis of the development of the solutions. The macro commands are not intended to be used as a means of teaching the mathematics—they are used only as a quick means of checking.

If there ever were to be a perfect union in computational mathematics, one between partial differential equations and powerful software, Maple would be close to it. This text is an attempt to join the two together.

Many years ago, I recall sitting in a partial differential equations class when the professor was discussing a heat-flow boundary value problem. Using a piece of chalk at the blackboard, he was making a seemingly desperate attempt to get his students to visualize the spatial-time development of the three-dimensional surface temperature of a plate that was allowed to cool down to a surrounding equilibrium temperature. You can imagine the frustration that he, and many professors before him, experienced at doing this task. Now, with the powerful computational tools and graphics capabilities at hand, this era of difficulty is over.

This text presents the formal mathematical concepts needed to develop solutions to boundary value problems, and it demonstrates the capabilities of Maple software as being a powerful computational tool. The graphics and animation commands allow for accurate visualization of the spatial-time development of the solutions on the computer screen—what students could only imagine many years ago can now be viewed in real time.

The text is targeted for use by senior-graduate level students and practitioners in the disciplines of physics, mathematics, and engineering. Typically, these people have already had some exposure to courses in basic physics, calculus, linear algebra, and ordinary differential equations. The need for previous exposure to the Maple software is not necessary. In Chapter 0, we provide an introduction to some simple Maple commands, which is all that is necessary for the reader to move on successfully into the text. In addition, we also review those important yet simple concepts from linear algebra and ordinary differential equations that are essential to understanding the development of solutions to partial differential equations.

The basic approach to teaching this material is very traditional. The main goal is to teach the fundamental mathematical procedures for developing solutions and to use the computer only as a tool. We do not want the computer to do all the work for us—this would defeat our purpose here. For example, in Chapter 1 we spend time developing solutions to first- and second-order differential equations in a manner similar to that found in a typical course in ordinary differential equations. There are simple Maple commands that can solve such problems with a single line of code. We do not use that approach here. Instead, we present the material in a traditional way (read: before computers) so that, first and foremost, the student learns the formal mathematics needed to develop and understand the solution. Traditionalist professors of mathematics would certainly welcome this more fundamental approach.

What is the purpose of using Maple here? Basically, we make use of the language to perform the tedious tasks of integration and graphics. The Maple code for doing these tasks is so intuitive and easy to remember that students, practitioners, and professors become experts almost immediately. Thus, use of this powerful computer software frees up our resources so that we can spend more time being mathematically creative.

There are ten chapters in the text and each one stands out as a self-contained unit on presenting the fundamental mathematical concepts followed by the equivalent Maple code for developing solutions to example problems. Each chapter looks at example problems and develops the mathematical solution to the problem first before presenting the Maple solution. In this manner, the student first learns the fundamental formal mathematical procedures for developing a solution. After seeing the equivalent Maple solution, the student then makes an easy transition in learning to use Maple as a powerful computational tool. Eventually, the student can interact directly with the software to solve the exercise problems.

Chapter 1 is dedicated to ordinary linear differential equations. Traditionally, before one can understand what a partial differential equation is, one must first understand what an ordinary differential equation is. We examine first- and second-order differential equations, introducing the important concept of basis vectors. Closed form solutions in addition to series solutions of differential equations are presented here.

Chapter 2 is dedicated to Sturm-Liouville eigenvalue problems and generalized Fourier series. These concepts are introduced very early in the text because they are so important to the development of solutions to boundary value problems in all of the later chapters. This very early introduction to Sturm-Liouville problems and series expansions in terms of sets of orthonormal eigenvectors, for both rectangular and cylindrical coordinate systems, makes this text singularly different from most others.

Chapters 3, 4, and 5 deal with three of the most famous partial differential equations—the diffusion or heat equation in one spatial dimension, the wave equation in one spatial dimension, and the Laplace equation in two spatial dimensions. Chapters 6 and 7 expand coverage of the diffusion and wave equation to two spatial dimensions. Chapter 8 examines the nonhomogeneous versions of the diffusion and wave equations in a single spatial dimension. Chapter 9 considers partial differential equations over infinite and semi-infinite domains, and Chapter 10 examines Laplace transform methods of solution.

Each chapter contains an extensive set of example problems in addition to an exhaustive array of exercise problems that present a challenge to understanding the material.

I would like to thank my many colleagues at Rutgers University who were very encouraging in the development of this text. Also, special thanks to the first edition editor, Charles B. Glaser, of Academic Press, and to the second edition editor, Lauren Schultz Yuhasz, of Elsevier, for their support and cooperation.

George A. Articolo

CHAPTER 0
Basic Review

0.1 Preparation for Maple Worksheets

This book combines the traditional presentation of fundamental mathematical concepts with the contemporary computational benefits of Maple software.

Each chapter begins with a basic preview of the relevant mathematics needed to understand and solve problems. The mathematics is presented in a style that is typical of the approach found in most traditional textbooks that deal with partial differential equations. Following the presentation of basic mathematical concepts, the corresponding Maple worksheets are constructed showing the development of the solutions to typical problems using Maple.

This chapter introduces some of the basic operational procedures of the Maple language as seen in the worksheets. This brief introduction provides enough information about the code so the reader can set up new worksheets to solve new problems. More insight into the finer details of Maple can be found in the many excellent texts dedicated to Maple (see references).

Much effort was put into using a minimal number of commands and avoiding procedures that might not be familiar to those who have never used Maple before. In most instances, generic commands like those used in traditional-style textbooks are used. Our main purpose of this book is to teach mathematics, not new code.

Some of the Maple macro commands are not used in this book, since they do not support learning mathematics. For example, in the solution to ordinary differential equations, Maple has a command "dsolve" that practically solves the entire problem. Instead, the solution is developed in a manner that is typical of that found in traditional-style mathematics texts. The procedures presented here are the fundamental basis of the methods that constitute the Maple "dsolve" command and thus provide an explanation of the code behind the command. It is our belief that in doing things this way, our primary focus is on learning the mathematics and using the Maple code only for computational ease and to use some of the Maple macro commands only as a means of checking our solutions. If we were to use the all-inclusive macro commands exclusively, we would be guilty of raising a generation of mathematicians who do not understand the basics.

Throughout the text, we use standard arithmetic operations such as addition, subtraction, multiplication, division, and exponentiation. In addition, we also use more sophisticated operations like differentiation, integration, summation, and factorization.

The easiest way to learn the commands for these operations is to set up examples and visualize the resulting Maple output. In the following examples, we see how the Maple command line is constructed. At the left side of the command line is the "prompt" symbol, >. In the middle of the line is the command equality statement, := (a colon followed by an equal sign). The command equality statement declares the value of what is to the left of the statement as in typical algebra. The extreme right end of the command line has either a semicolon or a colon. When a colon is used, the line is not printed out, whereas with a semicolon, the line is printed.

We now illustrate with examples. Observations of these examples provide enough learning experience to deal with almost anything in the text. It should be noted that all of the Maple material was developed using Release 12.

If we want to write out the function $f(x) = x^2$ as a Maple command, we use the Maple input prompt > (this is found in the Insert menu), and we write

> f(x):=x^2:

Note that there is no printout of the preceding command because of the colon at the end. To get a printout, we replace the colon with a semicolon:

> f(x):=x^2;

$$f(x) := x^2 \qquad (0.1)$$

Similarly, for $g(x) = x^3$ we write

> g(x):=x^3;

$$g(x) := x^3 \qquad (0.2)$$

The sum of $f(x)$ and $g(x)$ is

> f(x)+g(x);

$$x^2 + x^3 \qquad (0.3)$$

The product

> f(x)*g(x);

$$x^5 \qquad (0.4)$$

The quotient

> f(x)/g(x);

$$\frac{1}{x} \qquad (0.5)$$

The derivative of $f(x)$ with respect to x:

`> diff(f(x),x);`

$$2x \tag{0.6}$$

The indefinite integral of $g(x)$ written symbolically:

`> Int(g(x),x);`

$$\int x^3 \, dx \tag{0.7}$$

The evaluated indefinite integral of $g(x)$

`> int(g(x),x);`

$$\frac{1}{4} x^4 \tag{0.8}$$

The definite integral of $f(x)$ over the finite closed interval $[1, 4]$ written symbolically:

`> Int(f(x),x=1..4);`

$$\int_1^4 x^2 \, dx \tag{0.9}$$

The evaluated definite integral of $f(x)$ over the interval $[1, 4]$:

`> int(f(x),x=1..4);`

$$21 \tag{0.10}$$

Factorization:

`> factor(x^2−x−2);`

$$(x+1)(x-2) \tag{0.11}$$

Substitution:

`> g(2):=subs(x=2,g(x));`

$$g(2) := 8 \tag{0.12}$$

Summation:

`> S:=Sum(n*x,n=1..3);`

$$S := \sum_{n=1}^{3} nx \tag{0.13}$$

Evaluation of the previous command:

> S:=value(%);

$$S := 6x \qquad (0.14)$$

We must be aware of three other important items when using Maple worksheets. When we want to declare new values of variables in an entire problem, we wipe out all the previous declarations of these values by using the simple command

> restart:

When we want to use special computational packages that facilitate the use of Maple for specific applications, we must bring these packages into the worksheet area by using a specific command. For example, to implement the graphics capability of Maple, we bring the "plot" package into the worksheet area by using the command

> with(plots):

To implement the integral transform commands into the worksheet, such as Fourier and Laplace and their corresponding inverses, we bring the "transform" package into the worksheet using the Maple command

> with(inttrans):

Generally, the commands to bring in special packages are made at the very beginning of the Maple worksheet.

The preceding command-operations cover most of those in the text that use the Maple code. There are other commands that are also valuble and will become apparent in the development of the solutions of particular problems. Note the minimal number of operations and the adherence to traditional style. Mastery of the preceding concepts, at the very beginning, will set aside any problems we might have later with the code, allowing us to focus primarily on the mathematics. Please be aware that different versions or releases of Maple have different characteristics in entering commands and that the Maple help section should be read and used to resolve any difficulties.

0.2 Preparation for Linear Algebra

A linear vector space consists of a set of vectors or functions and the standard operations of addition, subtraction, and scalar multiplication. In solving ordinary and partial differential equations, we assume the solution space to behave like an ordinary linear vector space. A primary concern is whether or not we have enough of the correct vectors needed to span the solution space completely. We now investigate these notions as they apply directly to two-dimensional vector spaces and differential equations.

We use the simple example of the very familiar two-dimensional Euclidean vector space R2; this is the familiar (x, y) plane. The two standard vectors in the (x, y) plane are traditionally denoted as i and j. The vector i is a unit vector along the x-axis, and the vector j is a unit vector along the y-axis. Any point in the (x, y) plane can be reached by some linear combination, or superposition, of the two standard vectors i and j. We say the vectors "span" the space. The fact that only two vectors are needed to span the two-dimensional space R2 is not coincidental; three vectors would be redundant. One reason for this has to do with the fact that the two vectors i and j are "linearly independent"—that is, one cannot be written as a multiple of the other. The other reason has to do with the fact that in an n-dimensional Euclidean space, the minimum number of vectors needed to span the space is n.

A more formal mathematical definition of linear independence between two vectors or functions $v1$ and $v2$ reads as "The two vectors $v1$ and $v2$ are linearly independent if and only if the only solution to the linear equation

$$c1\,v1 + c2\,v2 = 0$$

is that both $c1$ and $c2$ are zero." Otherwise, the vectors are said to be linearly dependent.

In the simple case of the two-dimensional (x, y) space R2, linear independence can be geometrically understood to mean that the two vectors do not lie along the same direction (noncolinear). In fact, any set of two noncolinear vectors could also span the vector space of the (x, y) plane. There are an infinite number of sets of vectors that will do the job. One common connection between all sets, however, is that all the sets can be shown to be linearly dependent; that is, all the sets can be shown to be reducible to linear combinations of the standard i and j vectors.

For example, the two vector sets

$$S1 = \{i, j\}$$

and

$$S2 = \{i + j, 2\,i - 3\,j\}$$

are both linearly independent sets of vectors that span the two-dimensional (x, y) space. Note that the vectors within each set are linearly independent, but the vectors between sets are linearly dependent.

A set of vectors $S = \{v1, v2, v3, \ldots, vn\}$ that are linearly independent and that span the space is called a set of "basis" vectors for that particular vector space. Thus, for the two-dimensional Euclidean space R2, the vectors i and j form a basis, and for the three-dimensional Euclidean space R3, vectors i, j, and k form a basis. The number of vectors in a basis is called the "dimension" of the vector space.

A set of basis vectors is fundamental to a particular vector space because any vector in that space can then be written as a unique superposition of those basis vectors. These concepts are important to us when we consider the solution space of both ordinary and partial differential equations. Another important concept in linear algebra is that of the inner product of two vectors in that particular vector space.

For the Euclidean space R3, if we let u and v be two different vectors in this space with components

$$u = [u1, u2, u3]$$

and

$$v = [v1, v2, v3]$$

then the inner product of these two vectors is given as

$$ip(u, v) = u1\, v1 + u2\, v2 + u3\, v3$$

Thus, the inner product is the sum of the product of the components of the two vectors. The inner product is sometimes also referred to as the "dot product."

If we take the square root of an inner product of a vector with itself, then we are evaluating the length of the vector, commonly called the "norm."

$$norm(u) = \sqrt{ip(u, u)}$$

Different vector spaces have different inner products. For example, we consider the vector space C[a, b] of all functions that are continuous over the finite closed interval [a, b]. Let $f(x)$ and $g(x)$ be two different vectors in this space. The inner product of these two vectors over the interval, with respect to the weight function $w(x)$, is defined as the definite integral:

$$ip(f, g) = \int_a^b f(x)\, g(x)\, w(x)\, dx$$

From the basic definition of a definite integral, we see the inner product to be an (infinite) sum of the product of the components of the two vectors.

Similarly, in the space of continuous functions, if we take the square root of the inner product of a vector with itself, then we evaluate the length or norm of the vector to be

$$norm(f) = \sqrt{\int_a^b f(x)^2\, w(x)\, dx}$$

As an example, consider the two functions $f(x) = \sin(x)$ and $g(x) = \cos(x)$ over the finite closed interval $[0, \pi]$ with a weight function $w(x) = 1$. The length or norm of $f(x)$ is the definite integral

$$norm(f) = \sqrt{\int_0^\pi \sin(x)^2 \, dx}$$

which evaluates to

$$norm(f) = \sqrt{\frac{\pi}{2}}$$

Similarly, for $g(x)$ the norm is the definite integral

$$norm(g) = \sqrt{\int_0^\pi \cos(x)^2 \, dx}$$

which evaluates to

$$norm(g) = \sqrt{\frac{\pi}{2}}$$

If we evaluate the inner product of the two functions $f(x)$ and $g(x)$, we get the definite integral

$$ip(f, g) = \int_0^\pi \cos(x) \sin(x) \, dx$$

which evaluates to

$$ip(f, g) = 0$$

If the inner product between two vectors is zero, we say the two vectors are "orthogonal" to each other. Orthogonal vectors can also be shown to be linearly independent.

If we divide a vector by its length or norm, then we "normalize" the vector. For the preceding $f(x)$ and $g(x)$, the corresponding normalized vectors are

$$f(x) = \sqrt{\frac{2}{\pi}} \sin(x)$$

and

$$g(x) = \sqrt{\frac{2}{\pi}} \cos(x)$$

A set that consists of vectors that are both normal and orthogonal is said to be an "orthonormal" set. For orthonormal sets, the inner product of two vectors in the set gives the value 1 if the vectors are alike or the value 0 if the vectors are not alike.

Two vectors $\varphi_n(x)$ and $\varphi_m(x)$, which are indexed by the positive integers n and m, are orthonormal with respect to the weight function $w(x)$ over the interval $[a, b]$ if the following relation holds:

$$\int_a^b \varphi_n(x)\varphi_m(x)\,w(x)\,dx = \delta(n, m)$$

Here, $\delta(n, m)$ is the familiar Kronecker delta function whose value is 0 if $n \neq m$ and is 1 if $n = m$.

Orthonormal sets play a big role in the development of solutions to partial differential equations.

0.3 Preparation for Ordinary Differential Equations

An ordinary linear homogeneous differential equation of the second order has the form

$$a2(x)\left(\frac{d^2}{dx^2}y(x)\right) + a1(x)\left(\frac{d}{dx}y(x)\right) + a0(x)\,y(x) = 0$$

Here, the coefficients $a2(x)$, $a1(x)$, and $a0(x)$ are functions of the single independent variable x, and y is the dependent variable of the differential equation. We say the differential equation is "normal" over some finite interval I if the leading coefficient $a2(x)$ is never zero over that interval.

Recall that the second derivative of a function is a measure of its concavity, the first derivative is a measure of its slope, and the zero derivative is a measure of its magnitude. Thus, the solution $y(x)$ to the above second-order differential equation is that function whose concavity multiplied by $a2(x)$, plus the slope multiplied by $a1(x)$, plus the magnitude multiplied by $a0(x)$ must all add up to zero. Finding solutions to such differential equations is standard material for a course in differential equations.

For now, we state some fundamental theorems about the solution space of ordinary differential equations.

Theorem 0.1: On any interval I, over which the nth-order linear ordinary homogeneous differential is normal, the solution space is of finite dimension n and there exist n linearly independent solution vectors $y1(x), y2(x), y3(x), \ldots, yn(x)$.

Theorem 0.2: If $y1(x)$ and $y2(x)$ are two solutions to a linear second-order differential equation over some interval I, and the Wronskian of these two solutions does not equal zero anywhere over this interval, then the two solutions are linearly independent and form a set of "basis" vectors.

From differential equations, the second-order Wronskian of the two vectors $y1(x)$ and $y2(x)$ is defined as

$$W(y1(x), y2(x)) = y1(x)\left(\frac{d}{dx}y2(x)\right) - y2(x)\left(\frac{d}{dx}y1(x)\right)$$

Similar to what we do in linear algebra, with a set of basis vectors in hand, we can span the solution space and write any solution vector as a linear combination of these basis vectors.

As a simple example, one that is analogous to the two-dimensional Euclidean space R2, we consider the solution space of the linear second-order homogeneous differential equation

$$\frac{d^2}{dx^2}y(x) + y(x) = 0$$

The preceding differential equation is referred to as an "Euler"-type differential equation. As can easily be verified, two solution vectors are the Euler functions $y1(x) = \cos(x)$ and $y2(x) = \sin(x)$. Are the vectors linearly independent? If we evaluate the Wronskian of this set, we get

$$W(y1(x), y2(x)) = \cos(x)^2 + \sin(x)^2$$

Since the Wronskian is never equal to zero and the differential equation is normal everywhere, the two vectors form a basis for the solution space of the differential equation. Thus, the set

$$S1 = \{\cos(x), \sin(x)\}$$

is a basis for the solution space of this particular differential equation.

In terms of this basis, we can span the solution space and write the general solution to the preceding differential equation as

$$y(x) = C1\cos(x) + C2\sin(x)$$

where $C1$ and $C2$ are arbitrary constants.

It can be verified that another equivalent basis set is

$$S2 = \{e^{ix}, e^{-ix}\}$$

Similar to the Euclidean spaces discussed earlier, the two sets $S1$ and $S2$ contain two vectors that are linearly independent; however, the sets themselves are linearly dependent. This follows from the familiar Euler formulas

$$\sin(x) = \frac{e^{ix} - e^{-ix}}{2i}$$

and

$$\cos(x) = \frac{e^{ix} + e^{-ix}}{2}$$

With a set of basis vectors in hand, we can write any solution to a linear differential equation as a linear superposition of these basis vectors.

0.4 Preparation for Partial Differential Equations

Partial differential equations differ from ordinary differential equations in that the equation has a single dependent variable and more than one independent variable. We focus on three main types of partial differential equations in this text, all linear.

1. The heat or diffusion equation (first-order derivative in time t, second-order derivative in distance x)

$$\frac{\partial}{\partial t} u(x, t) = k \left(\frac{\partial^2}{\partial x^2} u(x, t) \right)$$

2. The wave equation (second-order derivative in time t, second-order derivative in distance x)

$$\frac{\partial^2}{\partial t^2} u(x, t) = c^2 \left(\frac{\partial^2}{\partial x^2} u(x, t) \right)$$

3. The Laplace equation (second-order derivative in both distance variables x and y)

$$\frac{\partial^2}{\partial x^2} u(x, y) + \frac{\partial^2}{\partial y^2} u(x, y) = 0$$

We note that in all three cases, we have a single dependent variable u and more than one independent variable. The terms c and k are constants.

For the particular types of partial differential equations we will be looking at, all are characterized by a linear operator, and all of them are solved by the method of separation of variables. A dramatic difference between ordinary and partial differential equations is the dimension of the solution space. For ordinary differential equations, the dimension of the solution space is finite; it is equal to the order of the differential equation. For partial differential equations with spatial boundary conditions, the dimension of the solution space is infinite. Thus, a basis for the solution space of a partial differential equation consists of an infinite number of vectors. As an example, consider the diffusion equation

$$\frac{\partial}{\partial t} u(x, t) = k \left(\frac{\partial^2}{\partial x^2} u(x, t) \right)$$

subject to a given set of spatial boundary conditions. By separation of variables, we assume a solution in the form of a product

$$u(x, t) = X(x)\, T(t)$$

After substitution of the assumed solution into the partial differential equation, we end up with two ordinary differential equations: one whose independent variable is x and one whose independent variable is t.

From the imposition of the given spatial boundary conditions, we find an infinite number of x-dependent solutions that take on the form of eigenfunctions that are indexed by positive integers n and written as

$$X_n(x)$$

for $n = 0, 1, 2, 3, \ldots$.

Similarly, the t-dependent solution can also be indexed by the integer n, and we write the t-dependent solution as

$$T_n(t)$$

Thus, for a given value of n, one solution to the homogeneous partial differential equation, which satisfies the boundary conditions, is given as

$$u_n(x, t) = X_n(x)\, T_n(t)$$

for $n = 0, 1, 2, 3, \ldots$.

Since the partial differential equation operator is linear, any superposition of solutions for all allowed values of n satisfies the partial differential equation and the given boundary conditions. Thus, the set of vectors

$$S = \{u_n(x, t)\}$$

for $n = 0, 1, 2, 3, \ldots$, forms a basis for the solution space of the partial differential equation. Since there are an infinite number of indexed solutions, we say the basis of the solution space is "infinite." Similar to what we do for ordinary differential equations, we can write the general solution to the problem as a superposition of the allowed basis vectors—that is,

$$u(x, t) = \sum_{n=0}^{\infty} u_n(x, t)$$

The following chapters provide the steps for solving partial differential equations with boundary conditions.

CHAPTER 1

Ordinary Linear Differential Equations

1.1 Introduction

We discuss ordinary linear differential equations in general by initially focusing on the second-order equation. We begin by considering a normal, linear, second-order, nonhomogeneous differential equation on some interval I.

$$a2(t) \left(\frac{d^2}{dt^2} y(t) \right) + a1(t) \left(\frac{d}{dt} y(t) \right) + a0(t) y(t) = f(t)$$

By "normal" we mean that the leading coefficient $a2(t)$ is not equal to zero anywhere on the interval I. The coefficients $a2(t)$, $a1(t)$, and $a0(t)$ are, in general, functions of the independent variable t. The solution $y(t)$ denotes the single dependent variable y to be a function of the single independent variable t.

The function $f(t)$ is generally referred to as the "driving" or external "source" function. If we set $f(t) = 0$, we get the corresponding homogeneous differential equation

$$a2(t) \left(\frac{d^2}{dt^2} y(t) \right) + a1(t) \left(\frac{d}{dt} y(t) \right) + a0(t) y(t) = 0$$

Basis Vectors

From theorems covered in Chapter 0, the dimension of the solution space of an ordinary linear differential equation is equal to the order of the differential equation. Thus, for a second-order equation, the dimension of the solution space is two, and a set of basis vectors of the system consists of two linearly independent solutions to the corresponding homogeneous differential equation.

We define the linear differential equation operator L acting on $y(t)$ as

$$L(y) = a2(t) \left(\frac{d^2}{dt^2} y(t) \right) + a1(t) \left(\frac{d}{dt} y(t) \right) + a0(t) y(t)$$

A basis of the system consists of two solution vectors $y1(t)$ and $y2(t)$, which are linearly independent and each of which satisfies the corresponding homogeneous equations $L(y1) = 0$ and $L(y2) = 0$. There are an infinite number of legitimate sets of basis vectors of the system. However, all of the sets can be shown to be linearly dependent on each other.

The test for linear independence of the two vectors $y1(t)$ and $y2(t)$ on the interval I is that the Wronskian $W(y1(t), y2(t))$ does not equal zero at any point on the interval. Recall, the Wronskian $W(y1(t), y2(t))$ is given as

$$W(y1(t), y2(t)) = y1(t)\left(\frac{d}{dt}y2(t)\right) - y2(t)\left(\frac{d}{dt}y1(t)\right)$$

If $y1(t)$ and $y2(t)$ are solutions of the homogeneous differential equation—that is, $L(y1) = 0$ and $L(y2) = 0$—and their Wronskian $W(y1(t), y2(t))$ does not vanish at any point on the interval, then these two vectors form a basis of the system on that interval.

If L is normal on the interval I and if $y1(t)$ and $y2(t)$ are the basis vectors of the system, then we say these two vectors form a "complete" set. This is equivalent to saying that they completely "span" the solution space to the homogeneous differential equation. Thus, the general solution to the homogeneous equation $y_h(t)$ is given as the linear superposition of the basis vectors

$$y_h(t) = C1\, y1(t) + C2\, y2(t)$$

In the preceding, $C1$ and $C2$ are arbitrary constants.

If we denote a particular solution to the nonhomogeneous differential equation as $y_p(t)$, then the general solution to the nonhomogeneous equation can be written as a sum of the preceding homogeneous solution plus the particular solution

$$y(t) = y_h(t) + y_p(t)$$

We will eventually show that a particular solution to the corresponding nonhomogeneous differential equation can be constructed from a set of basis vectors of the system.

1.2 First-Order Linear Differential Equations

We consider the first-order linear nonhomogeneous differential equation that is normal on an interval I and that has the form

$$a1(t)\left(\frac{d}{dt}y(t)\right) + a0(t)\, y(t) = f(t)$$

The corresponding first-order homogeneous equation can be written as

$$a1(t)\left(\frac{d}{dt}y(t)\right) + a0(t)\, y(t) = 0$$

The single basis vector solution to this homogeneous differential equation is

$$y1(t) = e^{\int \left(-\frac{a0(t)}{a1(t)} \right) dt}$$

DEMONSTRATION: We seek the basis vector solution to the first-order linear homogeneous differential equation

$$t^2 \left(\frac{d}{dt} y(t) \right) + 3ty(t) = 0$$

SOLUTION: We identify the coefficients of the differential equation to be $a1(t) = t^2$ and $a0(t) = 3t$. Note that the differential equation is not normal at the origin. The single system basis vector is given by the integral

$$y1(t) = e^{\int \left(-\frac{3}{t} \right) dt}$$

This evaluates to be

$$y1(t) = \frac{1}{t^3}$$

for $t \neq 0$. We can check this answer by substituting it back into the differential equation.

We now focus on the solution to the corresponding nonhomogeneous differential equation. We assume the solution to the nonhomogeneous differential equation to be a nonlinear multiple of this basis vector, and we set

$$y(t) = u(t) y1(t)$$

Here, $u(t)$ is some unknown, variable function of t [we are forcing a condition of linear independence between $y(t)$ and $y1(t)$]. Substituting this assumed solution into the differential equation, we get

$$a1(t) \left(\frac{d}{dt} u(t) \right) y1(t) + a1(t)u(t) \left(\frac{d}{dt} y1(t) \right) + a0(t)u(t)y1(t) = f(t)$$

Recognizing that $y1(t)$ is a solution to the homogeneous differential equation—that is, $L(y1) = 0$—the last two preceding terms vanish, and we get a first-order equation in $u(t)$ that reads

$$\frac{d}{dt} u(t) = \frac{f(t)}{a1(t) \, y1(t)}$$

A simple integration of the preceding yields

$$u(t) = \int \frac{f(t)}{a1(t) \, y1(t)} dt$$

Thus, a particular solution to the nonhomogeneous differential equation can be written as the product of $u(t)$ and $y1(t)$ shown next:

$$y_p(t) = y1(t) \left(\int \frac{f(t)}{a1(t)\,y1(t)}\,dt \right)$$

If we define the first-order Green's function $G1(t, s)$ as

$$G1(t, s) = \frac{y1(t)}{a1(s)\,y1(s)}$$

then we can express our particular solution in terms of the preceding Green's function as

$$y_p(t) = \int G1(t, s)\,f(s)\,ds \tag{1.1}$$

Writing the solution in terms of the Green's function illustrates its general usefulness in that once the Green's function for a particular differential equation is evaluated, the solution can accommodate any type of driving or source function $f(t)$.

DEMONSTRATION: We seek the solution to the first-order linear nonhomogeneous differential equation

$$t^2 \left(\frac{d}{dt} y(t) \right) + 3\,ty(t) = f(t)$$

SOLUTION: From the preceding, we identify $a1(t) = t^2$ and $a0(t) = 3\,t$. The single basis vector is

$$y1(t) = e^{\int \left(-\frac{a0(t)}{a1(t)} \right) dt}$$

which evaluates to

$$y1(t) = \frac{1}{t^3}$$

From knowledge of the basis vector, we constuct the first-order Green's function

$$G1(t, s) = \frac{y1(t)}{a1(s)\,y1(s)}$$

which evaluates to

$$G1(t, s) = \frac{s}{t^3}$$

The particular solution is the integral whose integrand is the product of the Green's function and the source function written in terms of the dummy variable s:

$$y_p(t) = \int \frac{sf(s)}{t^3} \, ds$$

Thus, the general solution is

$$y(t) = \frac{C1}{t^3} + \int \frac{sf(s)}{t^3} \, ds \tag{1.2}$$

for $t \neq 0$. Here, $C1$ is an arbitrary constant.

Care must be used in evaluating the preceding integral to ensure that we eventually get an answer that is dependent only on t. The correct procedure is to first perform the integration with respect to the dummy variable s and then substitute t back in for s. This is illustrated in the worked-out examples that follow.

As the preceding solution stands, we can appreciate its generality in that it can accommodate any source function $f(t)$. This is in dramatic contrast with the cumbersome method of undetermined coefficients whereby one makes a trial-and-error guess at the solution.

We have completed the development of the solution to all first-order linear differential equations. These solutions will be useful later in the development of solutions to partial differential equations. We now consider some example problems using Maple.

EXAMPLE 1.2.1: We seek a solution to the first-order linear nonhomogeneous differential equation

$$\frac{d}{dt} y(t) + y(t) = e^t$$

SOLUTION: We identify the coefficients and the driving function

> restart: a1(t):=1;a0(t):=1;

$$a1(t) := 1$$
$$a0(t) := 1 \tag{1.3}$$

> f(t):=exp(t);f(s):=subs(t=s,f(t)):

$$f(t) := e^t \tag{1.4}$$

System basis vector

> y1(t):=exp(int(−a0(t)/a1(t),t));

$$y1(t) := e^{-t} \tag{1.5}$$

First-order Green's function

> G1(t,s):=simplify(y1(t)/(subs(t=s,a1(t)*y1(t))));

$$G1(t, s) := e^{-t+s} \tag{1.6}$$

Particular solution (integrate first with respect to s and then substitute s with t)

> y[p](t):=Int(G1(t,s)*f(s),s);

$$y_p(t) := \int e^{-t+s} e^s ds \tag{1.7}$$

> y[p](t):=subs(s=t, value(%));

$$y_p(t) := \frac{1}{2} e^t \tag{1.8}$$

General solution: here, $C1$ is an arbitrary constant.

> y(t):=C1*y1(t)+y[p](t);

$$y(t) := C1e^{-t} + \frac{1}{2} e^t \tag{1.9}$$

Check: Using Maple dsolve command.

> restart:ode:=diff(y(t),t)+y(t)=exp(t):dsolve(ode);

$$y(t) = \frac{1}{2} e^t + e^{-t} \_C1 \tag{1.10}$$

EXAMPLE 1.2.2: We seek the solution to the first-order linear differential equation

$$t \left(\frac{\mathrm{d}}{\mathrm{d}t} y(t) \right) + y(t) = t$$

SOLUTION: We identify the coefficients and the driving function

> restart:a1(t):=t;a0(t):=1;

$$a1(t) := t$$
$$a0(t) := 1 \tag{1.11}$$

> f(t):=t;f(s):=subs(t=s,f(t)):

$$f(t) := t \tag{1.12}$$

System basis vector

> y1(t):=exp(int(−a0(t)/a1(t),t));

$$y1(t) := \frac{1}{t} \tag{1.13}$$

First-order Green's function

> G1(t,s):=simplify(y1(t)/(subs(t=s,a1(t)*y1(t))));

$$G1(t, s) := \frac{1}{t} \tag{1.14}$$

Particular solution (integrate first with respect to s and then substitute s with t)

> y[p](t):=Int(G1(t,s)*f(s),s);

$$y_p(t) := \int \frac{s}{t} \mathrm{d}s \tag{1.15}$$

> y[p](t):=subs(s=t,value(%));

$$y_p(t) := \frac{1}{2} t \tag{1.16}$$

General solution: here $C1$ is an arbitrary constant, and we note that the solution is valid everywhere except at $t = 0$.

> y(t):=C1*y1(t)+y[p](t);

$$y(t) := \frac{C1}{t} + \frac{1}{2} t \tag{1.17}$$

Check: Using Maple dsolve command.

>restart:ode:=t*diff(y(t),t)+y(t)=t:dsolve(ode);

$$y(t) = \frac{1}{2} t + \frac{\_C1}{t} \tag{1.18}$$

1.3 First-Order Initial-Value Problem

Many problems arise in partial differential equations whereby we are confronted with a linear first-order differential equation with an initial condition constraint. A typical problem that we confront is the following nonhomogeneous differential equation, which is generally written in standard form as

$$a1(t) \left(\frac{\mathrm{d}}{\mathrm{d}t} y(t) \right) + a0(t) y(t) = f(t)$$

We seek a solution that satisfies the initial (time $t = 0$) condition $y(0) = y_0$. We now write our basis vector in terms of the following definite integral

$$y1(t) = e^{\int_0^t \left(-\frac{a0(s)}{a1(s)}\right) ds}$$

In Section 1.2, the first-order Green's function was shown to be

$$G1(t, s) = \frac{y1(t)}{a1(s)y1(s)}$$

In terms of the Green's function, the particular solution is

$$y_p(t) = \int_0^t G1(t, s) f(s) \, ds$$

and our final general solution to the initial value problem becomes

$$y(t) = C1 \, y1(t) + \int_0^t G1(t, s) \, f(s) \, ds$$

From the initial condition constraint $y(0) = y_0$, the arbitrary constant $C1$ is evaluated to be

$$C1 = y_0$$

Thus, the final solution, which satisfies the initial condition constraint, is given as

$$y(t) = y_0 e^{\int_0^t \left(-\frac{a0(s)}{a1(s)}\right) ds} + \int_0^t G1(t, s) f(s) \, ds \tag{1.19}$$

Because we forced the initial condition, there is no arbitrary constant in the solution, and we see that the preceding form of the solution can accommodate any initial condition and any driving function $f(t)$.

EXAMPLE 1.3.1: We consider an object whose rate of thermal cooling obeys Newton's law whereby the rate of change of the temperature of the body is proportional to the difference between the temperature of the body and the temperature of its surroundings. Consider a specific problem whereby the initial temperature of the body is 100°C, the surrounding temperature is 20°C, and the thermal coefficient of diffusivity is $k = 0.2/$ sec. We seek $y(t)$: the temperature of the object as a function of the time t.

SOLUTION: The defining differential equation of the system (see Exercise 1.13 at the end of the chapter) is

$$\frac{d}{dt} y(t) = -k(y(t) - 20)$$

Inserting the value for k, we rewrite the preceding in the standard form

$$\frac{d}{dt}y(t) + \frac{2y(t)}{10} = 4$$

We can identify the coefficients of the differential equation as

> restart:a1(t):=1;a0(t):=2/10;

$$a1(t) := 1$$

$$a0(t) := \frac{1}{5} \tag{1.20}$$

Initial condition

> y(0):=100;

$$y(0) := 100 \tag{1.21}$$

The driving function is

> f(t):=4;f(s):=subs(t=s,f(t)):

$$f(t) := 4 \tag{1.22}$$

System basis vector

> y1(t):=exp(int(subs(t=s,−a0(t)/a1(t)),s=0..t));

$$y1(t) := e^{-\frac{1}{5}t} \tag{1.23}$$

First-order Green's function

> G1(t,s):=simplify(y1(t)/(subs(t=s,a1(t)*y1(t))));

$$G1(t, s) := e^{-\frac{1}{5}t + \frac{1}{5}s} \tag{1.24}$$

Particular solution

> y[p](t):=Int(G1(t,s)*f(s),s=0..t);

$$y_p(t) := \int_0^t 4\, e^{-\frac{1}{5}t + \frac{1}{5}s}\, ds \tag{1.25}$$

> y[p](t):=value(%);

$$y_p(t) := -20\, e^{-\frac{1}{5}t} + 20 \tag{1.26}$$

Final solution

> y(t):=simplify(eval(y(0)*y1(t)+y[p](t)));

$$y(t) := 80\,e^{-\frac{1}{5}t} + 20 \tag{1.27}$$

> plot(y(t),t=0..20,thickness=10);

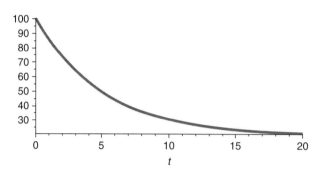

Figure 1.1

Figure 1.1 depicts the decay of the temperature of the body until it approaches the surrounding temperature. We can provide an animated view of the temperature decay as follows:

ANIMATION

> y(x,t):=y(t)*(Heaviside(x+1/2)−Heaviside(x−1/2));

$$y(x, t) := \left(80\,e^{-\frac{1}{5}t} + 20\right)\left(\text{Heaviside}\left(x + \frac{1}{2}\right) - \text{Heaviside}\left(x - \frac{1}{2}\right)\right) \tag{1.28}$$

> with(plots):animate(y(x,t),x=−1..1,t=0..30,thickness=2);

From Figure 1.1, we can see the actual real-time decay of the temperature until it finally reaches its surrounding temperature.

Check: Using Maple dsolve command with initial conditions.

> restart:ode:=diff(y(t),t)+2/10*y(t)=4;

$$ode := \frac{d}{dt}y(t) + \frac{1}{5}\,y(t) = 4 \tag{1.29}$$

> ics:=y(0)=100;

$$ics := y(0) = 100 \tag{1.30}$$

> dsolve({ode,ics});

$$y(t) = 20 + 80\,e^{-\frac{1}{5}t} \tag{1.31}$$

1.4 Second-Order Linear Differential Equations with Constant Coefficients

We now consider second-order linear nonhomogeneous differential equations with constant coefficients on some interval I. The equation, written in standard form, reads

$$a2\left(\frac{d^2}{dt^2}y(t)\right)+a1\left(\frac{d}{dt}y(t)\right)+a0\,y(t)=f(t)$$

Since the order of the differential equation is two, we must first find a set of two basis vectors of the corresponding homogeneous differential equation

$$a2\left(\frac{d^2}{dt^2}y(t)\right)+a1\left(\frac{d}{dt}y(t)\right)+a0\,y(t)=0$$

Since the coefficients $a2$, $a1$, and $a0$ are time invariant (constants), the method of undetermined coefficients can be used to find a set of system basis vectors. We assume a solution of the form

$$y(t)=e^{rt}$$

Substitution of this into the homogeneous equation provides us with the "characteristic" equation

$$a2\,r^2+a1\,r+a0=0$$

The roots of the characteristic equation are given as

$$r1=\frac{-a1+\sqrt{a1^2-4\,a2\,a0}}{2a2}$$

$$r2=\frac{-a1-\sqrt{a1^2-4\,a2\,a0}}{2a2}$$

Thus, our two solution vectors become

$$y1(t)=e^{\frac{\left(-a1+\sqrt{a1^2-4\,a2\,a0}\right)t}{2a2}}$$

and

$$y2(t)=e^{\frac{\left(-a1-\sqrt{a1^2-4\,a2\,a0}\right)t}{2a2}} \tag{1.32}$$

It can be shown that if the discriminant given here is not equal to 0, then the roots $r1$ and $r2$ are distinct and the two solutions are linearly independent; thus, for constant coefficient second-order differential equations, the preceding solution vectors constitute a set of system basis vectors.

DEMONSTRATION: We seek a set of basis vectors for the second-order linear homogeneous differential equation with constant coefficients as shown:

$$\frac{d^2}{dt^2}y(t) + 3\left(\frac{d}{dt}y(t)\right) + 2\,y(t) = 0$$

SOLUTION: We identify the coefficients of the differential equation $a2 = 1$, $a1 = 3$, and $a0 = 2$. The characteristic equation is

$$(r+2)(r+1) = 0$$

The roots of the characteristic equation are given as $r1 = -2$ and $r2 = -1$. Thus, a set of system basis vectors is

$$y1(t) = e^{-2t}$$

and

$$y2(t) = e^{-t}$$

We now consider a very special case of a differential equation with constant coefficients.

The Euler Differential Equation

One of the most frequently occurring ordinary differential equations, which arises in the solution of partial differential equations in the rectangular coordinate system, is the Euler differential equation. This differential equation is a special case of a second-order linear equation with constant coefficients. The general homogeneous form of this equation reads

$$\frac{d^2}{dt^2}y(t) + \lambda y(t) = 0$$

Note that the equation lacks a first-order derivative term. We now consider some example problems of the Euler differential equation.

EXAMPLE 1.4.1: Find a set of basis vectors for the Euler differential equation with a positive coefficient.

$$\frac{d^2}{dt^2}y(t) + \mu^2 y(t) = 0$$

SOLUTION: We identify the coefficients

> restart:a2:=1;a1:=0;a0:=mu^2;

$$a2 := 1$$

$$a1 := 0 \qquad\qquad (1.33)$$

$$a0 := \mu^2$$

Characteristic equation

```
> eq:=factor(a2*r^2+a1*r+a0=0);
```

$$eq := r^2 + \mu^2 = 0 \qquad (1.34)$$

The roots of the characteristic equation are given as

```
> con:=solve(eq,r):r1:=con[1];r2:=con[2];
```

$$r1 := I\mu$$
$$r2 := -I\mu \qquad (1.35)$$

System basis vectors

```
> y1(t):=evalc(Im(exp(r1*t)));
```

$$y1(t) := \sin(\mu t) \qquad (1.36)$$

```
> y2(t):=evalc(Re(exp(r2*t)));
```

$$y2(t) := \cos(\mu t) \qquad (1.37)$$

General solution: here *C1* and *C2* are arbitrary constants.

```
> y(t):=C1*y1(t)+C2*y2(t);
```

$$y(t) := C1\sin(\mu t) + C2\cos(\mu t) \qquad (1.38)$$

Check: Using Maple dsolve command.

```
> restart:ode:=diff(y(t),t,t)+mu^2*y(t)=0:dsolve(ode);
```

$$y(t) = \_C1\sin(\mu t) + \_C2\cos(\mu t) \qquad (1.39)$$

EXAMPLE 1.4.2: Find a set of basis vectors for the Euler differential equation with negative coefficient.

$$\frac{d^2}{dt^2}y(t) - \mu^2 y(t) = 0$$

SOLUTION: We identify the coefficients

```
> restart:a2:=1;a1:=0;a0:=-mu^2;
```

$$a2 := 1$$
$$a1 := 0$$
$$a0 := -\mu^2 \qquad (1.40)$$

Characteristic equation

```
> eq:=factor(a2*r^2+a1*r+a0=0);
```

$$eq := -(\mu - r)(\mu + r) = 0 \tag{1.41}$$

The roots of the characteristic equation are given as

```
> con:=solve(eq,r):r1:=con[1];r2:=con[2];
```

$$r1 := \mu$$

$$r2 := -\mu \tag{1.42}$$

System basis vectors

```
> y1(t):=exp(r1*t);
```

$$y1(t) := e^{\mu t} \tag{1.43}$$

```
> y2(t):=exp(r2*t);
```

$$y2(t) := e^{-\mu t} \tag{1.44}$$

General solution: here *C1* and *C2* are arbitrary constants.

```
> y(t):=C1*y1(t)+C2*y2(t);
```

$$y(t) := C1 e^{\mu t} + C2 e^{-\mu t} \tag{1.45}$$

Check: Using Maple dsolve command.

```
> restart:ode:=diff(y(t),t,t)−mu^2*y(t)=0:dsolve(ode);
```

$$y(t) = \_C1 e^{\mu t} + \_C2 e^{-\mu t} \tag{1.46}$$

In application problems in partial differential equations, it is more convenient to express this set of basis vectors in the (linearly dependent) equivalent form:

```
> y1(t):=sinh(mu*t);
```

$$y1(t) := \sinh(\mu t) \tag{1.47}$$

```
> y1(t):=cosh(mu*t);
```

$$y2(t) := \cosh(\mu t) \tag{1.48}$$

We now consider the following example, which occurs in the solution of the time-dependent portion of the wave partial differential equation that we will look at later.

EXAMPLE 1.4.3: Find the basis vectors for the equation

$$\frac{d^2}{dt^2} y(t) + \gamma \left(\frac{d}{dt} y(t) \right) + c^2 \lambda \, y(t) = 0$$

SOLUTION: We identify the coefficients

> restart:a2:=1;a1:=gamma;a0:=c^2*lambda;

$$a2 := 1$$
$$a1 := \gamma$$
$$a0 := c^2 \lambda \tag{1.49}$$

Characteristic equation

> eq:=factor(a2*r^2+a1*r+a0=0);

$$eq := r^2 + \gamma r + c^2 \lambda = 0 \tag{1.50}$$

The roots of the characteristic equation are given as

> con:=solve(eq,r):r1:=con[1];r2:=con[2];

$$r1 := -\frac{1}{2} \gamma + \frac{1}{2} \sqrt{\gamma^2 - 4c^2 \lambda}$$

$$r2 := -\frac{1}{2} \gamma - \frac{1}{2} \sqrt{\gamma^2 - 4c^2 \lambda} \tag{1.51}$$

System basis vectors

> y1(t):=exp(r1*t);

$$y1(t) := e^{\left(-\frac{1}{2}\gamma + \frac{1}{2}\sqrt{\gamma^2 - 4c^2\lambda} \right)t} \tag{1.52}$$

> y2(t):=exp(r2*t);

$$y2(t) := e^{\left(-\frac{1}{2}\gamma - \frac{1}{2}\sqrt{\gamma^2 - 4c^2\lambda} \right)t} \tag{1.53}$$

General solution: here *C1* and *C2* are arbitrary constants.

> y(t):=C1*y1(t)+C2*y2(t);

$$y(t) := C1 e^{\left(-\frac{1}{2}\gamma + \frac{1}{2}\sqrt{\gamma^2 - 4c^2\lambda} \right)t} + C2 e^{\left(-\frac{1}{2}\gamma - \frac{1}{2}\sqrt{\gamma^2 - 4c^2\lambda} \right)t} \tag{1.54}$$

Check: Using Maple dsolve command.

> restart:ode:=diff(y(t),t,t)+gamma*diff(y(t),t)+c^2*lambda*y(t)=0:dsolve(ode);

$$y(t) = \_C1\,e^{\left(-\frac{1}{2}\gamma+\frac{1}{2}\sqrt{\gamma^2-4c^2\lambda}\right)t} + \_C2\,e^{\left(-\frac{1}{2}\gamma-\frac{1}{2}\sqrt{\gamma^2-4c^2\lambda}\right)t} \qquad (1.55)$$

For the case where γ is very small, the discriminant given is negative, and we end up with complex roots. For this situation, it is more convenient to express this set of basis vectors in the (linearly dependent) equivalent form:

> y1(t):=exp(−gamma/2*t)*cos(sqrt(lambda*c^2−gamma^2/4)*t);

$$y1(t) := e^{-\frac{1}{2}\gamma t}\cos\left(\frac{1}{2}\sqrt{-\gamma^2+4c^2\lambda}\,t\right) \qquad (1.56)$$

> y2(t):=exp(−gamma/2*t)*sin(sqrt(lambda*c^2−gamma^2/4)*t);

$$y2(t) := e^{-\frac{1}{2}\gamma t}\sin\left(\frac{1}{2}\sqrt{-\gamma^2+4c^2\lambda}\,t\right) \qquad (1.57)$$

In the preceding paragraphs, we considered the special case of differential equations with constant coefficients. In general, the coefficients of linear differential equations are not constants and are functionally dependent on the independent variable. We now look at the general case of variable coefficients.

1.5 Second-Order Linear Differential Equations with Variable Coefficients

We now consider second-order linear nonhomogeneous differential equations with variable coefficients on a finite interval I. The equation written in standard form reads as

$$a2(t)\left(\frac{d^2}{dt^2}y(t)\right) + a1(t)\left(\frac{d}{dt}y(t)\right) + a0(t)y(t) = f(t)$$

We note that the generalized coefficients $a2(t)$, $a1(t)$, and $a0(t)$ are not constant. They are functionally dependent on the independent variable t; thus, the method of undetermined coefficients cannot be used here to find the basis vectors.

Since the order of the differential equation is two, we must first find two basis vectors of the corresponding homogeneous differential equation.

$$a2(t)\left(\frac{d^2}{dt^2}y(t)\right) + a1(t)\left(\frac{d}{dt}y(t)\right) + a0(t)y(t) = 0$$

Different types of variable coefficient differential equations demand their own peculiar technique for solution. We now consider a prominent example of such an equation.

The Cauchy-Euler Differential Equation

We consider the special case of the Cauchy-Euler differential equation, which will occur later in our study of partial differential equations in the cylindrical coordinate system.

The nonhomogeneous Cauchy-Euler differential equation has the form

$$a2\, t^2 \left(\frac{d^2}{dt^2} y(t) \right) + a1\, t \left(\frac{d}{dt} y(t) \right) + a0\, y(t) = f(t)$$

where $a2$, $a1$, and $a0$ are constants, and the t-dependence of each term has been extracted explicitly. This differential equation is easily recognized by the fact that the power of the coefficient of the independent variable t in each term is equal to the order of differentiation of the term $y(t)$, which it multiplies.

We now consider finding a set of basis vectors for the corresponding homogeneous Cauchy-Euler equation

$$a2\, t^2 \left(\frac{d^2}{dt^2} y(t) \right) + a1\, t \left(\frac{d}{dt} y(t) \right) + a0\, y(t) = 0$$

We substitute an assumed solution of the form

$$y(t) = t^m$$

into the homogeneous differential equation and get the characteristic equation

$$a2\, m(m - 1) + a1\, m + a0 = 0$$

Solving for the roots of this equation, we get

$$m1 = \frac{a2 - a1 + \sqrt{a2^2 - 2\, a2\, a1 + a1^2 - 4\, a2\, a0}}{2\, a2}$$

and

$$m2 = \frac{a2 - a1 - \sqrt{a2^2 - 2\, a2\, a1 + a1^2 - 4\, a2\, a0}}{2\, a2}$$

Thus, the solution vectors to the Cauchy-Euler dfferential equation are

$$y1(t) = t^{\frac{a2 - a1 + \sqrt{a2^2 - 2\, a2\, a1 + a1^2 - 4\, a2\, a0}}{2a2}}$$

and

$$y2(t) = t^{\frac{a2 - a1 - \sqrt{a2^2 - 2\, a2\, a1 + a1^2 - 4\, a2\, a0}}{2a2}}$$

If the discriminant—the term under the square root—is not equal to zero, then the roots $m1$ and $m2$ are distinct, and the two solutions can be shown to be linearly independent. Thus, the two solutions constitute a set of basis vectors for the Cauchy-Euler differential equation.

EXAMPLE 1.5.1: Find a set of basis vectors for the Cauchy-Euler differential equation whose characteristic equation has negative real roots:

$$2t^2 \left(\frac{d^2}{dt^2} y(t) \right) + 5t \left(\frac{d}{dt} y(t) \right) + y(t) = 0$$

SOLUTION: We identify the coefficients

> restart:a2:=2;a1:=5;a0:=1;

$$a2 := 2$$
$$a1 := 5 \tag{1.58}$$
$$a0 := 1$$

Characteristic equation

> eq:=a2*m*(m−1)+a1*m+a0=0;

$$eq := 2m(m-1) + 5m + 1 = 0 \tag{1.59}$$

Solving for the roots of this equation, we get two distinct roots

> con:=solve(eq,m):m1:=con[1];m2:=con[2];

$$m1 := -\frac{1}{2}$$
$$m2 := -1 \tag{1.60}$$

System basis vectors

> y1(t):=t^m1;

$$y1(t) := \frac{1}{\sqrt{t}} \tag{1.61}$$

> y2(t):=t^m2;

$$y2(t) := \frac{1}{t} \tag{1.62}$$

General solution: here $C1$ and $C2$ are arbitrary constants.

> y(t):=C1*y1(t)+C2*y2(t);

$$y(t) := \frac{C1}{\sqrt{t}} + \frac{C2}{t} \tag{1.63}$$

Check: Using Maple dsolve command.

\> restart:ode:=2*t^2*diff(y(t),t,t)+5*t*diff(y(t),t)+y(t)=0:dsolve(ode);

$$y(t) = \frac{\_C1}{\sqrt{t}} + \frac{\_C2}{\sqrt{t}} \tag{1.64}$$

EXAMPLE 1.5.2: Find the basis vectors for the following Cauchy-Euler differential equation whose characteristic equation has complex roots:

$$t^2 \left(\frac{d^2}{dt^2} y(t) \right) + t \left(\frac{d}{dt} y(t) \right) + 4 y(t) = 0$$

SOLUTION: We identify the coefficients

\> restart:a2:=1;a1:=1;a0:=4;

$$a2 := 1$$
$$a1 := 1 \tag{1.65}$$
$$a0 := 4$$

Characteristic equation

\> eq:=a2*m*(m−1)+a1*m+a0=0;

$$eq := m(m-1) + m + 4 = 0 \tag{1.66}$$

Solving for the roots of this equation, we get two distinct roots

\> con:=solve(eq,m):m1:=con[1];m2:=con[2];

$$m1 := 2\mathrm{I} \tag{1.67}$$
$$m2 := -2\mathrm{I}$$

System basis vectors

\> y1(t):=t^m1;

$$y1(t) := t^{2\mathrm{I}} \tag{1.68}$$

\> y2(t):=t^m2;

$$y2(t) := t^{-2\mathrm{I}} \tag{1.69}$$

General solution: here *C1* and *C2* are arbitrary constants.

\> y(t):=C1*y1(t)+C2*y2(t);

$$y(t) := C1t^{2\mathrm{I}} + C2t^{-2\mathrm{I}} \tag{1.70}$$

For convenience later on, we express the preceding set of basis vectors in the (linearly dependent) equivalent form:

> y1(t):=sin(2*ln(t));

$$yl(t) := \sin(2\ln(t)) \tag{1.71}$$

> y2(t):=cos(2*ln(t));

$$y2(t) := \cos(2\ln(t)) \tag{1.72}$$

General solution: here C1 and C2 are arbitrary constants.

> y(t):=C1*y1(t)+C2*y2(t);

$$y(t) := C1\sin(2\ln(t)) + C2\cos(2\ln(t)) \tag{1.73}$$

Check: Using Maple dsolve command.

> restart:ode:=t^2*diff(y(t),t,t)+t*diff(y(t),t)+4*y(t)=0:dsolve(ode);

$$y(t) = \_C1\sin(2\ln(t)) + \_C2\cos(2\ln(t)) \tag{1.74}$$

1.6 Finding a Second Basis Vector by the Method of Reduction of Order

In some cases, a second linearly independent solution vector does not always become readily available. Generally, this occurs if any of the preceding discriminants vanish; in this case, we do not get two distinct roots. However, if we know one solution vector for the second-order linear differential equation, then the method of reduction of order can be used to find a second linearly independent solution vector.

Let $yl(t)$ be one solution to the homogeneous differential equation:

$$a2(t)\left(\frac{d^2}{dt^2}yl(t)\right) + al(t)\left(\frac{d}{dt}yl(t)\right) + a0(t)yl(t) = 0$$

Using operator notation, we write the preceding equation as $L(yl) = 0$. We now assume that a second solution $y2(t)$ can be expressed as a nonlinear multiple of $yl(t)$:

$$y2(t) = u(t)yl(t)$$

Substituting this into the preceding homogeneous differential equation yields

$$a2(t)\left(\frac{d^2}{dt^2}u(t)\right)y1(t)+2a2(t)\left(\frac{d}{dt}u(t)\right)\left(\frac{d}{dt}y1(t)\right)+a2(t)u(t)\left(\frac{d^2}{dt^2}y1(t)\right)$$

$$+a1(t)\left(\frac{d}{dt}u(t)\right)y1(t)+a1(t)u(t)\left(\frac{d}{dt}y1(t)\right)+a0(t)u(t)y1(t)=0$$

Collecting terms, we get

$$\left(a2(t)\left(\frac{d^2}{dt^2}y1(t)\right)+a1(t)\left(\frac{d}{dt}y1(t)\right)+a0(t)y1(t)\right)u(t)+a2(t)\left(\frac{d^2}{dt^2}u(t)\right)y1(t)$$

$$+2a2(t)\left(\frac{d}{dt}u(t)\right)\left(\frac{d}{dt}y1(t)\right)+a1(t)\left(\frac{d}{dt}u(t)\right)y1(t)=0$$

Since $y1(t)$ satisfies the homogeneous equation—that is, $L(y1)=0$—then the first term above vanishes, and we end up with the following homogeneous differential equation

$$a2(t)\left(\frac{d^2}{dt^2}u(t)\right)y1(t)+\left(a1(t)y1(t)+2a2(t)\left(\frac{d}{dt}y1(t)\right)\right)\left(\frac{d}{dt}u(t)\right)=0$$

This is basically a first-order linear differential equation in terms of the first derivative of $u(t)$. We have reduced the order of the differential equation. From the solution procedure for all first-order linear equations in Section 1.2, we solve for $u(t)$ and get

$$u(t)=\int e^{\int\left(-\frac{a1(t)}{a2(t)}-\frac{2\left(\frac{d}{dt}y1(t)\right)}{y1(t)}\right)dt}dt$$

Simplifying the preceding yields

$$u(t)=\int\frac{e^{-\left(\int\frac{a1(t)}{a2(t)}dt\right)}}{y1(t)^2}dt$$

Thus, our second solution $y2(t)$ is given as

$$y2(t)=y1(t)\left(\int\frac{e^{-\left(\int\frac{a1(t)}{a2(t)}dt\right)}}{y1(t)^2}dt\right)$$

This solution can be shown to be linearly independent of $y1(t)$. Thus, from one solution $y1(t)$, the method of reduction of order allows us to generate a second basis vector $y2(t)$.

DEMONSTRATION: Use the method of reduction of order to determine a set of basis vectors for the Cauchy-Euler differential equation

$$t^2\left(\frac{d^2}{dt^2}y(t)\right)-3t\left(\frac{d}{dt}y(t)\right)+3y(t)=0$$

SOLUTION: By inspection, we see that one solution vector is

$$y1(t) = t$$

We identify the coefficients of the differential equation $a2(t) = t^2$, $a1(t) = -3t$, and $a0(t) = 3$. Using the preceding previous procedure, with knowledge of one vector, we generate a second linearly independent basis vector from the integral

$$y2(t) = t \left(\int \frac{e^{\int \frac{3t}{t^2} dt}}{t^2} dt \right)$$

After evaluation of the interior integral, we have

$$y2(t) = t \left(\int t \, dt \right)$$

which evaluates to

$$y2(t) = \frac{t^3}{2}$$

With the preceding set of basis vectors, the general solution to the homogeneous differential equation is

$$y(t) = C1t + \frac{C2t^3}{2} \tag{1.75}$$

Here, $C1$ and $C2$ are arbitrary constants.

The ability to generate a second basis vector from knowledge of a single solution vector to a second-order differential equation is very important.

EXAMPLE 1.6.1: We seek a set of basis vectors for the following differential equation whose characteristic equation has repeated roots. Use the method of reduction of order to find a second basis vector.

$$\frac{d^2}{dt^2} y(t) + 2 \left(\frac{d}{dt} y(t) \right) + y(t) = 0$$

SOLUTION: We identify the coefficients

> restart:a2:=1;a1:=2;a0:=1;

$$a2 := 1$$

$$a1 := 2 \tag{1.76}$$

$$a0 := 1$$

We assume a solution

> y(t):=exp(r*t);

$$y(t) := e^{rt} \tag{1.77}$$

Characteristic equation

> eq:=factor(a2*diff(y(t),t,t)+a1*diff(y(t),t)+a0*y(t))/exp(r*t)=0;

$$eq := (r+1)^2 = 0 \tag{1.78}$$

The roots of the characteristic equation are given as

> con:=solve(eq,r):r1:=con[1];r2:=con[2];

$$r1 := -1 \tag{1.79}$$

$$r2 := -1$$

This is a situation whereby the discriminant is 0, and we get two equal roots from the characteristic equation. The situation of two equal roots does not provide us with the opportunity to get two linearly independent solutions. To get a second linearly independent solution vector, we declare one basis vector to be

> y1(t):=exp(r1*t);

$$y1(t) := e^{-t} \tag{1.80}$$

Using the preceding method of reduction of order, we obtain a second basis vector

> y2(t):=y1(t)*int(exp(-int(a1/a2,t))/y1(t)^2,t);

$$y2(t) := e^{-t} t \tag{1.81}$$

General solution: here *C1* and *C2* are arbitrary constants.

> y(t):=C1*y1(t)+C2*y2(t);

$$y(t) := C1e^{-t} + C2e^{-t}t \tag{1.82}$$

Check: Using Maple dsolve command.

> restart:ode:=diff(y(t),t,t)+2*diff(y(t),t)+y(t)=0:dsolve(ode);

$$y(t) = \_C1e^{-t} + \_C2e^{-t}t \tag{1.83}$$

EXAMPLE 1.6.2: We seek the solution to the following second-order linear differential equation with trigonometric coefficients:

$$\frac{d^2}{dt^2}y(t) + (\tan(t) - 2\cot(t))\left(\frac{d}{dt}y(t)\right) = 0$$

SOLUTION: We identify the coefficients

> restart:a2(t):=1;a1(t):=tan(t)−2*cot(t);a0(t):=0;

$$a2(t) := 1$$
$$a1(t) := \tan(t) - 2\cot(t) \tag{1.84}$$
$$a0(t) := 0$$

By inspection, we see that one solution is

> y1(t):=1;

$$y1(t) := 1 \tag{1.85}$$

Using the method of reduction of order, we generate a second basis vector

> y2(t):=y1(t)*int(exp(−int(a1(t)/a2(t),t))/y1(t)^2,t);

$$y2(t) := \frac{1}{3}\sin(t)^3 \tag{1.86}$$

General solution: here *C1* and *C2* are arbitrary constants.

> y(t):=C1*y1(t)+C2*y2(t);

$$y(t) := C1 + \frac{1}{3}C2\sin(t)^3 \tag{1.87}$$

Check: Using Maple dsolve command.

> restart:ode:=diff(y(t),t,t)+(tan(t)−2*cot(t))*diff(y(t),t)=0:dsolve(ode);

$$y(t) = \_C1 + \sin(t)^3\_C2 \tag{1.88}$$

1.7 The Method of Variation of Parameters—Second-Order Green's Function

We now consider finding a particular solution to a second-order linear nonhomogeneous differential equation on an interval *I*. By the method of variation of parameters, we will construct this solution from a known set of basis vectors of the system. Recall that the basis vectors of the second-order system are two linearly independent solutions of the corresponding homogeneous system.

The second-order linear nonhomogeneous differential equation reads as

$$a2(t)\left(\frac{d^2}{dt^2}y(t)\right) + a1(t)\left(\frac{d}{dt}y(t)\right) + a0(t)y(t) = f(t)$$

Similar to what we did for the first-order system, we assume our particular solution to the second-order nonhomogeneous equation to be a nonlinear superposition of the two basis vectors $y1(t)$ and $y2(t)$—that is, we set

$$y(t) = u1(t)y1(t) + u2(t)y2(t) \tag{1.89}$$

Here, $u1(t)$ and $u2(t)$ are two unknown functions of the independent variable t for which we now seek their solutions. Taking the first derivative of the preceding using the product rule yields

$$\frac{d}{dt}y(t) = u1(t)\left(\frac{d}{dt}y1(t)\right) + y1(t)\left(\frac{d}{dt}u1(t)\right) + u2(t)\left(\frac{d}{dt}y2(t)\right) + y2(t)\left(\frac{d}{dt}u2(t)\right)$$

To simplify our solution, we set

$$y1(t)\left(\frac{d}{dt}u1(t)\right) + y2(t)\left(\frac{d}{dt}u2(t)\right) = 0$$

Taking the second derivative of the remaining term and substituting into the preceding nonhomogeneous differential equation and collecting terms yields

$$\left(a2(t)\left(\frac{d^2}{dt^2}y1(t)\right) + a1(t)\left(\frac{d}{dt}y1(t)\right) + a0(t)y1(t)\right)u1(t) + \left(a2(t)\left(\frac{d^2}{dt^2}y2(t)\right)\right.$$

$$+ a1(t)\left(\frac{d}{dt}y2(t)\right) + a0(t)y2(t)\right)u2(t) + a2(t)\left(\left(\frac{d}{dt}u1(t)\right)y1(t) + \left(\frac{d}{dt}u2(t)\right)y2(t)\right) = f(t)$$

Since $y1(t)$ and $y2(t)$ are both solutions to the homogeneous equation—that is, $L(y1) = 0$ and $L(y2) = 0$—the first two preceding terms vanish, and we arrive at

$$a2(t)\left(\left(\frac{d}{dt}u1(t)\right)y1(t) + \left(\frac{d}{dt}u2(t)\right)y2(t)\right) = f(t)$$

Thus, we finally end up with a system of two differential equations

$$\left(\frac{d}{dt}u1(t)\right)y1(t) + \left(\frac{d}{dt}u2(t)\right)y2(t) = 0$$

and

$$\left(\frac{d}{dt}u1(t)\right)\left(\frac{d}{dt}y1(t)\right) + \left(\frac{d}{dt}u2(t)\right)\left(\frac{d}{dt}y2(t)\right) = \frac{f(t)}{a2(t)}$$

Solving these two equations simultaneously, for the derivatives of $u1(t)$ and $u2(t)$, we arrive at the two first-order linear nonhomogeneous differential equations:

$$\frac{d}{dt}u1(t) = -\frac{f(t)y2(t)}{a2(t)\left(y1(t)\left(\frac{d}{dt}y2(t)\right) - y2(t)\left(\frac{d}{dt}y1(t)\right)\right)}$$

and

$$\frac{d}{dt}u2(t) = -\frac{f(t)y1(t)}{a2(t)\left(y1(t)\left(\frac{d}{dt}y2(t)\right) - y2(t)\left(\frac{d}{dt}y1(t)\right)\right)}$$

Integrating each of the preceding yields

$$u1(t) = \int\left(-\frac{f(t)y2(t)}{\left(-y2(t)\left(\frac{d}{dt}y1(t)\right) + \left(\frac{d}{dt}y2(t)\right)y1(t)\right)a2(t)}\right)dt$$

and

$$u2(t) = \int -\frac{f(t)y1(t)}{\left(-y2(t)\left(\frac{d}{dt}y1(t)\right) + \left(\frac{d}{dt}y2(t)\right)y1(t)\right)a2(t)}dt$$

Recognizing the denominator in the integrands as the Wronskian

$$W(y1(t), y2(t)) = -y2(t)\left(\frac{d}{dt}y1(t)\right) + \left(\frac{d}{dt}y2(t)\right)y1(t)$$

we get the following solutions

$$u1(t) = -\left(\int\frac{f(t)y2(t)}{a2(t)W(y1(t), y2(t))}dt\right)$$

and

$$u2(t) = \int\frac{f(t)y1(t)}{a2(t)W(y1(t), y2(t))}dt$$

Thus, our particular solution to the differential equation becomes

$$y_p(t) = y2(t)\left(\int\frac{f(t)y1(t)}{a2(t)W(y1(t), y2(t))}dt\right) - y1(t)\left(\int\frac{f(t)y2(t)}{a2(t)W(y1(t), y2(t))}dt\right)$$

Using the dummy variable s in the integrands, a more compact form of the particular solution can be rewritten as

$$y_p(t) = \int\frac{(y1(s)y2(t) - y1(t)y2(s))f(s)}{a2(s)W(y1(s), y2(s))}ds$$

Similar to what we did for first-order differential equations, we now define $G2(t, s)$ as the second-order Green's function

$$G2(t, s) = \frac{y1(s)y2(t) - y1(t)y2(s)}{a2(s)W(y1(s), y2(s))} \tag{1.90}$$

Of course, this result is valid only on an interval I, where the term $a2(t)$ does not vanish—that is, on an interval where the equation is normal. Further, since the vectors form a basis, they are linearly independent, and the Wronskian is nonzero on this interval.

In terms of the second-order Green's function given earlier, the particular solution takes on the form

$$y_p(t) = \int G2(t, s) f(s) \mathrm{d}s \qquad (1.91)$$

Again, we see that this form of the solution can accommodate any type of driving or source function $f(t)$. This procedure, by way of the method of variation of parameters, leads to an integral solution whereby the Green's function acts as the kernel of the integrand.

The beauty of the preceding form just derived becomes apparent when we compare it with the alternate procedure of undetermined coefficients. In that procedure, we have to make a tentative guess at the solution. The guess, of course, depends on the character of the source function $f(t)$ and whether or not this function is linearly dependent on the system basis vectors. With the preceding method, there is no need to guess, and we get immediate solutions.

DEMONSTRATION: We seek the particular solution to the second-order linear nonhomogeneous differential equation

$$\frac{\mathrm{d}^2}{\mathrm{d}t^2} y(t) + 2 \left(\frac{\mathrm{d}}{\mathrm{d}t} y(t) \right) + y(t) = te^{-t}$$

SOLUTION: We identify the coefficients of the differential equation $a2(t) = 1$, $a1(t) = 2$, and $a0(t) = 1$, and the source term

$$f(t) = te^{-t}$$

Earlier, from Example 1.6.1, we found a set of basis vectors to be

$$y1(t) = e^{-t}$$

and

$$y2(t) = te^{-t}$$

We note the linear dependence of the source term on one of the basis vectors. Evaluation of the Wronskian of the basis yields

$$W(e^t, te^t) = e^{-2t}$$

Evaluation of the second-order Green's function from the formula

$$G2(t, s) = \frac{y1(s)y2(t) - y1(t)y2(s)}{a2(s)W(y1(s), y2(s))}$$

yields

$$G2(t, s) = \frac{e^{-s}te^{-t} - e^{-t}se^{-s}}{e^{-2s}}$$

The particular solution is the integral whose integrand is the product of the Green's function and the source function written in terms of the dummy variable s:

$$y_p(t) = \int \frac{(e^{-s}te^{-t} - e^{-t}se^{-s})se^{-s}}{e^{-2s}}ds \qquad (1.92)$$

Integrating first with respect to the dummy variable s and then substituting t for s yields

$$y_p(t) = \frac{t^3 e^{-t}}{6}$$

Thus, the general solution to the problem is the sum of the homogeneous plus the particular solution. Here $C1$ and $C2$ are arbitary constants.

$$y(t) = C1e^{-t} + C2\,te^{-t} + \frac{t^3 e^{-t}}{6} \qquad (1.93)$$

The example just described is especially interesting because the source term $f(t)$ is linearly dependent on one of the basis vectors. The ease with which we found this solution indicates the method of variation of paramaters to be far more convenient than what we would have encountered with the method of undetermined coefficients.

EXAMPLE 1.7.1: Use the method of variation of parameters to generate the particular solution to the differential equation that is not normal at the origin.

$$t\left(\frac{d^2}{dt^2}y(t)\right) + \frac{d}{dt}y(t) = t + 1$$

SOLUTION: We identify the coefficients

> restart:a2(t):=t;a1(t):=1;a0(t):=0;

$$a2(t) := t$$

$$a1(t) := 1$$

$$a0(t) := 0 \qquad (1.94)$$

Source term

> f(t):=t+1;

$$f(t) := t + 1 \qquad (1.95)$$

From earlier methods, we see that one basis vector is a constant and a second basis vector can be found using methods of Section 1.6. Thus, two possible basis vectors are

> y1(t):=1;

$$y1(t) := 1 \tag{1.96}$$

> y2(t):=ln(t);

$$y2(t) := \ln(t) \tag{1.97}$$

Wronskian

> W(y1(t),y2(t)):=y1(t)*diff(y2(t),t)−y2(t)*diff(y1(t),t);

$$W(1, \ln(t)) := \frac{1}{t} \tag{1.98}$$

> y1(s):=subs(t=s,y1(t)):y2(s):=subs(t=s,y2(t)):W(y1(s),y2(s)):=subs(t=s,W(y1(t),y2(t))):a2(s)
  :=subs(t=s,a2(t)):f(s):=subs(t=s,f(t)):

Green's function

> G2(t,s):=(y1(s)*y2(t)−y1(t)*y2(s))/(a2(s)*W(y1(s),y2(s)));

$$G2(t, s) := \ln(t) - \ln(s) \tag{1.99}$$

Particular solution (integrate first with respect to s and then substitute s with t)

> y[p](t):=Int(G2(t,s)*f(s),s);

$$y_p(t) := \int (\ln(t) - \ln(s))(s+1)ds \tag{1.100}$$

> y[p](t):=subs(s=t,value(%));

$$y_p(t) := \frac{1}{4}t^2 + t \tag{1.101}$$

General solution: here C1 and C2 are arbitrary constants.

> y(t):=C1*y1(t)+C2*y2(t)+y[p](t);

$$y(t) := C1 + C2\ln(t) + \frac{1}{4}t^2 + t \tag{1.102}$$

This solution is valid everywhere except at the point $t = 0$. This follows, since the differential equation fails to be normal at the origin.

Check: Using Maple dsolve command.

> restart:ode:=t*diff(y(t),t,t)+diff(y(t),t)=t+1:dsolve(ode);

$$y(t) = \frac{1}{4}t^2 + \_C1\ln(t) + t + \_C2 \tag{1.103}$$

EXAMPLE 1.7.2: We seek the solution to the nonhomogeneous Euler differential equation

$$\frac{d^2}{dt^2}y(t) + y(t) = \sec(t)$$

SOLUTION: We identify the coefficients

> restart:a2(t):=1;a1(t):=0;a0(t):=1;

$$a2(t) := 1$$
$$a1(t) := 0 \tag{1.104}$$
$$a0(t) := 1$$

Source function

> f(t):=sec(t);

$$f(t) := \sec(t) \tag{1.105}$$

System basis vectors from earlier result

> y1(t):=sin(t);

$$y1(t) := \sin(t) \tag{1.106}$$

> y2(t):=cos(t);

$$y2(t) := \cos(t) \tag{1.107}$$

Wronskian

> W(y1(t),y2(t)):=simplify(y1(t)*diff(y2(t),t)−y2(t)*diff(y1(t),t));

$$W(\sin(t), \cos(t)) := -1 \tag{1.108}$$

> y1(s):=subs(t=s,y1(t)):y2(s):=subs(t=s,y2(t)):W(y1(s),y2(s)):=subs(t=s,W(y1(t),y2(t))):a2(s)
:=subs(t=s,a2(t)):f(s):=subs(t=s,f(t)):

Green's function

> G2(t,s):=(y1(s)*y2(t)−y1(t)*y2(s))/(a2(s)*W(y1(s),y2(s)));

$$G2(t, s) := -\sin(s)\cos(t) + \sin(t)\cos(s) \tag{1.109}$$

Particular solution (integrate first with respect to s and then substitute s with t)

> y[p](t):=Int(G2(t,s)*f(s),s);

$$y_p(t) := \int (-\sin(s)\cos(t) + \sin(t)\cos(s))\sec(s)ds \tag{1.110}$$

> y[p](t):=subs(s=t,value(%));

$$y_p(t) := \cos(t) \ln(\cos(t)) + \sin(t)t \tag{1.111}$$

General solution: here *C1* and *C2* are arbitrary constants.

> y(t):=C1*y1(t)+C2*y2(t)+y[p](t);

$$y(t) := C1 \sin(t) + C2 \cos(t) + \cos(t) \ln(\cos(t)) + \sin(t)t \tag{1.112}$$

Check: Using Maple dsolve command.

> restart:ode:=diff(y(t),t,t)+y(t)=sec(t):dsolve(ode);

$$y(t) = \sin(t)\_C2 + \cos(t)\_C1 + t\sin(t) + \ln(\cos(t))\cos(t) \tag{1.113}$$

EXAMPLE 1.7.3: We are given one solution to the Cauchy-Euler homogeneous differential equation shown here. From this single basis vector, we are asked to find a complete solution to the diffferential equation.

$$t^2 \left(\frac{d^2}{dt^2} y(t) \right) - 2t \left(\frac{d}{dt} y(t) \right) + 2y(t) = t \ln(t)$$

SOLUTION: We identify the coefficients

> restart:a2(t):=t^2;a1(t):=−2*t;a0(t):=2;

$$a2(t) := t^2$$
$$a1(t) := -2t$$
$$a0(t) := 2 \tag{1.114}$$

By inspection, we see one basis vector to be

> y1(t):=t;

$$y1(t) := t \tag{1.115}$$

Source term

> f(t):=t*ln(t);

$$f(t) := t \ln(t) \tag{1.116}$$

A second basis vector by the method of reduction of order is

> y2(t):=y1(t)*int(exp(−int(a1(t)/a2(t),t))/y1(t)^2,t);

$$y2(t) := t^2 \tag{1.117}$$

Wronskian

>W(y1(t),y2(t)):=simplify(y1(t)*diff(y2(t),t)−y2(t)*diff(y1(t),t));

$$W(t, t^2) := t^2 \tag{1.118}$$

>y1(s):=subs(t=s,y1(t)):y2(s):=subs(t=s,y2(t)):W(y1(s),y2(s)):=subs(t=s,W(y1(t),y2(t))):a2(s)
  :=subs(t=s,a2(t)):f(s):=subs(t=s,f(t)):

Green's function

> G2(t,s):=(y1(s)*y2(t)−y1(t)*y2(s))/(a2(s)*W(y1(s),y2(s)));

$$G2(t, s) := \frac{st^2 - ts^2}{s^4} \tag{1.119}$$

Particular solution (integrate first with respect to s and then substitute s with t)

> y[p](t):=Int(G2(t,s)*f(s),s);

$$y_p(t) := \int \frac{(st^2 - ts^2) \ln(s)}{s^3} ds \tag{1.120}$$

> y[p](t):=subs(s=t,value(%));

$$y_p(t) := -t \ln(t) - t - \frac{1}{2} t \ln(t)^2 \tag{1.121}$$

General solution: here $C1$ and $C2$ are arbitrary constants.

> y(t):=C1*y1(t)+C2*y2(t)+y[p](t);

$$y(t) := C1\, t + C2\, t^2 - t \ln(t) - t - \frac{1}{2}\, t \ln(t)^2 \tag{1.122}$$

This solution is valid everywhere except at the origin.

Check: Using Maple dsolve command.

> restart:ode:=t^2*diff(y(t),t,t)−2*t*diff(y(t),t)+2*y(t)=t*ln(t):expand(dsolve(ode));

$$y(t) = t^2\,\_C2 + t\,\_C1 - t \ln(t) - t - \frac{1}{2}\, \ln(t)^2 t \tag{1.123}$$

This example demonstrates a significant solution procedure in that, from knowledge of only a single solution vector, we were able to construct a basis in addition to the complete solution to the nonhomogeneous differential equation.

1.8 Initial-Value Problem for Second-Order Differential Equations

Often in partial differential equations, we have to find solutions to time-dependent, second-order linear differential equations with initial conditions as constraints. We now investigate the initial-value problem for the second-order differential equation.

Consider the following linear nonhomogeneous differential equation with constant coefficients, which is descriptive of the motion of a simple harmonic oscillator with damping. Here, m is the mass, γ is the damping coefficient, and k is the spring constant.

$$m\left(\frac{d^2}{dt^2}y(t)\right) + \gamma\left(\frac{d}{dt}y(t)\right) + ky(t) = f(t)$$

We assume damping in the system to be small ($\gamma^2 < 4km$) and we set

$$\omega = \frac{\sqrt{4km - \gamma^2}}{2m}$$

From Section 1.4, we evaluate a set of basis vectors

$$y1(t) = e^{-\frac{\gamma t}{2m}}\sin(\omega t)$$

and

$$y2(t) = e^{-\frac{\gamma t}{2m}}\cos(\omega t)$$

Evaluation of the Wronskian and the subsequent second-order Green's function yields

$$G2(t,s) = \frac{e^{-\frac{\gamma(t-s)}{2m}}\sin(\omega t - \omega s)}{m\omega}$$

Thus, we can write our general solution to the differential equation as

$$y(t) = C1\,e^{-\frac{\gamma t}{2m}}\sin(\omega t) + C2\,e^{-\frac{\gamma t}{2m}}\cos(\omega t) + \int_0^t \frac{e^{-\frac{\gamma(t-s)}{2m}}\sin(\omega t - \omega s)f(s)}{m\omega}\,ds$$

Since the differential equation is of order two, we have two initial conditions, and, at time $t = 0$, the two initial conditions on the problem are the initial position

$$y(0) = y_0$$

and the initial speed

$$v(0) = v_0$$

Substituting these initial conditions, we evaluate the arbitrary constants *C1* and *C2*, and we arrive at the final solution

$$y(t) = \frac{(2\,v_0 m + y_0 \gamma)e^{-\frac{\gamma t}{2m}}\sin(\omega t)}{2\omega m} + y_0 e^{-\frac{\gamma t}{2m}}\cos(\omega t) + \int_0^t \frac{e^{-\frac{\gamma(t-s)}{2m}}\sin(\omega t - \omega s)\,f(s)}{m\omega}\,ds$$

(1.124)

This form of the solution can accommodate any set of initial conditions and any source function $f(t)$. Note that there are no arbitrary constants in the preceding equations because the initial conditions have already been incorporated into the solution.

EXAMPLE 1.8.1: We consider a mass $m = 1$ kg attached to a simple spring with a spring constant $k = 65/4$ N/m, which is immersed in a damping medium with a damping coefficient $\gamma = 1$ N/m/sec. An applied force $f(t)$ in units of N acts on the system and is given following. The initial position of the mass is 10 m and the initial speed is 20 m/sec. From Newton's second-law, the dynamic equation of motion of the mass reads as

$$\frac{d^2}{dt^2}y(t) + \frac{d}{dt}y(t) + \frac{65y(t)}{4} = f(t)$$

SOLUTION: We identify the coefficients

> restart:a2:=1;a1:=1;a0:=65/4;

$$a2 := 1$$
$$a1 := 1$$
$$a0 := \frac{65}{4}$$

(1.125)

The driving function

> f(t):=10*t*exp(−t/2);

$$f(t) := 10t\,e^{-\frac{1}{2}t}$$

(1.126)

The damped angular oscillation frequency of the spring-mass system is

> omega:=sqrt(4*a2*a0−a1^2)/(2*a2);

$$\omega := 4$$

(1.127)

Initial conditions

> y(0):=10;

$$y(0) := 10$$

(1.128)

> v(0):=20;

$$v(0) := 20 \qquad (1.129)$$

System basis vectors

> y1(t):=exp(−a1*t/(2*a2))*sin(omega*t);

$$y1(t) := e^{-\frac{1}{2}t}\sin(4t) \qquad (1.130)$$

> y2(t):=exp(−a1*t/(2*a2))*cos(omega*t);

$$y2(t) := e^{-\frac{1}{2}t}\cos(4t) \qquad (1.131)$$

Wronskian

> W(y1(t),y2(t)):=simplify(y1(t)*diff(y2(t),t)−y2(t)*diff(y1(t),t));

$$W\left(e^{-\frac{1}{2}t}\sin(4t),\, e^{-\frac{1}{2}t}\cos(4t)\right) := -4e^{-t} \qquad (1.132)$$

> y1(s):=subs(t=s,y1(t)):y2(s):=subs(t=s,y2(t)):W(y1(s),y2(s)):=subs(t=s,W(y1(t),y2(t))):f(s):=
  subs(t=s,f(t)):

Green's function

> G2(t,s):=(y1(s)*y2(t)−y1(t)*y2(s))/(a2*W(y1(s),y2(s)));

$$G2(t, s) := -\frac{1}{4}\frac{e^{-\frac{1}{2}s}\sin(4s)\, e^{-\frac{1}{2}t}\cos(4t) - e^{-\frac{1}{2}t}\sin(4t)\, e^{-\frac{1}{2}s}\cos(4s)}{e^{-s}} \qquad (1.133)$$

> G2(t,s):=simplify(combine(G2(t,s),trig));

$$G2(t, s) := -\frac{1}{4}\,\sin(4s - 4t)e^{\frac{1}{2}s-\frac{1}{2}t} \qquad (1.134)$$

Particular solution

> y[p](t):=Int(G2(t,s)*f(s),s=0..t);

$$y_p(t) := \int_0^t \left(-\frac{5}{2}\sin(4s - 4t)\ e^{\frac{1}{2}s-\frac{1}{2}t}s\, e^{-\frac{1}{2}s}\right) ds \qquad (1.135)$$

> y[p](t):=combine(value(%),trig);

$$y_p(t) := -\frac{5}{32}\,e^{-\frac{1}{2}t}\,\sin(4t) + \frac{5}{8}te^{-\frac{1}{2}t} \qquad (1.136)$$

Substituting the initial conditions into the general solution, we get the final solution

> y(t):=(2*v(0)*a2+y(0)*a1)/(2*omega*a2)*y1(t)+y(0)*y2(t)+y[p](t);

$$y(t) := \frac{195}{32} e^{-\frac{1}{2}t} \sin(4t) + 10 e^{-\frac{1}{2}t} \cos(4t) + \frac{5}{8} t e^{-\frac{1}{2}t} \tag{1.137}$$

> plot(y(t),t=0..10,thickness=10);

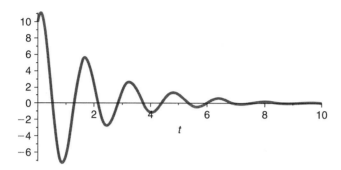

Figure 1.2

Figure1.2 depicts the motion of the mass as a function of time. We can provide an animated view of the preceding example as follows:

ANIMATION

>y(x,t):=y(t)*(Heaviside(x+1/2)−Heaviside(x−1/2));

$$y(x, t) := \left(\frac{195}{32} e^{-\frac{1}{2}t} \sin(4t) + 10 e^{-\frac{1}{2}t} \cos(4t) + \frac{5}{8} t e^{-\frac{1}{2}t} \right)$$
$$\left(\text{Heaviside} \left(x + \frac{1}{2} \right) - \text{Heaviside} \left(x - \frac{1}{2} \right) \right) \tag{1.138}$$

> with(plots):animate(y(x,t),x=−1..1,t=0..10,thickness=3,frames=100);

From the animated display, we can observe the actual real-time motion of the mass on the spring.

Check: Using Maple dsolve command with initial conditions.

> restart:ode:=diff(y(t),t,t)+diff(y(t),t)+65/4*y(t)=10*t*exp(−t/2);

$$ode := \frac{d^2}{dt^2} y(t) + \frac{d}{dt} y(t) + \frac{65}{4} y(t) = 10 t e^{-\frac{1}{2}t} \tag{1.139}$$

> ics:=y(0)=10,D(y)(0)=20;

$$ics := y(0) = 10, \ D(y)(0) = 20 \tag{1.140}$$

`> dsolve({ode,ics});`

$$y(t) = \frac{195}{32}e^{-\frac{1}{2}t}\sin(4t) + 10\,e^{-\frac{1}{2}t}\cos(4t) + \frac{5}{8}te^{-\frac{1}{2}t} \tag{1.141}$$

1.9 Frobenius Method of Series Solutions to Ordinary Differential Equations

We consider linear second-order differential equations, with variable coefficients, whose method of solution differs from those of the constant coefficients and Cauchy-Euler-type equations. Specifically, we now look at the Frobenius method for finding the series solution to the differential equation. The solution is valid about a point that can, at most, be a "regular singular point" of the differential equation.

Consider the second-order homogeneous differential equation with variable coefficients over a finite closed interval $I = [0, b]$, where the point $x = 0$ is, at most, a regular singular point of the differential equation

$$a2(x)\left(\frac{d^2}{dx^2}y(x)\right) + a1(x)\left(\frac{d}{dx}y(x)\right) + a0(x)\,y(x) = 0$$

The method of Frobenius is based on a collection of theorems that state that if the point $x = 0$ is, at most, a regular singular point of the differential equation, then we can develop a solution that has the form of a Taylor series expansion about the origin, and this series has convergence characteristics that allow for term-by-term differentiation of the series. Thus, we seek a Frobenius series solution expansion about the origin of the form

$$y(x) = \sum_{n=0}^{\infty} c(n)x^{n+r}$$

Here, r is a constant that is to be determined. Assuming conditions are in place that allow for the valid interchange in the order of operations between differentiation and summation (we must be sensitive to this because the series just given is an infinite series), we can calculate the first- and second-order derivatives of the series solution. The first derivative is

$$\frac{d}{dx}y(x) = \sum_{n=0}^{\infty} c(n)x^{n+r-1}(n+r)$$

and the second derivative is

$$\frac{d^2}{dx^2}y(x) = \sum_{n=0}^{\infty} c(n)x^{n+r-2}(n+r)(n+r-1)$$

Combining the derivative terms in the differential equation and assuming the validity of the interchange between the multiplication operation and the summation operation, we obtain the following homogeneous series equation:

$$\sum_{n=0}^{\infty}(a2(x)c(n)(n+r)(n+r-1)x^{n+r-2}+a1(x)c(n)(n+r)x^{n+r-1}+a0(x)c(n)x^{n+r})=0$$

Linear Independence of Unlike Powers of x

The solution to the preceding series equation is driven by the fact that terms of unlike powers of x are linearly independent; that is, we cannot express x raised to some integer power as a linear multiple of x raised to a different power. Thus, the set of vectors $1, x, x^2, x^3, x^4 \ldots$ is linearly independent and any sum of coefficients multiplying unlike powers of x set equal to zero can only be satisfied if each of the coefficients is set equal to zero. In the Frobenius series solution to differential equations, we encounter sums of coefficients multiplying terms of x raised to different powers, and since these sums must equal zero—as shown in the preceding series—the only way we can satisfy these sums is that we set the coefficients of such terms equal to zero. Doing the preceding gives rise to what we later encounter as "indicial" equations and "recursion" formulas.

To facilitate the solution of the preceding series equation, we establish the three partitioned series from the preceding:

$$S2 = \sum_{n=0}^{\infty} a2(x)c(n)(n+r)(n+r-1)x^{n+r-2}$$

$$S1 = \sum_{n=0}^{\infty} a1(x)c(n)(n+r)x^{n+r-1}$$

$$S0 = \sum_{n=0}^{\infty} a0(x)c(n)x^{n+r}$$

Thus, the preceding homogeneous series equation can be written in terms of the three partitioned series $S2$, $S1$, and $S0$, respectively, as

$$S2 + S1 + S0 = 0$$

In order to develop the solution further, we must look at examples for specific values of the differential equation terms $a2(x)$, $a1(x)$, and $a0(x)$.

1.10 Series Sine and Cosine Solutions to the Euler Differential Equation

We again consider the Euler-type differential equation that we already solved using an earlier method. This is an exercise in demonstrating the Frobenius method of solution. We now use simple Maple commands to generate a Frobenius series solution. This allows us to compare the solutions obtained by the two different methods.

EXAMPLE 1.10.1: We seek the Frobenius series solution basis vectors about the origin to the Euler differential equation

$$\frac{d^2}{dx^2}y(x) + y(x) = 0$$

SOLUTION: We assume a series solution

$$y(x) = \sum_{n=0}^{\infty} c(n)x^{n+r}$$

For the Euler differential equation, we identify the coefficients

> restart:a2(x):=1;a1(x):=0;a0(x):=1;

$$a2(x) := 1$$
$$a1(x) := 0$$
$$a0(x) := 1 \tag{1.142}$$

We evaluate the three partitioned series

> S2:=Sum((simplify(a2(x)*c(n)*(n+r)*(n+r−1)*x^(n+r−2)),n=0..infinity));

$$S2 := \sum_{n=0}^{\infty} c(n)(n+r)(n+r-1)x^{n+r-2} \tag{1.143}$$

> S1:=Sum((simplify(a1(x)*c(n)*(n+r)*x^(n+r−1)),n=0..infinity));

$$S1 := \sum_{n=0}^{\infty} 0 \tag{1.144}$$

> S0:=Sum((simplify(a0(x)*c(n)*x^(n+r)),n=0..infinity));

$$S0 := \sum_{n=0}^{\infty} c(n)x^{n+r} \tag{1.145}$$

Adding all of the preceding terms yields the homogeneous series equation

> S:=S2+S0=0;

$$S := \sum_{n=0}^{\infty} c(n)(n+r)(n+r-1)x^{n+r-2} + \sum_{n=0}^{\infty} c(n)x^{n+r} = 0 \qquad (1.146)$$

We now take steps in shifting the summation indices so as to preserve the product terms that have the lowest power on x. In this case, the first series term has the lowest power $n+r-2$; thus, we must shift the summation index $n = n-2$ on the preceding second series, leaving the other series intact.

Doing so, the partitioned series with shifted summation indices becomes

> S2:=Sum(subs(n=n,a2(x)*c(n)*(n+r)*(n+r−1)*x^(n+r−2)),n=0..infinity);

$$S2 := \sum_{n=0}^{\infty} c(n)(n+r)(n+r-1)x^{n+r-2} \qquad (1.147)$$

> S0:=Sum(subs(n=n−2,a0(x)*c(n)*x^(n+r)),n=2..infinity);

$$S0 := \sum_{n=2}^{\infty} c(n-2)x^{n+r-2} \qquad (1.148)$$

The homogeneous series equation now reads (with $S1 = 0$)

> S:=S2+S0=0;

$$S := \sum_{n=0}^{\infty} c(n)(n+r)(n+r-1)x^{n+r-2} + \sum_{n=2}^{\infty} c(n-2)x^{n+r-2} = 0 \qquad (1.149)$$

Note that by shifting the summation indices, all the preceding series have x terms raised to the same lowest power. We now take steps to obtain a general summation term whose summing index is the same for all terms. Those terms that cannot be swept into the general summation give rise to what we call "residual" terms. These terms eventually give rise to what we call the "indicial" equations. Those terms remaining in the generalized sum give rise to the "recursion" formula.

Extracting the first two terms from the preceding first series and sweeping like terms under a general sum with the starting summation index $n = 2$, we get the final homogeneous series equation corresponding to the Euler differential equation

> S:=c(0)*r*(r−1)*x^(r−2)+c(1)*r*(r+1)*x^(r−1)+Sum((a2(x)c(n)*(n+r)*(n+r−1)+a0(x)*
>   c(n−2))*x^(n+r−2),n=2..infinity)=0;

$$S := c(0)r(r-1)x^{r-2} + c(1)r(r+1)x^{r-1} + \sum_{n=2}^{\infty} (c(n)(n+r)(n+r-1) + c(n-2))x^{n+r-2} = 0$$

$$(1.150)$$

The beginning terms that are not in the general summation constitute the residual equation terms shown following:

> resid:=c(0)*r*(r−1)*x^(r−2)+c(1)*r*(r+1)*x^(r−1);

$$resid := c(0)r(r-1)x^{r-2} + c(1)r(r+1)x^{r-1} \qquad (1.151)$$

As stated earlier, since terms of unlike powers of x are linearly independent, then each of the coefficients of unlike powers of x in the preceding sum must equal zero. Setting the coefficient terms in the preceding infinite sum equal to zero gives rise to the general recursion formula

> recur:=a2(x)*c(n)*(n+r)*(n+r−1)+a0(x)*c(n−2)=0;

$$recur := c(n)(n+r)(n+r-1) + c(n-2) = 0 \qquad (1.152)$$

Likewise, setting each of the two coefficient terms in the residual equation equal to zero gives rise to the two indicial equations

> ind1:=c(0)*r*(r−1)=0;

$$ind1 := c(0)r(r-1) = 0 \qquad (1.153)$$

> ind2:=c(1)*r*(r+1)=0;

$$ind2 := c(1)r(r+1) = 0 \qquad (1.154)$$

We now seek nontrivial solutions to the indicial equations. Since both of these terms must vanish, ideally, we solve for those values of r that will make both vanish simultaneously, leaving c(0) and c(1) arbitrary.

> solve(ind1,r);

$$0, 1 \qquad (1.155)$$

> solve(ind2,r);

$$0, -1 \qquad (1.156)$$

The simplest solution to the preceding is to choose $r = 0$, leaving c(0) and c(1) arbitrary. With two arbitrary constants, we are able to generate two linearly independent solutions from only a single recursion formula. Substituting this solution into the recursion formula yields

> r:=0:recur:=recur;

$$recur := c(n)n(n-1) + c(n-2) = 0 \qquad (1.157)$$

An alternative generating form of the recursion formula reads

```
> c(n):=simplify(solve(recur,c(n)));
```

$$c(n) := -\frac{c(n-2)}{n(n-1)} \tag{1.158}$$

Note that the preceding holds only for the summation index $2 \leq n$. We now evaluate the first eight terms in the expansion.

```
> for k from 2 to 8 do
> c(k):=−c(k−2)/(k*(k−1))
> od:
```

The first seven terms of the solution read

```
> y(x):=c(0)*x^r+c(1)*x^(1+r)+eval(sum(c(n)*x^(n+r),n=2..6));
```

$$y(x) := c(0) + c(1)x - \frac{1}{2}c(0)\,x^2 - \frac{1}{6}c(1)x^3 + \frac{1}{24}c(0)x^4 + \frac{1}{120}c(1)x^5 - \frac{1}{720}\,c(0)x^6 \tag{1.159}$$

Substituting $c(0)$ and $c(1)$ by the arbitrary contstants C1 and C2, respectively, and collecting terms, we get

```
> y(x):=subs({c(0)=C1,c(1)=C2},%):y(x):=collect(y(x),{C1,C2});
```

$$y(x) := \left(1 - \frac{1}{2}x^2 + \frac{1}{24}\,x^4 - \frac{1}{720}\,x^6\right)C1 + \left(x - \frac{1}{6}\,x^3 + \frac{1}{120}\,x^5\right)C2 \tag{1.160}$$

Since the constants $C1$ and $C2$ are both arbitrary, we see that the preceding result contains two linearly independent solutions to the differential equation. Thus, a set of truncated system basis vectors is

```
> y1(x):=coeff(y(x),C1);
```

$$y1(x) := 1 - \frac{1}{2}x^2 + \frac{1}{24}x^4 - \frac{1}{720}x^6 \tag{1.161}$$

```
> y2(x):=coeff(y(x),C2);
```

$$y2(x) := x - \frac{1}{6}x^3 + \frac{1}{120}x^5 \tag{1.162}$$

If we compare the preceding two series solutions with the Maclaurin series expansions of the sine and cosine functions, we have

```
> ys:=sin(x)=series(sin(x),x,7);
```

$$ys := \sin(x) = x - \frac{1}{6}x^3 + \frac{1}{120}x^5 + O(x^7) \tag{1.163}$$

> yc:=cos(x)=series(cos(x),x,8);

$$yc := \cos(x) = 1 - \frac{1}{2}x^2 + \frac{1}{24}x^4 - \frac{1}{720}x^6 + O(x^8) \tag{1.164}$$

It is obvious that the two basis vectors obtained from the preceding series solution method are identical to the basis vectors $\sin(x)$ and $\cos(x)$ that we would have obtained from earlier methods. The system basis vectors will play a significant role in developing solutions to partial differential equations in the rectangular-cartesian coordinate system. A plot of the two basis vectors is shown in Figure 1.3.

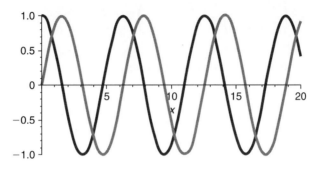

Figure 1.3

General solution: here $C1$ and $C2$ are arbitrary constants.

> y(t):=C1*sin(x)+C2*cos(x);

$$y(t) := C1 \sin(x) + C2 \cos(x) \tag{1.165}$$

Check: Using Maple dsolve command.

> restart:ode:=diff(y(x),x,x)+y(x)=0:expand(dsolve(ode));

$$y(x) = \_C1 \sin(x) + \_C2 \cos(x) \tag{1.166}$$

> plot({sin(x),cos(x)},x=0..20,thickness=10);

Check: Using Maple dsolve command for a series solution.

> restart:ode:=diff(y(x),x,x)+y(x)=0;

$$ode := \frac{d^2}{dx^2} y(x) + y(x) = 0 \tag{1.167}$$

> expand(dsolve(ode),y(x),series);

$$y(x) = \_C1 \sin(x) + \_C2 \cos(x) \tag{1.168}$$

1.11 Frobenius Series Solution to the Bessel Differential Equation

We now consider the Bessel differential equation of order m. This equation is a second-order linear differential equation that occurs often in problems in the cylindrical coordinate system. The point $x = 0$ is a regular singular point.

EXAMPLE 1.11.1: We seek the Frobenius series solution to the Bessel differential equation of order m (m is a positive number) about the origin

$$x^2 \left(\frac{d^2}{dx^2} y(x) \right) + x \left(\frac{d}{dx} y(x) \right) + (x^2 - m^2) y(x) = 0$$

SOLUTION: Assumed series solution about the origin

$$y(x) = \sum_{n=0}^{\infty} c(n) x^{n+r}$$

For the Bessel differential equation, we identify the coefficients

> restart:a2(x):=x^2;a1(x):=x;a01(x):=x^2;a02(x):=−m^2;

$$a2(x) := x^2$$
$$a1(x) := x$$
$$a01(x) := x^2$$
$$a02(x) := -m^2 \tag{1.169}$$

The homogeneous series equation is

> Sum((a2(x)*c(n)*(n+r)*(n+r−1)*x^(n+r−2)+a1(x)*c(n)*(n+r)*x^(n+r−1)+a01(x)*
>   c(n)*x^(n+r))+a02(x)*c(n)*x^(n+r),n=0..infinity)=0;

$$\sum_{n=0}^{\infty} x^2 c(n)(n+r)(n+r-1)x^{n+r-2} + xc(n)(n+r)x^{n+r-1} + x^2 c(n)x^{n+r} - m^2 c(n)x^{n+r} = 0$$

$$\tag{1.170}$$

We evaluate the partitioned series

> S2:=Sum(simplify(a2(x)*c(n)*(n+r)*(n+r−1)*x^(n+r−2)),n=0..infinity);

$$S2 := \sum_{n=0}^{\infty} x^{n+r} c(n)(n+r)(n+r-1) \tag{1.171}$$

> S1:=Sum(simplify(a1(x)*c(n)*(n+r)*x^(n+r−1)),n=0..infinity);

$$S1 := \sum_{n=0}^{\infty} x^{n+r} c(n)(n+r) \tag{1.172}$$

> S01:=Sum(simplify(a01(x)*c(n)*x^(n+r)),n=0..infinity);

$$S01 := \sum_{n=0}^{\infty} x^{2+n+r} c(n) \tag{1.173}$$

> S02:=Sum(simplify(a02(x)*c(n)*x^(n+r)),n=0..infinity);

$$S02 := \sum_{n=0}^{\infty} \left(-m^2 c(n) x^{n+r} \right) \tag{1.174}$$

Collecting all of the preceding, we get the homogeneous series equation

> S:=S2+S1+S01+S02=0;

$$S := \sum_{n=0}^{\infty} x^{n+r} c(n)(n+r)(n+r-1) + \sum_{n=0}^{\infty} x^{n+r} c(n)(n+r) + \sum_{n=0}^{\infty} x^{2+n+r} c(n)$$

$$+ \sum_{n=0}^{\infty} \left(-m^2 c(n) x^{n+r} \right) = 0 \tag{1.175}$$

We now take steps in shifting the summation indices so as to preserve the product terms that have the lowest power on x. In this case, the lowest power is $n+r$; thus, we must shift the summation index $n = n - 2$ on the third preceding series, leaving the other series intact.

Doing so, the partitioned series with shifted summation indices becomes

> S2:=Sum(subs(n=n,simplify(a2(x)*c(n)*(n+r)*(n+r−1)*x^(n+r−2))),n=0..infinity);

$$S2 := \sum_{n=0}^{\infty} x^{n+r} c(n)(n+r)(n+r-1) \tag{1.176}$$

> S1:=Sum(subs(n=n,simplify(a1(x)*c(n)*(n+r)*x^(n+r−1))),n=0..infinity);

$$S1 := \sum_{n=0}^{\infty} x^{n+r} c(n)(n+r) \tag{1.177}$$

> S01:=Sum(subs(n=n−2,simplify(a01(x)*c(n)*x^(n+r))),n=2..infinity);

$$S01 := \sum_{n=2}^{\infty} x^{n+r} c(n-2) \tag{1.178}$$

> S02:=Sum(subs(n=n,simplify(a02(x)*c(n)*x^(n+r))),n=0..infinity);

$$S02 := \sum_{n=0}^{\infty} (-m^2 c(n) x^{n+r}) \tag{1.179}$$

Adding all of the preceding, we get the homogeneous series equation

> S:=S2+S1+S01+S02=0;

$$S := \sum_{n=0}^{\infty} x^{n+r} c(n)(n+r)(n+r-1) + \sum_{n=0}^{\infty} x^{n+r} c(n)(n+r) + \sum_{n=2}^{\infty} x^{n+r} c(n-2)$$

$$+ \sum_{n=0}^{\infty} (-m^2 c(n) x^{n+r}) = 0 \tag{1.180}$$

Note that the preceding series all have x terms raised to the same lowest power. We now take steps to obtain a general summation term whose summing index $n = 2$ is the same for all terms. Those terms that cannot be swept into the general summation give rise to residual terms, which eventually give rise to what we call the "indicial" equations. Those terms remaining in the generalized sum give rise to the "recursion" formula.

Extracting the first two terms from the preceding first, second, and fourth series, and sweeping like terms under a general sum with the starting summation index $n = 2$, we get the final homogeneous series equation corresponding to the Bessel differential equation

> S:=c(0)*(r*r−1)+r−m^2)*x^r+c(1)*(r*(r+1)+(r+1)−m^2)*x^(r+1)+Sum((c(n)*(n+r)*
  (n+r−1)+c(n)*(n+r)+c(n−2)−c(n)*m^2)*x^(n+r),n=2..infinity)=0;

$$S := c(0)(r(r-1) + r - m^2) x^r + c(1)(r(r+1) + r + 1 - m^2) x^{r+1} + \sum_{n=2}^{\infty} (c(n)(n+r)(n+r-1)$$

$$+ c(n)(n+r) + c(n-2) - c(n)m^2) x^{n+r} = 0 \tag{1.181}$$

The beginning terms that are not in the general summation constitute the residual equation terms

> resid:=c(0)*(r*(r−1)+r−m^2)*x^r+c(1)*(r*(r+1)+(r+1)−m^2)*x^(r+1);

$$resid := c(0)(r(r-1) + r - m^2) x^r + c(1)(r(r+1) + r + 1 - m^2) x^{r+1} \tag{1.182}$$

As stated earlier, since terms of unlike powers of x are linearly independent, then each of the coefficients of unlike powers of x in the preceding summation must equal zero. Setting the coefficient terms in the preceding infinite sum equal to zero gives rise to the general recursion formula

> recur:=c(n)*(n+r)*(n+r−1)+c(n)*(n+r)+c(n−2)−c(n)*m^2=0;

$$recur := c(n)(n+r)(n+r-1) + c(n)(n+r) + c(n-2) - c(n)m^2 = 0 \qquad (1.183)$$

Likewise, setting each of the two coefficient terms in the residual equation equal to zero gives rise to the two indicial equations

> ind1:=c(0)*(r*(r−1)+r−m^2)=0;

$$ind1 := c(0)(r(r-1) + r - m^2) = 0 \qquad (1.184)$$

> ind2:=c(1)*(r*(r+1)+r+1−m^2)=0;

$$ind2 := c(1)(r(r+1) + r + 1 - m^2) = 0 \qquad (1.185)$$

Since both of the preceding terms must vanish, ideally, we solve for those values of r that will make both indicial equations vanish simultaneously:

> solve(ind1,r);

$$m, \ -m \qquad (1.186)$$

> solve(ind2,r);

$$-m - 1, \ m - 1 \qquad (1.187)$$

We see that it is impossible to choose a value of r that will make both of the preceding terms vanish simultaneously. In choosing a solution for r, we purposely avoid values that will give negative exponents in the series solution, since this gives rise to singularities. Thus, from the four solutions shown, the only two possible choices are

1. $r = m$, c(0) arbitrary, $c(1) = 0$

2. $r = m - 1$, c(1) arbitrary, $c(0) = 0$

We choose the solution $r = m$, $c(1) = 0$, c(0) arbitrary. Setting $r = m$ in the recursion formula yields

> r:=m:recur:=subs(r=m,recur);

$$recur := c(n)(n+m)(n+m-1) + c(n)(n+m) + c(n-2) - c(n)m^2 = 0 \qquad (1.188)$$

From this, we evaluate the recursion relation

> c(n):=simplify(solve(recur,c(n)));

$$c(n) := -\frac{c(n-2)}{n(n+2\,m)} \qquad (1.189)$$

The preceding holds for $2 \leq n$. We now evaluate terms in the series. From the preceding choices, we set

> c(1):=0;

$$c(1) := 0 \tag{1.190}$$

> for k from 2 to 12 do
> c(k):=−c(k−2)/(k*(k+2*m))
> od:

The first few terms in the expansion are

> y(x):=c(0)*x^r+c(1)*x^(1+r)+eval(sum(c(n)*x^(n+r),n=2..5));

$$y(x) := c(0) \, x^m - \frac{1}{2} \frac{c(0) \, x^{2+m}}{2+2m} + \frac{1}{8} \frac{c(0) \, x^{4+m}}{(2+2m)(4+2m)} \tag{1.191}$$

Thus, one truncated solution to the Bessel differential equation of order m is

> y(x):=collect(y(x),c(0));

$$y(x) := \left(x^m - \frac{1}{2} \frac{x^{2+m}}{2+2m} + \frac{1}{8} \frac{x^{4+m}}{(2+2m)(4+2m)} \right) c(0) \tag{1.192}$$

For the special case when m is an integer, the convention used in most math textbooks is to set the arbitary coefficient c(0) to have the value

$$c(0) = \frac{1}{2^m m!}$$

Thus, for m a positive number, the preceding terms are the first three terms of the Bessel function of the first kind of order m, which, in general, can be expressed as

$$J(m, x) = \sum_{n=0}^{\infty} \frac{(-1)^n x^{m+2n}}{2^{m+2n} n!(n+m)!}$$

For reasons that will become apparent when we study partial differential equations, the term m is generally a positive integer. We now plot Bessel functions of the first kind of integer order zero, one, and two in Figure 1.4.

> plot({BesselJ(0,x),BesselJ(1,x),BesselJ(2,x)},x=0..10,thickness=10);

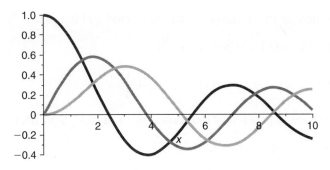

Figure 1.4

EXAMPLE 1.11.2: Now that we have one solution to the Bessel differential equation of order m, we seek a second basis vector using the method of reduction of order. For illustration purposes and ease of calculation here, we focus on the special case for $m = 0$, and we consider only the first three terms in the series expansion of the first basis vector $y1(x)$.

SOLUTION: From Section 1.6, with knowledge of one solution vector $y1(x)$, a second basis vector $y2(x)$ can be found from the equation

$$y2(x) = y1(x) \left(\int \frac{e^{-\left(\int \frac{a1(x)}{a2(x)} dx \right)}}{y1(x)^2} dx \right)$$

For the Bessel differential equation, we identify the coefficients

> restart: a2(x):=x^2;a1(x):=x;

$$a2(x) := x^2 \tag{1.193}$$

$$a1(x) := x$$

The first few terms for $y1(x)$ for $m = 0$ are

> y1(x):=1−x^2/4+x^4/64;

$$y1(x) := 1 - \frac{1}{4}x^2 + \frac{1}{64}x^4 \tag{1.194}$$

Evaluation of the preceding integrand gives

> v(x):=(1/y1(x)^2)*exp(−int(a1(x)/a2(x),x));

$$v(x) := \frac{1}{\left(1 - \frac{1}{4}x^2 + \frac{1}{64}x^4 \right)^2 x} \tag{1.195}$$

Taking just the first three terms of the preceding integrand and integrating yields

> u(x):=int(convert(series(v(x),x,3),polynom),x);

$$u(x) := \ln(x) + \frac{1}{4}x^2 \tag{1.196}$$

Thus, our truncated series expansion of the second basis vector $y2(x)$ becomes

> y2(x):=simplify(expand(y1(x)*u(x)));

$$y2(x) := \ln(x) + \frac{1}{4}x^2 - \frac{1}{4}x^2\ln(x) - \frac{1}{16}x^4 + \frac{1}{64}x^4\ln(x) + \frac{1}{256}x^6 \tag{1.197}$$

If we compare this second linearly independent solution with $Y_0(x)$, the Bessel function of the second kind of order $m = 0$, we have

> Y[0](x):=expand((Pi/2)*(convert(series(BesselY(0,x),x,5),polynom)));

$$Y_0(x) := -\ln(2) + \ln(x) + \gamma + \frac{1}{4}x^2\ln(2) - \frac{1}{4}x^2\ln(x) + \frac{1}{4}x^2 - \frac{1}{4}x^2\gamma - \frac{1}{64}x^4\ln(2)$$

$$+ \frac{1}{64}x^4\ln(x) - \frac{3}{128}x^4 + \frac{1}{64}x^4\gamma \tag{1.198}$$

Except for the conventions used in establishing certain constants ($\gamma = $ Euler's constant), we see that our second series solution for $y2(x)$ is equivalent to the Bessel function of the second kind of order zero. Because $x = 0$ is a regular singular point of the differential equation, we note that the preceding function fails to exist at the point $x = 0$.

We plot Bessel functions of the second kind of integer orders zero, one, and two in Figure 1.5.

> plot({BesselY(0,x),BesselY(1,x),BesselY(2,x)},x=0..8,y=−4..2,thickness=10);

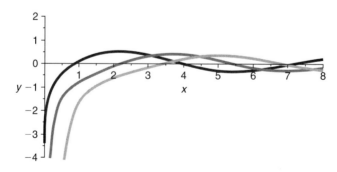

Figure 1.5

The significant feature of the plot of Figure 1.5 shows that the Bessel functions of the second kind, of integer order, all fail to exist at $x = 0$. The preceding system basis vectors will play a significant role when we consider partial differential equations in the cylindrical coordinate system.

Chapter Summary

First-order linear nonhomogeneous differential equation

$$a1(t)\left(\frac{d}{dt}y(t)\right) + a0(t)y(t) = f(t)$$

System basis vector

$$y1(t) = e^{\int\left(-\frac{a0(t)}{a1(t)}\right)dt}$$

Particular solution

$$y_p(t) = \int_0^t G1(t,s)f(s)ds$$

First-order Green's function $G1(t,s)$

$$G1(t,s) = \frac{y1(t)}{a1(s)y1(s)}$$

Second-order linear nonhomogeneous differential equation

$$a2(t)\left(\frac{d^2}{dt^2}y(t)\right) + a1(t)\left(\frac{d}{dt}y(t)\right) + a0(t)y(t) = f(t)$$

Second basis vector $y2(t)$ as evaluated from the first $y1(t)$

$$y2(t) = y1(t)\left(\int \frac{e^{-\left(\int \frac{a1(s)}{a2(s)}ds\right)}}{y1(s)^2}ds\right)$$

Particular solution in terms of the second-order Green's function

$$y_p(t) = \int_0^t G2(t,s)f(s)ds$$

Second-order Green's function $G2(t,s)$

$$G2(t,s) = \frac{y1(s)y2(t) - y1(t)y2(s)}{a2(s)W(y1(s),y2(s))}$$

Series solution to the second-order linear ordinary differential equation

$$a2(x)\left(\frac{d^2}{dx^2}y(x)\right)+a1(x)\left(\frac{d}{dx}y(x)\right)+a0(x)y(x)=0$$

Assumed Frobenius solution about the origin

$$y(x)=\sum_{n=0}^{\infty}c(n)x^{n+r}$$

Equivalent series equation

$$\sum_{n=0}^{\infty}\left(a2(x)c(n)(n+r)(n+r-1)x^{n+r-2}+a1(x)c(n)(n+r)x^{n+r-1}+a0(x)c(n)x^{n+r}\right)=0$$

From the preceding series equation, we develop the indicial equations and the recursion formula for the final solution.

Most often occurring differential equations

1. The Euler differential equation

$$\frac{d^2}{dx^2}y(x)+\lambda\,y(x)=0$$

A set of basis vectors for the Euler equation

$$y1(x)=\sin\left(\sqrt{\lambda}\,x\right)$$

and

$$y2(x)=\cos\left(\sqrt{\lambda}\,x\right)$$

2. The Cauchy-Euler differential equation

$$x^2\left(\frac{d^2}{dx^2}y(x)\right)+x\left(\frac{d}{dx}y(x)\right)+\lambda y(x)=0$$

A set of basis vectors for the Cauchy-Euler equation

$$y1(x)=\sin\left(\sqrt{\lambda}\,\ln(x)\right)$$

and

$$y2(x)=\cos\left(\sqrt{\lambda}\,\ln(x)\right)$$

3. The Bessel differential equation of order m

$$x^2 \left(\frac{d^2}{dx^2} y(x) \right) + x \left(\frac{d}{dx} y(x) \right) + \left(x^2 - m^2 \right) y(x) = 0$$

A set of basis vectors for the Bessel equation

$$y1(x) = J(m, x)$$

and

$$y2(x) = Y(m, x)$$

These procedures have shown that we can find the general solution to both first- and second-order linear ordinary nonhomogeneous differential equations. For the case of second-order nonhomogeneous differential equations, we demonstrated that from the knowledge of only a single basis vector, we were able to generate the complete solution to the differential equation. We have stressed the procedure of the method of variation of parameters and the subsequent evaluation of the corresponding Green's function. These solutions will play a big role in the development of solutions to some partial differential equations later.

Exercises

Consider the following first-order linear nonhomogeneous differential equations. In all cases, evaluate one basis vector $y1(x)$ and the first-order Green's function $G1(x, s)$, and then write out the general solution. When possible, use the Maple command dsolve to verify the solution.

1.1.

$$\frac{d}{dx} y(x) + \frac{y(x)}{x} = 2 + x^2$$

1.2.

$$x \left(\frac{d}{dx} y(x) \right) + 2 y(x) = e^x + \ln(x)$$

1.3.

$$\left(x^2 + 1 \right) \left(\frac{d}{dx} y(x) \right) - 2 xy(x) = x^2 + 1$$

1.4.

$$\frac{d}{dx} y(x) + \tan(x) y(x) = \sec(x)$$

1.5.

$$x\left(\frac{d}{dx}y(x)\right) + (1+x)y(x) = 3e^{-x}$$

Consider the following initial-value problems. In all cases, evaluate one basis vector $y1(t)$ and the first-order Green's function $G1(t, s)$, and then evaluate the final solution.

1.6. $y(0) = 4$

$$\frac{d}{dt}y(t) + 2\,y(t) = t + \sin(t)$$

1.7. $y(0) = 3$

$$\frac{d}{dt}y(t) + ty(t) = t^3$$

1.8. $y(0) = -5$

$$\frac{d}{dt}y(t) + \frac{4y(t)}{t+2} = 3$$

1.9. $y(0) = 7$

$$\left(2t^2 + 1\right)\left(\frac{d}{dt}y(t)\right) - 10ty(t) = 4t$$

1.10. $y(0) = 1$

$$\frac{d}{dt}y(t) + y(t) = e^{-2t}$$

The following problems demonstrate applications of first-order initial value problems. Generate all solutions and develop a graph of the solution and an animated display as done in the examples in Section 1.3.

1.11.   A tank is initially filled with 40 gal of a salt solution containing 2 lb of salt per gal. Fresh brine containing 3 lb of salt per gal runs into the tank at 4 gal per min, and the mixture is kept uniform by constant stirring. (a) If the mixture runs out of the tank at the same rate it runs in, find the amount of salt $q(t)$ at any time in the tank. (b) Evaluate the time at which the amount of salt in the tank reaches 100 lb. Hint: This is an initial value problem with $q(0) = 80$ lb, and the differential equation is

$$\frac{d}{dt}q(t) = 12 - \frac{q(t)}{10}$$

1.12.   Solve Exercise 1.11 for the case where the initial concentration of salt in the container is 5 lb of salt per gal.

1.13. According to Newton's law of cooling, a body loses heat at a rate proportional to the instantaneous temperature difference between the body and its surroundings. Consider a body that loses heat in accordance with this law, where the initial temperature of the body is 70°C, the surrounding temperature is 20°C, and the temperature decay coefficient is $k = 3/10$ per min. (a) Find an expression for the temperature $T(t)$ at any time t. (b) At what time will the body temperature be 30°C? Hint: This is an initial condition problem with $T(0) = 70°C$, and the differential equation is

$$\frac{d}{dt}T(t) = -k(T(t) - 20)$$

1.14. Solve Exercise 1.13 for the case where $k = 5/10$/min and $T(0) = 100°C$.

1.15. A body of mass m undergoes freefall due to gravity in a viscous medium, where the viscous force is proportional to the body speed. We would like to calculate the speed $v(t)$ of a body whose mass is $m = 5$ kg in freefall if its initial speed is 50 m/sec upward and the damping coefficient is $\gamma = 1$ N/m/sec. The acceleration due to gravity is $g = 9.8$ m/sec/sec in the MKS system. Hint: This is an initial value problem with $v(0) = 50$ m/sec, and the differential equation is

$$m\left(\frac{d}{dt}v(t)\right) = -mg - \gamma\, v(t)$$

1.16. Solve Exercise 1.15 for the time at which the speed is zero.

1.17. Solve Exercise 1.15 for the case where the damping coefficient is doubled.

1.18. We consider a resistance-inductance (RL) circuit with an impressed voltage $e(t)$, where the electric current in the circuit at any time t is $i(t)$. The circuit resistance $R = 20$ ohms and the inductance $L = 100$ henries. At time $t = 0$, the circuit switch is closed, and the impressed voltage is $e(t) = 20$ volts. Solve for the current in the circuit. Hint: This is an initial value problem with $i(0) = 0$ amps, and the differential equation is

$$L\left(\frac{d}{dt}i(t)\right) + Ri(t) = e(t)$$

1.19. Solve Exercise 1.18 for the case where the impressed voltage is $e(t) = 40 \sin(t)$ volts and the initial current is $i(0) = 10$ amps.

1.20. We consider a resistance-capacitance (RC) circuit with an impressed voltage $e(t)$, where the electric charge on the capacitor at any time t is $q(t)$. The circuit resistance $R = 3$ ohms, and the capacitance $C = 2$ farads. At time $t = 0$, the circuit switch is closed and the capacitor has an initial charge of 200 coulombs. If the impressed voltage is $e(t) = 20$ volts, solve for the charge in the circuit. Hint: This is an initial value problem with $q(0) = 200$ coulombs, and the differential equation is

$$R\left(\frac{d}{dt}q(t)\right) + \frac{q(t)}{C} = e(t)$$

1.21.   Solve Exercise 1.20 where the impressed voltage is $e(t) = 10te^{-t}$ volts and the initial charge is $q(0) = 0$ coulombs. Find the time at which the charge in the circuit is a maximum.

Consider the following second-order linear differential equations with constant coefficients. Evaluate two basis vectors $y1(t)$ and $y2(t)$ and, from these, evaluate the second-order Green's function $G2(t, s)$. With the Green's function, evaluate a particular solution. When possible, use the Maple dsolve command to verify the solution.

1.22.

$$\frac{d^2}{dt^2}y(t) + 4\left(\frac{d}{dt}y(t)\right) + 3y(t) = 3e^t$$

1.23.

$$\frac{d^2}{dt^2}y(t) - 5\left(\frac{d}{dt}y(t)\right) + 4y(t) = 10\sinh(t)$$

1.24.

$$\frac{d^2}{dt^2}y(t) + 2\left(\frac{d}{dt}y(t)\right) - 3y(t) = 2 + 3e^t$$

1.25.

$$\frac{d^2}{dt^2}y(t) + 4\left(\frac{d}{dt}y(t)\right) + 13y(t) = \sin(t)$$

Consider the following Cauchy-Euler second-order differential equations with variable coefficients. Evaluate two basis vectors $y1(t)$ and $y2(t)$ and, from these, evaluate the second-order Green's function $G2(t, s)$.

1.26.

$$t^2\left(\frac{d^2}{dt^2}y(t)\right) + 7t\left(\frac{d}{dt}y(t)\right) + 9y(t) = f(t)$$

1.27.

$$t^2\left(\frac{d^2}{dt^2}y(t)\right) + t\left(\frac{d}{dt}y(t)\right) + y(t) = f(t)$$

1.28.

$$t^2 \left(\frac{d^2}{dt^2} y(t) \right) - 4t \left(\frac{d}{dt} y(t) \right) + 6y(t) = f(t)$$

1.29.

$$2t^2 \left(\frac{d^2}{dt^2} y(t) \right) + 5t \left(\frac{d}{dt} y(t) \right) + y(t) = f(t)$$

1.30. In Exercise 1.26, let $f(t) = \ln(t)$. Use the evaluated Green's function to find a particular solution. Use the Maple dsolve command to verify the solution.

1.31. In Exercise 1.27, let $f(t) = t$. Use the evaluated Green's function to find a particular solution. Use the Maple dsolve command to verify the solution.

1.32. In Exercise 1.28, let $f(t) = t^4 e^t$. Use the evaluated Green's function to find a particular solution. Use the Maple dsolve command to verify the solution.

1.33. In Exercise 1.29, let $f(t) = t^2 - t$. Use the evaluated Green's function to find a particular solution. Use the Maple dsolve command to verify the solution.

1.34. The simple spring-mass system consists of a mass m that is attached to an ideal spring with a spring constant k. The system is immersed in a viscous damping medium where γ is the damping coefficient. The external applied force acting on the system is denoted $f(t)$, and the displacement of the mass from its equilibrium position is denoted $y(t)$. The differential equation that describes the system is

$$m \left(\frac{d^2}{dt^2} y(t) \right) + \gamma \left(\frac{d}{dt} y(t) \right) + ky(t) = f(t)$$

We look at the case where $m = 1$ kg, $\gamma = 0$ N/m/sec, and $k = 9$ N/m. (a) Evaluate a set of system basis vectors $y1(t)$ and $y2(t)$, and evaluate the second-order Green's function $G2(t, s)$. (b) If the initial conditions are $y(0) = 2$ m, $v(0) = 10$ m/sec, and the driving force is $f(t) = 10 \sin(t)$ N, evaluate the time-dependent solution $y(t)$ for the problem. (c) Graph the solution and develop the animation as done in Example 1.8.1. This is a simple second-order initial value problem.

1.35. Do Exercise 1.34 for the case where $m = 1$ kg, $\gamma = 1$ N/m/sec, $k = 37/4$ N/m, and the driving force is $f(t) = 20te^{-t}$ N. All other conditions are the same. Develop all graphics and animation.

1.36. Do Exercise 1.34 for the case where $m = 1$ kg, $\gamma = 1$ N/m/sec, $k = 82/4$ N/m, and the driving force is $f(t) = 10\,\text{Heaviside}(t)$ N. [Note: Heaviside(t) denotes the unit step function.] Develop all graphics and animation.

1.37.   The series RLC circuit consists of a resistor R, an inductor L, and a capacitor C connected in series. If the electric charge on the capacitor in the system is denoted as $q(t)$, and the driving voltage is $e(t)$, then the differential equation that describes the time dependence of the charge is given as

$$L\left(\frac{d^2}{dt^2}q(t)\right) + R\left(\frac{d}{dt}q(t)\right) + \frac{q(t)}{C} = e(t) \qquad (1.199)$$

We consider the case where $L = 1$ henry, $R = 1$ ohms, and $C = 4/50$ farads and the driving voltage is $e(t) = 10\,te^{-t}$ volts. (a) Evaluate the system basis vectors $q1(t)$ and $q2(t)$ and the second-order Green's function $G2(t, s)$. (b) If the initial charge $q(0) = 0$ coulombs and the initial current $i(0) = 0$ amps, evaluate the time dependence of the charge $q(t)$ on the capacitor. (c) Graph the solution and develop the animation as done in Example 1.8.1. This is a simple second-order initial value problem.

1.38.   Do Exercise 1.37 for the case where $L = 1$ henry, $R = 1$ ohms, and $C = 4/10$ farads, and the driving voltage is $e(t) = 10\sin(t)$ volts. All other conditions are the same. Develop all graphics and animation.

For the following differential equations, you are given one basis vector $y1(t)$. You are asked to evaluate a second basis vector $y2(t)$ using the method of reduction of order. You are then asked to evaluate the second-order Green's function $G2(t, s)$.

1.39.   Given one basis vector $y1(t) = e^{3t}$, find the second-order Green's function $G2(t, s)$.

$$\frac{d^2}{dt^2}y(t) - 6\left(\frac{d}{dt}y(t)\right) + 9y(t) = f(t)$$

1.40.   Given one basis vector $y1(t) = t$, find the second-order Green's function $G2(t, s)$.

$$t^2\left(\frac{d^2}{dt^2}y(t)\right) - t\left(\frac{d}{dt}y(t)\right) + y(t) = f(t)$$

1.41.   Given one basis vector $y1(t) = t - 1$, find the second-order Green's function $G2(t, s)$.

$$\left(2t - t^2\right)\left(\frac{d^2}{dt^2}y(t)\right) + 2(t - 1)\left(\frac{d}{dt}y(t)\right) - 2y(t) = f(t)$$

1.42.   Given one basis vector $y1(t) = t^3$, find the second-order Green's function $G2(t, s)$.

$$t^2\left(\frac{d^2}{dt^2}y(t)\right) - 6y(t) = f(t)$$

1.43.   Given one basis vector $y1(t) = t$, find the second-order Green's function $G2(t, s)$.

$$t^3 \left(\frac{d^2}{dt^2} y(t) \right) + t \left(\frac{d}{dt} y(t) \right) - y(t) = f(t) \qquad (1.200)$$

1.44.   Let $f(t) = te^{3t}$ in Exercise 1.39. Use the Green's function that you evaluated to find a particular solution. Check your answer by substituting the solution back into the differential equation.

1.45.   Let $f(t) = \frac{1}{t}$ in Exercise 1.40. Use the Green's function that you evaluated to find a particular solution. Use the Maple dsolve command to verify the solution.

1.46.   Let $f(t) = t^3 \ln(t)$ in Exercise 1.42. Use the Green's function that you evaluated to find a particular solution. Use the Maple dsolve command to verify the solution.

1.47.   Let $f(t) = t$ in Exercise 1.43. Use the Green's function that you evaluated to find a particular solution. Use the Maple dsolve command to verify the solution.

The following linear differential equations are to be solved by using the method of Frobenius for finding a series solution about the origin ($x = 0$). For the Exercises 1.48 through 1.51, (a) evaluate the indicial equations and the corresponding recursion formula. (b) From the indicial equation, obtain a set of basis vectors from different choices of roots of the indicial equation. If you are able to obtain only one solution, then use the method of reduction of order to find a second basis vector.

1.48.

$$\frac{d^2}{dx^2} y(x) + x \left(\frac{d}{dx} y(x) \right) + y(x) = 0$$

1.49.

$$\frac{d^2}{dx^2} y(x) - x \left(\frac{d}{dx} y(x) \right) - x^2 y(x) = 0$$

1.50.

$$x \left(\frac{d^2}{dx^2} y(x) \right) + 2 \left(\frac{d}{dx} y(x) \right) + xy(x) = 0$$

1.51.

$$x \left(\frac{d^2}{dx^2} y(x) \right) + 3 \left(\frac{d}{dx} y(x) \right) - y(x) = 0$$

1.52. The Bessel differential equation solved in Section 1.11 comes about when one considers the diffusion and wave equations in the cylindrical coordinate system. For an integer m, the Bessel function of the first kind of order m reads as

$$J(m, x) = \sum_{n=0}^{\infty} \frac{(-1)^n x^{m+2n}}{2^{m+2n} n!(m+n)!}$$

From the preceding series, verify the following

(a)

$$J(0, 0) = 1$$

(b)

$$J(0, -x) = J(0, x)$$

(c)

$$\frac{d}{dx} J(0, x) = -J(1, x)$$

(d)

$$x\left(\frac{\partial}{\partial x} J(m, x)\right) = -m J(m, x) + x J(m-1, x)$$

(e)

$$\int x^m J(m-1, x) dx = x^m J(m, x)$$

(f)

$$2 m J(m, x) = x(J(m-1, x) + J(m+1, x))$$

CHAPTER 2

Sturm-Liouville Eigenvalue Problems and Generalized Fourier Series

2.1 Introduction

In Chapter 1 we examined both first- and second-order linear homogeneous and nonhomogeneous differential equations. We established the significance of the dimension of the solution space and the basis vectors. With a set of basis vectors, we could span the entire solution space of the homogeneous equation by way of a linear superposition of these basis vectors. We also showed how the method of variation of parameters utilized the basis vectors in the construction of the Green's function. With the Green's function in hand, we were then able to evaluate the solution to the corresponding nonhomogeneous differential equation.

2.2 The Regular Sturm-Liouville Eigenvalue Problem

We will now focus on second-order differential equations, over finite intervals, when the independent variable is the spatial variable x. These problems will play a prominent role in the type of partial differential equations that we will look at later.

The problems that we examine here fall under the general category of what are called "regular" Sturm-Liouville eigenvalue problems. These are basically second-order linear homogeneous differential equations whose constraints are in the form of two-point, homogeneous, nonmixed boundary conditions.

To begin, we first define the generalized Sturm-Liouville differential operator L acting on some function $y(x)$ over a finite interval $I = \{x \mid a < x < b\}$ as follows:

$$L(y) = \mathrm{D}(p(x)\mathrm{D}(y)) + q(x)y \tag{2.1}$$

Here, D is the standard differentiation operator defined as

$$D(y) = \frac{d}{dx} y(x)$$

The "regular" Sturm-Liouville eigenvalue problem consists of two parts:

1. The homogeneous Sturm-Liouville differential equation

$$L(\varphi) + \lambda\, w(x)\varphi = 0$$

2. The "regular" boundary conditions that are homogeneous, nonmixed (single point, spatially separated) conditions at the end points

$$\kappa_1 \varphi(a) + \kappa_2 \varphi_x(a) = 0$$

and

$$\kappa_3 \varphi(b) + \kappa_4 \varphi_x(b) = 0$$

The coefficients $\kappa_1, \kappa_2, \kappa_3, \kappa_4$ are all real constants. Restrictive conditions on the system are that over the finite interval I, the functions $p(x)$, $p'(x)$, $q(x)$, and $w(x)$ are real continuous functions and that $p(x)$ and $w(x)$ are positive over this interval I. In some situations, the preceding conditions can be relaxed and modified somewhat and the problem becomes "singular." We shall consider singular problems later.

The Sturm-Liouville differential equation given earlier is linear and homogeneous, and the boundary conditions are homogeneous and nonmixed. The allowed values of λ_n for which the preceding differential equation satisfies the boundary conditions are called the "eigenvalues," and the corresponding solutions $\varphi_n(x)$ are called the "eigenfunctions."

We state, without proof, some of the important features of "regular" Sturm-Liouville problems:

1. There exist an infinite number of eigenvalues λ_n that can be ordered in magnitude and that can be indexed by the positive integers $n = 0, 1, 2, 3, \ldots.$

2. All eigenvalues are real.

3. Corresponding to each eigenvalue λ_n there exists a unique eigenfunction $\varphi_n(x)$.

4. The eigenfunctions form a "complete" set with respect to any piecewise smooth function $f(x)$ over the finite interval $I = \{x \mid a < x < b\}$. This means that, over this interval, the function can be represented as a generalized Fourier series expansion in terms of the eigenfunctions shown here:

$$f(x) = \sum_{n=0}^{\infty} F(n)\varphi_n(x)$$

where the terms $F(n)$ are the properly evaluated Fourier coefficients.

5. The infinite series given converges to the average of the left- and right-hand limits of the function at any point in the interval; that is,

$$\sum_{n=0}^{\infty} F(n)\varphi_n(x) = \frac{f(x+0) + f(x-0)}{2}$$

6. The eigenfunctions corresponding to different eigenvalues are orthogonal relative to the "weight function" $w(x)$ over the interval I. If the orthogonal eigenfunctions are further normalized, then we can express the statement of orthonormality of the eigenfunctions in terms of the inner product (see Chapter 0), with respect to the weight function $w(x)$, as

$$\int_a^b \varphi_n(x)\varphi_m(x)w(x)\mathrm{d}x = \delta(n, m)$$

where we have used the Kronecker delta function $\delta_{n,m}$, which, by definition, has the value 1 for $n = m$ and the value 0 for $n \neq m$.

2.3 Green's Formula and the Statement of Orthonormality

In terms of the standard differentiation operator D, the Sturm-Liouville operator L acting on some function $y(x)$ reads

$$L(y) = \mathrm{D}(p(x)\mathrm{D}(y)) + q(x)y$$

It should be remarked that any second-order linear differential operator

$$a2(x)\mathrm{D}^2 + a1(x)\mathrm{D} + a0(x)$$

can be written in Sturm-Liouville form (see Exercise 2.1) by multiplying the operator by the integrating factor

$$r(x) = \frac{e^{\int \frac{a1(x)}{a2(x)}\mathrm{d}x}}{a2(x)}$$

We now provide a formal proof of the statement of orthogonality of the eigenfunctions. We consider the Sturm-Liouville operator acting on the two different functions u and v:

$$L(u) = \mathrm{D}(p(x)\mathrm{D}(u)) + q(x)u$$

and

$$L(v) = \mathrm{D}(p(x)\mathrm{D}(v)) + q(x)v$$

We evaluate the Green's formula, which is shown next for the two functions u and v.

$$\int_a^b (uL(v) - vL(u))dx = \int_a^b (u(Dp(x)D(v) + q(x)v) - v(Dp(x)D(u) + q(x)u))dx$$

Expansion and simplification of the integrand on the right-hand side yields

$$\int_a^b (uL(v) - vL(u))dx = \int_a^b D(p(x)(uD(v) - vD(u)))dx$$

Since the integrand on the right is an exact differential, a simple integration yields

$$\int_a^b (uL(v) - vL(u))dx = p(b)(u(b)v_x(b) - v(b)u_x(b)) - p(a)(u(a)v_x(a) - v(a)u_x(a))$$

If we impose the regular Sturm-Liouville boundary conditions from Section 2.2 on the functions u and v, then the right-hand term in the preceding Green's formula vanishes, and we obtain

$$\int_a^b uL(v)dx = \int_a^b vL(u)dx$$

When the preceding statement holds, we say the operator L, for the appropriate boundary conditions, is "self-adjoint."

If we now write the Sturm-Liouville differential equation for the two different eigenfunctions $\varphi_n(x)$ and $\varphi_m(x)$, we have

$$L(\varphi_n) + \lambda_n w(x)\varphi_n = 0$$

and

$$L(\varphi_m) + \lambda_m w(x)\varphi_m = 0$$

Rearranging these equations and integrating over the interval yields

$$\int_a^b (\varphi_n L(\varphi_m) - \varphi_m L(\varphi_n)) dx = (\lambda_n - \lambda_m)\left(\int_a^b \varphi_n(x)\varphi_m(x)w(x)dx\right)$$

If the eigenfunctions satisfy the regular Sturm-Liouville boundary conditions from Section 2.2, then Green's formula shows the left-hand side to be zero and, thus, the right-hand side is also zero. This yields the statement of orthogonality of the eigenfunctions:

$$(\lambda_n - \lambda_m) \left(\int_a^b \varphi_n(x)\varphi_m(x)w(x)\mathrm{d}x \right) = 0$$

The integral here denotes the inner product (see Chapter 0) of the two different eigenfunctions with respect to the weight function $w(x)$ over the interval $I = \{x \mid a < x < b\}$. This result shows that if the eigenvalues are not equal—that is, $\lambda_n \neq \lambda_m$—then the eigenfunctions, corresponding to the different eigenvalues, are orthogonal with respect to the weight function $w(x)$ over the interval I.

The Statement of Orthonormality

To normalize a vector, we simply divide the vector by its length or norm. From Chapter 0, recall that the norm is the square root of the inner product of the vector with itself with respect to the weight function $w(x)$—that is,

$$norm = \sqrt{\int_a^b \varphi_n(x)^2 w(x)\mathrm{d}x}$$

If we normalize the eigenfunctions—that is, divide each by its length (norm)—then the preceding condition of orthogonality yields an equivalent "statement of orthonormality," which, in terms of the Kronecker delta function, reads

$$\int_a^b \varphi_n(x)\varphi_m(x)w(x)\mathrm{d}x = \delta(n, m) \tag{2.2}$$

This statement indicates the two different eigenfunctions to be orthonormal (a combination of being orthogonal and normalized) with respect to the weight function $w(x)$ over the interval I. This result will play a significant role in the development of the generalized Fourier series.

Types of Boundary Conditions

Before we consider different examples of Sturm-Liouville operators with various types of nonmixed homogeneous boundary conditions, we first classify the regular boundary conditions that we will look at in three types. Let point a be the boundary point.

1. **Type 1 or Dirichlet** condition (vanishing of the zero derivative at the boundary)

$$y(a) = 0$$

2. **Type 2 or Neumann** condition (vanishing of the first derivative at the boundary)

$$y_x(a) = 0$$

3. **Type 3 or Robin** condition (linear dependence between zero and first derivative at boundary)

$$y_x(a) + hy(a) = 0$$

The constant h in the Robin condition has the dimension of the reciprocal of displacement. When we consider actual boundary value problems, we will be able to attach physical significance to each of these conditions. We now provide an illustration of a typical Sturm-Liouville eigenvalue problem.

DEMONSTRATION: We seek the eigenvalues and corresponding eigenfunctions for the Euler differential equation shown:

$$\frac{d^2}{dx^2} y(x) + \lambda y(x) = 0$$

over the interval $I = \{x \mid 0 < x < 1\}$ with the boundary conditions

$$y_x(0) = 0$$

and

$$y(1) = 0$$

From our earlier classification of boundary conditions, we have a type 2 condition at the left and a type 1 condition at the right.

By comparison, we see the Euler differential equation to be a special case of a Sturm-Liouville differential equation

$$L(y) + \lambda w(x)y = 0$$

where L is the Sturm-Liouville operator

$$L(y) = D(p(x)D(y)) + q(x)y$$

For the Euler differential equation, we identify the coefficients $p(x) = 1$, $q(x) = 0$, and $w(x) = 1$.

SOLUTION: To solve for the eigenvalues, we consider three possible cases for values of λ. We first consider the case for $\lambda < 0$. We set $\lambda = -\mu^2$. From Chapter 1, a set of basis vectors for this case is

$$y1(x) = \sinh(\mu x)$$

and

$$y2(x) = \cosh(\mu x)$$

and the general solution can be written in terms of this basis as

$$y(x) = C1 \sinh(\mu x) + C2 \cosh(\mu x)$$

Substituting the boundary conditions at $x = 0$ and at $x = 1$, we get

$$C1\mu = 0$$

and

$$C2\mu \sinh(1) = 0$$

The only solution to the preceding is the trivial solution $C1 = 0$ and $C2 = 0$.

We next consider the case for $\lambda = 0$. A set of system basis vectors is

$$y1(x) = 1$$

and

$$y2(x) = x$$

and the general solution in terms of this basis is

$$y(x) = C1 + C2 x$$

Substituting the boundary condition at $x = 0$ and at $x = 1$ gives

$$C2 = 0$$

and

$$C1 + C2 = 0$$

The only solution to the preceding is the trivial solution $C1 = 0$ and $C2 = 0$. From a geometric standpoint, we expected this result because it is impossible to have a nontrivial line that will satisfy the boundary conditions at both the left and right end points of the interval.

Finally, we consider the case for $\lambda > 0$. We set $\lambda = \mu^2$. From Chapter 1, a set of basis vectors for this case is

$$y1(x) = \sin(\mu x)$$

and

$$y2(x) = \cos(\mu x)$$

The general solution here is

$$y(x) = C1\sin(\mu x) + C2\cos(\mu x)$$

Substituting the boundary conditions at $x = 0$ and at $x = 1$, we get

$$C1\mu = 0$$

and

$$C1\sin(\mu) + C2\cos(\mu) = 0$$

The only nontrivial solutions to the preceding are that $C1 = 0$, $C2$ is arbitrary, and μ has values that are the roots of the eigenvalue equation

$$\cos(\mu) = 0$$

Thus, μ takes on the special values

$$\mu_n = \frac{(2n-1)\pi}{2}$$

for $n = 1, 2, 3, \ldots$.

The allowed eigenvalues $\lambda_n = \mu_n^2$ are

$$\lambda_n = \frac{(2n-1)^2\pi^2}{4}$$

and the corresponding eigenfunctions are

$$\varphi_n(x) = \cos\left(\frac{(2n-1)\pi x}{2}\right)$$

for $n = 1, 2, 3, \ldots$.

These eigenfunctions are not normalized. We must first find the norm of each eigenfunction by evaluating the square root of the inner product of the eigenfunctions with respect to the weight function $w(x) = 1$ over the interval.

The norm is

$$norm = \sqrt{\int_0^1 \cos\left(\frac{(2n-1)\pi x}{2}\right)^2 dx}$$

Evaluating this integral yields

$$norm = \frac{\sqrt{2}}{2}$$

Dividing each eigenfunction by its norm yields the orthonormal set of eigenfunctions

$$\varphi_n(x) = \sqrt{2}\cos\left(\frac{(2n-1)\pi x}{2}\right)$$

for $n = 1, 2, 3, \ldots$.

The statement of orthonormality for this set of eigenfunctions reads

$$\int_0^1 2\cos\left(\frac{(2n-1)\pi x}{2}\right)\cos\left(\frac{(2m-1)\pi x}{2}\right) dx = \delta(n, m)$$

for $n, m = 1, 2, 3, \ldots$.

2.4 The Generalized Fourier Series Expansion

Since the eigenfunctions form a "complete" set with respect to any piecewise smooth function $f(x)$ over the interval $I = \{x \mid a < x < b\}$, then we can expand $f(x)$ as a generalized Fourier series. In terms of the orthonormalized eigenfunctions, the generalized expansion formula for $f(x)$ reads

$$f(x) = \sum_{n=0}^{\infty} F(n)\varphi_n(x)$$

where the $F(n)$ are the generalized Fourier coefficients of the function $f(x)$. We proceed to evaluate these coefficients in a formal manner by taking advantage of the statement of orthonormality given earlier.

If we take the inner product of both sides of the preceding equation, with respect to the weight function $w(x)$ over the interval, and we assume it is valid to interchange the order of the summation and integration operators (see exercises on conditions for this), then we get

$$\int_a^b f(x)\varphi_m(x)w(x)dx = \sum_{n=0}^{\infty} F(n)\left(\int_a^b \varphi_n(x)\varphi_m(x)w(x)dx\right)$$

Because of the statement of orthonormality, we can write the integral on the right in terms of the Kronecker delta function as

$$\int_a^b f(x)\varphi_m(x)w(x)dx = \sum_{n=0}^{\infty} F(n)\delta(n,m)$$

The sum on the right is easy to evaluate, since $\delta(n,m)$ is 0 for $n \neq m$ and 1 for $n = m$. Evaluating the sum (only one term survives), we get

$$F(m) = \int_a^b f(x)\varphi_m(x)w(x)dx$$

Thus, we see that the Fourier coefficients are evaluated from the inner product of $f(x)$ and the corresponding orthonormal eigenfunction, with respect to the weight function $w(x)$, over the interval I.

Convergence of the Fourier Series

Before we provide some examples, we first consider some important aspects on the convergence of a series. We consider two important questions dealing with the convergence of the Fourier series: (1) Will the series converge? (2) To what value does the series converge? Both of these questions are addressed by the following two theorems (from references, see Berg and McGregor, page 157).

Theorem 2.4.1 (Pointwise convergence): Let $f(x)$ be piecewise smooth on the closed interval $[a, b]$, and let the set $\{\varphi_n(x)\}$ be the eigenfunctions of a regular Sturm-Liouville problem over that interval. Then for each value of x in the corresponding open interval (a, b), the Fourier series, relative to these eigenfunctions, converges "pointwise" in accordance with

$$\sum_{n=0}^{\infty} F(n)\varphi_n(x) = \frac{f(x+0) + f(x-0)}{2}$$

Here, the $F(n)$ are the properly evaluated Fourier coefficients over the interval, and the term on the right is the average of the left- and right-hand limits of $f(x)$ at the point. For piecewise smooth functions, the quality of convergence of the series depends on the value of x in the interval.

Theorem 2.4.2 (Uniform convergence): Let $f(x)$ be defined and continuous with continuous first and second derivatives on the closed interval $[a, b]$, and let $f(x)$ satisfy the same boundary conditions as the eigenfunctions $\{\varphi_n(x)\}$, which are the solutions to the regular Sturm-Liouville problem over that interval. Then the Fourier series expansion of $f(x)$, relative to these eigenfunctions, converges "uniformly" to $f(x)$ on the closed interval.

Functions that satisfy the more rigid conditions for uniform convergence are more well behaved and have Fourier series expansions that converge more rapidly and with better fidelity to the actual function. Uniform convergence obviously implies pointwise convergence at all points on the closed interval, and, in addition, the convergence is independent of the value of x on that interval. For more details on testing a series for uniform convergence, see Section 3.7, which discusses the Weierstrass M-test for the uniform convergence of a series. For more formal definitions of the concepts of pointwise and uniform convergence, please be advised that these concepts are best discussed in more advanced texts in Fourier series (see references). The more advanced concepts of periodic entensions and Gibb's phenomena are best discussed in these more advanced texts.

We now provide illustrations of Fourier series expansions of three different functions in terms of those eigenfunctions derived in the problem in Section 2.3. This exercise demonstrates how the quality of convergence of the resulting series depends on the character of the function and how well it adheres to the boundary conditions on the problem.

DEMONSTRATION: We seek the generalized series expansion of a piecewise smooth function $f(x)$ over the interval $I = \{x \mid 0 < x < 1\}$, in terms of the orthonormal eigenfunctions evaluated in Section 2.3.

SOLUTION: In terms of the evaluated orthonormal eigenfunctions, the expansion reads

$$f(x) = \sum_{n=1}^{\infty} F(n) \cos\left(\frac{(2n-1)\pi x}{2}\right) \sqrt{2}$$

From the preceding, the Fourier coefficients are determined from the integral

$$F(n) = \int_{0}^{1} f(x) \cos\left(\frac{(2n-1)\pi x}{2}\right) \sqrt{2}\, dx$$

We evaluate the series for three different functions $f(x)$, each having a different character over the interval.

Case 1 We let $f1(x) = x$. For this case, $f1(x)$ does not satisfy either of the boundary conditions at the left or right end points of the interval. The Fourier coefficient integral reads

$$F(n) = \int_{0}^{1} x\sqrt{2} \cos\left(\frac{(2n-1)\pi x}{2}\right) dx$$

Evaluation of this integral yields

$$F(n) = -\frac{2\sqrt{2}(2\pi(-1)^n n - \pi(-1)^n + 2)}{(2n-1)^2 \pi^2} \tag{2.3}$$

Thus, the Fourier series expansion of $f1(x)$ over the interval is

$$f1(x) = \sum_{n=1}^{\infty} \left(-\frac{4(2\pi n(-1)^n - \pi(-1)^n + 2)\cos\left(\frac{(2n-1)\pi x}{2}\right)}{(2n-1)^2\pi^2} \right)$$

A plot of both $f1(x)$ and the first five terms of the series expansion is shown in Figure 2.1.

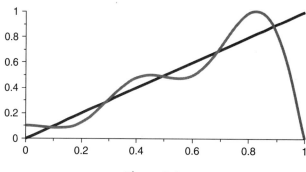

Figure 2.1

Because the function $f1(x)$ does not satisfy either of the boundary conditions, the series converges pointwise. The series converges very slowly, as an alternating $\frac{1}{n}$ series, and the quality of convergence appears to be poor.

Case 2 We let $f2(x) = 1 - x$. For this case, $f2(x)$ does not satisfy the boundary condition at the left, but it does satisfy the condition at the right end point of the interval. The Fourier coefficient integral reads

$$F(n) = \int_{0}^{1} (1-x)\sqrt{2}\cos\left(\frac{(2n-1)\pi x}{2}\right) dx$$

Evaluation of this integral yields

$$F(n) = \frac{4\sqrt{2}}{(2n-1)^2\pi^2} \tag{2.4}$$

Thus, the Fourier series expansion of $f2(x)$ over the interval is

$$f2(x) = \sum_{n=1}^{\infty} \frac{8\cos\left(\frac{(2n-1)\pi x}{2}\right)}{(2n-1)^2\pi^2}$$

A plot of both $f2(x)$ and the first five terms of the series expansion is shown in Figure 2.2.

The function $f2(x)$ satisfies the boundary condition at the right but not at the left end point, and the series converges pointwise. The series converges very rapidly as a $\frac{1}{n^2}$ series, and, since the function satisfies the boundary condition at one of the end points, the quality of convergence appears to be much better than that for case 1.

Case 3 We let $f3(x) = 1 - x^2$. For this case, $f3(x)$ satisfies both boundary conditons at the left and right end points of the interval. The Fourier coefficient integral reads

$$F(n) = \int_0^1 (1 - x^2)\sqrt{2}\cos\left(\frac{(2n-1)\pi x}{2}\right) dx$$

Evaluation of this integral yields

$$F(n) = \frac{16\sqrt{2}(-1)^{n+1}}{(2n-1)^3\pi^2}$$

Thus, the Fourier series expansion of $f3(x)$ over the interval is

$$f3(x) = \sum_{n=1}^{\infty} \frac{32(-1)^{n+1}\cos\left(\frac{(2n-1)\pi x}{2}\right)}{(2n-1)^3\pi^2}$$

A plot of both $f3(x)$ and the first five terms of the series expansion is shown in Figure 2.3.

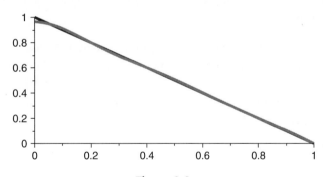

Figure 2.2

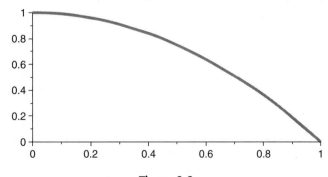

Figure 2.3

Because the function $f3(x)$ satisfies both of the boundary conditions at the left and right end points, the series converges uniformly to $f(x)$. The series converges very rapidly as a $\frac{1}{n^3}$ series, and the quality of convergence is considerably much better than either of the two cases already discussed.

2.5 Examples of Regular Sturm-Liouville Eigenvalue Problems

We will now look at examples of regular Sturm-Liouville differential equations with various combinations of the three types of boundary conditions discussed earlier. All of the examples are special cases of the Sturm-Liouville differential equation

$$L(y) + \lambda\, w(x)y = 0$$

where L is the Sturm-Liouville operator

$$L(y) = D(p(x)D(y)) + q(x)y$$

We focus on three types of differential equations: Euler, Cauchy-Euler, and Bessel. Each one of these differential equations is characterized by a different set of Sturm-Liouville coefficients: $p(x)$, $q(x)$, and $w(x)$. Subject to a particular set of boundary conditions, we will generate the eigenvalues, the corresponding eigenfunctions, and the statement of orthonormality. We will then provide an example of a generalized Fourier series expansion of a given function in terms of the particular eigenfunctions.

In solving for the allowed eigenvalues and corresponding eigenfunctions, we would ordinarily consider three possibilities for values of λ: $\lambda < 0$, $\lambda = 0$, and $\lambda > 0$. However, to make our task a little simpler, we will not consider the case for $\lambda < 0$ because it can be shown, by way of the Rayleigh quotient (see Exercises 2.24 through 2.26), that for the particular Sturm-Liouville problems we will be considering, λ must be greater than or equal to zero.

EXAMPLE 2.5.1: Consider the Euler operator with Dirichlet conditions. We seek the eigenvalues and corresponding orthonormal eigenfunctions for the Euler differential equation [Sturm-Liouville type for $p(x) = 1$, $q(x) = 0$, $w(x) = 1$] over the interval $I = \{x|0 < x < b\}$. The boundary conditions are type 1 at the left and type 1 at the right end points.

Euler differential equation

$$\frac{d^2}{dx^2}y(x) + \lambda y(x) = 0$$

Boundary conditions

$$y(0) = 0 \quad \text{and} \quad y(b) = 0$$

SOLUTION: We consider two possibilities for values of λ. We first consider $\lambda = 0$. For this case, the system basis vectors are

> restart:y1(x):=1;y2(x):=x;

$$y1(x) := 1$$

$$y2(x) := x \tag{2.5}$$

General solution

> y(x):=C1*y1(x)+C2*y2(x);

$$y(x) := C1 + C2x \tag{2.6}$$

Substitution into the boundary conditions yields

> eval(subs(x=0,y(x)))=0;

$$C1 = 0 \tag{2.7}$$

> eval(subs(x=b,C1=0,y(x)))=0;

$$C2\,b = 0 \tag{2.8}$$

The only solution to the preceding is the trivial solution. We next consider $\lambda > 0$. We set $\lambda = \mu^2$ and, for this case, the system basis vectors are

> y1(x):=sin(mu*x);y2(x):=cos(mu*x);

$$y1(x) := \sin(\mu x)$$

$$y2(x) := \cos(\mu x) \tag{2.9}$$

General solution

> y(x):=C1*y1(x)+C2*y2(x);

$$y(x) := C1\sin(\mu x) + C2\cos(\mu x) \tag{2.10}$$

Substituting into the boundary conditions, we get

> eval(subs(x=0,y(x)))=0;

$$C2 = 0 \tag{2.11}$$

> eval(subs(x=b,y(x)))=0;

$$C1\sin(\mu b) + C2\cos(\mu b) = 0 \tag{2.12}$$

The only nontrivial solutions to the preceding occur when $C2 = 0$, $C1$ is arbitrary, and μ satisfies the following eigenvalue equation:

> sin(mu*b)=0;

$$\sin(\mu b) = 0 \tag{2.13}$$

Thus, μ takes on the values

> mu[n]:=n*Pi/b;

$$\mu_n := \frac{n\pi}{b} \tag{2.14}$$

for $n = 1, 2, 3, \ldots$.

Allowed eigenvalues are $\lambda_n = \mu_n^2$

> lambda[n]:=(n*Pi/b)^2;

$$\lambda_n := \frac{n^2\pi^2}{b^2} \tag{2.15}$$

Nonnormalized eigenfunctions are

> phi[n](x):=sin(mu[n]*x);

$$\phi_n(x) := \sin\left(\frac{n\pi x}{b}\right) \tag{2.16}$$

Normalization

Evaluating the norm from the inner product of the eigenfunctions with respect to the weight function $w(x) = 1$ over the interval yields

> w(x):=1:unprotect(norm):norm:=sqrt(Int(phi[n](x)^2*w(x),x=0..b));norm:=expand (value(%)):

$$norm := \sqrt{\int_0^b \sin\left(\frac{n\pi x}{b}\right)^2 dx} \tag{2.17}$$

Substitution of the eigenvalue equation simplifies the norm

> norm:=radsimp(subs({sin(n*Pi)=0,cos(n*Pi)=(−1)^n},norm));

$$norm := \frac{1}{2}\sqrt{2}\sqrt{b} \tag{2.18}$$

Orthonormal eigenfunctions

> phi[n](x):=phi[n](x)/norm;phi[m](x):=subs(n=m,phi[n](x)):

$$\phi_n(x) := \frac{\sin\left(\frac{n\pi x}{b}\right)\sqrt{2}}{\sqrt{b}} \tag{2.19}$$

Statement of orthonormality

> Int(phi[n](x)*phi[m](x)*w(x),x=0..b)=delta(n,m);

$$\int_0^b \frac{2\sin\left(\frac{n\pi x}{b}\right)\sin\left(\frac{m\pi x}{b}\right)}{b}\,dx = \delta(n,m) \tag{2.20}$$

Generalized Fourier series expansion

> f(x):=Sum(F(n)*phi[n](x),n=1..infinity);f(x):='f(x)':

$$f(x) := \sum_{n=1}^{\infty} \frac{F(n)\sin\left(\frac{n\pi x}{b}\right)\sqrt{2}}{\sqrt{b}} \tag{2.21}$$

Fourier coefficients

> F(n):=Int(f(x)*phi[n](x)*w(x),x=0..b);F(n):='F(n)':

$$F(n) := \int_0^b \frac{f(x)\sin\left(\frac{n\pi x}{b}\right)\sqrt{2}}{\sqrt{b}}\,dx \tag{2.22}$$

This is the generalized series expansion of $f(x)$ in terms of the "complete" set of eigenfunctions for the particular Sturm-Liouville operator and given boundary conditions over the interval.

DEMONSTRATION: Develop the generalized series expansion for $f(x) = x$ over the interval $I = \{x \mid 0 < x < 1\}$ in terms of the preceding eigenfunctions. We assign the system values

> a:=0;b:=1;f(x):=x;

$$a := 0$$
$$b := 1$$
$$f(x) := x \tag{2.23}$$

SOLUTION: We evaluate the Fourier coefficients

> F(n):=eval(Int(f(x)*phi[n](x)*w(x),x=a..b));F(n):=value(%):

$$F(n) := \int_0^1 x\sin(n\pi x)\sqrt{2}\,dx \tag{2.24}$$

> F(n):=subs({sin(n*Pi)=0,cos(n*Pi)=(−1)^n},F(n));

$$F(n) := -\frac{\sqrt{2}(-1)^n}{n\pi} \tag{2.25}$$

> Series:=eval(Sum(F(n)*phi[n](x),n=1..infinity));

$$Series := \sum_{n=1}^{\infty} \left(-\frac{2(-1)^n \sin(n\pi x)}{n\pi} \right) \tag{2.26}$$

First five terms of expansion

> Series:=sum(F(n)*phi[n](x),n=1..5):

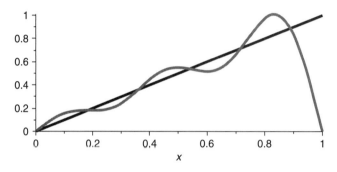

Figure 2.4

The curves of Figure 2.4 depict the function $f(x)$ and its Fourier series approximation in terms of the orthonormal eigenfunctions for the particular operator and boundary conditions given earlier. Note that $f(x)$ satisfies the given boundary conditions at the left but fails to do so at the right end point. The convergence is pointwise.

EXAMPLE 2.5.2: Consider the Euler operator with Dirichlet and Neumann conditions. We seek the eigenvalues and corresponding orthonormal eigenfunctions for the Euler differential equation [Sturm-Liouville type for $p(x) = 1$, $q(x) = 0$, $w(x) = 1$] over the interval $I = \{x \mid 0 < x < b\}$. The boundary conditions are type 1 at the left and type 2 at the right.

Euler differential equation

$$\frac{d^2}{dx^2}y(x) + \lambda y(x) = 0$$

Boundary conditions

$$y(0) = 0 \quad \text{and} \quad y_x(b) = 0$$

SOLUTION: We consider two possibilities for values of λ. We first consider $\lambda = 0$. For this case, the system basis vectors are

> restart:y1(x):=1;y2(x):=x;

$$y1(x) := 1$$

$$y2(x) := x \tag{2.27}$$

General solution

> y(x):=C1*y1(x)+C2*y2(x);

$$y(x) := C1 + C2x \tag{2.28}$$

Substituting the boundary conditions yields

> eval(subs(x=0,y(x)))=0;

$$C1 = 0 \tag{2.29}$$

> eval(subs(x=b,diff(y(x),x)))=0;

$$C2 = 0 \tag{2.30}$$

The only solution to the preceding is the trivial solution. We next consider $\lambda > 0$. We set $\lambda = \mu^2$ and, for this case, the system basis vectors are

> y1(x):=sin(mu*x);y2(x):=cos(mu*x);

$$y1(x) := \sin(\mu x)$$

$$y2(x) := \cos(\mu x) \tag{2.31}$$

General solution

> y(x):=C1*y1(x)+C2*y2(x);

$$y(x) := C1 \sin(\mu x) + C2 \cos(\mu x) \tag{2.32}$$

Substituting the boundary conditions yields

> eval(subs(x=0,y(x)))=0;

$$C2 = 0 \tag{2.33}$$

> eval(subs(x=b,diff(y(x),x)))=0;

$$C1 \cos(\mu b)\mu - C2 \sin(\mu b)\mu = 0 \tag{2.34}$$

The only nontrivial solutions occur when $C2 = 0$, $C1$ is arbitrary, and μ satisfies the following eigenvalue equation:

> cos(mu*b)=0;

$$\cos(\mu b) = 0 \tag{2.35}$$

Thus, μ takes on values

> mu[n]:=(2*n−1)*Pi/(2*b);

$$\mu_n := \frac{1}{2} \frac{(2n - 1)\pi}{b} \tag{2.36}$$

for $n = 1, 2, 3, \ldots$.

Allowed eigenvalues are $\lambda_n = \mu_n^2$

> lambda[n]:=mu[n]^2;

$$\lambda_n := \frac{1}{4} \frac{(2n - 1)^2 \pi^2}{b^2} \tag{2.37}$$

Nonnormalized eigenfunctions are

> phi[n](x):=sin(mu[n]*x);

$$\varphi_n(x) := \sin\left(\frac{1}{2} \frac{(2n - 1)\pi x}{b}\right) \tag{2.38}$$

Normalization

Evaluating the norm from the inner product of the eigenfunctions with respect to the weight function $w(x) = 1$ over the interval yields

> w(x):=1:unprotect(norm):norm:=sqrt(Int(phi[n](x)^2*w(x),x=0..b));norm:=expand (value(%)):

$$norm := \sqrt{\int_0^b \sin\left(\frac{1}{2} \frac{(2n - 1)\pi x}{b}\right)^2 dx} \tag{2.39}$$

Substitution of the eigenvalue equation simplifies the norm

> norm:=radsimp(subs({sin(n*Pi)=0,cos(n*Pi)=(−1)^n},norm));

$$norm := \frac{1}{2} \sqrt{2}\sqrt{b} \tag{2.40}$$

Orthonormal eigenfunctions

> phi[n](x):=phi[n](x)/norm;phi[m](x):=subs(n=m,phi[n](x)):

$$\varphi_n(x) := \frac{\sin\left(\frac{1}{2}\frac{(2n-1)\pi x}{b}\right)\sqrt{2}}{\sqrt{b}} \tag{2.41}$$

Statement of orthonormality

> Int(phi[n](x)*phi[m](x)*w(x),x=0..b)=delta(n,m);

$$\int_0^b \frac{2\sin\left(\frac{1}{2}\frac{(2n-1)\pi x}{b}\right)\sin\left(\frac{1}{2}\frac{(2m-1)\pi x}{b}\right)}{b}\,dx = \delta(n,m) \tag{2.42}$$

Generalized Fourier series expansion

> f(x):=Sum(F(n)*phi[n](x),n=1..infinity);f(x):='f(x)':

$$f(x) := \sum_{n=1}^{\infty} \frac{F(n)\sin\left(\frac{1}{2}\frac{(2n-1)\pi x}{b}\right)\sqrt{2}}{\sqrt{b}} \tag{2.43}$$

Fourier coefficients

> F(n):=Int(f(x)*phi[n](x)*w(x),x=0..b);F(n):='F(n)':

$$F(n) := \int_0^b \frac{f(x)\sin\left(\frac{1}{2}\frac{(2n-1)\pi x}{b}\right)\sqrt{2}}{\sqrt{b}}\,dx \tag{2.44}$$

This is the generalized series expansion of $f(x)$ in terms of the "complete" set of eigenfunctions for the particular Sturm-Liouville operator and boundary conditions over the interval.

DEMONSTRATION: Develop the generalized series expansion for $f(x) = x$ over the interval $I = \{x \mid 0 < x < 1\}$ in terms of the preceding eigenfunctions. We assign the system values

> a:=0;b:=1;f(x):=x;

$$a := 0$$
$$b := 1$$
$$f(x) := x \tag{2.45}$$

SOLUTION: We evaluate the Fourier coefficients

> F(n):=eval(Int(f(x)*phi[n](x)*w(x),x=a..b));F(n):=value(%):

$$F(n) := \int_0^1 x\sin\left(\frac{1}{2}(2n-1)\pi x\right)\sqrt{2}\,dx \tag{2.46}$$

> F(n):=subs({sin(n*Pi)=0,cos(n*Pi)=(−1)^n,sin((2*n+1)/2*Pi)=(−1)^n,
  cos((2*n+1)/2*Pi)=0},F(n));

$$F(n) := -\frac{4\sqrt{2}(-1)^n}{\pi^2(4n^2 - 4n + 1)} \tag{2.47}$$

> Series:=eval(Sum(F(n)*phi[n](x),n=1..infinity));

$$Series := \sum_{n=1}^{\infty} \left(-\frac{8(-1)^n \sin\left(\frac{1}{2}(2n-1)\pi x\right)}{\pi^2(4n^2 - 4n + 1)} \right) \tag{2.48}$$

First five terms of expansion

> Series:=eval(sum(F(n)*phi[n](x),n=1..5)):

> plot({Series,f(x)},x=0..b,thickness=10);

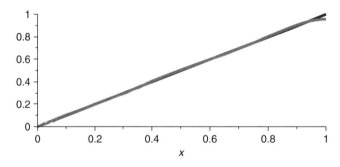

Figure 2.5

The two curves of Figure 2.5 depict the function $f(x)$ and its Fourier series approximation in terms of the orthonormal eigenfunctions for the particular operator and boundary conditions given earlier. Note that $f(x)$ satisfies the given boundary conditions at the left but fails to do so at the right end point. The convergence is pointwise.

EXAMPLE 2.5.3: Consider the Euler operator with Neumann conditions. We seek the eigenvalues and corresponding orthonormal eigenfunctions for the Euler differential equation [Sturm-Liouville type for $p(x) = 1$, $q(x) = 0$, $w(x) = 1$] over the interval $I = \{x \mid 0 < x < b\}$. The boundary conditions are type 2 at the left and type 2 at the right end points.

Euler differential equation

$$\frac{d^2}{dx^2} y(x) + \lambda y(x) = 0$$

Boundary conditions

$$y_x(0) = 0 \quad \text{and} \quad y_x(b) = 0$$

SOLUTION: We consider two possibilities for values of λ. We first consider $\lambda = 0$. For this case, the system basis vectors are

> restart:y1(x):=1;y2(x):=x;

$$y1(x) := 1$$

$$y2(x) := x \tag{2.49}$$

General solution

> y(x):=C1*y1(x)+C2*y2(x);

$$y(x) := C1 + C2x \tag{2.50}$$

Substituting into the boundary conditions yields

> eval(subs(x=0,diff(y(x),x)))=0;

$$C2 = 0 \tag{2.51}$$

> eval(subs(x=b,diff(y(x),x)))=0;

$$C2 = 0 \tag{2.52}$$

The only nontrivial solution to the above occurs when $C2 = 0$ and $C1$ is arbitrary. Thus, our eigenvalue and corresponding eigenfunction for $\lambda = 0$ are

> lambda[0]:=0;

$$\lambda_0 := 0 \tag{2.53}$$

> phi[0](x):=1;

$$\phi_0(x) := 1 \tag{2.54}$$

We next consider $\lambda > 0$. We set $\lambda = \mu^2$. For this case, the system basis vectors are

> y1(x):=sin(mu*x);y2(x):=cos(mu*x);

$$y1(x) := \sin(\mu x)$$

$$y2(x) := \cos(\mu x) \tag{2.55}$$

General solution

> y(x):=C1*y1(x)+C2*y2(x);

$$y(x) := C1 \sin(\mu x) + C2 \cos(\mu x) \tag{2.56}$$

Substituting into the boundary conditions yields

> eval(subs(x=0,diff(y(x),x)))=0;

$$C1\,\mu = 0 \tag{2.57}$$

> eval(subs(x=b,diff(y(x),x)))=0;

$$C1\cos(\mu\,b)\mu - C2\sin(\mu\,b)\mu = 0 \tag{2.58}$$

The only nontrivial solutions to the above occur when $C1 = 0$, $C2$ is arbitrary, and μ satisfies the following eigenvalue equation:

> sin(mu*b)=0;

$$\sin(\mu\,b) = 0 \tag{2.59}$$

Thus, μ takes on values

> mu[n]:=n*Pi/b;

$$\mu_n := \frac{n\pi}{b} \tag{2.60}$$

for $n = 1, 2, 3, \ldots$.

Allowed eigenvalues are $\lambda_n = \mu_n^2$

> lambda[n]:=mu[n]^2;

$$\lambda_n := \frac{n^2\pi^2}{b^2} \tag{2.61}$$

Nonnormalized eigenfunctions are

> phi[n](x):=cos(mu[n]*x);

$$\phi_n(x) := \cos\left(\frac{n\pi x}{b}\right) \tag{2.62}$$

Normalization

Evaluating the norm from the inner product of the eigenfunctions with respect to the weight function $w(x) = 1$ over the interval, yields, for $n = 0$

> w(x):=1:norm0:=sqrt(Int(phi[0](x)^2*w(x),x=0..b));

$$norm0 := \sqrt{\int_0^b 1\,dx} \tag{2.63}$$

> norm0:=value(%);

$$norm0 := \sqrt{b} \tag{2.64}$$

For $n = 1, 2, 3, \ldots$, we get

> norm1:=sqrt(Int(phi[n](x)^2*w(x),x=0..b));norm1:=expand(value(%)):

$$norm1 := \sqrt{\int_0^b \cos\left(\frac{n\pi x}{b}\right)^2 dx} \tag{2.65}$$

Substitution of the eigenvalue equation simplifies the norm

> norm1:=radsimp(subs({sin(n*Pi)=0,cos(n*Pi)=(−1)^n},norm1));

$$norm1 := \frac{1}{2}\sqrt{2}\sqrt{b} \tag{2.66}$$

Orthonormal eigenfunctions

> phi[0](x):=phi[0](x)/norm0;

$$\phi_0(x) := \frac{1}{\sqrt{b}} \tag{2.67}$$

> phi[n](x):=phi[n](x)/norm1;phi[m](x):=subs(n=m,phi[n](x)):

$$\phi_n(x) := \frac{\cos\left(\frac{n\pi x}{b}\right)\sqrt{2}}{\sqrt{b}} \tag{2.68}$$

Statement of orthonormality

> Int(phi[n](x)*phi[m](x)*w(x),x=0..b)=delta(n,m);

$$\int_0^b \frac{2\cos\left(\frac{n\pi x}{b}\right)\cos\left(\frac{m\pi x}{b}\right)}{b} dx = \delta(n, m) \tag{2.69}$$

Generalized Fourier series expansion

> f(x):=F(0)*phi[0](x)+Sum(F(n)*phi[n](x),n=1..infinity);f(x):='f(x)':

$$f(x) := \frac{F(0)}{\sqrt{b}} + \sum_{n=1}^{\infty} \frac{F(n)\cos\left(\frac{n\pi x}{b}\right)\sqrt{2}}{\sqrt{b}} \tag{2.70}$$

Fourier coefficients

> F(n):=Int(f(x)*phi[n](x)*w(x),x=0..b);F(n):=`F(n)`:

$$F(n) := \int_0^b \frac{f(x)\, \cos\left(\frac{n\pi x}{b}\right)\sqrt{2}}{\sqrt{b}}\,dx \tag{2.71}$$

> F(0):=Int(f(x)*phi[0](x)*w(x),x=0..b);

$$F(0) := \int_0^b \frac{f(x)}{\sqrt{b}}\,dx \tag{2.72}$$

This is the generalized series expansion of $f(x)$ in terms of the "complete" set of eigenfunctions for the particular Sturm-Liouville operator and boundary conditions over the interval.

DEMONSTRATION: Develop the generalized series expansion for $f(x) = x$ over the interval $I = \{x \mid 0 < x < 1\}$ in terms of the preceding eigenfunctions. We assign the system values

> a:=0;b:=1;f(x):=x;

$$a := 0$$
$$b := 1$$
$$f(x) := x \tag{2.73}$$

SOLUTION: We evaluate the Fourier coefficients

> F(n):=eval(Int(f(x)*phi[n](x)*w(x),x=a..b));F(n):=value(%):

$$F(n) := \int_0^1 x\cos(n\pi x)\sqrt{2}\,dx \tag{2.74}$$

> F(n):=simplify(subs({sin(n*Pi)=0,cos(n*Pi)=(-1)^n},F(n)));

$$F(n) := \frac{\sqrt{2}\,(-1+(-1)^n)}{n^2\pi^2} \tag{2.75}$$

> F(0):=eval(Int(f(x)*phi[0](x)*w(x),x=a..b));

$$F(0) := \int_0^1 x\,dx \tag{2.76}$$

> F(0):=value(%);

$$F(0) := \frac{1}{2} \tag{2.77}$$

> Series:=eval(F(0)*phi[0](x)+Sum(F(n)*phi[n](x),n=1..infinity));

$$Series := \frac{1}{2} + \sum_{n=1}^{\infty} \frac{2\left(-1+(-1)^n\right)\cos(n\pi x)}{n^2\pi^2} \tag{2.78}$$

First five terms of expansion

> Series:=eval(F(0)*phi[0](x)+sum(F(n)*phi[n](x),n=1..5)):

> plot({Series,f(x)},x=a..b,thickness=10);

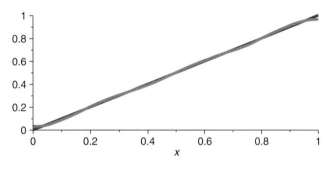

Figure 2.6

The two curves of Figure 2.6 depict the function $f(x)$ and its Fourier series approximation in terms of the orthonormal eigenfunctions for the particular operator and boundary conditions given earlier. Note that $f(x)$ does not satisfy either of the boundary conditions imposed on the eigenfunctions at the end points. The convergence is pointwise.

EXAMPLE 2.5.4: Consider the Euler operator with Dirichlet and Robin conditions. We seek the eigenvalues and corresponding orthonormal eigenfunctions for the Euler differential equation [Sturm-Liouville type for $p(x) = 1$, $q(x) = 0$, $w(x) = 1$] over the interval $I = \{x \mid 0 < x < b\}$. The boundary conditions are type 1 at the left and type 3 at the right end points.

Euler differential equation

$$\frac{d^2}{dx^2}y(x) + \lambda y(x) = 0$$

Boundary conditions ($h > 0$)

$$y(0) = 0 \quad \text{and} \quad y_x(b) + hy(b) = 0$$

SOLUTION: We consider two possibilities for values of λ. We first consider $\lambda = 0$. For this case, the system basis vectors are

> restart:y1(x):=1;y2(x):=x;

$$y1(x) := 1$$
$$y2(x) := x \tag{2.79}$$

General solution

> y(x):=C1*y1(x)+C2*y2(x);

$$y(x) := C1 + C2x \tag{2.80}$$

Substituting into the boundary conditions yields

> eval(subs(x=0,y(x)))=0;

$$C1 = 0 \tag{2.81}$$

> eval(subs(x=b,diff(y(x),x)+h*y(x)))=0;

$$C2 + h(C1 + C2b) = 0 \tag{2.82}$$

The only solution to the preceding is the trivial solution. We next consider $\lambda > 0$. We set $\lambda = \mu^2$, and for this case, the system basis vectors are

> y1(x):=sin(mu*x);y2(x):=cos(mu*x);

$$y1(x) := \sin(\mu x)$$
$$y2(x) := \cos(\mu x) \tag{2.83}$$

General solution

> y(x):=C1*y1(x)+C2*y2(x);

$$y(x) := C1 \sin(\mu x) + C2 \cos(\mu x) \tag{2.84}$$

Substituting into the boundary conditions yields

> eval(subs(x=0,y(x)))=0;

$$C2 = 0 \tag{2.85}$$

> eval(subs(x=b,diff(y(x),x)+h*y(x)))=0;

$$C1 \cos(\mu b)\mu - C2 \sin(\mu b)\mu + h(C1 \sin(\mu b) + C2 \cos(\mu b)) = 0 \tag{2.86}$$

The only nontrivial solutions to the preceding are that $C2 = 0$, $C1$ is arbitrary, and μ must satisfy the following eigenvalue equation:

> h*sin(mu*b)+mu*cos(mu*b)=0;

$$h \sin(\mu\, b) + \mu \cos(\mu\, b) = 0 \qquad (2.87)$$

We indicate these roots as μ_n for $n = 1, 2, 3, \ldots$.

Allowed eigenvalues are $\lambda_n = \mu_n^2$

> lambda[n]=mu[n]^2;

$$\lambda_n = \mu_n^2 \qquad (2.88)$$

Nonnormalized eigenfunctions are

> phi[n](x):=sin(sqrt(lambda[n])*x);

$$\phi_n(x) := \sin\left(\sqrt{\lambda_n}\,x\right) \qquad (2.89)$$

Normalization

Evaluating the norm from the inner product of the eigenfunctions with respect to the weight function $w(x) = 1$ over the interval yields

> w(x):=1:unprotect(norm):norm:=sqrt(Int(phi[n](x)^2*w(x),x=0..b));norm:=value(%):

$$norm := \sqrt{\int_0^b \sin\left(\sqrt{\lambda_n}\,x\right)^2 dx} \qquad (2.90)$$

Substitution of the eigenvalue equation simplifies the norm

> norm:=radsimp(subs(sin(sqrt(lambda[n])*b)=−sqrt(lambda[n])/
  h*cos(sqrt(lambda[n])*b),norm));

$$norm := \frac{1}{2} \frac{\sqrt{2}\sqrt{\left(\cos(\sqrt{\lambda_n}b)^2 + bh\right)h}}{h} \qquad (2.91)$$

Orthonormal eigenfunctions

> phi[n](x):=phi[n](x)/norm;phi[m](x):=subs(n=m,phi[n](x)):

$$\phi_n(x) := \frac{\sin\left(\sqrt{\lambda_n}x\right)\sqrt{2h}}{\sqrt{\left(\cos(\sqrt{\lambda_n}b)^2 + bh\right)h}} \qquad (2.92)$$

Statement of orthonormality

> Int(phi[n](x)*phi[m](x)*w(x),x=0..b)=delta(n,m);

$$\int_0^b \frac{2 \sin\left(\sqrt{\lambda_n}x\right) h^2 \sin\left(\sqrt{\lambda_m}x\right)}{\sqrt{\left(\cos\left(\sqrt{\lambda_n}b\right)^2 + bh\right) h} \sqrt{\left(\cos\left(\sqrt{\lambda_m}b\right)^2 + bh\right) h}} dx = \delta(n, m) \qquad (2.93)$$

Generalized Fourier series expansion

> f(x):=Sum(F(n)*phi[n](x),n=1..infinity);f(x):='f(x)':

$$f(x) := \sum_{n=1}^{\infty} \frac{F(n) \sin\left(\sqrt{\lambda_n}\, x\right) \sqrt{2}\, h}{\sqrt{\left(\cos\left(\sqrt{\lambda_n}b\right)^2 + bh\right) h}} \qquad (2.94)$$

Fourier coefficients

> F(n):=Int(f(x)*phi[n](x)*w(x),x=0..b);F(n):='F(n)':

$$F(n) := \int_0^b \frac{f(x) \sin\left(\sqrt{\lambda_n}\, x\right) \sqrt{2}\, h}{\sqrt{\left(\cos\left(\sqrt{\lambda_n}b\right)^2 + bh\right) h}} dx \qquad (2.95)$$

This is the generalized series expansion of $f(x)$ in terms of the "complete" set of eigenfunctions for the particular Sturm-Liouville operator and boundary conditions over the interval.

DEMONSTRATION: Develop the generalized series expansion for $f(x) = x$ over the interval $I = \{x \mid 0 < x < 1\}$ in terms of the preceding eigenfunctions for $h = 1$. We assign the system values

> a:=0;b:=1;h:=1;f(x):=x;

$$a := 0$$
$$b := 1$$
$$h := 1$$
$$f(x) := x \qquad (2.96)$$

SOLUTION: We evaluate the Fourier coefficients

> F(n):=Int(eval(f(x)*phi[n](x))*w(x),x=a..b);F(n):=value(%):

$$F(n) := \int_0^1 \frac{x \sin\left(\sqrt{\lambda_n}\, x\right) \sqrt{2}}{\sqrt{\cos\left(\sqrt{\lambda_n}\right)^2 + 1}} dx \qquad (2.97)$$

Substitution of the eigenvalue equation simplifies the preceding equation:

> F(n):=simplify(subs(sin(sqrt(lambda[n])*b)=−sqrt(lambda[n])/
h*cos(sqrt(lambda[n])*b),F(n)));

$$F(n) := -\frac{2\sqrt{2}\,\cos\left(\sqrt{\lambda_n}\right)}{\sqrt{\cos\left(\sqrt{\lambda_n}\right)^2 + 1}\,\sqrt{\lambda_n}} \qquad (2.98)$$

> Series:=simplify(Sum(F(n)*phi[n](x),n=1..infinity));

$$Series := \sum_{n=1}^{\infty}\left(-\frac{4\,\cos\left(\sqrt{\lambda_n}\right)\sin\left(\sqrt{\lambda_n}x\right)}{\left(\cos\left(\sqrt{\lambda_n}\right)^2 + 1\right)\sqrt{\lambda_n}}\right) \qquad (2.99)$$

Evaluation of the eigenvalues from the roots of the eigenvalue equation yields

> tan(sqrt(lambda[n])*b)=−sqrt(lambda[n])/h;

$$\tan\left(\sqrt{\lambda_n}\right) = -\sqrt{\lambda_n} \qquad (2.100)$$

> plot({tan(v),−v/(b*h)},v=0..20,y=−20..0,thickness=10);

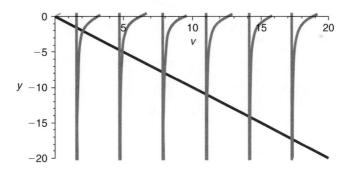

Figure 2.7

If we set $v = \sqrt{\lambda}\,b$, then the eigenvalues are found from the intersection points of the curves shown in Figure 2.7. We evaluate a few of these eigenvalues from the Maple fsolve command for the roots of the eigenvalue equation to be given as

> lambda[1]:=(1/b*(fsolve((tan(v)+v/(h*b)),v=1..3)))^2;

$$\lambda_1 := 4.115858365 \qquad (2.101)$$

> lambda[2]:=(1/b*(fsolve((tan(v)+v/(h*b)),v=3..6)))^2;

$$\lambda_2 := 24.13934203 \qquad (2.102)$$

> lambda[3]:=(1/b*(fsolve((tan(v)+v/(h*b)),v=6..9)))^2;

$$\lambda_3 := 63.65910654 \qquad (2.103)$$

> lambda[4]:=(1/b*(fsolve((tan(v)+v/(h*b)),v=9..13)))^2;

$$\lambda_4 := 122.8891618 \qquad (2.104)$$

> lambda[5]:=(1/b*(fsolve((tan(v)+v/(h*b)),v=13..16)))^2;

$$\lambda_5 := 201.8512584 \qquad (2.105)$$

First five terms of expansion

> Series:=eval(sum(F(n)*phi[n](x),n=1..5)):

> plot({Series,f(x)},x=a..b,thickness=10);

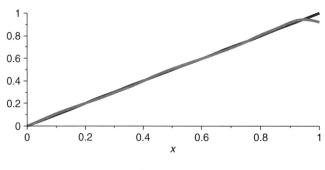

Figure 2.8

The two curves of Figure 2.8 depict the function $f(x)$ and its Fourier series approximation in terms of the orthonormal eigenfunctions for the particular operator and boundary conditions given earlier. Note that $f(x)$ satisfies the given boundary conditions at the left but fails to do so at the right end point. The convergence is pointwise.

EXAMPLE 2.5.5: Consider the Euler operator with Neumann and Robin conditions. We seek the eigenvalues and corresponding orthonormal eigenfunctions for the Euler differential equation [Sturm-Liouville type for $p(x) = 1$, $q(x) = 0$, $w(x) = 1$] over the interval $I = \{x \mid 0 < x < b\}$. The boundary conditions are type 2 at the left and type 3 at the right end points.

Euler differential equation

$$\frac{d^2}{dx^2}y(x) + \lambda y(x) = 0$$

Boundary conditions ($h > 0$)

$$y_x(0) = 0 \quad \text{and} \quad y_x(b) + hy(b) = 0$$

SOLUTION: We consider two possibilities for values of λ. We first consider $\lambda = 0$. For this case, the system basis vectors are

> restart:y1(x):=1;y2(x):=x;

$$y1(x) := 1$$
$$y2(x) := x \tag{2.106}$$

General solution

> y(x):=C1*y1(x)+C2*y2(x);

$$y(x) := C1 + C2x \tag{2.107}$$

Substituting into the boundary conditions yields

> eval(subs(x=0,diff(y(x),x)))=0;

$$C2 = 0 \tag{2.108}$$

> eval(subs(x=b,diff(y(x),x)+h*y(x)))=0;

$$C2 + h(C1 + C2b) = 0 \tag{2.109}$$

The only solution to the preceding is the trivial solution. We next consider $\lambda > 0$. We set $\lambda = \mu^2$. For this case, the system basis vectors are

> y1(x):=sin(mu*x); y2(x):=cos(mu*x);

$$y1(x) := \sin(\mu x)$$
$$y2(x) := \cos(\mu x) \tag{2.110}$$

General solution

> y(x):=C1*y1(x)+C2*y2(x);

$$y(x) := C1 \sin(\mu x) + C2 \cos(\mu x) \tag{2.111}$$

Substituting into the boundary conditions yields

> eval(subs(x=0,diff(y(x),x)))=0;

$$C1 \mu = 0 \tag{2.112}$$

> eval(subs(x=b,diff(y(x),x)+h*y(x)))=0;

$$C1\cos(\mu\,b)\mu - C2\sin(\mu\,b)\mu + h(C1\sin(\mu\,b) + C2\cos(\mu\,b)) = 0 \qquad (2.113)$$

The only nontrivial solutions to the preceding occur when $C1 = 0$, $C2$ is arbitrary, and μ satisfies the following eigenvalue equation:

> mu*sin(mu*b)−h*cos(mu*b)=0;

$$\mu\sin(\mu\,b) - h\cos(\mu\,b) = 0 \qquad (2.114)$$

We identify these roots as μ_n for $n = 1, 2, 3, \ldots$.

Allowed eigenvalues are $\lambda_n = \mu_n^2$

> lambda[n]=mu[n]^2;

$$\lambda_n = \mu_n^2 \qquad (2.115)$$

Nonnormalized eigenfunctions are

> phi[n](x):=cos(sqrt(lambda[n])*x);

$$\phi_n(x) := \cos\left(\sqrt{\lambda_n}x\right) \qquad (2.116)$$

Normalization

Evaluating the norm from the inner product of the eigenfunctions with respect to the weight function $w(x) = 1$ over the interval yields

> w(x):=1:unprotect(norm):norm:=sqrt(Int(phi[n](x)^2*w(x),x=0..b));norm:=value(%):

$$norm := \sqrt{\int_0^b \cos\left(\sqrt{\lambda_n}x\right)^2 dx} \qquad (2.117)$$

Substitution of the eigenvalue equation simplifies the preceding equation:

> norm:=radsimp(subs(cos(sqrt(lambda[n])*b)=sqrt(lambda[n])/
    h*sin(sqrt(lambda[n])*b),norm));

$$norm := \frac{1}{2}\frac{\sqrt{2}\sqrt{\left(\sin(\sqrt{\lambda_n}b)^2 + bh\right)h}}{h} \qquad (2.118)$$

Orthonormal eigenfunctions

> phi[n](x):=phi[n](x)/norm;phi[m](x):=subs(n=m,phi[n](x)):

$$\phi_n(x) := \frac{\cos\left(\sqrt{\lambda_n}x\right)\sqrt{2h}}{\sqrt{\left(\sin\left(\sqrt{\lambda_n}b\right)^2 + bh\right)h}} \tag{2.119}$$

Statement of orthonormality

> Int(phi[n](x)*phi[m](x)*w(x),x=0..b)=delta(n,m);;

$$\int_0^b \frac{2\cos\left(\sqrt{\lambda_n}x\right)h^2\cos\left(\sqrt{\lambda_m}x\right)}{\sqrt{\left(\sin\left(\sqrt{\lambda_n}b\right)^2 + bh\right)h}\sqrt{\left(\sin\left(\sqrt{\lambda_m}b\right)^2 + bh\right)h}}\, dx = \delta(n,m) \tag{2.120}$$

Generalized Fourier series expansion

> f(x):=Sum(F(n)*phi[n](x),n=1..infinity);f(x):='f(x)':

$$f(x) := \sum_{n=1}^{\infty} \frac{F(n)\cos\left(\sqrt{\lambda_n}x\right)\sqrt{2h}}{\sqrt{\left(\sin\left(\sqrt{\lambda_n}b\right)^2 + bh\right)h}} \tag{2.121}$$

Fourier coefficients

> F(n):=Int(f(x)*phi[n](x)*w(x),x=0..b);F(n):='F(n)':

$$F(n) := \int_0^b \frac{f(x)\cos\left(\sqrt{\lambda_n}x\right)\sqrt{2h}}{\sqrt{\left(\sin\left(\sqrt{\lambda_n}b\right)^2 + bh\right)h}}\, dx \tag{2.122}$$

This is the generalized series expansion of $f(x)$ in terms of the "complete" set of eigenfunctions for the particular Sturm-Liouville operator and boundary conditions over the interval.

DEMONSTRATION: Develop the generalized series expansion for $f(x) = 1 - x$ over the interval $I = \{x \mid 0 < x < 1\}$ in terms of the preceding eigenfunctions for $h = 1$. We assign the system values

> a:=0;b:=1;h:=1;f(x):=1−x;

$$a := 0$$
$$b := 1$$
$$h := 1$$
$$f(x) := 1 - x \tag{2.123}$$

SOLUTION: We evaluate the Fourier coefficients

> F(n):=Int(eval(f(x)*phi[n](x))*w(x),x=a..b);F(n):=value(%):

$$F(n) := \int_0^1 \frac{(1-x)\cos\left(\sqrt{\lambda_n}x\right)\sqrt{2}}{\sqrt{\sin\left(\sqrt{\lambda_n}\right)^2+1}} dx \tag{2.124}$$

> F(n):=radsimp(subs(tan(sqrt(lambda[n])*b)=h/sqrt(lambda[n]),F(n)));

$$F(n) := \frac{\left(1-\cos\left(\sqrt{\lambda_n}\right)\right)\sqrt{2}}{\lambda_n\sqrt{2-\cos\left(\sqrt{\lambda_n}\right)^2}} \tag{2.125}$$

> Series:=simplify(Sum(F(n)*phi[n](x),n=1..infinity));

$$Series := \sum_{n=1}^{\infty} \frac{2\left(-1+\cos\left(\sqrt{\lambda_n}\right)\right)\cos\left(\sqrt{\lambda_n}x\right)}{\lambda_n\left(-2+\cos\left(\sqrt{\lambda_n}\right)^2\right)} \tag{2.126}$$

Evaluation of the eigenvalues from the roots of the eigenvalue equation yields

> tan(sqrt(lambda[n])*b)=h/sqrt(lambda[n]);

$$\tan\left(\sqrt{\lambda_n}\right) = \frac{1}{\sqrt{\lambda_n}} \tag{2.127}$$

> plot({tan(v),(h*b)/v},v=0..20,y=0..3/2,thickness=10);

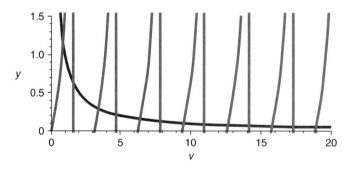

Figure 2.9

If we set $v = \sqrt{\lambda}b$, then the eigenvalues are found from the intersection points of the curves shown in Figure 2.9. We evaluate a few of these eigenvalues from the Maple fsolve command for the roots of the eigenvalue equation to be given as

> lambda[1]:=(1/b*(fsolve((tan(v)−(h*b)/v),v=0..1)))^2;

$$\lambda_1 := 0.7401738844 \qquad (2.128)$$

> lambda[2]:=(1/b*(fsolve((tan(v)−(h*b)/v),v=1..4)))^2;

$$\lambda_2 := 11.73486183 \qquad (2.129)$$

> lambda[3]:=(1/b*(fsolve((tan(v)−(h*b)/v),v=4..7)))^2;

$$\lambda_3 := 41.43880785 \qquad (2.130)$$

> lambda[4]:=(1/b*(fsolve((tan(v)−(h*b)/v),v=7..10)))^2;

$$\lambda_4 := 90.80821420 \qquad (2.131)$$

> lambda[5]:=(1/b*(fsolve((tan(v)−(h*b)/v),v=10..13)))^2;

$$\lambda_5 := 159.9032889 \qquad (2.132)$$

First five terms of expansion

> Series:=eval(sum(F(n)*phi[n](x),n=1..5)):

> plot({Series,f(x)},x=a..b,thickness=10);

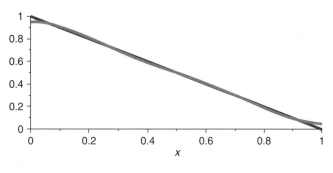

Figure 2.10

The two curves of Figure 2.10 depict the function $f(x)$ and its Fourier series approximation in terms of the orthonormal eigenfunctions for the particular operator and boundary conditions given earlier. Note that $f(x)$ does not satisfy either of the given boundary conditions at the left or right end points. The convergence is pointwise.

EXAMPLE 2.5.6: Consider the Euler operator with Robin and Dirichlet conditions. We seek the eigenvalues and corresponding orthonormal eigenfunctions for the Euler differential equation [Sturm-Liouville type for $p(x) = 1, q(x) = 0, w(x) = 1$] over the interval

$I = \{x \mid 0 < x < b\}$. The boundary conditions are type 3 at the left and type 1 at the right end points.

Euler differential equation

$$\frac{d^2}{dx^2}y(x) + \lambda y(x) = 0$$

Boundary conditions ($h > 0$)

$$y_x(0) - hy(0) = 0 \quad \text{and} \quad y(b) = 0$$

SOLUTION: We consider two possibilities for values of λ. We first consider $\lambda = 0$. For this case, the system basis vectors are

> restart:y1(x):=1;y2(x):=x;

$$y1(x) := 1$$
$$y2(x) := x$$

(2.133)

General solution

> y(x):=C1*y1(x)+C2*y2(x);

$$y(x) := C1 + C2x \tag{2.134}$$

Substituting into the boundary conditions yields

> eval(subs(x=0,diff(y(x),x)−h*y(x)))=0;

$$C2 - hC1 = 0 \tag{2.135}$$

> eval(subs(x=b,y(x)))=0;

$$C1 + C2b = 0 \tag{2.136}$$

The only solution to the above is the trivial solution. We next consider $\lambda > 0$. We set $\lambda = \mu^2$, and, for this case, the system basis vectors are

> y1(x):=sin(mu*x);y2(x):=cos(mu*x);

$$y1(x) := \sin(\mu x)$$
$$y2(x) := \cos(\mu x)$$

(2.137)

General solution

> y(x):=C1*y1(x)+C2*y2(x);

$$y(x) := C1 \sin(\mu x) + C2 \cos(\mu x) \tag{2.138}$$

Substituting into the boundary conditions yields

> eval(subs(x=0,diff(y(x),x)−h*y(x)))=0;

$$C1\mu - hC2 = 0 \tag{2.139}$$

> eval(subs(x=b,y(x)))=0;

$$C1 \sin(\mu b) + C2 \cos(\mu b) = 0 \tag{2.140}$$

The only nontrivial solutions to the preceding occur when both $C1$ and $C2$ are arbitrary and μ satisfies the following eigenvalue equation:

> h*sin(mu*b)+mu*cos(mu*b)=0;

$$h \sin(\mu b) + \mu \cos(\mu b) = 0 \tag{2.141}$$

We identify these roots as μ_n for $n = 1, 2, 3, \ldots$.

Allowed eigenvalues are $\lambda_n = \mu_n^2$

> lambda[n]=mu[n]^2;

$$\lambda_n = \mu_n^2 \tag{2.142}$$

Nonnormalized eigenfunctions are

> phi[n](x):=h*sin(sqrt(lambda[n])*x)+sqrt(lambda[n])*cos(sqrt(lambda[n])*x);

$$\phi_n(x) := h \sin\left(\sqrt{\lambda_n}x\right) + \sqrt{\lambda_n} \cos\left(\sqrt{\lambda_n}x\right) \tag{2.143}$$

Normalization

Evaluating the norm from the inner product of the eigenfunctions with respect to the weight function $w(x) = 1$ over the interval yields

> w(x):=1:unprotect(norm):norm:=sqrt(Int(phi[n](x)^2*w(x),x=0..b));norm:=value(%):

$$norm := \sqrt{\int_0^b \left(h \sin\left(\sqrt{\lambda_n}x\right) + \sqrt{\lambda_n} \cos\left(\sqrt{\lambda_n}x\right)\right)^2 dx} \tag{2.144}$$

Substitution of the eigenvalue equation simplifies the preceding equation:

```
> norm:=radsimp(subs(sin(sqrt(lambda[n])*b)=-sqrt(lambda[n])/
  h*cos(sqrt(lambda[n])*b),norm));
```

$$norm := \frac{1}{2} \frac{\sqrt{2}\sqrt{-\left(-2h^2 + h^2 \cos\left(\sqrt{\lambda_n}b\right)^2 - h^3b + \lambda_n \cos\left(\sqrt{\lambda_n}b\right)^2 - \lambda_n bh\right)h}}{h} \tag{2.145}$$

Orthonormal eigenfunctions

```
> phi[n](x):=phi[n](x)/norm;phi[m](x):=subs(n=m,phi[n](x)):
```

$$\phi_n(x) := \frac{\left(h \sin\left(\sqrt{\lambda_n}x\right) + \sqrt{\lambda_n} \cos\left(\sqrt{\lambda_n}x\right)\right)\sqrt{2h}}{\sqrt{-\left(-2h^2 + h^2 \cos\left(\sqrt{\lambda_n}b\right)^2 - h^3b + \lambda_n \cos\left(\sqrt{\lambda_n}b\right)^2 - \lambda_n bh\right)h}} \tag{2.146}$$

Statement of orthonormality

```
> Int(phi[n](x)*phi[m](x)*w(x),x=0..b)=delta(n,m);;
```

$$\int_0^b \left(2\left(h \sin\left(\sqrt{\lambda_n}x\right) + \sqrt{\lambda_n} \cos\left(\sqrt{\lambda_n}x\right)\right)h^2\left(h \sin\left(\sqrt{\lambda_m}x\right) + \sqrt{\lambda_m} \cos\left(\sqrt{\lambda_m}x\right)\right)\right) \Bigg/$$

$$\left(\sqrt{-\left(-2h^2 + h^2 \cos\left(\sqrt{\lambda_n}b\right)^2 - h^3b + \lambda_n \cos\left(\sqrt{\lambda_n}b\right)^2 - \lambda_n bh\right)h}\right.$$

$$\left.\sqrt{-\left(-2h^2 + h^2 \cos\left(\sqrt{\lambda_m}b\right)^2 - h^3b + \lambda_m \cos\left(\sqrt{\lambda_m}b\right)^2 - \lambda_m bh\right)h}\right) dx = \delta(n,m)$$

$$\tag{2.147}$$

Generalized Fourier series expansion

```
> f(x):=Sum(F(n)*phi[n](x),n=1..infinity);f(x):='f(x)':
```

$$f(x) := \sum_{n=1}^{\infty} \frac{F(n)\left(h \sin\left(\sqrt{\lambda_n}x\right) + \sqrt{\lambda_n} \cos\left(\sqrt{\lambda_n}x\right)\right)\sqrt{2h}}{\sqrt{-\left(-2h^2 + h^2 \cos\left(\sqrt{\lambda_n}b\right)^2 - h^3b + \lambda_n \cos\left(\sqrt{\lambda_n}b\right)^2 - \lambda_n bh\right)h}} \tag{2.148}$$

Fourier coefficients

> F(n):=Int(f(x)*phi[n](x)*w(x),x=0..b);F(n):='F(n)':

$$F(n) := \int_0^b \frac{f(x)\left(h\sin\left(\sqrt{\lambda_n}x\right) + \sqrt{\lambda_n}\cos\left(\sqrt{\lambda_n}x\right)\right)\sqrt{2h}}{\sqrt{-\left(-2h^2 + h^2\cos\left(\sqrt{\lambda_n}b\right)^2 - h^3b + \lambda_n\cos\left(\sqrt{\lambda_n}b\right)^2 - \lambda_n bh\right)h}}\,dx \quad (2.149)$$

This is the generalized series expansion of $f(x)$ in terms of the "complete" set of eigenfunctions for the particular Sturm-Liouville operator and boundary conditions over the interval.

DEMONSTRATION: Develop the generalized series expansion for $f(x) = 1 - x$ over the interval $I = \{x \mid 0 < x < 1\}$ in terms of the preceding eigenfunctions for $h = 1$. We assign the system values

> a:=0;b:=1;h:=1;f(x):=1−x;

$$a := 0$$
$$b := 1$$
$$h := 1$$
$$f(x) := 1 - x \quad (2.150)$$

SOLUTION: We evaluate the Fourier coefficients

> F(n):=Int(eval(f(x)*phi[n](x))*w(x),x=a..b);F(n):=value(%):

$$F(n) := \int_0^1 \frac{(1-x)\left(\sin\left(\sqrt{\lambda_n}x\right) + \sqrt{\lambda_n}\cos\left(\sqrt{\lambda_n}x\right)\right)\sqrt{2}}{\sqrt{3 - \cos\left(\sqrt{\lambda_n}\right)^2 - \lambda_n\cos\left(\sqrt{\lambda_n}\right)^2 + \lambda_n}}\,dx \quad (2.151)$$

> F(n):=radsimp(subs(sin(sqrt(lambda[n])*b)=−sqrt(lambda[n])/
  h*cos(sqrt(lambda[n])*b),F(n)));

$$F(n) := \frac{2\sqrt{2}}{\sqrt{\lambda_n}\sqrt{3 - \cos\left(\sqrt{\lambda_n}\right)^2 - \lambda_n\cos\left(\sqrt{\lambda_n}\right)^2 + \lambda_n}} \quad (2.152)$$

> Series:=simplify(Sum(F(n)*phi[n](x),n=1..infinity));

$$Series := \sum_{n=1}^{\infty} \left(-\frac{4\left(\sin\left(\sqrt{\lambda_n}x\right) + \sqrt{\lambda_n}\cos\left(\sqrt{\lambda_n}x\right)\right)}{\sqrt{\lambda_n}\left(-3 + \cos\left(\sqrt{\lambda_n}\right)^2 + \lambda_n\cos\left(\sqrt{\lambda_n}\right)^2 - \lambda_n\right)} \right) \quad (2.153)$$

Evaluation of the eigenvalues from the roots of the eigenvalue equation yields

> tan(sqrt(lambda[n])*b)=−sqrt(lambda[n])/h;

$$\tan\left(\sqrt{\lambda_n}\right) = -\sqrt{\lambda_n} \tag{2.154}$$

> plot({tan(v),−v/(h*b)},v=0..20,y=−20..0,thickness=10);

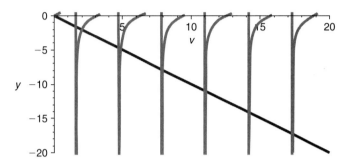

Figure 2.11

If we set $v = \sqrt{\lambda}b$, then the eigenvalues are found from the intersection points of the curves shown in Figure 2.11. We evaluate a few of these eigenvalues from the Maple fsolve command for the roots of the eigenvalue equation to be given as

> lambda[1]:=(1/b*(fsolve((tan(v)+v/(b*h)),v=1..3)))^2;

$$\lambda_1 := 4.115858365 \tag{2.155}$$

> lambda[2]:=(1/b*(fsolve((tan(v)+v/(b*h)),v=3..6)))^2;

$$\lambda_2 := 24.13934203 \tag{2.156}$$

> lambda[3]:=(1/b*(fsolve((tan(v)+v/(b*h)),v=6..9)))^2;

$$\lambda_3 := 63.65910654 \tag{2.157}$$

> lambda[4]:=(1/b*(fsolve((tan(v)+v/(b*h)),v=10..12)))^2;

$$\lambda_4 := 122.8891618 \tag{2.158}$$

> lambda[5]:=(1/b*(fsolve((tan(v)+v/(b*h)),v=13..15)))^2;

$$\lambda_5 := 201.8512584 \tag{2.159}$$

First five terms of expansion

> Series:=eval(sum(F(n)*phi[n](x),n=1..5)):
> plot({Series,f(x)},x=a..b,thickness=10);

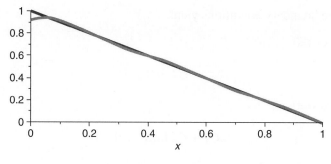

Figure 2.12

The two curves shown in Figure 2.12 depict the function $f(x)$ and its Fourier series approximation in terms of the orthonormal eigenfunctions for the particular operator and boundary conditions given earlier. Note that $f(x)$ satisfies the boundary condition at the right but not at the left end point. The convergence is pointwise.

EXAMPLE 2.5.7: Consider the Cauchy-Euler operator with Dirichlet conditions. We seek the eigenvalues and corresponding orthonormal eigenfunctions for the Cauchy-Euler differential equation [Sturm-Liouville type for $p(x) = x$, $q(x) = 0$, $w(x) = \frac{1}{x}$] over the interval $I = \{x \mid 1 < x < b\}$. The boundary conditions are type 1 at the left and type 1 at the right end points.

Cauchy-Euler differential equation

$$x^2\left(\frac{d^2}{dx^2}y(x)\right) + x\left(\frac{d}{dx}y(x)\right) + \lambda y(x) = 0$$

Boundary conditions

$$y(1) = 0 \quad \text{and} \quad y(b) = 0$$

SOLUTION: We consider two possibilities for values of λ. We first consider $\lambda = 0$. For this case, the system basis vectors are

> restart:y1(x):=1;y2(x):=ln(x);

$$y1(x) := 1$$

$$y2(x) := \ln(x) \tag{2.160}$$

General solution

> y(x):=C1*y1(x)+C2*y2(x);

$$y(x) := C1 + C2\ln(x) \tag{2.161}$$

Substituting into the boundary conditions yields

> eval(subs(x=1,y(x)))=0;

$$C1 = 0 \tag{2.162}$$

> eval(subs(x=b,y(x)))=0;

$$C1 + C2 \ln(b) = 0 \tag{2.163}$$

The only solution to the preceding is the trivial solution. We next consider $\lambda > 0$. We set $\lambda = \mu^2$, and, for this case, the system basis vectors are

> y1(x):=sin(mu*ln(x));y2(x):=cos(mu*ln(x));

$$y1(x) := \sin(\mu \ln(x))$$
$$y2(x) := \cos(\mu \ln(x)) \tag{2.164}$$

General solution

> y(x):=C1*y1(x)+C2*y2(x);

$$y(x) := C1 \sin(\mu \ln(x)) + C2 \cos(\mu \ln(x)) \tag{2.165}$$

Substituting into the boundary conditions yields

> eval(subs(x=1,y(x)))=0;

$$C2 = 0 \tag{2.166}$$

> eval(subs(x=b,y(x)))=0;

$$C1 \sin(\mu \ln(b)) + C2 \cos(\mu \ln(b)) = 0 \tag{2.167}$$

The only nontrivial solutions to the preceding occur when $C2 = 0$, $C1$ is arbitrary, and μ satisfies the following eigenvalue equation:

> sin(mu*ln(b))=0;

$$\sin(\mu \ln(b)) = 0 \tag{2.168}$$

Thus, the values for μ are

> mu[n]:=n*Pi/ln(b);

$$\mu_n := \frac{n\pi}{\ln(b)} \tag{2.169}$$

for $n = 1, 2, 3, \ldots$.

Allowed eigenvalues are $\lambda_n = \mu_n^2$

> lambda[n]:=(n*Pi/ln(b))^2;

$$\lambda_n := \frac{n^2\pi^2}{\ln(b)^2} \qquad (2.170)$$

Nonnormalized eigenfunctions are

> phi[n](x):=(sin(mu[n]*ln(x)));

$$\phi_n(x) := \sin\left(\frac{n\pi \ln(x)}{\ln(b)}\right) \qquad (2.171)$$

Normalization

Evaluating the norm from the inner product of the eigenfunctions with respect to the weight function $w(x) = \frac{1}{x}$ over the interval yields

> w(x):=1/x:unprotect(norm):norm:=sqrt(Int(phi[n](x)^2*w(x),x=1..b));norm:=value(%):

$$norm := \sqrt{\int_1^b \frac{\sin\left(\frac{n\pi \ln(x)}{\ln(b)}\right)^2}{x}\,dx} \qquad (2.172)$$

Substitution of the eigenvalue equation simplifies the norm

> norm:=radsimp(subs({sin(n*Pi)=0,cos(n*Pi)=(−1)^n},norm));

$$norm := \frac{1}{2}\sqrt{2}\,\sqrt{\ln(b)} \qquad (2.173)$$

Orthonormal eigenfunctions

> phi[n](x):=phi[n](x)/norm;phi[m](x):=subs(n=m,phi[n](x)):

$$\phi_n(x) := \frac{\sin\left(\frac{n\pi \ln(x)}{\ln(b)}\right)\sqrt{2}}{\sqrt{\ln(b)}} \qquad (2.174)$$

Statement of orthonormality

> Int(phi[n](x)*phi[m](x)*w(x),x=1..b)=delta(n,m);;

$$\int_1^b \frac{2\sin\left(\frac{n\pi \ln(x)}{\ln(b)}\right)\sin\left(\frac{m\pi \ln(x)}{\ln(b)}\right)}{\ln(b)x}\,dx = \delta(n,m) \qquad (2.175)$$

Generalized Fourier series expansion

> f(x):=Sum(F(n)*phi[n](x),n=1..infinity);f(x):='f(x)':

$$f(x) := \sum_{n=1}^{\infty} \frac{F(n) \sin\left(\frac{n\pi \ln(x)}{\ln(b)}\right) \sqrt{2}}{\sqrt{\ln(b)}} \qquad (2.176)$$

Fourier coefficients

> F(n):=Int(f(x)*phi[n](x)*w(x),x=1..b);F(n):='F(n)':

$$F(n) := \int_{1}^{b} \frac{f(x) \sin\left(\frac{n\pi \ln(x)}{\ln(b)}\right) \sqrt{2}}{\sqrt{\ln(b)}x} \, dx \qquad (2.177)$$

This is the generalized series expansion of $f(x)$ in terms of the "complete" set of eigenfunctions for the particular Sturm-Liouville operator and boundary conditions over the interval.

DEMONSTRATION: Develop the generalized series expansion for $f(x) = x - 1$ over the interval $I = \{x \mid 1 < x < 2\}$ in terms of the preceding eigenfunctions. We assign the system values

> a:=1;b:=2;f(x):=x−1;

$$a := 1$$
$$b := 2$$
$$f(x) := x - 1 \qquad (2.178)$$

SOLUTION: We evaluate the Fourier coefficients

> F(n):=eval(Int(f(x)*phi[n](x)*w(x),x=a..b));F(n):=simplify(value(%)):

$$F(n) := \int_{1}^{2} \frac{(x-1) \sin\left(\frac{n\pi \ln(x)}{\ln(2)}\right) \sqrt{2}}{\sqrt{\ln(2)}x} \, dx \qquad (2.179)$$

> F(n):=combine(%,trig):F(n):=subs(cos(n*Pi)=(−1)^n,sin(n*Pi)=0,F(n));

$$F(n) := \frac{-\sqrt{\ln(2)}\sqrt{2}n^2\pi^2(-1)^n - \ln(2)^{5/2}\sqrt{2} + \ln(2)^{5/2}\sqrt{2}(-1)^n}{n^3\pi^3 + n\pi \ln(2)^2} \qquad (2.180)$$

> Series:=eval(Sum(F(n)*phi[n](x),n=1..infinity));

$$Series := \sum_{n=1}^{\infty} \frac{\left(-\sqrt{\ln(2)}\,\sqrt{2}\,n^2\pi^2(-1)^n - \ln(2)^{5/2}\sqrt{2} + \ln(2)^{5/2}\sqrt{2}(-1)^n\right)\sin\left(\frac{n\pi\ln(x)}{\ln(2)}\right)\sqrt{2}}{\left(n^3\pi^3 + n\pi\ln(2)^2\right)\sqrt{\ln(2)}}$$

(2.181)

First few terms of expansion

> Series:=eval(sum(F(n)*phi[n](x),n=1..5)):

> plot({Series,f(x)},x=a..b,thickness=10);

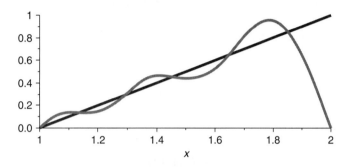

Figure 2.13

The two curves in Figure 2.13 depict the function $f(x)$ and its Fourier series approximation in terms of the orthonormal eigenfunctions for the particular operator and boundary conditions given earlier. Note that $f(x)$ satisfies the given boundary conditions at the left but fails to do so at the right end point. The convergence is pointwise.

EXAMPLE 2.5.8: Consider the Cauchy-Euler operator with Neumann and Dirichlet conditions. We seek the eigenvalues and corresponding orthonormal eigenfunctions for the Cauchy-Euler differential equation [Sturm-Liouville type for $p(x) = x$, $q(x) = 0$, $w(x) = \frac{1}{x}$] over the interval $I = \{x \mid 1 < x < b\}$. The boundary conditions are type 2 at the left and type 1 at the right end points.

Cauchy-Euler differential equation

$$x^2\left(\frac{d^2}{dx^2}y(x)\right) + x\left(\frac{d}{dx}y(x)\right) + \lambda y(x) = 0$$

Boundary conditions

$$y_x(1) = 0 \quad \text{and} \quad y(b) = 0$$

SOLUTION: We consider two possibilities for values of λ. We first consider $\lambda = 0$. For this case, the system basis vectors are

> restart:y1(x):=1;y2(x):=ln(x);

$$y1(x) := 1$$

$$y2(x) := \ln(x) \tag{2.182}$$

General solution

> y(x):=C1*y1(x)+C2*y2(x);

$$y(x) := C1 + C2 \ln(x) \tag{2.183}$$

Substituting into the boundary conditions yields

> eval(subs(x=1,diff(y(x),x)))=0;

$$C2 = 0 \tag{2.184}$$

> eval(subs(x=b,y(x)))=0;

$$C1 + C2 \ln(b) = 0 \tag{2.185}$$

The only solution to the preceding is the trivial solution. We next consider $\lambda > 0$. We set $\lambda = \mu^2$, and, for this case, the system basis vectors are

> y1(x):=sin(mu*ln(x));y2(x):=cos(mu*ln(x));

$$y1(x) := \sin(\mu \ln(x))$$

$$y2(x) := \cos(\mu \ln(x)) \tag{2.186}$$

General solution

> y(x):=C1*y1(x)+C2*y2(x);

$$y(x) := C1 \sin(\mu \ln(x)) + C2 \cos(\mu \ln(x)) \tag{2.187}$$

Substituting into the boundary conditions yields

> eval(subs(x=1,diff(y(x),x)))=0;

$$C1\mu = 0 \tag{2.188}$$

> eval(subs(x=b,y(x)))=0;

$$C1 \sin(\mu \ln(b)) + C2 \cos(\mu \ln(b)) = 0 \tag{2.189}$$

The only nontrivial solutions to the preceding occur when $C1 = 0$, $C2$ is arbitrary, and μ satisfies the following eigenvalue equation:

> cos(mu*ln(b))=0;

$$\cos(\mu \ln(b)) = 0 \tag{2.190}$$

Thus, μ takes on the values

> mu[n]:=(2*n−1)*Pi/(2*ln(b));

$$\mu_n := \frac{1}{2} \frac{(2n-1)\pi}{\ln(b)} \tag{2.191}$$

for $n = 1, 2, 3, \ldots$.

Allowed eigenvalues are $\lambda_n = \mu_n^2$

> lambda[n]:=mu[n]^2;

$$\lambda_n := \frac{1}{4} \frac{(2n-1)^2 \pi^2}{\ln(b)^2} \tag{2.192}$$

Nonnormalized eigenfunctions are

> phi[n](x):=cos(mu[n]*ln(x));

$$\phi_n(x) := \cos\left(\frac{1}{2} \frac{(2n-1)\pi \ln(x)}{\ln(b)}\right) \tag{2.193}$$

Normalization

Evaluating the norm from the inner product of the eigenfunctions with respect to the weight function $w(x) = \frac{1}{x}$ over the interval yields

> w(x):=1/x:unprotect(norm):norm:=sqrt(Int(phi[n](x)^2*w(x),x=1..b));
>   norm:=expand(value(%)):

$$norm := \sqrt{\int_1^b \frac{\cos\left(\frac{1}{2} \frac{(2n-1)\pi \ln(x)}{\ln(b)}\right)^2}{x} \, dx} \tag{2.194}$$

Substitution of the eigenvalue equation simplifies the norm

> norm:=radsimp(subs({sin(n^Pi)=0,cos(n^Pi)=(−1)^n},norm));

$$norm := \frac{1}{2}\sqrt{2}\sqrt{\ln(b)} \tag{2.195}$$

Orthonormal eigenfunctions

> phi[n](x):=phi[n](x)/norm;phi[m](x):=subs(n=m,phi[n](x)):

$$\phi_n(x) := \frac{\cos\left(\frac{1}{2}\frac{(2n-1)\pi \ln(x)}{\ln(b)}\right)\sqrt{2}}{\sqrt{\ln(b)}} \tag{2.196}$$

Statement of orthonormality

> Int(phi[n](x)*phi[m](x)*w(x),x=1..b)=delta(n,m);

$$\int_1^b \frac{2\cos\left(\frac{1}{2}\frac{(2n-1)\pi \ln(x)}{\ln(b)}\right)\cos\left(\frac{1}{2}\frac{(2m-1)\pi \ln(x)}{\ln(b)}\right)}{\ln(b)x}\,dx = \delta(n,m) \tag{2.197}$$

Generalized Fourier series expansion

> f(x):=Sum(F(n)*phi[n](x),n=1..infinity);f(x):='f(x)':

$$f(x) := \sum_{n=1}^{\infty} \frac{F(n)\cos\left(\frac{1}{2}\frac{(2n-1)\pi \ln(x)}{\ln(b)}\right)\sqrt{2}}{\sqrt{\ln(b)}} \tag{2.198}$$

Fourier coefficients

> F(n):=Int(f(x)*phi[n](x)*w(x),x=1..b);F(n):='F(n)':

$$F(n) := \int_1^b \frac{f(x)\cos\left(\frac{1}{2}\frac{(2n-1)\pi \ln(x)}{\ln(b)}\right)\sqrt{2}}{\sqrt{\ln(b)}x}\,dx \tag{2.199}$$

This is the generalized series expansion of $f(x)$ in terms of the "complete" set of eigenfunctions for the particular Sturm-Liouville operator and boundary conditions over the interval.

DEMONSTRATION: Develop the generalized series expansion for $f(x) = 2x - x^2$ over the interval $I = \{x \mid 1 < x < 2\}$ in terms of the preceding eigenfunctions. We assign the system values

> a:=1;b:=2;f(x):=2*x−x^2;

$$a := 1$$
$$b := 2$$
$$f(x) := 2x - x^2 \tag{2.200}$$

SOLUTION: We evaluate the Fourier coefficients

> F(n):=eval(Int(f(x)*phi[n](x)*w(x),x=a..b));F(n):=simplify(value(%)):

$$F(n) := \int_1^2 \frac{(2x - x^2) \cos\left(\frac{1}{2}\frac{(2n-1)\pi \ln(x)}{\ln(2)}\right) \sqrt{2}}{\sqrt{\ln(2)}\, x}\, dx \qquad (2.201)$$

> F(n):=combine(%,trig):F(n):=factor(simplify(subs(cos(n*Pi)=(−1)^n,sin(n*Pi)=0,F(n))));

$$F(n) := -\frac{96\ln(2)^{5/2}\sqrt{2}\left(2\pi n(-1)^n - \pi(-1)^n + \ln(2)\right)}{\left(16\ln(2)^2 - 4\pi^2 n + \pi^2 + 4\pi^2 n^2\right)\left(4\pi^2 n^2 - 4\pi^2 n + 4\ln(2)^2 + \pi^2\right)} \qquad (2.202)$$

> Series:=eval(Sum(F(n)*phi[n](x),n=1..infinity));

$$Series := \sum_{n=1}^{\infty} \left(-\frac{192\ln(2)^2\left(2\pi n(-1)^n - \pi(-1)^n + \ln(2)\right)\cos\left(\frac{1}{2}\frac{(2n-1)\pi \ln(x)}{\ln(2)}\right)}{\left(16\ln(2)^2 - 4\pi^2 n + \pi^2 + 4\pi^2 n^2\right)\left(4\pi^2 n^2 - 4\pi^2 n + 4\ln(2)^2 + \pi^2\right)} \right) \qquad (2.203)$$

First five terms of expansion

> Series:=eval(sum(F(n)*phi[n](x),n=1..5)):

> plot({Series,f(x)},x=a..b,thickness=10);

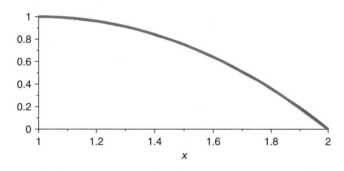

Figure 2.14

The two curves in Figure 2.14 depict the function $f(x)$ and its Fourier series approximation in terms of the orthonormal eigenfunctions for the particular operator and boundary conditions given earlier. Note that $f(x)$ satisfies the given boundary conditions at both the left and the right end points. The convergence is uniform.

EXAMPLE 2.5.9: Consider the Cauchy-Euler operator with Dirichlet and Robin conditions. We seek the eigenvalues and corresponding orthonormal eigenfunctions for the Cauchy-Euler differential equation [Sturm-Liouville type for $p(x) = x$, $q(x) = 0$, $w(x) = \frac{1}{x}$] over the interval $I = \{x \mid 1 < x < b\}$. The boundary conditions are type 3 at the left and type 1 at the right end points.

Cauchy-Euler differential equation

$$x^2 \left(\frac{d^2}{dx^2} y(x) \right) + x \left(\frac{d}{dx} y(x) \right) + \lambda y(x) = 0$$

Boundary conditions ($h > 0$)

$$y(1) = 0 \quad \text{and} \quad y_x(b) + h y(b) = 0$$

SOLUTION: We consider two possibilities for values of λ. We first consider $\lambda = 0$. For this case, the system basis vectors are

> restart:y1(x):=1;y2(x):=ln(x);

$$y1(x) := 1$$
$$y2(x) := \ln(x) \tag{2.204}$$

General solution

> y(x):=C1*y1(x)+C2*y2(x);

$$y(x) := C1 + C2 \ln(x) \tag{2.205}$$

Substituting into the boundary conditions yields

> eval(subs(x=1,y(x)))=0;

$$C1 = 0 \tag{2.206}$$

> eval(subs(C1=0,x=b,diff(y(x),x)+h*y(x)))=0;

$$\frac{C2}{b} + h C2 \ln(b) = 0 \tag{2.207}$$

The only solution to the preceding is the trivial solution. We next consider $\lambda > 0$. We set $\lambda = \mu^2$, and, for this case, the system basis vectors are

> y1(x):=sin(mu*ln(x));y2(x):=cos(mu*ln(x));

$$y1(x) := \sin(\mu \ln(x))$$
$$y2(x) := \cos(\mu \ln(x)) \tag{2.208}$$

General solution

> y(x):=C1*y1(x)+C2*y2(x);

$$y(x) := C1\sin(\mu\ln(x)) + C2\cos(\mu\ln(x)) \tag{2.209}$$

Substituting into the boundary conditions yields

> eval(subs(x=1,y(x)))=0;

$$C2 = 0 \tag{2.210}$$

> eval(subs(C2=0,x=b,diff(y(x),x)+h*y(x)))=0;

$$\frac{C1\cos(\mu\ln(b))\mu}{b} + hC1\sin(\mu\ln(b)) = 0 \tag{2.211}$$

The only nontrivial solutions to the preceding are that $C2 = 0$, $C1$ is arbitrary, and μ must satisfy the following eigenvalue equation:

> h*sin(mu*ln(b))+mu/b*cos(mu*ln(b))=0;

$$h\sin(\mu\ln(b)) + \frac{\mu\cos(\mu\ln(b))}{b} = 0 \tag{2.212}$$

We indicate these roots as μ_n for $n = 1, 2, 3, \ldots$.

Allowed eigenvalues are $\lambda_n = \mu_n^2$

> lambda[n]=mu[n]^2;

$$\lambda_n = \mu_n^2 \tag{2.213}$$

Nonnormalized eigenfunctions are

> phi[n](x):=sin(sqrt(lambda[n])*ln(x));

$$\phi_n(x) := \sin\left(\sqrt{\lambda_n}\ln(x)\right) \tag{2.214}$$

Normalization

Evaluating the norm from the inner product of the eigenfunctions with respect to the weight function $w(x) = \frac{1}{x}$ over the interval yields

> w(x):=1/x:unprotect(norm):norm:=sqrt(Int(phi[n](x)^2*w(x),x=1..b));norm:=value(%):

$$norm := \sqrt{\int_1^b \frac{\sin\left(\sqrt{\lambda_n}\ln(x)\right)^2}{x}\,dx} \tag{2.215}$$

Substitution of the eigenvalue equation simplifies the norm

> norm:=simplify(subs(sin(sqrt(lambda[n])*ln(b))=−sqrt(lambda[n])/
  (h*b)*cos(sqrt(lambda[n])*ln(b)),norm));

$$norm := \frac{1}{2}\sqrt{2}\sqrt{\frac{\cos\left(\sqrt{\lambda_n}\,\ln(b)\right)^2 + \ln(b)hb}{hb}} \tag{2.216}$$

Orthonormal eigenfunctions

> phi[n](x):=phi[n](x)/norm;phi[m](x):=subs(n=m,phi[n](x)):

$$\phi_n(x) := \frac{\sin\left(\sqrt{\lambda_n}\,\ln(x)\right)\sqrt{2}}{\sqrt{\dfrac{\cos\left(\sqrt{\lambda_n}\,\ln(b)\right)^2 + \ln(b)hb}{hb}}} \tag{2.217}$$

Statement of orthonormality

> Int(phi[n](x)*phi[m](x)*w(x),x=1..b)=delta(n,m);

$$\int_1^b \frac{2\sin\left(\sqrt{\lambda_n}\,\ln(x)\right)\sin\left(\sqrt{\lambda_n}\,\ln(x)\right)}{\sqrt{\dfrac{\cos\left(\sqrt{\lambda_n}\,\ln(b)\right)^2 + \ln(b)hb}{hb}}\sqrt{\dfrac{\cos\left(\sqrt{\lambda_n}\,\ln(b)\right)^2 + \ln(b)hb}{hb}}x}\,dx = \delta(n,m) \tag{2.218}$$

Generalized Fourier series expansion

> f(x):=Sum(F(n)*phi[n](x),n=1..infinity);f(x):='f(x)':

$$f(x) := \sum_{n=1}^{\infty} \frac{F(n)\sin\left(\sqrt{\lambda_n}\,\ln(x)\right)\sqrt{2}}{\sqrt{\dfrac{\cos\left(\sqrt{\lambda_n}\,\ln(b)\right)^2 + \ln(b)hb}{hb}}} \tag{2.219}$$

Fourier coefficients

> F(n):=Int(f(x)*phi[n](x)*w(x),x=1..b);F(n):='F(n)':

$$F(n) := \int_1^b \frac{f(x)\sin\left(\sqrt{\lambda_n}\,\ln(x)\right)\sqrt{2}}{\sqrt{\dfrac{\cos\left(\sqrt{\lambda_n}\,\ln(b)\right)^2 + \ln(b)hb}{hb}}x}\,dx \tag{2.220}$$

This is the generalized series expansion of $f(x)$ in terms of the "complete" set of eigenfunctions for the particular Sturm-Liouville operator and boundary conditions over the interval.

DEMONSTRATION: Develop the generalized series expansion for $f(x) = x - 1$ over the interval $I = \{x \mid 1 < x < 2\}$ in terms of the preceding eigenfunctions for $h = 1$. We assign the system values

> a:=1;b:=2;h:=1;f(x):=x−1;

$$a := 1$$
$$b := 2$$
$$h := 1$$
$$f(x) := x - 1 \tag{2.221}$$

SOLUTION: We evaluate the Fourier coefficients

> F(n):=eval(expand((Int(f(x)*phi[n](x)*w(x),x=a..b))));F(n):=simplify(value(%)):

$$F(n) := \frac{\sqrt{2}\left(\displaystyle\int_{1}^{2}\left(\sin\left(\sqrt{\lambda_n}\,\ln(x)\right) - \frac{\sin\left(\sqrt{\lambda_n}\,\ln(x)\right)}{x}\right)dx\right)}{\sqrt{\frac{1}{2}\cos\left(\sqrt{\lambda_n}\,\ln(2)\right)^2 + \ln(2)}} \tag{2.222}$$

Substitution of the eigenvalue equation simplifies the preceding equation

> F(n):=(eval(subs(sin(sqrt(lambda[n])*ln(b))=−sqrt(lambda[n])/
  (b*h)*cos(sqrt(lambda[n])*ln(b)),F(n))));

$$F(n) := \frac{1}{\sqrt{2\cos\left(\sqrt{\lambda_n}\,\ln(2)\right)^2 + 4\ln(2)}\,\sqrt{\lambda_n}\,(1+\lambda_n)}\left(2\left(-2\lambda_n\cos\left(\frac{1}{2}\sqrt{\lambda_n}\,\ln(2)\right)^2 + \lambda_n\right.\right.$$
$$\left.\left. + 2\cos\left(\frac{1}{2}\sqrt{\lambda_n}\,\ln(2)\right)^2 + 4\sin\left(\frac{1}{2}\sqrt{\lambda_n}\,\ln(2)\right)\sqrt{\lambda_n}\cos\left(\frac{1}{2}\sqrt{\lambda_n}\,\ln(2)\right) - 2\right)\sqrt{2}\right) \tag{2.223}$$

> Series:=eval(Sum(F(n)*phi[n](x),n=1..infinity));

$$Series := \sum_{n=1}^{\infty}\left(4\left(-2\lambda_n\cos\left(\frac{1}{2}\sqrt{\lambda_n}\,\ln(2)\right)^2 + \lambda_n + 2\cos\left(\frac{1}{2}\sqrt{\lambda_n}\,\ln(2)\right)^2\right.\right.$$
$$\left.\left. + 4\sin\left(\frac{1}{2}\sqrt{\lambda_n}\,\ln(2)\right)\sqrt{\lambda_n}\cos\left(\frac{1}{2}\sqrt{\lambda_n}\,\ln(2)\right) - 2\right)\sin\left(\sqrt{\lambda_n}\,\ln(x)\right)\right) \Bigg/$$
$$\left(\sqrt{2\cos\left(\sqrt{\lambda_n}\,\ln(2)\right)^2 + 4\ln(2)}\,\sqrt{\lambda_n}\,(1+\lambda_n)\sqrt{\frac{1}{2}\cos\left(\sqrt{\lambda_n}\,\ln(2)\right)^2 + \ln(2)}\right) \tag{2.224}$$

Evaluation of the eigenvalues from the roots of the eigenvalue equation yields

> tan(sqrt(lambda[n])*ln(b))=−sqrt(lambda[n])/(b*h);

$$\tan\left(\sqrt{\lambda_n}\ln(2)\right) = -\frac{1}{2}\sqrt{\lambda_n} \tag{2.225}$$

> plot({tan(v),−v/(b*h*ln(b))},v=0..20,y=−20..0,thickness=10);

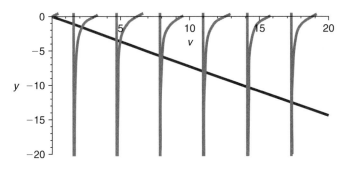

Figure 2.15

If we set $v = \sqrt{\lambda}\ln(b)$, then the eigenvalues are found from the intersection points of the curves shown in Figure 2.15. We evaluate a few of these eigenvalues from the Maple fsolve command for the roots of the eigenvalue equation to be given as

> lambda[1]:=evalf(1/ln(b)*(fsolve((tan(v)+v/(h*b*ln(b))),v=1..3)))^2;

$$\lambda_1 := 9.573189425 \tag{2.226}$$

> lambda[2]:=evalf(1/ln(b)*(fsolve((tan(v)+v/(h*b*ln(b))),v=3..6)))^2;

$$\lambda_2 := 51.69543396 \tag{2.227}$$

> lambda[3]:=evalf(1/ln(b)*(fsolve((tan(v)+v/(h*b*ln(b))),v=6..9)))^2;

$$\lambda_3 := 134.0427844 \tag{2.228}$$

> lambda[4]:=evalf(1/ln(b)*(fsolve((tan(v)+v/(h*b*ln(b))),v=9..13)))^2;

$$\lambda_4 := 257.3521722 \tag{2.229}$$

> lambda[5]:=evalf(1/ln(b)*(fsolve((tan(v)+v/(h*b*ln(b))),v=13..16)))^2;

$$\lambda_5 := 421.7143609 \tag{2.230}$$

First five terms of expansion

> Series:=eval(sum(F(n)*phi[n](x),n=1..5)):

> plot({Series,f(x)},x=a..b,thickness=10);

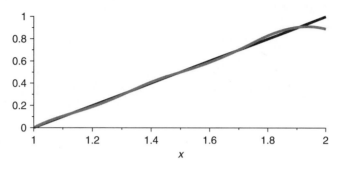

Figure 2.16

The two curves in Figure 2.16 depict the function $f(x)$ and its Fourier series approximation in terms of the orthonormal eigenfunctions for the particular operator and boundary conditions given earlier. Note that $f(x)$ satisfies the given boundary condition at the left but not at the right end point. The convergence is pointwise.

2.6 Nonregular or Singular Sturm-Liouville Eigenvalue Problems

If the Sturm-Liouville problem does not conform to the conditions of a "regular" format as given in Section 2.1, then the problem is "nonregular" or "singular." There are a variety of reasons why the problem may not have the regular format. We now consider a few examples that will be of use to us later when solving problems in partial differential equations. The first example shown here is of particular interest because it gives rise to what we recognize as being the familiar "Fourier" series.

EXAMPLE 2.6.1: Consider the Euler operator with periodic boundary conditions. We seek the eigenvalues and corresponding orthonormal eigenfunctions for the Euler differential equation [Sturm-Liouville type with $p(x) = 1$, $q(x) = 0$, $w(x) = 1$] over the symmetric interval $I = \{x \mid -b < x < b\}$. The boundary conditions are of the "periodic" type; this condition is nonregular because the boundary conditions are "mixed." A mixed condition comes about because each boundary condition involves two different spatial points. From Green's formula in Section 2.2, it can be shown that for periodic boundary conditions and $p(x) = 1$, the eigenfunctions continue to be orthogonal over the interval I (see Exercise 2.32). In addition, we need only consider the cases for λ greater than or equal to zero.

Euler differential equation

$$\frac{d^2}{dx^2} y(x) + \lambda y(x) = 0$$

Boundary conditions (periodic)

$$y(-b) = y(b) \quad \text{and} \quad y_x(-b) = y_x(b)$$

SOLUTION: We consider two possibilities for values of λ. We first consider $\lambda = 0$. For this case, the system basis vectors are

> restart:y1(x):=1;y2(x):=x;

$$y1(x) := 1$$
$$y2(x) := x \tag{2.231}$$

General solution

> y(x):=C1*y1(x)+C2*y2(x);

$$y(x) := C1 + C2x \tag{2.232}$$

Substituting into the boundary conditions yields

> eval(subs(x=−b,y(x)))=eval(subs(x=b,y(x)));

$$C1 - C2b = C1 + C2b \tag{2.233}$$

> eval(subs(x=−b,diff(y(x),x)))=eval(subs(x=b,diff(y(x),x)));

$$C2 = C2 \tag{2.234}$$

The only nontrivial solution to the preceding is that $C2 = 0$ and $C1$ is arbitrary. Thus, for $\lambda = 0$, we have

> lambda[0]:=0;

$$\lambda_0 := 0 \tag{2.235}$$

> phi[0](x):=1;

$$\phi_0(x) := 1 \tag{2.236}$$

We now consider the case for $\lambda > 0$. We set $\lambda = \mu^2$, and, for this case, the system basis vectors are

> y1(x):=sin(mu*x);y2(x):=cos(mu*x);

$$yI(x) := \sin(\mu x)$$

$$y2(x) := \cos(\mu x) \tag{2.237}$$

General solution

> y(x):=C1*y1(x)+C2*y2(x);

$$y(x) := CI \sin(\mu x) + C2 \cos(\mu x) \tag{2.238}$$

Substituting into the boundary conditions yields

> eval(subs(x=−b,y(x)))=eval(subs(x=b,y(x)));

$$-CI \sin(\mu b) + C2 \cos(\mu b) = CI \sin(\mu b) + C2 \cos(\mu b) \tag{2.239}$$

> eval(subs(x=−b,diff(y(x),x)))=eval(subs(x=b,diff(y(x),x)));

$$CI \cos(\mu b)\mu + C2 \sin(\mu b)\mu = CI \cos(\mu b)\mu - C2 \sin(\mu b)\mu \tag{2.240}$$

The only nontrivial solutions are that both CI and $C2$ be arbitrary and independent and that μ satisfies the following eigenvalue equation:

> sin(mu*b)=0;

$$\sin(\mu b) = 0 \tag{2.241}$$

The values of μ that satisfy this equation are

> mu[n]:=(n*Pi/b);

$$\mu_n := \frac{n\pi}{b} \tag{2.242}$$

for $n = 1, 2, 3, \ldots$.

Allowed eigenvalues are $\lambda_n = \mu_n^2$

> lambda[n]:=mu[n]^2;

$$\lambda_n := \frac{n^2\pi^2}{b^2} \tag{2.243}$$

Nonnormalized eigenfunctions are

> phi[n](x):=cos(mu[n]*x);psi[n](x):=sin(mu[n]*x);

$$\phi_n(x) := \cos\left(\frac{n\pi x}{b}\right)$$

$$\psi_n(x) := \sin\left(\frac{n\pi x}{b}\right) \tag{2.244}$$

This multiple set of eigenfunctions comes about because the boundary conditions are not regular. For "regular" Sturm-Liouville problems, such degeneracy does not occur.

Normalization

Evaluating the norm from the inner product of the eigenfunctions with respect to the weight function $w(x) = 1$ over the interval yields, for $n = 0$,

> w(x):=1:norm0:=sqrt(Int(phi[0](x)^2*w(x),x=−b..b));

$$norm0 := \sqrt{\int_{-b}^{b} 1 dx} \tag{2.245}$$

> norm0:=value(%);

$$norm0 := \sqrt{2}\sqrt{b} \tag{2.246}$$

For $n = 1, 2, 3, \ldots$, substitution of the eigenvalue equation simplifies the norms as follows:

> norm1:=sqrt(Int(phi[n](x)^2*w(x),x=−b..b));norm1:=value(%);

$$norm1 := \sqrt{\int_{-b}^{b} \cos\left(\frac{n\pi x}{b}\right)^2 dx} \tag{2.247}$$

> norm1:=radsimp(subs({sin(n*Pi)=0,cos(n*Pi)=(−1)^n},norm1));

$$norm1 := \sqrt{b} \tag{2.248}$$

> norm2:=sqrt(Int(expand(psi[n](x)^2*w(x)),x=−b..b));norm2:=value(%);

$$norm2 := \sqrt{\int_{-b}^{b} \sin\left(\frac{n\pi x}{b}\right)^2 dx} \tag{2.249}$$

> norm2:=radsimp(subs({sin(n*Pi)=0,cos(n*Pi)=(−1)^n},norm2));

$$norm2 := \sqrt{b} \tag{2.250}$$

Orthonormal eigenfunctions

> phi[0](x):=phi[0](x)/norm0;

$$\phi_0(x) := \frac{1}{2}\frac{\sqrt{2}}{\sqrt{b}} \tag{2.251}$$

> phi[n](x):=phi[n](x)/norm1;phi[m](x):=subs(n=m,phi[n](x)):

$$\phi_n(x) := \frac{\frac{\cos\left(\frac{n\pi x}{b}\right)}{b}}{\sqrt{b}} \tag{2.252}$$

> psi[n](x):=psi[n](x)/norm2;psi[m](x):=subs(n=m,psi[n](x)):

$$\psi_n(x) := \frac{\frac{\sin\left(\frac{n\pi x}{b}\right)}{b}}{\sqrt{b}} \tag{2.253}$$

Statements of orthonormality

> Int(phi[n](x)*phi[m](x)*w(x),x=−b..b)=delta(n,m);

$$\int_{-b}^{b} \frac{\cos\left(\frac{n\pi x}{b}\right)\cos\left(\frac{m\pi x}{b}\right)}{b}\,\mathrm{d}x = \delta(n, m) \tag{2.254}$$

> Int(psi[n](x)*psi[m](x)*w(x),x=−b..b)=delta(n,m);

$$\int_{-b}^{b} \frac{\sin\left(\frac{n\pi x}{b}\right)\sin\left(\frac{m\pi x}{b}\right)}{b}\,\mathrm{d}x = \delta(n, m) \tag{2.255}$$

Generalized Fourier series expansion

> f(x):=A(0)*phi[0](x)+Sum(A(n)*phi[n](x)+B(n)*psi[n](x),n=1..infinity);f(x):=`f(x)`:

$$f(x) := \frac{1}{2}\frac{A(0)\sqrt{2}}{\sqrt{b}} + \sum_{n=1}^{\infty}\left(\frac{A(n)\cos\left(\frac{n\pi x}{b}\right)}{\sqrt{b}} + \frac{B(n)\sin\left(\frac{n\pi x}{b}\right)}{\sqrt{b}}\right) \tag{2.256}$$

Fourier coefficients

For $n = 1, 2, 3, \ldots$,

> A(n):=Int(f(x)*phi[n](x)*w(x),x=−b..b);A(n):=`A(n)`:

$$A(n) := \int_{-b}^{b} \frac{f(x)\cos\left(\frac{n\pi x}{b}\right)}{\sqrt{b}}\,\mathrm{d}x \tag{2.257}$$

> B(n):=Int(f(x)*psi[n](x)*w(x),x=−b..b);B(n):=`B(n)':

$$B(n) := \int_{-b}^{b} \frac{f(x)\sin\left(\frac{n\pi x}{b}\right)}{\sqrt{b}} \, dx \qquad (2.258)$$

For $n = 0$,

> A(0):=int(f(x)*phi[0](x)*w(x),x=−b..b);

$$A(0) := \int_{-b}^{b} \frac{1}{2}\frac{f(x)\sqrt{2}}{\sqrt{b}} \, dx \qquad (2.259)$$

This is the generalized series expansion of $f(x)$ in terms of the "complete" set of orthonormal eigenfunctions for the particular Sturm-Liouville operator with periodic boundary conditions over the interval. This particular series is recognized as being the familiar "Fourier" series. Note that since the expansion is in terms of the "orthonormal" eigenfunctions, it may appear different from formats found in other textbooks.

DEMONSTRATION: Develop the Fourier series expansion for $f(x) = 1 - x^2$ over the interval $I = \{x \mid -1 < x < 1\}$ in terms of the preceding eigenfunctions. We assign the system values

> a:=−1;b:=1;f(x):=1−x^2;

$$a := -1$$
$$b := 1$$
$$f(x) := 1 - x^2 \qquad (2.260)$$

SOLUTION: We evaluate the Fourier coefficients

> A(n):=eval(Int(f(x)*phi[n](x)*w(x),x=a..b));A(n):=value(%):

$$A(n) := \int_{-1}^{1} (1 - x^2)\cos(n\pi x) \, dx \qquad (2.261)$$

> A(n):=subs({sin(n*Pi)=0,cos(n*Pi)=(−1)^n},A(n));

$$A(n) := -\frac{4(-1)^n}{n^2\pi^2} \qquad (2.262)$$

> B(n):=eval(Int(f(x)*psi[n](x)*w(x),x=a..b));B(n):=value(%):

$$B(n) := \int_{-1}^{1} (1 - x^2)\sin(n\pi x) \, dx \qquad (2.263)$$

> B(n):=subs({sin(n*Pi)=0,cos(n*Pi)=(−1)^n},B(n));

$$B(n) := 0 \tag{2.264}$$

> A(0):=eval(Int(f(x)*phi[0](x)*w(x),x=a..b));

$$A(0) := \int_{-1}^{1} \frac{1}{2}(1 - x^2)\sqrt{2}\,dx \tag{2.265}$$

> A(0):=value(%);

$$A(0) := \frac{2}{3}\sqrt{2} \tag{2.266}$$

> Series:=eval(A(0)*phi[0](x)+Sum(A(n)*phi[n](x)+B(n)*psi[n](x),n=1..infinity));

$$Series := \frac{2}{3} + \sum_{n=1}^{\infty} \left(\frac{-4(-1)^n \cos(n\pi x)}{n^2 \pi^2} \right) \tag{2.267}$$

First five terms of expansion

> Series:=A(0)*phi[0](x)+sum(A(n)*phi[n](x)+B(n)*psi[n](x),n=1..5):

> plot({Series,f(x)},x=a..b,thickness=10);

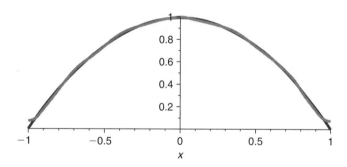

Figure 2.17

The two curves in Figure 2.17 depict the actual function $f(x)$ and its Fourier series approximation in terms of the orthonormal eigenfunctions for the particular operator and boundary conditions given earlier. Note that $f(x)$ does not satisfy the given periodic boundary conditions.

EXAMPLE 2.6.2: Consider the Bessel operator with Dirichlet conditions. We seek the eigenvalues and corresponding orthonormal eigenfunctions for the Bessel differential equation of order m [Sturm-Liouville type for $p(x) = x$, $q(x) = -\frac{m^2}{x}$, $w(x) = x$] over the interval $I = \{x \mid 0 < x < b\}$.

The boundary conditions are that the solution be finite at the origin and there is a type 1 condition at the point $x = b$. The Sturm-Liouville problem involving the Bessel operator is nonregular over this interval I for the simple reason that $p(x)$ and $w(x)$ both vanish at $x = 0$ and $q(x)$ is undefined at $x = 0$; thus, $x = 0$ is a singular point. From Green's formula in Section 2.2, since $p(0) = 0$, we continue to obtain a condition for orthogonality over this interval if, in addition to requiring the solution to be finite at the origin, we further require that at $x = b$, the solution vanishes (see Exercise 2.33).

Bessel differential equation of order m

$$x^2 \left(\frac{d^2}{dx^2} y(x) \right) + x \left(\frac{d}{dx} y(x) \right) - m^2 y(x) + \lambda^2 x^2 y(x) = 0$$

Boundary conditions

$$|y(0)| < \infty \quad \text{and} \quad y(b) = 0$$

SOLUTION: We only consider the case for $\lambda > 0$. Note that for $\lambda = 0$, the Bessel differential equation reduces to a Cauchy-Euler differential equation and we do not get nontrivial solutions for the given boundary conditions. The system basis vectors are

> restart:y1(x):=BesselJ(m,lambda*x);y2(x):=BesselY(m,lambda*x);

$$y1(x) := \text{BesselJ}(m, \lambda x)$$

$$y2(x) := \text{BesselY}(m, \lambda x) \tag{2.268}$$

General solution

> y(x):=C1*y1(x)+C2*y2(x);

$$y(x) := C1 \, \text{BesselJ}(m, \lambda x) + C2 \, \text{BesselY}(m, \lambda x) \tag{2.269}$$

Substituting the boundary condition at the origin indicates that, since the Bessel function of the second kind $[Y(m, \mu x)]$ is not finite at the origin, we must set $C2 = 0$. Substituting the remaining condition at $x = b$ yields

> eval(subs({x=b,C2=0},y(x)))=0;

$$C1 \, \text{BesselJ}(m, \lambda b) = 0 \tag{2.270}$$

The only nontrivial solutions to the preceding are that $C1$ be arbitrary and that λ satisfy the following eigenvalue equation:

> BesselJ(m,lambda[m,n]*b)=0;

$$\text{BesselJ}(m, \lambda_{m,n} b) = 0 \tag{2.271}$$

for $n = 1, 2, 3, \ldots$.

Nonnormalized eigenfunctions are

> phi[m,n](x):=BesselJ(m,lambda[m,n]*x);

$$\phi_{m,n}(x) := \text{BesselJ}(m, \lambda_{m,n}x) \tag{2.272}$$

Normalization

Evaluating the norm from the inner product of the eigenfunctions with respect to the weight function $w(x) = x$ over the interval yields

> w(x):=x:unprotect(norm):norm:=sqrt(Int(phi[m,n](x)^2*w(x),x=0..b));

$$norm := \sqrt{\int_0^b \text{BesselJ}\left(m, \lambda_{m,n}x\right)^2 x\mathrm{d}x} \tag{2.273}$$

Orthonormal eigenfunctions

> phi[m,n](x):=phi[m,n](x)/norm;phi[m,p](x):=subs(n=p,phi[m,n](x)):

$$\phi_{m,n}(x) := \frac{\text{BesselJ}\left(m, \lambda_{m,n}x\right)}{\sqrt{\int_0^b \text{BesselJ}\left(m, \lambda_{m,n}x\right)^2 x\,\mathrm{d}x}} \tag{2.274}$$

Statement of orthonormality

> Int(phi[m,n](x)*phi[m,p](x)*w(x),x=0..b)=delta(n,p);

$$\int_0^b \frac{\text{BesselJ}\left(m, \lambda_{m,n}x\right)\text{BesselJ}\left(m, \lambda_{m,p}x\right)x}{\sqrt{\int_0^b \text{BesselJ}\left(m, \lambda_{m,n}x\right)^2 x\,\mathrm{d}x}\sqrt{\int_0^b \text{BesselJ}\left(m, \lambda_{m,n}x\right)^2 x\,\mathrm{d}x}}\mathrm{d}x = \delta(n, p) \tag{2.275}$$

Generalized Fourier series expansion

> f(x):=Sum(F(n)*phi[m,n](x),n=1..infinity);f(x):='f(x)':

$$f(x) := \sum_{n=1}^{\infty} \frac{F(n)\text{BesselJ}\left(m, \lambda_{m,n}x\right)}{\sqrt{\int_0^b \text{BesselJ}\left(m, \lambda_{m,n}x\right)^2 x\,\mathrm{d}x}} \tag{2.276}$$

Fourier coefficients

```
> F(n):=Int(f(x)*phi[m,n](x)*w(x),x=0..b);F(n):=`F(n)`:
```

$$F(n) := \int_0^b \frac{f(x) \text{BesselJ}\left(m, \lambda_{m,n} x\right) x}{\sqrt{\int_0^b \text{BesselJ}\left(m, \lambda_{m,n} x\right)^2 x \, dx}} \, dx \tag{2.277}$$

This is the Fourier-Bessel series expansion of $f(x)$ in terms of the "complete" set of eigenfunctions for the particular Sturm-Liouville operator and boundary conditions over the interval. Here, $J(m, \lambda x)$ denotes the Bessel function of the first kind of order m.

DEMONSTRATION: Develop the Fourier-Bessel series expansion for $f(x) = x$ over the interval $I = \{x \mid 0 < x < 1\}$ in terms of the Bessel functions of the first kind of order $m = 1$. We assign the system values

```
> a:=0;b:=1;m:=1;f(x):=x;
```

$$a := 0$$
$$b := 1$$
$$m := 1$$
$$f(x) := x \tag{2.278}$$

SOLUTION: Evaluation of the eigenvalues from the roots of the eigenvalue equation yields

```
> BesselJ(m,lambda[m,n]*b)=0;
```

$$\text{BesselJ}\left(1, \lambda_{1,n}\right) = 0 \tag{2.279}$$

```
> plot(BesselJ(m,v),v=0..20,thickness=10);
```

If we set $v = \lambda b$, then the eigenvalues are found from the intersection points of the curve with the v-axis shown in Figure 2.18. We evaluate a few of these eigenvalues from the Maple fsolve command for the roots of the eigenvalue equation to be given as

```
> lambda[1,1]:=(1/b)*fsolve(BesselJ(m,v)=0,v=1..4);
```

$$\lambda_{1,1} := 3.831705970 \tag{2.280}$$

```
> lambda[1,2]:=(1/b)*fsolve(BesselJ(m,v)=0,v=4..9);
```

$$\lambda_{1,2} := 7.015586670 \tag{2.281}$$

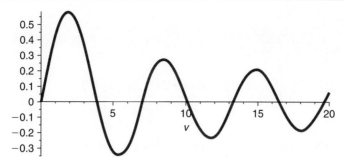

Figure 2.18

> lambda[1,3]:=(1/b)*fsolve(BesselJ(m,v)=0,v=9..12);

$$\lambda_{1,3} := 10.17346814 \tag{2.282}$$

> lambda[1,4]:=(1/b)*fsolve(BesselJ(m,v)=0,v=12..16);

$$\lambda_{1,4} := 13.32369194 \tag{2.283}$$

> lambda[1,5]:=(1/b)*fsolve(BesselJ(m,v)=0,v,16..19);

$$\lambda_{1,5} := 16.47063005 \tag{2.284}$$

Normalization

> norm:=sqrt(Int(w(x)*(BesselJ(m,lambda[m,n]*x))^2,x=0..b));

$$norm := \sqrt{\int_0^1 x\,\mathrm{BesselJ}\left(1, \lambda_{1,n}x\right)^2 dx} \tag{2.285}$$

Substitution of the eigenvalue equation simplifies the norm

> norm:=radsimp(subs(BesselJ(1,lambda[1,n])=0,value(%)));

$$norm := \frac{1}{2}\sqrt{2}\,\mathrm{BesselJ}\left(0, \lambda_{1,n}\right) \tag{2.286}$$

Orthonormal eigenfunctions

> phi[m,n](x):=BesselJ(m,lambda[m,n]*x)/norm;

$$\phi_{1,n}(x) := \frac{\mathrm{BesselJ}\left(1, \lambda_{1,n}x\right)\sqrt{2}}{\mathrm{BesselJ}\left(0, \lambda_{1,n}\right)} \tag{2.287}$$

Fourier coefficients

> F(n):=Int(f(x)*phi[m,n](x)*w(x),x=a..b);

$$F(n) := \int_0^1 \frac{x^2 \text{BesselJ}\left(1, \lambda_{1,n}x\right) \sqrt{2}}{\text{BesselJ}\left(0, \lambda_{1,n}\right)}\, dx \qquad (2.288)$$

Substitution of the eigenvalue equation simplifies the preceding equation

> F(n):=subs(BesselJ(1,lambda[1,n])=0,value(%));

$$F(n) := -\frac{\sqrt{2}}{\lambda_{1,n}} \qquad (2.289)$$

> Series:=Sum(F(n)*phi[m,n](x),n=1..infinity);

$$Series := \sum_{n=1}^{\infty} \left(-\frac{2\text{BesselJ}\left(1, \lambda_{1,n}x\right)}{\lambda_{1,n}\text{BesselJ}\left(0, \lambda_{1,n}\right)} \right) \qquad (2.290)$$

First five terms in expansion

> Series:=sum(F(n)*phi[m,n](x),n=1..5):

> plot({Series,f(x)},x=a..b,thickness=10);

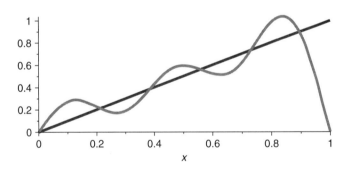

Figure 2.19

The two curves in Figure 2.19 depict the function $f(x)$ and its series expansion approximation in terms of the Bessel functions of the first kind of order $m = 1$, which satisfy the given boundary conditions over the interval. Note that $f(x)$ satisfies the given boundary conditions at the left but fails to do so at the right end point.

EXAMPLE 2.6.3: Consider the Bessel operator with Neumann conditions. We seek the eigenvalues and corresponding orthonormal eigenfunctions for the Bessel differential equation of order $m = 0$ [Sturm-Liouville type for $p(x) = x$, $q(x) = 0$, $w(x) = x$] over the interval $I = \{x\,|\,0 < x < b\}$. The boundary conditions are that the solution be finite at the origin and that there is a type 2 condition at the point $x = b$. The Sturm-Liouville problem involving the Bessel operator is nonregular over this interval. From Green's formula in Section 2.2, since $p(0) = 0$, we continue to obtain a condition for orthogonality over this interval if, in addition to requiring the solution to be finite at the origin, we further require that at $x = b$, its first derivative vanishes (see Exercise 2.34).

Bessel differential equation of order $m = 0$

$$x^2 \left(\frac{d^2}{dx^2} y(x) \right) + x \left(\frac{d}{dx} y(x) \right) + \lambda^2 x^2 y(x) = 0$$

Boundary conditions

$$|y(0)| < \infty \quad \text{and} \quad y_x(b) = 0$$

SOLUTION: We consider two possibilities for values of λ. We first consider the case for $\lambda = 0$. The system basis vectors are

> restart:y1(x):=1;y2(x):=ln(x);

$$y1(x) := 1$$

$$y2(x) := \ln(x) \tag{2.291}$$

General solution

> y(x):=C1*y1(x)+C2*y2(x);

$$y(x) := C1 + C2 \ln(x) \tag{2.292}$$

Since $\ln(x)$ fails to exist at $x = 0$, we set $C2 = 0$, giving

> y(x):=C1;

$$y(x) := C1 \tag{2.293}$$

Substituting into the boundary condition at the periphery yields

> eval(subs(x=a,diff(y(x),x)))=0;

$$0 = 0 \tag{2.294}$$

Thus, $C1$ is arbitrary and the eigenvalue and eigenfunction corresponding to $\lambda = 0$ are given as

> lambda[0]:=0;

$$\lambda_0 := 0 \tag{2.295}$$

> phi[0](x):=1;

$$\phi_0(x) := 1 \tag{2.296}$$

We next consider the case for $\lambda > 0$. The system basis vectors are

> y1(x):=BesselJ(0,lambda*x);y2(x):=BessselY(0,lambda*x);

$$y1(x) := \text{BesselJ}(0, \lambda x)$$

$$y2(x) := \text{BesselY}(0, \lambda x) \tag{2.297}$$

General solution

> y(x):=C1*y1(x)+C2*y2(x);

$$y(x) := C1 \, \text{BesselJ}(0, \lambda x) + C2 \, \text{BesselY}(0, \lambda x) \tag{2.298}$$

Substituting the boundary condition at the origin indicates that, since the Bessel function of the second kind $[Y(m, \mu x)]$ is not finite at the origin, we must set $C2 = 0$. Substituting the remaining condition at $x = b$ yields

> eval(subs({x=b,C2=0},diff(y(x),x)))=0;

$$-C1 \, \text{BesselJ}(1, \lambda b)\lambda = 0 \tag{2.299}$$

The only nontrivial solutions to the above occur when $C1$ is arbitrary and λ satisfies the following eigenvalue equation

$$\text{BesselJ}\,(1, \lambda_n b) = 0 \tag{2.300}$$

for $n = 1, 2, 3, \ldots$.

Nonnormalized eigenfunctions are

> phi[n](x):=BesselJ(0,lambda[n]*x);

$$\varphi_n(x) := \text{BesselJ}\,(0, \lambda_n x) \tag{2.301}$$

Normalization

Evaluating the norm from the inner product of the eigenfunctions, with respect to the weight function $w(x) = x$ over the interval, yields, for $n = 0$,

> w(x):=x:norm0:=sqrt(Int(phi[0](x)^2*w(x),x=0..b));

$$norm0 := \sqrt{\int_0^b x \, dx} \tag{2.302}$$

> norm0:=radsimp(value(%));

$$norm0 := \frac{1}{2}\sqrt{2}b \tag{2.303}$$

For $n = 1, 2, 3, \ldots,$

> norm1:=sqrt(Int(phi[n](x)^2*w(x),x=0..b));norm1:=value(%):

$$norm1 := \sqrt{\int_0^b (\text{BesselJ}\,(0, \lambda_n x))^2 \, x \, dx} \qquad (2.304)$$

Substitution of the eigenvalue equation simplifies the norm

> norm1:=radsimp(subs(BesselJ(1,lambda[n]*b)=0,norm1));

$$norm1 := \sqrt{2}\, b\, \text{BesselJ}\,(0, \lambda_n b) \qquad (2.305)$$

Orthonormal eigenfunctions

> phi[0](x):=phi[0](x)/norm0;

$$\phi_0(x) := \frac{\sqrt{2}}{b} \qquad (2.306)$$

> phi[n](x):=phi[n](x)/norm1;phi[p](x):=subs(n=p,phi[n](x)):

$$\phi_n(x) := \frac{\text{BesselJ}\,(0, \lambda_n x)\,\sqrt{2}}{b\,\text{BesselJ}\,(0, \lambda_n b)} \qquad (2.307)$$

Statement of orthonormality

> Int(phi[n](x)*phi[p](x)*w(x),x=0..b)=delta(n,p);

$$\int_0^b \frac{2\,\text{BesselJ}\,(0, \lambda_n x)\,\text{BesselJ}\,\left(0, \lambda_p x\right)\,x}{b^2\,\text{BesselJ}\,(0, \lambda_n b)\,\text{BesselJ}\,\left(0, \lambda_p b\right)} \, dx = \delta(n, p) \qquad (2.308)$$

Generalized Fourier series expansion

> f(x):=F(0)*phi[0](x)+Sum(F(n)*phi[n](x),n=1..infinity);f(x):='f(x)':

$$f(x) := \frac{F(0)\sqrt{2}}{b} + \sum_{n=1}^{\infty} \frac{F(n)\,\text{BesselJ}\,(0, \lambda_n x)\,\sqrt{2}}{b\,\text{BesselJ}\,(0, \lambda_n b)} \qquad (2.309)$$

Fourier coefficients

> F(n):=Int(f(x)*phi[n](x)*w(x),x=0..b);F(n):='F(n)':

$$F(n) := \int_0^b \frac{f(x)\text{BesselJ}\,(0, \lambda_n x)\,\sqrt{2}x}{b\,\text{BesselJ}\,(0, \lambda_n b)} \, dx \qquad (2.310)$$

> F(0):=Int(f(x)*phi[0](x)*w(x),x=0..b);F(0):='F(0)':

$$F(0) := \int_0^b \frac{f(x)\sqrt{2x}}{b}\,dx \qquad\qquad (2.311)$$

This is the Fourier-Bessel series expansion of $f(x)$ in terms of the "complete" set of orthonormal eigenfunctions for the particular Sturm-Liouville operator and boundary conditions over the interval.

DEMONSTRATION: Develop the Fourier-Bessel series expansion for $f(x) = 1 - x^2$ over the interval $I = \{x \mid 0 < x < 1\}$ in terms of the preceding eigenfunctions. We assign the system values

> a:=0;b:=1;f(x):=1−x^2;

$$a := 0$$
$$b := 1$$
$$f(x) := 1 - x^2 \qquad\qquad (2.312)$$

SOLUTION: Evaluation of the eigenvalues from the roots of the eigenvalue equation yields

> BesselJ(1,lambda[n]*b)=0;

$$\text{BesselJ}\,(1, \lambda_n) = 0 \qquad\qquad (2.313)$$

> plot(BesselJ(1,v),v=0..20,thickness=10);

If we set $v = \lambda b$, then the eigenvalues are found from the intersection points of the curve with the v-axis shown in Figure 2.20. We evaluate a few of these eigenvalues from the Maple fsolve command for the roots of the eigenvalue equation to be given as

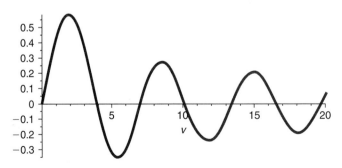

Figure 2.20

> lambda[1]:=(1/b)*fsolve(BesselJ(1,v)=0,v=1..4);

$$\lambda_1 := 3.831705970 \qquad (2.314)$$

> lambda[2]:=(1/b)*fsolve(BesselJ(1,v)=0,v=4..9);

$$\lambda_2 := 7.015586670 \qquad (2.315)$$

> lambda[3]:=(1/b)*fsolve(BesselJ(1,v)=0,v=9..12);

$$\lambda_3 := 10.17346814 \qquad (2.316)$$

> lambda[4]:=(1/b)*fsolve(BesselJ(1,v)=0,v=12..16);

$$\lambda_4 := 13.32369194 \qquad (2.317)$$

> lambda[5]:=(1/b)*fsolve(BesselJ(1,v)=0,v=16..19);

$$\lambda_5 := 16.47063005 \qquad (2.318)$$

Fourier coefficients

> F(n):=eval(Int(f(x)*phi[n](x)*w(x),x=a..b));

$$F(n) := \int_0^1 \frac{\left(1-x^2\right) \mathrm{BesselJ}\,(0,\lambda_n x)\,\sqrt{2}x}{\mathrm{BesselJ}\,(0,\lambda_n)}\, dx \qquad (2.319)$$

Substitution of the eigenvalue equation simplifies the preceding equation

> F(n):=subs(BesselJ(1,lambda[n]*b)=0,value(%));

$$F(n) := -\frac{2\sqrt{2}}{\lambda_n^2} \qquad (2.320)$$

> F(0):=eval(Int(f(x)*phi[0](x)*w(x),x=a..b));

$$F(0) := \int_0^1 \left(1-x^2\right)\sqrt{2}x\, dx \qquad (2.321)$$

> F(0):=value(%);

$$F(0) := \frac{1}{4}\sqrt{2} \qquad (2.322)$$

> Series:=eval(F(0)*phi[0](x)+Sum(F(n)*phi[n](x),n=1..infinity));

$$Series := \frac{1}{2} + \sum_{n=1}^{\infty} \left(-\frac{4\mathrm{BesselJ}\,(0,\lambda_n x)}{\lambda_n^2\,\mathrm{BesselJ}\,(0,\lambda_n)} \right) \qquad (2.323)$$

First five terms in the expansion

> Series:=F(0)*phi[0](x)+sum(F(n)*phi[n](x),n=1..5):

> plot({Series,f(x)},x=a..b,thickness=10);

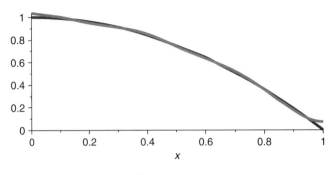

Figure 2.21

The two curves in Figure 2.21 depict the function $f(x)$ and its Fourier series expansion approximation in terms of the Bessel functions of the first kind of order zero, which satisfy the given boundary conditions. Note that $f(x)$ satisfies the given boundary condition at the left but fails to do so at the right end point.

Chapter Summary

Sturm-Liouville operator acting on $y(x)$

$$L(y) = D(p(x)D(y)) + q(x)y$$

The Sturm-Liouville differential equation

$$L(\varphi) + \lambda\, w(x)\varphi = 0$$

Regular homogeneous boundary conditions

$$\kappa_1 \varphi(a) + \kappa_2 \varphi_x(a) = 0$$

and

$$\kappa_3 \varphi(b) + \kappa_4 \varphi_x(b) = 0$$

Generalized Fourier series expansion

$$f(x) = \sum_{n=0}^{\infty} F(n)\varphi_n(x)$$

Statement of orthonormality with respect to the weight function $w(x)$

$$\int_a^b \phi_n(x)\,\phi_m(x)\,w(x)\,\mathrm{d}x = \delta(n,m)$$

Generalized Fourier coefficients

$$F(n) = \int_a^b f(x)\,\phi_n(x)\,w(x)\,\mathrm{d}x$$

The generalized Fourier series expansions above are based on the orthonormalized eigenfunction solutions to Sturm-Liouville boundary value problems. These generalized series expansions will play an important role in the solutions to boundary value problems in partial differential equations.

Exercises

2.1.   From Section 2.2, any second-order linear differential equation of the form

$$a_2(x)\left(\frac{\mathrm{d}^2}{\mathrm{d}x^2}y(x)\right) + a_1(x)\left(\frac{\mathrm{d}}{\mathrm{d}x}y(x)\right) + a_0(x)y(x) = 0$$

can be converted into self-adjoint form by multiplying by the integrating factor

$$r(x) = \frac{e^{\displaystyle\int \frac{a_1(x)}{a_2(x)}\mathrm{d}x}}{a_2(x)}$$

Consider the following differential equations shown below. In all cases, evaluate the integrating factor, identify $p(x)$, $q(x)$, and the weight function $w(x)$, and rewrite the equations in the Sturm-Liouville (self-adjoint) form

$$(D(p(x)D) + q(x) + \lambda\,w(x))y(x) = 0$$

(a)   The Euler equation

$$\frac{\mathrm{d}^2}{\mathrm{d}x^2}y(x) + \lambda\,y(x) = 0$$

(b)   The Cauchy-Euler equation

$$x^2\left(\frac{\mathrm{d}^2}{\mathrm{d}x^2}y(x)\right) + x\left(\frac{\mathrm{d}}{\mathrm{d}x}y(x)\right) + \lambda\,y(x) = 0$$

(c)   The Laguerre equation from quantum mechanics

$$x\left(\frac{d^2}{dx^2}y(x)\right)+(1-x)\left(\frac{d}{dx}y(x)\right)+\lambda\,y(x)=0$$

(d)   The Hermite equation from quantum mechanics

$$\frac{d^2}{dx^2}y(x)-2x\left(\frac{d}{dx}y(x)\right)+2\lambda\,y(x)=0$$

(e)   The Tchebycheff equation

$$(1-x^2)\left(\frac{d^2}{dx^2}y(x)\right)-x\left(\frac{d}{dx}y(x)\right)+\lambda\,y(x)=0$$

(f)   The Bessel differential equation of order m

$$x^2\left(\frac{d^2}{dx^2}y(x)\right)+x\left(\frac{d}{dx}y(x)\right)-m^2y(x)+\lambda^2x^2y(x)=0$$

In the following problems, you will be asked to generate a set of eigenfunctions for a Sturm-Liouville problem over an interval I with a particular set of boundary conditions. You are also asked to develop a generalized Fourier series expansion for three functions over the interval. The first, $f1(x)$, should not satisfy any of the boundary conditions. The second, $f2(x)$, should satisfy one of the boundary conditions (left or right). The third, $f3(x)$, should satisfy both boundary conditions (left and right). Pay particular attention to the quality of the convergence of each of the series depending on the boundary condition behavior of the particular function $f(x)$ over that interval.

For problems 2.2 through 2.21, you are asked to do the following:

(a)   Evaluate the eigenvalues and corresponding eigenfunctions.
(b)   Normalize the eigenfunctions with the respective weight function.
(c)   Write out the statement of orthonormality.
(d)   Write out the generalized Fourier series expansion for function $f(x)$.
(e)   Write out the integral for the Fourier coefficients F(n).

2.2.   Consider the Sturm-Liouville problem with the Euler operator over the interval $I=\{x\,|\,0<x<1\}$.

$$\frac{d^2}{dx^2}\varphi(x)+\lambda\,\varphi(x)=0$$

with boundary conditions

$$\varphi(0)=0,\ \varphi(1)=0$$

Evaluate the generalized Fourier series expansion for the following:

$$f1(x) = 1$$
$$f2(x) = x$$
$$f3(x) = x(1-x)$$

With the first five terms in each series above, generate the curve of $f(x)$ and its corresponding series on the same set of axes.

2.3.   Consider the Sturm-Liouville problem with the Euler operator over the interval $I = \{x \mid 0 < x < 1\}$.

$$\frac{d^2}{dx^2}\varphi(x) + \lambda\,\varphi(x) = 0$$

with boundary conditions

$$\varphi(0) = 0,\ \varphi_x(1) = 0$$

Evaluate the generalized Fourier series expansion for the following:

$$f1(x) = \left(x - \frac{1}{2}\right)^2$$
$$f2(x) = x$$
$$f3(x) = x\left(1 - \frac{x}{2}\right)$$

With the first five terms in each preceding series, generate the curve of $f(x)$ and its corresponding series on the same set of axes.

2.4.   Consider the Sturm-Liouville problem with the Euler operator over the interval $I = \{x \mid 0 < x < 1\}$.

$$\frac{d^2}{dx^2}\varphi(x) + \lambda\,\varphi(x) = 0$$

with boundary conditions

$$\varphi_x(0) = 0,\ \varphi(1) = 0$$

Evaluate the generalized Fourier series expansion for the following:

$$f1(x) = x$$
$$f2(x) = 1$$
$$f3(x) = 1 - x^2$$

With the first five terms in each series above, generate the curve of $f(x)$ and its corresponding series on the same set of axes.

2.5. Consider the Sturm-Liouville problem with the Euler operator over the interval $I = \{x \mid 0 < x < 1\}$.

$$\frac{d^2}{dx^2}\varphi(x) + \lambda\,\varphi(x) = 0$$

with boundary conditions

$$\varphi_x(0) = 0, \ \varphi_x(1) = 0$$

Evaluate the generalized Fourier series expansion for the following:

$$f1(x) = x$$

$$f2(x) = 1 - x^2$$

$$f3(x) = x^2\left(1 - \frac{2x}{3}\right)$$

With the first five terms in each series above, generate the curve of $f(x)$ and its corresponding series on the same set of axes.

2.6. Consider the Sturm-Liouville problem with the Euler operator over the interval $I = \{x \mid 0 < x < 1\}$.

$$\frac{d^2}{dx^2}\varphi(x) + \lambda\,\varphi(x) = 0$$

with boundary conditions

$$\varphi(0) = 0, \ \varphi_x(1) + \varphi(1) = 0$$

Evaluate the generalized Fourier series expansion for the following:

$$f1(x) = 1$$

$$f2(x) = x$$

$$f3(x) = x\left(1 - \frac{2x}{3}\right)$$

With the first five terms in each series above, generate the curve of $f(x)$ and its corresponding series on the same set of axes.

2.7. Consider the Sturm-Liouville problem with the Euler operator over the interval $I = \{x \mid 0 < x < 1\}$.

$$\frac{d^2}{dx^2}\varphi(x) + \lambda\,\varphi(x) = 0$$

with boundary conditions

$$\varphi_x(0) - \varphi(0) = 0, \ \varphi(1) = 0$$

Evaluate the generalized Fourier series expansion for the following:

$$f1(x) = 1$$
$$f2(x) = 1 - x$$
$$f3(x) = -\frac{2x^2}{3} + \frac{x}{3} + \frac{1}{3}$$

With the first five terms in each preceding series, generate the curve of $f(x)$ and its corresponding series on the same set of axes.

2.8.   Consider the Sturm-Liouville problem with the Euler operator over the interval $I = \{x \,|\, 0 < x < 1\}$.

$$\frac{d^2}{dx^2}\varphi(x) + \lambda\,\varphi(x) = 0$$

with boundary conditions

$$\varphi_x(0) = 0, \ \varphi_x(1) + \varphi(1) = 0$$

Evaluate the generalized Fourier series expansion for the following:

$$f1(x) = x$$
$$f2(x) = 1$$
$$f3(x) = 1 - \frac{x^2}{3}$$

With the first five terms in each preceding series, generate the curve of $f(x)$ and its corresponding series on the same set of axes.

2.9.   Consider the Sturm-Liouville problem with the Euler operator over the interval $I = \{x \,|\, 0 < x < 1\}$.

$$\frac{d^2}{dx^2}\varphi(x) + \lambda\,\varphi(x) = 0$$

with boundary conditions

$$\varphi_x(0) - \varphi(0) = 0, \ \varphi_x(1) = 0$$

Evaluate the generalized Fourier series expansion for the following:

$$f1(x) = x$$

$$f2(x) = 1$$

$$f3(x) = -\frac{(x-1)^2}{3} + 1$$

With the first five terms in each preceding series, generate the curve of $f(x)$ and its corresponding series on the same set of axes.

2.10.   Consider the Sturm-Liouville problem with the Cauchy-Euler operator over the interval $I = \{x \mid 1 < x < 2\}$.

$$x^2 \left(\frac{d^2}{dx^2} \varphi(x) \right) + x \left(\frac{d}{dx} \varphi(x) \right) + \lambda \varphi(x) = 0$$

with boundary conditions

$$\varphi(1) = 0, \ \varphi(2) = 0$$

Evaluate the generalized Fourier series expansion for the following:

$$f1(x) = 1$$

$$f2(x) = x - 1$$

$$f3(x) = (x-1)(2-x)$$

With the first five terms in each preceding series, generate the curve of $f(x)$ and its corresponding series on the same set of axes.

2.11.   Consider the Sturm-Liouville problem with the Cauchy-Euler operator over the interval $I = \{x \mid 1 < x < 2\}$.

$$x^2 \left(\frac{d^2}{dx^2} \varphi(x) \right) + x \left(\frac{d}{dx} \varphi(x) \right) + \lambda \varphi(x) = 0$$

with boundary conditions

$$\varphi_x(1) = 0, \ \varphi(2) = 0$$

Evaluate the generalized Fourier series expansion for the following:

$$f1(x) = x - 1$$

$$f2(x) = 1$$

$$f3(x) = x(2-x)$$

With the first five terms in each preceding series, generate the curve of $f(x)$ and its corresponding series on the same set of axes.

2.12. Consider the Sturm-Liouville problem with the Cauchy-Euler operator over the interval $I = \{x \mid 1 < x < 2\}$.

$$x^2 \left(\frac{d^2}{dx^2} \varphi(x) \right) + x \left(\frac{d}{dx} \varphi(x) \right) + \lambda \, \varphi(x) = 0$$

with boundary conditions

$$\varphi(1) = 0, \ \varphi_x(2) = 0$$

Evaluate the generalized Fourier series expansion for the following:

$$f1(x) = \left(x - \frac{3}{2} \right)^2$$

$$f2(x) = x - 1$$

$$f3(x) = \frac{(x-1)(3-x)}{2}$$

With the first five terms in each preceding series, generate the curve of $f(x)$ and its corresponding series on the same set of axes.

2.13. Consider the Sturm-Liouville problem with the Cauchy-Euler operator over the interval $I = \{x \mid 1 < x < 2\}$.

$$x^2 \left(\frac{d^2}{dx^2} \varphi(x) \right) + x \left(\frac{d}{dx} \varphi(x) \right) + \lambda \, \varphi(x) = 0$$

with boundary conditions

$$\varphi_x(1) = 0, \ \varphi_x(2) = 0$$

Evaluate the generalized Fourier series expansion for the following:

$$f1(x) = x - 1$$

$$f2(x) = x(2 - x)$$

$$f3(x) = 3x^2 - \frac{2x^2}{3} - 4x + \frac{5}{3}$$

With the first five terms in each preceding series, generate the curve of $f(x)$ and its corresponding series on the same set of axes.

2.14. Consider the Sturm-Liouville problem with the Cauchy-Euler operator over the interval $I = \{x \mid 1 < x < 2\}$.

$$x^2 \left(\frac{d^2}{dx^2} \varphi(x) \right) + x \left(\frac{d}{dx} \varphi(x) \right) + \lambda \varphi(x) = 0$$

with boundary conditions

$$\varphi(1) = 0, \ \varphi_x(2) + \varphi(2) = 0$$

Evaluate the generalized Fourier series expansion for the following:

$$f1(x) = 1$$
$$f2(x) = x - 1$$
$$f3(x) = \frac{(x-1)(5-2x)}{3}$$

With the first five terms in each preceding series, generate the curve of $f(x)$ and its corresponding series on the same set of axes.

2.15. Consider the Sturm-Liouville problem with the Cauchy-Euler operator over the interval $I = \{x \mid 1 < x < 2\}$.

$$x^2 \left(\frac{d^2}{dx^2} \varphi(x) \right) + x \left(\frac{d}{dx} \varphi(x) \right) + \lambda \varphi(x) = 0$$

with boundary conditions

$$\varphi_x(1) - \varphi(1) = 0, \ \varphi(2) = 0$$

Evaluate the generalized Fourier series expansion for the following:

$$f1(x) = 1$$
$$f2(x) = 2 - x$$
$$f3(x) = -\frac{2x^2}{3} + \frac{5x}{3} - \frac{2}{3}$$

With the first five terms in each preceding series, generate the curve of $f(x)$ and its corresponding series on the same set of axes.

2.16. Consider the Sturm-Liouville problem with the Cauchy-Euler operator over the interval $I = \{x \mid 1 < x < 2\}$.

$$x^2 \left(\frac{d^2}{dx^2} \varphi(x) \right) + x \left(\frac{d}{dx} \varphi(x) \right) + \lambda \varphi(x) = 0$$

with boundary conditions

$$\varphi_x(1) = 0, \; \varphi_x(2) + \varphi(2) = 0$$

Evaluate the generalized Fourier series expansion for the following:

$$f1(x) = x - 1$$
$$f2(x) = 1$$
$$f3(x) = \frac{2}{3} - \frac{x^2}{3} + \frac{2x}{3}$$

With the first five terms in each preceding series, generate the curve of $f(x)$ and its corresponding series on the same set of axes.

2.17. Consider the Sturm-Liouville problem with the Cauchy-Euler operator over the interval $I = \{x \mid 1 < x < 2\}$.

$$x^2 \left(\frac{d^2}{dx^2} \varphi(x) \right) + x \left(\frac{d}{dx} \varphi(x) \right) + \lambda \varphi(x) = 0$$

with boundary conditions

$$\varphi_x(1) - \varphi(1) = 0, \; \varphi_x(2) = 0$$

Evaluate the generalized Fourier series expansion for the following:

$$f1(x) = x - 1$$
$$f2(x) = 1$$
$$f3(x) = -\frac{x^2}{3} + \frac{4x}{3} - \frac{1}{3}$$

With the first five terms in each preceding series, generate the curve of $f(x)$ and its corresponding series on the same set of axes.

2.18. Consider the Sturm-Liouville problem with the Euler operator over the interval $I = \{x \mid -1 < x < 1\}$.

$$\frac{d^2}{dx^2} \varphi(x) + \lambda \varphi(x) = 0$$

with periodic boundary conditions

$$\varphi(-1) = \varphi(1), \; \varphi_x(-1) = \varphi_x(1)$$

Evaluate the generalized Fourier series expansion for the following:

$$f1(x) = x$$

$$f2(x) = 1 - x^2$$

$$f(3) = |x|$$

With the first five terms in each preceding series, generate the curve of $f(x)$ and its corresponding series on the same set of axes.

2.19.   Consider the Sturm-Liouville problem with the Bessel operator of order zero over the interval $I = \{x \mid 0 < x < 1\}$.

$$x^2 \left(\frac{d^2}{dx^2} \varphi(x) \right) + x \left(\frac{d}{dx} \varphi(x) \right) + \lambda^2 x^2 \varphi(x) = 0$$

with boundary conditions

$$\varphi(0) < \infty, \ \varphi(1) = 0$$

Evaluate the generalized Fourier series expansion for the following:

$$f1(x) = x^2$$

$$f2(x) = 1$$

$$f3(x) = 1 - x^2$$

With the first five terms in each preceding series, generate the curve of $f(x)$ and its corresponding series on the same set of axes.

2.20.   Consider the Sturm-Liouville problem with the Bessel operator of order zero over the interval $I = \{x \mid 0 < x < 1\}$.

$$x^2 \left(\frac{d^2}{dx^2} \varphi(x) \right) + x \left(\frac{d}{dx} \varphi(x) \right) + \lambda^2 x^2 \varphi(x) = 0$$

with boundary conditions

$$|\varphi(0)| < \infty, \ \varphi_x(1) = 0$$

Evaluate the generalized Fourier series expansion for the following:

$$f1(x) = 1$$

$$f2(x) = x^2$$

$$f3(x) = x^2 - \frac{x^4}{2}$$

With the first five terms in each preceding series, generate the curve of $f(x)$ and its corresponding series on the same set of axes.

2.21. Consider the Sturm-Liouville problem with the Bessel operator of order zero over the interval $I = \{x \mid 0 < x < 1\}$.

$$x^2 \left(\frac{d^2}{dx^2} \varphi(x) \right) + x \left(\frac{d}{dx} \varphi(x) \right) + \lambda^2 x^2 \varphi(x) = 0$$

with boundary conditions

$$|\varphi(0)| < \infty, \varphi_x(1) + \varphi(1) = 0$$

Evaluate the generalized Fourier series expansion for the following:

$$f1(x) = 1$$
$$f2(x) = x^2$$
$$f3(x) = 1 - \frac{x^2}{3}$$

With the first five terms in each preceding series, generate the curve of $f(x)$ and its corresponding series on the same set of axes.

2.22. Consider the non-self-adjoint boundary value problem

$$\frac{d^2}{dx^2} y(x) + 4 \left(\frac{d}{dx} y(x) \right) + (4 + 9\lambda) y(x) = 0$$

with boundary conditions

$$y(0) = 0, \, y(1) = 0$$

(a) Transform into the self-adjoint form and identify the weight function $w(x)$.
(b) Solve for the eigenvalues and the corresponding eigenfuctions.
(c) Normalize the eigenfunctions and write out the statement of orthonormality.
(d) Evaluate the generalized Fourier series expansion of the function $f(x) = x - x^2$ over the interval $I = \{x \mid 0 < x < 1\}$.

2.23. Consider the non-self-adjoint boundary value problem

$$\frac{d^2}{dx^2} y(x) + 2 \left(\frac{d}{dx} y(x) \right) + (1 - \lambda) y(x) = 0$$

with boundary conditions

$$y(0) = 0, \, y(1) = 0$$

(a) Transfer into the self-adjoint form and identify the weight function $w(x)$.

(b) Solve for the eigenvalues and the corresponding eigenfuctions.

(c) Normalize the eigenfunctions and write out the statement of orthonormality.

(d) Evaluate the generalized Fourier series expansion of the function $f(x) = 1 - x$ over the interval $I = \{x \mid 0 < x < 1\}$.

The Rayleigh Quotient

The Rayleigh quotient is useful in determining the sign of the eigenvalues. It can be derived from the expanded form of the Sturm-Liouville equation:

$$\left(\frac{d}{dx}p(x)\right)\left(\frac{d}{dx}\varphi(x)\right) + p(x)\left(\frac{d^2}{dx^2}\varphi(x)\right) + q(x)\varphi(x) + \lambda w(x)\varphi(x) = 0$$

If we multiply both sides of the above by $\varphi(x)$ and integrate over the interval $I = \{x \mid a < x < b\}$, we get

$$\int_a^b \left(\varphi(x)\left(\left(\frac{d}{dx}p(x)\right)\left(\frac{d}{dx}\varphi(x)\right) + p(x)\left(\frac{d^2}{dx^2}\varphi(x)\right)\right) + q(x)\varphi(x)^2\right)dx$$

$$+ \lambda\left(\int_a^b \varphi(x)^2 w(x)dx\right) = 0$$

Integrating by parts and solving for λ, we get

$$\lambda = \frac{-p(b)\,\varphi(b)\,\varphi_x(b) + p(a)\,\varphi(a)\,\varphi_x(a) + \int_a^b \left(p(x)\left(\frac{d}{dx}\varphi(x)\right)^2 - q(x)\varphi(x)^2\right)dx}{\int_a^b \varphi(x)^2 w(x)\,dx}$$

We recognize the denominator here to be the norm of the eigenfunction. It is obvious that if the following conditions hold

$$0 \le -p(b)\,\varphi(b)\,\varphi_x(b) + p(a)\,\varphi(a)\,\varphi_x(a)$$

in addition to

$$0 \le p(x) \quad \text{and} \quad q(x) \le 0$$

then the term on the right-hand side must be positive or equal to zero; thus, the eigenvalues λ must be greater than or equal to zero—that is, there are no negative eigenvalues. We took advantage of this earlier when we evaluated the eigenvalues in the example problems.

2.24. For the Sturm-Liouville problems 2.7 through 2.9 for the Euler operator, identify $q(x)$ and $p(x)$ and show that for the given boundary conditions, we have $0 \leq \lambda$.

2.25. For the Sturm-Liouville problems 2.15 through 2.17 for the Cauchy-Euler operator, identify $q(x)$ and $p(x)$ and show that for the given boundary conditions, we have $0 \leq \lambda$.

2.26. For the Sturm-Liouville problems 2.19 through 2.20 for the Bessel operator, identify $q(x)$ and $p(x)$ and show that for the given boundary conditions, we have $0 \leq \lambda$.

Statement of Orthogonality

From Section 2.3, we showed that for the Sturm-Liouville operator

$$L = D(p(x)D) + q(x)$$

acting upon two functions $u(x)$ and $v(x)$ over the interval $I = \{x \mid a < x < b\}$, Green's formula reads

$$\int_a^b (uL(v) - vL(u)) dx = p(b)(u(b)v_x(b) - v(b)u_x(b)) - p(a)(u(a)v_x(a) - v(a)u_x(a))$$

If $\varphi_n(x)$ and $\varphi_m(x)$ are both solutions to the Sturm-Liouville differential equations

$$L\varphi_m(x) + \lambda_m w(x)\varphi_m(x) = 0$$
$$L\varphi_n(x) + \lambda_n w(x)\varphi_n(x) = 0$$

then it can be shown that for regular boundary conditions, the right-hand side of Green's formula vanishes and we get

$$(\lambda_n - \lambda_m)\left(\int_a^b \varphi_n(x)\,\varphi_m(x)w(x)\,dx \right) = 0$$

For $\lambda_m \neq \lambda_n$, the preceding is the equivalent statement of orthogonality of the eigenfunctions with respect to the weight function $w(x)$. In the following exercises, you are asked to consider the specific boundary conditions for the given problem and to show that the right-hand side of Green's formula does indeed reduce to zero.

2.27. Consider Exercise 2.4. Show that for the given boundary conditions, the right-hand side of Green's formula vanishes, thus establishing the orthogonality of the eigenfunctions.

2.28. Consider Exercise 2.6. Show that for the given boundary conditions, the right-hand side of Green's formula vanishes, thus establishing the orthogonality of the eigenfunctions.

2.29. Consider Exercise 2.10. Show that for the given boundary conditions, the right-hand side of Green's formula vanishes, thus establishing the orthogonality of the eigenfunctions.

2.30.　Consider Exercise 2.12. Show that for the given boundary conditions, the right-hand side of Green's formula vanishes, thus establishing the orthogonality of the eigenfunctions.

2.31.　Consider Exercise 2.14. Show that for the given boundary conditions, the right-hand side of Green's formula vanishes, thus establishing the orthogonality of the eigenfunctions.

2.32.　Consider Exercise 2.18. Show that for the given periodic boundary conditions, the right-hand side of Green's formula vanishes, thus establishing the orthogonality of the eigenfunctions.

2.33.　Consider Exercise 2.19. Show that for the given boundary conditions, the right-hand side of Green's formula vanishes, thus establishing the orthogonality of the eigenfunctions.

2.34.　Consider Exercise 2.20. Show that for the given boundary conditions, the right-hand side of Green's formula vanishes, thus establishing the orthogonality of the eigenfunctions.

CHAPTER 3

The Diffusion or Heat Partial Differential Equation

3.1 Introduction

We begin by examining types of partial differential equations that occur in describing heat or diffusion phenomena. Heat or diffusion phenomena are treated similarly, since physicists liken the random motion of molecules under nonuniform temperature distributions to be the same as the diffusion of molecules under nonuniform concentration.

The partial differential equations that we examine here are said to be linear in that the partial differential operator obeys the defining characteristics of a linear operator. The operator L is defined to be linear if it satisfies the following, where C1 and C2 are constants:

$$L(C1u1 + C2u2) = C1L(u) + C2L(u2)$$

We examine the following partial differential equations that have linear operators. We shall see that many of the results of Chapters 1 and 2 will play an important role in the results of this chapter.

3.2 One-Dimensional Diffusion Operator in Rectangular Coordinates

It can be shown that heat or diffusion phenomena in one dimension can be described by the following partial differential equation in the rectangular coordinate system:

$$c(x)\rho(x)\left(\frac{\partial}{\partial t}u(x,t)\right) = \left(\frac{d}{dx}K(x)\right)\left(\frac{\partial}{\partial x}u(x,t)\right) + K(x)\left(\frac{\partial^2}{\partial x^2}u(x,t)\right) + q(x,t)$$

where, for the situation of heat phenomena, $u(x,t)$ denotes the spatial-time dependent temperature, $K(x)$ denotes the thermal conductivity, $c(x)$ denotes the specific heat, $\rho(x)$ denotes the mass density of the medium, and $q(x,t)$ denotes the time rate of the input source heat density into the medium. If the medium is uniform, then the coefficients are spatially invariant, and we can write them as constants. For constant coefficients, the preceding simplifies to

$$\frac{\partial}{\partial t}u(x,t) = k\left(\frac{\partial^2}{\partial x^2}u(x,t)\right) + h(x,t)$$

where k, the thermal diffusivity of the medium, and $h(x,t)$ are given as

$$k = \frac{K}{c\rho}$$

and

$$h(x,t) = \frac{q(x,t)}{c\rho}$$

We can write the preceding partial differential equation in terms of the linear heat or diffusion operator as

$$L(u) = h(x,t)$$

where the diffusion operator in rectangular coordinates reads

$$L(u) = \frac{\partial}{\partial t}u(x,t) - k\left(\frac{\partial^2}{\partial x^2}u(x,t)\right)$$

If $h(x,t) = 0$, then there are no heat sources in the system and this nonhomogeneous partial differential equation reduces to its corresponding homogeneous equation

$$L(u) = 0$$

This homogeneous equation can be written in the more familiar form

$$\frac{\partial}{\partial t}u(x,t) = k\left(\frac{\partial^2}{\partial x^2}u(x,t)\right)$$

This partial differential equation for diffusion or heat phenomena is characterized by the fact that it relates the first partial derivative with respect to the time variable with the second partial derivative with respect to the spatial variable.

Some motivation for the derivation of this equation can be provided in terms of some simple concepts from introductory calculus. Recall that the slope of a curve is found from the first derivative and the concavity of the curve is found from the second derivative with respect to the spatial variable x. In addition, the time rate of change, or speed, is given as the first derivative with respect to the time variable t.

We imagine the sideview profile of a dome-shaped pile of dry beach sand on a flat surface while it undergoes settling or diffusion. We let $u(x,t)$ denote the time and spatial dependent height of the pile of sand. For a typical sandpile—as the sand is being poured from a pail—the profile is concave down; thus, the second derivative with respect to the spatial variable is negative. As the sand settles from a peak, the concavity goes to zero; the speed of diffusion, or settling, of the pile height is negative; and the speed of drop of the pile height lessens as the pile gets flatter. Thus, the time-rate of change (speed) of the profile height is proportional to the spatial concavity of the profile, the identical behavior described by the preceding partial differential equation.

A typical diffusion problem reads as follows. We seek the temperature distribution $u(x, t)$ in a thin uniform rod over the finite interval $I = \{x \mid 0 < x < 1\}$. The lateral surface of the rod is insulated, so no heat can escape from the sides of the rod. Both the left and right ends of the rod are held at a fixed temperature of zero. The initial temperature distribution in the rod is given as $u(x, 0) = f(x)$. There are no external heat sources, and the thermal diffusivity of the rod is $k = 1/10$.

The homogeneous diffusion equation for this problem reads

$$\frac{\partial}{\partial t} u(x, t) = \frac{\frac{\partial^2}{\partial x^2} u(x, t)}{10}$$

Since both ends of the rod are at a fixed temperature of zero, the boundary conditions are type 1 at $x = 0$ and type 1 at $x = 1$:

$$u(0, t) = 0 \quad \text{and} \quad u(1, t) = 0$$

The initial condition is

$$u(x, 0) = f(x)$$

We will solve this specific problem later. For now, we develop the generalized procedure for the solution to this homogeneous partial differential equation over a finite domain with homogeneous boundary conditions using the method of separation of variables.

3.3 Method of Separation of Variables for the Diffusion Equation

The homogeneous diffusion equation for a uniform system in one dimension in rectangular coordinates is rewritten as

$$\frac{\partial}{\partial t} u(x, t) = k \left(\frac{\partial^2}{\partial x^2} u(x, t) \right)$$

Generally, one seeks a solution to this problem over the finite one-dimensional domain $I = \{x \mid a < x < b\}$ subject to the regular homogeneous boundary conditions

$$\kappa_1 u(a, t) + \kappa_2 u_x(a, t) = 0$$

and

$$\kappa_3 u(b, t) + \kappa_4 u_x(b, t) = 0$$

and the initial condition

$$u(x, 0) = f(x)$$

In the method of separation of variables, the character of the equation is such that we can assume a solution in the form of a product as follows:

$$u(x, t) = X(x)T(t)$$

Since the two independent variables in the partial differential equation are the spatial variable x and the time variable t, then the preceding assumed solution is written as a product of two functions: one that is exclusively a function of x and one that is an exclusive function of t.

Substituting this assumed solution into the partial differential equation yields

$$X(x)\left(\frac{d}{dt}T(t)\right) = k\left(\frac{d^2}{dx^2}X(x)\right)T(t)$$

Dividing both sides of this equation by the product solution gives

$$\frac{\frac{d}{dt}T(t)}{kT(t)} = \frac{\frac{d^2}{dx^2}X(x)}{X(x)}$$

Because the left-hand side is an exclusive function of t and the right-hand side an exclusive function of x, and x and t are independent, then the only way this can hold for all values of x and t is to set each side equal to a constant. Doing so, we get the following two ordinary differential equations in terms of the separation constant λ:

$$\frac{d}{dt}T(t) + k\lambda T(t) = 0$$

and

$$\frac{d^2}{dx^2}X(x) + \lambda X(x) = 0$$

The preceding differential equation in t is an ordinary first-order linear homogeneous differential equation for which we already have the solution from Section 1.2. The second differential equation in x is an ordinary second-order linear homogeneous equation for which we already have the solution from Section 1.4.

If we substitute the spatial boundary conditions into the assumed solution, we get

$$\kappa_1 X(a)T(t) + \kappa_2 X_x(a)T(t) = 0$$

and

$$\kappa_3 X(b)T(t) + \kappa_4 X_x(b)T(t) = 0$$

If the preceding condition is to hold for all times t, then the spatial function $X(x)$ must independently satisfy the boundary conditions.

3.4 Sturm-Liouville Problem for the Diffusion Equation

The second differential equation in the spatial variable x just given must be solved subject to the homogeneous spatial boundary conditions over the finite interval. The Sturm-Liouville problem for the diffusion equation consists of the ordinary differential equation

$$\frac{d^2}{dx^2}X(x) + \lambda\, X(x) = 0$$

along with the corresponding regular homogeneous boundary conditions, which read

$$\kappa_1 X(a) + \kappa_2 X_x(a) = 0$$

and

$$\kappa_3 X(b) + \kappa_4 X_x(b) = 0$$

We see this problem in the spatial variable x to be a "regular" Sturm-Liouville eigenvalue problem where the allowed values of λ are called the system "eigenvalues" and the corresponding solutions are called the "eigenfunctions." Note that the preceding ordinary differential equation is of the Euler type, and the weight function is $w(x) = 1$.

From Section 2.1, it was noted that for regular Sturm-Liouville problems over finite domains, there exists an infinite number of eigenvalues that can be indexed by the positive integers n. The indexed eigenvalues and corresponding eigenfunctions are given, respectively, as

$$\lambda_n, X_n(x)$$

for $n = 0, 1, 2, 3, \ldots$.

The eigenfunctions form a "complete" set with respect to any piecewise smooth function over the finite interval $I = \{x \mid a < x < b\}$. Further, the eigenfunctions can be normalized and the corresponding statement of orthonormality reads

$$\int_a^b X_n(x)X_m(x)\mathrm{d}x = \delta(n, m)$$

where the term on the right is the familiar Kronecker delta function, which was defined in Section 2.2. We now focus on the solution to the time-dependent differential equation, which reads

$$\frac{\mathrm{d}}{\mathrm{d}t}T(t) + k\lambda T(t) = 0$$

From Section 1.2 on linear first-order differential equations, the solution to this time-dependent differential equation in terms of the allowed eigenvalues is

$$T_n(t) = C(n)\mathrm{e}^{-k\lambda_n t}$$

where the coefficients $C(n)$ are unknown arbitrary constants.

By the method of separation of variables, we arrive at an infinite number of indexed solutions $u_n(x, t)$ $(n = 0, 1, 2, 3, \ldots)$ for the homogeneous diffusion partial differential equation, over a finite interval, given as

$$u_n(x, t) = X_n(x)C(n)\mathrm{e}^{-k\lambda_n t}$$

Because the differential operator is linear, then any superposition of solutions to the homogeneous equation is also a solution to the problem. Thus, we can write the general solution to the homogeneous partial differential equation as the infinite sum

$$u(x, t) = \sum_{n=0}^{\infty} X_n(x)C(n)\mathrm{e}^{-k\lambda_n t}$$

This equation is the eigenfunction expansion form of the solution to the diffusion partial differential equation. The terms of its sum are the "basis vectors" of the solution space of the partial differential equation. Thus, for the diffusion partial differential equation, there are an infinite number of basis vectors in the solution space and we say the dimension of the solution space is infinite. This compares dramatically with an ordinary differential equation whereby the dimension of the solution space is finite and equal to the order of the equation.

We demonstrate the discussed concepts with the example diffusion problem in rectangular coordinates given in Section 3.2.

DEMONSTRATION: We seek the temperature distribution $u(x, t)$ in a thin uniform rod over the finite interval $I = \{x \mid 0 < x < 1\}$. The lateral surface of the rod is insulated. Both the left and right ends of the rod are held at a fixed temperature of zero. The initial temperature distribution in the rod is given as $f(x)$. There are no external heat sources, and the thermal diffusivity of the rod is $k = 1/10$.

SOLUTION: The homogeneous diffusion equation for this problem reads

$$\frac{\partial}{\partial t} u(x, t) = \frac{\frac{\partial^2}{\partial x^2} u(x, t)}{10}$$

Since both ends of the rod are at a fixed temperature of zero, the boundary conditions are type 1 at $x = 0$ and type 1 at $x = 1$.

$$u(0, t) = 0 \quad \text{and} \quad u(1, t) = 0$$

The initial condition is

$$u(x, 0) = f(x)$$

From the method of separation of variables, we obtain the two ordinary differential equations

$$\frac{d}{dt} T(t) + \frac{\lambda T(t)}{10} = 0$$

and

$$\frac{d^2}{dx^2} X(x) + \lambda X(x) = 0$$

We first consider the spatial differential equation in x. This is an Euler-type differential equation with type 1 conditions at $x = 0$ and at $x = 1$:

$$X(0) = 0 \quad \text{and} \quad X(1) = 0$$

This same problem was considered in Example 2.5.1 in Chapter 2. The allowed eigenvalues and corresponding orthonormal eigenfunctions were evaluated as

$$\lambda_n = n^2 \pi^2$$

and

$$X_n(x) = \sqrt{2} \sin(n\pi x)$$

for $n = 1, 2, 3, \ldots$.

The corresponding statement of orthonormality with respect to the weight function $w(x) = 1$ over the interval I is

$$\int_0^1 2 \sin(n\pi x) \sin(m\pi x) dx = \delta(n, m)$$

When solving boundary value problems, the first choice of differential equation to be solved is generally the spatial differential equation. The reason for this is that the spatial equation, along with the spatial boundary conditions, determines the character of the system eigenvalues. These eigenvalues must be known before we can get a complete solution to the time-dependent differential equation.

We now consider the solution to the time-dependent differential equation. This is a first-order ordinary differential equation, which we solved in Section 1.2. The solution, for the allowed values of λ given earlier, reads

$$T_n(t) = C(n)e^{-\frac{n^2\pi^2 t}{10}}$$

Thus, the eigenfunction expansion for the solution to the problem reads

$$u(x, t) = \sum_{n=1}^{\infty} C(n)e^{-\frac{n^2\pi^2 t}{10}} \sqrt{2} \sin(n\pi x)$$

The unknown coefficients $C(n)$ for $n = 1, 2, 3, \ldots$ are to be determined from the initial condition function that is imposed on the problem.

3.5 Initial Conditions for the Diffusion Equation in Rectangular Coordinates

We now consider the initial conditions on the problem. Suppose that, at time $t = 0$, the rod has an initial temperature distribution

$$u(x, 0) = f(x)$$

Substituting this into the following general solution

$$u(x, t) = \sum_{n=0}^{\infty} X_n(x)C(n)e^{-k\lambda_n t}$$

at time $t = 0$ yields

$$f(x) = \sum_{n=0}^{\infty} X_n(x)C(n)$$

Since the eigenfunctions of the regular Sturm-Liouville problem form a complete set with respect to piecewise smooth functions over the finite interval I, then the preceding is the generalized Fourier series expansion of the function $f(x)$ in terms of the eigenfunctions. The terms $C(n)$ are the corresponding Fourier coefficients. We now evaluate these coefficients.

As we did in Section 2.3 for the generalized Fourier series expansion, we take the inner product of both sides of the preceding series with respect to the weight function $w(x) = 1$. Assuming validity of the interchange between the summation and integration operations, we get

$$\int_a^b f(x) X_m(x) \mathrm{d}x = \sum_{n=0}^{\infty} C(n) \left(\int_a^b X_m(x) X_n(x) \mathrm{d}x \right)$$

Taking advantage of the statement of orthonormality, this reduces to

$$\int_a^b f(x) X_m(x) \mathrm{d}x = \sum_{n=0}^{\infty} C(n) \delta(n, m)$$

Due to the mathematical character of the Kronecker delta function, only one term ($m = n$) in the sum survives, and the preceding sum reduces to

$$C(m) = \int_a^b f(x) X_m(x) \mathrm{d}x$$

Thus, we have evaluated the Fourier coefficients in the generalized expansion of $f(x)$. We can write the final generalized solution to the diffusion equation in one dimension, subject to the given homogeneous boundary conditions and initial conditions, as

$$u(x, t) = \sum_{n=0}^{\infty} X_n(x) e^{-k\lambda_n t} \left(\int_a^b f(s) X_n(s) \mathrm{d}s \right)$$

All of the previous operations are based on the assumption that the infinite series is uniformly convergent and that the formal interchange between the integration and the summation operators is valid. It can be shown that if the function $f(x)$ is piecewise smooth over the interval I and it satisfies the same boundary conditions as the eigenfunctions, then the preceding series is, indeed, uniformly convergent.

DEMONSTRATION: We now provide a demonstration of these concepts for the example problem given in Section 3.2 for the case where the initial temperature distribution $u(x, 0)$ is

$$f(x) = x(1 - x)$$

SOLUTION: From an earlier development, the unknown Fourier coefficients $C(n)$ are evaluated from the inner product between the orthonormalized eigenfunctions and $f(x)$ over the interval $I = \{x \mid 0 < x < 1\}$ as shown:

$$C(n) = \int_0^1 x(1 - x) \sqrt{2} \sin(n\pi x) \mathrm{d}x$$

Evaluation of this integral yields

$$C(n) = -\frac{2\sqrt{2}((-1)^n - 1)}{n^3\pi^3}$$

for $n = 1, 2, 3, \ldots$. Thus, the final series solution to the example problem reads

$$u(x, t) = \sum_{n=1}^{\infty}\left(-\frac{4((-1)^n - 1)e^{\frac{-n^2\pi^2 t}{10}}\sin(n\pi x)}{n^3\pi^3}\right)$$

The detailed development of the solution of this problem and its graphics are given in one of the Maple worksheet examples given later.

The Maple worksheets given later are expressed in terms of generalized values for length, diffusivity, and initial condition function $u(x, 0) = f(x)$. The reason for the generalization is so that the student or practitioner of the problem can change these values at will for different applications.

3.6 Example Diffusion Problems in Rectangular Coordinates

We now consider several examples of partial differential equations for heat or diffusion phenomena under various homogeneous boundary conditions over finite, one-dimensional intervals in the rectangular coordinate system. We note that all the spatial ordinary differential equations are of the Euler type.

EXAMPLE 3.6.1: We seek the temperature distribution $u(x, t)$ in a thin rod over the finite interval $I = \{x \,|\, 0 < x < 1\}$ whose lateral surface is insulated. Both the left and right ends of the rod are held at a fixed temperature of zero. The initial temperature distribution $f(x)$ is given following, and the diffusivity is $k = 1/10$.

SOLUTION: The homogeneous diffusion equation is

$$\frac{\partial}{\partial t}u(x, t) = k\left(\frac{\partial^2}{\partial x^2}u(x, t)\right)$$

The boundary conditions are type 1 at $x = 0$ and type 1 at $x = 1$

$$u(0, t) = 0 \quad \text{and} \quad u(1, t) = 0$$

The initial condition is

$$u(x, 0) = x(1 - x)$$

Ordinary differential equations obtained from the method of separation of variables:

$$\frac{d}{dt}T(t) + k\lambda T(t) = 0$$

and

$$\frac{d^2}{dx^2}X(x) + \lambda X(x) = 0$$

Boundary conditions on spatial equation

$$X(0) = 0 \quad \text{and} \quad X(1) = 0$$

Assignment of system parameters

> restart:with(plots):a:=0:b:=1:k:=1/10:

Allowed eigenvalues and orthonormal eigenfunctions are obtained from Example 2.5.1.

> lambda[n]:=(n*Pi/b)^2;

$$\lambda_n := n^2 \pi^2 \tag{3.1}$$

for $n = 1, 2, 3, \ldots$.

Orthonormal eigenfunctions

> X[n](x):=sqrt(2/b)*sin(n*Pi/b*x);X[m](x):=subs(n=m,X[n](x)):

$$X_n(x) := \sqrt{2}\sin(n\pi x) \tag{3.2}$$

Statement of orthonormality with respect to the weight function $w(x) = 1$

> w(x):=1:Int(X[n](x)*X[m](x)*w(x),x=a...b)=delta(n,m);

$$\int_0^1 2\sin(n\pi x)\sin(m\pi x)dx = \delta(n, m) \tag{3.3}$$

Time-dependent solution

> T[n](t):=C(n)*exp(−k*lambda[n]*t);u[n](x,t):=T[n](t)*X[n](x):

$$T_n(t) := C(n)e^{-\frac{1}{10}n^2\pi^2 t} \tag{3.4}$$

Eigenfunction expansion

> u(x,t):=Sum(u[n](x,t),n=1..infinity);

$$u(x, t) := \sum_{n=1}^{\infty} C(n)e^{-\frac{1}{10}n^2\pi^2 t}\sqrt{2}\sin(n\pi x) \tag{3.5}$$

Evaluation of Fourier coefficients for the specific initial condition

> f(x):=x*(1−x);

$$f(x) := x(1 - x) \tag{3.6}$$

> C(n):=Int(f(x)*X[n](x)*w(x),x=a..b);C(n):=expand(value(%)):

$$C(n) := \int_0^1 x(1 - x)\sqrt{2}\sin(n\pi x)dx \tag{3.7}$$

> C(n):=radsimp(subs({sin(n*Pi)=0,cos(n*Pi)=(−1)^n},C(n)));

$$C(n) := -\frac{2\sqrt{2}(-1 + (-1)^n)}{n^3\pi^3} \tag{3.8}$$

Generalized series terms

> u[n](x,t):=eval(T[n](t)*X[n](x));

$$u_n(x, t) := -\frac{4(-1 + (-1)^n)e^{-\frac{1}{10}n^2\pi^2 t}\sin(n\pi x)}{n^3\pi^3} \tag{3.9}$$

Series solution

> u(x,t):=Sum(u[n](x,t),n=1..infinity);

$$u(x, t) := \sum_{n=1}^{\infty}\left(-\frac{4(-1 + (-1)^n)e^{-\frac{1}{10}n^2\pi^2 t}\sin(n\pi x)}{n^3\pi^3}\right) \tag{3.10}$$

First few terms of sum

> u(x,t):=sum(u[n](x,t),n=1..3):

ANIMATION

> animate(u(x,t),x=a..b,t=0..5,color=red,thickness=10);

The preceding animation command illustrates the spatial-time-dependent solution for $u(x, t)$. The animation sequence shown in Figure 3.1 shows snapshots at times $t = 0, 1, 2, 3, 4$, and 5.

ANIMATION SEQUENCE

```
> u(x,0):=subs(t=0,u(x,t)):u(x,1):=subs(t=1,u(x,t)):
> u(x,2):=subs(t=2,u(x,t)):u(x,3):=subs(t=3,u(x,t)):
> u(x,4):=subs(t=4,u(x,t)):u(x,5):=subs(t=5,u(x,t)):
> plot({u(x,0),u(x,1),u(x,2),u(x,3),u(x,4),u(x,5),},x=a..b,thickness=10);
```

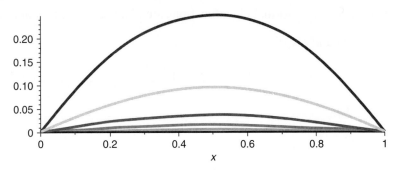

Figure 3.1

EXAMPLE 3.6.2: We seek the temperature distribution $u(x, t)$ in a thin rod over the finite interval $I = \{x \mid 0 < x < 1\}$ whose lateral surface is insulated. The left end is at a fixed temperature of zero and the right end is insulated. The initial temperature distribution $f(x)$ is given following, and the diffusivity is $k = 1/10$.

SOLUTION: The homogeneous diffusion equation is

$$\frac{\partial}{\partial t}u(x, t) = k\left(\frac{\partial^2}{\partial x^2}u(x, t)\right)$$

The boundary conditions are type 1 at $x = 0$ and type 2 at $x = 1$

$$u(0, t) = 0 \quad \text{and} \quad u_x(1, t) = 0$$

The initial condition is

$$u(x, 0) = x - \frac{x^2}{2}$$

Ordinary differential equations obtained from the method of separation of variables:

$$\frac{d}{dt}T(t) + k\lambda T(t) = 0$$

and

$$\frac{d^2}{dx^2}X(x) + \lambda X(x) = 0$$

Boundary conditions on spatial equation

$$X(0) = 0 \quad \text{and} \quad X_x(1) = 0$$

Assignment of system parameters

> restart:with(plots):a:=0:b:=1:k:=1/10:

Allowed eigenvalues and orthonormal eigenfunctions are obtained from Example 2.5.2.

> lambda[n]:=((2*n−1)*Pi/(2*b))^2;

$$\lambda_n := \frac{1}{4}(2n-1)^2\pi^2 \tag{3.11}$$

for $n = 1, 2, 3, \ldots$.

Orthonormal eigenfunctions

> X[n](x):=sqrt(2/b)*sin((2*n−1)*Pi/(2*b)*x);X[m](x):=subs(n=m,X[n](x)):

$$X_n(x) := \sqrt{2}\sin\left(\frac{1}{2}(2n-1)\pi x\right) \tag{3.12}$$

Statement of orthonormality with respect to the weight function $w(x) = 1$

> w(x):=1:Int(X[n](x)*X[m](x)*w(x),x=a..b)=delta(n,m);

$$\int_0^1 2\sin\left(\frac{1}{2}(2n-1)\pi x\right)\sin\left(\frac{1}{2}(2m-1)\pi x\right)dx = \delta(n, m) \tag{3.13}$$

Time-dependent solution

> T[n](t):=C(n)*exp(−k*lambda[n]*t);

$$T_n(t) := C(n)e^{-\frac{1}{40}(2n-1)^2\pi^2 t} \tag{3.14}$$

Generalized series terms

> u[n](x,t):=T[n](t)*X[n](x);

$$u_n(x, t) := C(n)e^{-\frac{1}{40}(2n-1)^2\pi^2 t}\sqrt{2}\sin\left(\frac{1}{2}(2n-1)\pi x\right) \tag{3.15}$$

Eigenfunction expansion

> u(x,t):=Sum(u[n](x,t),n=1..infinity);

$$u(x, t) := \sum_{n=1}^{\infty} C(n)e^{-\frac{1}{40}(2n-1)^2\pi^2 t}\sqrt{2}\sin\left(\frac{1}{2}(2n-1)\pi x\right) \tag{3.16}$$

Evaluation of Fourier coefficients for the specific initial condition

> f(x):=x−x^2/2;

$$f(x) := x - \frac{1}{2}x^2 \tag{3.17}$$

>C(n):=Int(f(x)*X[n](x)*w(x),x=a..b);C(n):=expand(value(%)):

$$C(n) := \int_0^1 \left(x - \frac{1}{2}x^2\right)\sqrt{2}\sin\left(\frac{1}{2}(2n-1)\pi x\right)dx \tag{3.18}$$

> C(n):=subs({sin(n*Pi)=0,cos(n*Pi)=(−1)^n},C(n));

$$C(n) := \frac{8\sqrt{2}}{\pi^3(8n^3 - 12n^2 + 6n - 1)} \tag{3.19}$$

Generalized series terms

> u[n](x,t):=eval(T[n](t)*X[n](x));

$$u_n(x, t) := \frac{16e^{-\frac{1}{40}(2n-1)^2\pi^2 t}\sin\left(\frac{1}{2}(2n-1)\pi x\right)}{\pi^3(8n^3 - 12n^2 + 6n - 1)} \tag{3.20}$$

Series solution

> u(x,t):=Sum(u[n](x,t),n=1..infinity);

$$u(x, t) := \sum_{n=1}^{\infty} \frac{16e^{-\frac{1}{40}(2n-1)^2\pi^2 t}\sin\left(\frac{1}{2}(2n-1)\pi x\right)}{\pi^3(8n^3 - 12n^2 + 6n - 1)} \tag{3.21}$$

First few terms of sum

> u(x,t):=sum(u[n](x,t),n=1..4):

ANIMATION

> animate(u(x,t),x=a..b,t=0..5,thickness=3);

The preceding animation command illustrates the spatial-time-dependent solution for $u(x, t)$. The animation sequence shown in Figure 3.2 shows snapshots at times $t = 0, 1, 2, 3, 4$, and 5.

ANIMATION SEQUENCE

> u(x,0):=subs(t=0,u(x,t)):u(x,1):=subs(t=1,u(x,t)):
> u(x,2):=subs(t=2,u(x,t)):u(x,3):=subs(t=3,u(x,t)):

> u(x,4):=subs(t=4,u(x,t)):u(x,5):=subs(t=5,u(x,t)):
> plot({u(x,0),u(x,1),u(x,2),u(x,3),u(x,4),u(x,5)},x=0..b,thickness=10);

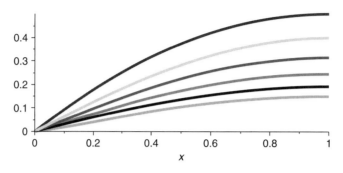

Figure 3.2

EXAMPLE 3.6.3: We again consider the temperature distribution $u(x, t)$ in a thin rod over the interval $I = \{x \mid 0 < x < 1\}$ whose lateral surface is insulated. Both the left and right ends of the rod are insulated. The initial temperature distribution $f(x)$ is given following, and the diffusivity is $k = 1/10$.

SOLUTION: The homogeneous diffusion equation is

$$\frac{\partial}{\partial t}u(x, t) = k\left(\frac{\partial^2}{\partial x^2}u(x, t)\right)$$

The boundary conditions are type 2 at $x = 0$ and type 2 at $x = 1$

$$u_x(0, t) = 0 \quad \text{and} \quad u_x(1, t) = 0$$

The initial condition is

$$u(x, 0) = x$$

Ordinary differential equations obtained from the method of separation of variables

$$\frac{d}{dt}T(t) + k\lambda T(t) = 0$$

and

$$\frac{d^2}{dx^2}X(x) + \lambda X(x) = 0$$

Boundary conditions on spatial equation

$$X_x(0) = 0 \quad \text{and} \quad X_x(1) = 0$$

Assignment of system parameters

> restart:with(plots):a:=0:b:=1:k:=1/10:

Allowed eigenvalues and orthonormal eigenfunctions are obtained from Example 2.5.3.

> lambda[0]:=0;

$$\lambda_0 := 0 \tag{3.22}$$

for $n = 0$.

Orthonormal eigenfunction

> X[0](x):=1/sqrt(b);

$$X_0(x) := 1 \tag{3.23}$$

For $n = 1, 2, 3, \ldots$,

> lambda[n]:=(n*Pi/b)^2;

$$\lambda_n := n^2\pi^2 \tag{3.24}$$

Orthonormal eigenfunctions

> X[n](x):=sqrt(2/b)*cos(n*Pi/b*x);X[m](x):=subs(n=m,X[n](x)):

$$X_n(x) := \sqrt{2}\cos(n\pi x) \tag{3.25}$$

Statement of orthonormality with respect to the weight function $w(x) = 1$

> w(x):=1:Int(X[n](x)*X[m](x)*w(x),x=a...b)=delta(n,m);

$$\int_0^1 2\cos(n\pi x)\cos(m\pi x)\mathrm{d}x = \delta(n, m) \tag{3.26}$$

Time-dependent solution

For $n = 1, 2, 3, \ldots$,

> T[n](t):=C(n)*exp(−k*lambda[n]*t);

$$T_n(t) := C(n)\mathrm{e}^{-\frac{1}{10}n^2\pi^2 t} \tag{3.27}$$

Generalized series terms

> u[n](x,t):=T[n](t)*X[n](x);

$$u_n(x, t) := C(n)\mathrm{e}^{-\frac{1}{10}n^2\pi^2 t}\sqrt{2}\cos(n\pi x) \tag{3.28}$$

For $n = 0$,

> T[0](t):=eval(subs(n=0,T[n](t)));u[0](x,t):=T[0](t)*X[0](x):

$$T_0(t) := C(0) \tag{3.29}$$

Eigenfunction expansion

> u(x,t):=u[0](x,t)+Sum(u[n](x,t),n=1..infinity);

$$u(x, t) := C(0) + \sum_{n=1}^{\infty} C(n) e^{-\frac{1}{10}n^2\pi^2 t} \sqrt{2} \cos(n\pi x) \tag{3.30}$$

Evaluation of Fourier coefficients for the specific initial condition

> f(x):=x;

$$f(x) := x \tag{3.31}$$

> C(n):=Int(f(x)*X[n](x)*w(x),x=a..b);C(n):=expand(value(%)):

$$C(n) := \int_0^1 x\sqrt{2}\cos(n\pi x)\,dx \tag{3.32}$$

> C(n):=radsimp(subs({sin(n*Pi)=0,cos(n*Pi)=(-1)^n},C(n)));

$$C(n) := \frac{\sqrt{2}(-1+(-1)^n)}{n^2\pi^2} \tag{3.33}$$

> C(0):=Int(f(x)*X[0](x)*w(x),x=a..b);C(0):=expand(value(%)):

$$C(0) := \int_0^1 x\,dx \tag{3.34}$$

> C(0):=value(%);

$$C(0) := \frac{1}{2} \tag{3.35}$$

> u[0](x,t):=eval(T[0](t)*X[0](x));

Generalized series terms

> u[n](x,t):=eval(T[n](t)*X[n](x));

$$u_n(x, t) := \frac{2(-1+(-1)^n)\,e^{-\frac{1}{10}n^2\pi^2 t}\cos(n\pi x)}{n^2\pi^2} \tag{3.36}$$

Series solution

```
> u(x,t):=eval(u[0](x,t)+Sum(u[n](x,t),n=1..infinity));
```

$$u(x, t) := \frac{1}{2} + \sum_{n=1}^{\infty} \frac{2(-1+(-1)^n)e^{-\frac{1}{10}n^2\pi^2 t}\cos(n\pi x)}{n^2\pi^2} \tag{3.37}$$

First few terms of sum

```
> u(x,t):=eval(u[0](x,t)+sum(u[n](x,t),n=1..3)):
```

ANIMATION

```
> animate(u(x,t),x=a..b,t=0..4,color=red,thickness=3);
```

The preceding animation command illustrates the spatial-time-dependent solution for $u(x, t)$. The animation sequence shown in Figure 3.3 shows snapshots at times $t = 0, 1, 2, 3, 4$, and 5.

ANIMATION SEQUENCE

```
> u(x,0):=subs(t=0,u(x,t)):u(x,1):=subs(t=1,u(x,t)):
> u(x,2):=subs(t=2,u(x,t)):u(x,3):=subs(t=3,u(x,t)):
> u(x,4):=subs(t=4,u(x,t)):u(x,5):=subs(t=5,u(x,t)):
> plot({u(x,0),u(x,1),u(x,2),u(x,3),u(x,4),u(x,5),},x=a..b,thickness=10);
```

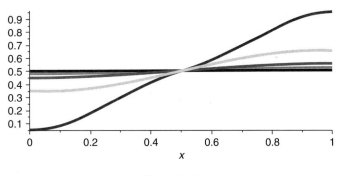

Figure 3.3

EXAMPLE 3.6.4: We consider the temperature distribution $u(x, t)$ in a thin rod whose lateral surface is insulated over the finite interval $I = \{x \mid 0 < x < 1\}$. The left end of the rod is insulated and the right end experiences a convection heat loss. The initial temperature distribution $f(x)$ is given following, and the diffusivity is $k = 1/5$.

SOLUTION: The homogeneous diffusion equation is

$$\frac{\partial}{\partial t} u(x, t) = k \left(\frac{\partial^2}{\partial x^2} u(x, t) \right)$$

The boundary conditions are type 2 at $x = 0$ and type 3 at $x = 1$

$$u_x(0, t) = 0 \quad \text{and} \quad u_x(1, t) + u(1, t) = 0$$

The initial condition is

$$u(x, 0) = 1 - \frac{x^3}{4}$$

Ordinary differential equations obtained from the method of separation of variables:

$$\frac{d}{dt} T(t) + k\lambda T(t) = 0$$

and

$$\frac{d^2}{dx^2} X(x) + \lambda X(x) = 0$$

Boundary conditions on spatial equation

$$X_x(0) = 0 \quad \text{and} \quad X_x(1) + X(1) = 0$$

Assignment of system parameters

> restart:with(plots):a:=0:b:=1:k:=1/5:

Allowed eigenvalues and orthonormal eigenfunctions are obtained from Example 2.5.5. The eigenvalues are the roots of the eigenvalue equation

> tan(sqrt(lambda[n])*b)=1/sqrt(lambda[n]);

$$\tan\left(\sqrt{\lambda_n}\right) = \frac{1}{\sqrt{\lambda_n}} \tag{3.38}$$

for $n = 1, 2, 3, \ldots$.

Orthonormal eigenfunctions

> X[n](x):=sqrt(2)*cos(sqrt(lambda[n])*x)/sqrt(((sin(sqrt(lambda[n])*b))^2+b));
  X[m](x):=subs(n=m,X[n](x)):

$$X_n(x) := \frac{\sqrt{2}\cos\left(\sqrt{\lambda_n}\,x\right)}{\sqrt{\sin\left(\sqrt{\lambda_n}\right)^2 + 1}} \tag{3.39}$$

Statement of orthonormality with respect to the weight function $w(x) = 1$

> w(x):=1:Int(X[n](x)*X[m](x)*w(x),x=a...b)=delta(n,m);

$$\int_0^1 \frac{2\cos\left(\sqrt{\lambda_n}x\right)\cos\left(\sqrt{\lambda_m}x\right)}{\sqrt{\sin\left(\sqrt{\lambda_n}\right)^2+1}\sqrt{\left(\sqrt{\lambda_m}\right)^2+1}}\,dx = \delta(n,m) \tag{3.40}$$

Time-dependent solution

> T[n](t):=C(n)*exp(−k*lambda[n]*t);u[n](x,t):=T[n](t)*X[n](x):

$$T_n(t) := C(n)e^{-\frac{1}{5}\lambda_n t} \tag{3.41}$$

Generalized series terms

> u[n](x,t):=T[n](t)*X[n](x);

$$u_n(x,t) := \frac{C(n)e^{-\frac{1}{5}\lambda_n t}\sqrt{2}\cos\left(\sqrt{\lambda_n}x\right)}{\sqrt{\sin\left(\sqrt{\lambda_n}\right)^2+1}} \tag{3.42}$$

Eigenfunction expansion

> u(x,t):=Sum(u[n](x,t),n=1..infinity);

$$u(x,t) := \sum_{n=1}^{\infty} \frac{C(n)e^{-\frac{1}{5}\lambda_n t}\sqrt{2}\cos\left(\sqrt{\lambda_n}x\right)}{\sqrt{\sin\left(\sqrt{\lambda_n}\right)^2+1}} \tag{3.43}$$

Evaluation of Fourier coefficients for the specific initial condition

> f(x):=1−x^3/4;

$$f(x) := 1 - \frac{1}{4}x^3 \tag{3.44}$$

> C(n):=Int(f(x)*X[n](x)*w(x),x=a..b);C(n):=simplify(value(%)):

$$C(n) := \int_0^1 \frac{\left(1-\frac{1}{4}x^3\right)\sqrt{2}\cos\left(\sqrt{\lambda_n}x\right)}{\sqrt{\sin\left(\sqrt{\lambda_n}\right)^2+1}}\,dx \tag{3.45}$$

Substitution of the eigenvalue equation simplifies this integral

> C(n):=simplify(subs(sin(sqrt(lambda[n])*b)=(1/sqrt(lambda[n]))*cos(sqrt(lambda[n])*b),
  C(n)));

$$C(n) := \frac{3}{2} \frac{\sqrt{2}\left(-1+2\cos\left(\sqrt{\lambda_n}\right)\right)}{\lambda_n^2 \sqrt{2-\cos\left(\sqrt{\lambda_n}\right)^2}} \tag{3.46}$$

Generalized series terms

> u[n](x,t):=eval(T[n](t)*X[n](x));

$$u_n(x, t) := \frac{3\left(-1+2\cos\left(\sqrt{\lambda_n}\right)\right) e^{-\frac{1}{5}\lambda_n t} \cos\left(\sqrt{\lambda_n}x\right)}{\lambda_n^2 \sqrt{2-\cos\left(\sqrt{\lambda_n}\right)^2} \sqrt{\sin\left(\sqrt{\lambda_n}\right)^2+1}} \tag{3.47}$$

Series solution

> u(x,t):=Sum(u[n](x,t),n=1..infinity);

$$u(x, t) := \sum_{n=1}^{\infty} \frac{3\left(-1+2\cos\left(\sqrt{\lambda_n}\right)\right) e^{-\frac{1}{5}\lambda_n t} \cos\left(\sqrt{\lambda_n}x\right)}{\lambda_n^2 \sqrt{2-\cos\left(\sqrt{\lambda_n}\right)^2} \sqrt{\sin\left(\sqrt{\lambda_n}\right)^2+1}} \tag{3.48}$$

Evaluation of the eigenvalues from the roots of the eigenvalue equation yields

> tan(sqrt(lambda[n]*b))=1/sqrt(lambda[n]);

$$\tan\left(\sqrt{\lambda_n}\right) = \frac{1}{\sqrt{\lambda_n}} \tag{3.49}$$

> plot({tan(v),1/v},v=0..20,y=0..3/2,thickness=10);

If we set $v = \sqrt{\lambda}b$, then the eigenvalues are found from the intersection points of the curves shown in Figure 3.4. We evaluate a few of these eigenvalues using the Maple fsolve command:

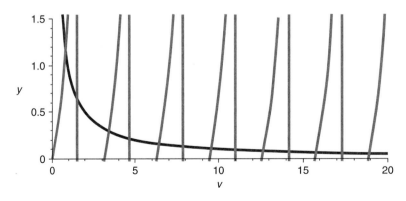

Figure 3.4

> lambda[1]:=(1/b*(fsolve((tan(v)−1/v),v=0..1)))^2;

$$\lambda_1 := 0.7401738844 \qquad (3.50)$$

> lambda[2]:=(1/b*(fsolve((tan(v)−1/v),v=1..4)))^2;

$$\lambda_2 := 11.73486183 \qquad (3.51)$$

> lambda[3]:=(1/b*(fsolve((tan(v)−1/v),v=4..7)))^2;

$$\lambda_3 := 41.43880785 \qquad (3.52)$$

First few terms of sum

> u(x,t):=eval(sum(u[n](x,t),n=1..3)):

ANIMATION

> animate(u(x,t),x=a..b,t=0..10,thickness=10);

The preceding animation command illustrates the spatial-time-dependent solution for $u(x, t)$. The animation sequence shown in Figure 3.5 shows snapshots at times $t = 0, 1, 2, 3, 4$, and 5.

ANIMATION SEQUENCE

> u(x,0):=subs(t=0,u(x,t)):u(x,1):=subs(t=1,u(x,t)):
> u(x,2):=subs(t=2,u(x,t)):u(x,3):=subs(t=3,u(x,t)):
> u(x,4):=subs(t=4,u(x,t)):u(x,5):=subs(t=5,u(x,t)):

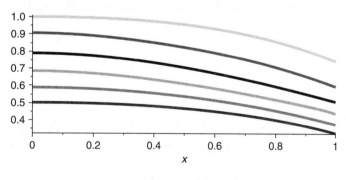

Figure 3.5

EXAMPLE 3.6.5: We seek the temperature distribution $u(x, t)$ in a thin rod over the finite interval $I = \{x \mid 0 < x < 1\}$. This rod experiences heat loss through the lateral surface into a surrounding medium of temperature zero. The coefficient β accounts for the heat loss. Both the left and right ends of the rod are held at the fixed temperature zero. The initial temperature distribution $f(x)$ is given following, $\beta = 1/5$, and the diffusivity is $k = 1/10$.

SOLUTION: The homogeneous diffusion equation is

$$\frac{\partial}{\partial t} u(x, t) = k\left(\frac{\partial^2}{\partial x^2} u(x, t)\right) - \beta u(x, t) \tag{3.53}$$

The boundary conditions are type 1 at $x = 0$ and type 1 at $x = 1$

$$u(0, t) = 0 \quad \text{and} \quad u(1, t) = 0$$

The initial condition is

$$u(x, 0) = x(1 - x)$$

Ordinary differential equations obtained from the method of separation of variables

$$\frac{d}{dt} T(t) + k\lambda T(t) = 0$$

and

$$\frac{d^2}{dx^2} X(x) + \left(\lambda - \frac{\beta}{k}\right) X(x) = 0$$

Boundary conditions on spatial equation

$$X(0) = 0 \quad \text{and} \quad X(1) = 0$$

Assignment of system parameters

> restart:with(plots):a:=0:b:=1:k:=1/10:beta:=1/5:

Allowed eigenvalues and orthonormal eigenfunctions are obtained from Example 2.5.1.

> lambda[n]:=(n*Pi/b)^2+beta/k;

$$\lambda_n := n^2\pi^2 + 2 \tag{3.54}$$

for $n = 1, 2, 3, \ldots$.

Orthonormal eigenfunctions

> X[n](x):=sqrt(2/b)*sin(n*Pi/b*x);X[m](x):=subs(n=m,X[n](x)):

$$X_n(x) := \sqrt{2}\sin(n\pi x) \tag{3.55}$$

Statement of orthonormality with respect to the weight function $w(x) = 1$

> w(x):=1:Int(X[n](x)*X[m](x)*w(x),x=a...b)=delta(n,m);

$$\int_0^1 2\sin(n\pi x)\sin(m\pi x)dx = \delta(n, m) \tag{3.56}$$

Time-dependent solution

> T[n](t):=C(n)*exp(−k*lambda[n]*t);

$$T_n(t) := C(n) e^{-\frac{1}{10}(n^2\pi^2+2)t} \tag{3.57}$$

Generalized series terms

> u[n](x,t):=T[n](t)*X[n](x);

$$u_n(x, t) := C(n) e^{-\frac{1}{10}(n^2\pi^2+2)t} \sqrt{2} \sin(n\pi x) \tag{3.58}$$

Eigenfunction expansion

> u(x,t):=Sum(u[n](x,t),n=1..infinity);

$$u(x, t) := \sum_{n=1}^{\infty} C(n) e^{-\frac{1}{10}(n^2\pi^2+2)t} \sqrt{2} \sin(n\pi x) \tag{3.59}$$

Evaluation of Fourier coefficients for the specific initial condition

> f(x):=x*(1−x);

$$f(x) := x(1 - x) \tag{3.60}$$

> C(n):=Int(f(x)*X[n](x)*w(x),x=a..b);C(n):=expand(value(%)):

$$C(n) := \int_0^1 x(1 - x)\sqrt{2} \sin(n\pi x)\, dx \tag{3.61}$$

> C(n):=simplify(subs({sin(n*Pi)=0,cos(n*Pi)=(−1)^n},C(n)));

$$C(n) := -\frac{2\sqrt{2}(-1 + (-1)^n)}{n^3\pi^3} \tag{3.62}$$

Generalized series terms

> u[n](x,t):=eval(T[n](t)*X[n](x));

$$u_n(x, t) := -\frac{4(-1 + (-1)^n) e^{-\frac{1}{10}(n^2\pi^2+2)t} \sin(n\pi x)}{n^3\pi^3} \tag{3.63}$$

Series solution

> u(x,t):=Sum(u[n](x,t),n=1..infinity);

$$u(x, t) := \sum_{n=1}^{\infty} \left(-\frac{4(-1 + (-1)^n) e^{-\frac{1}{10}(n^2\pi^2+2)t} \sin(n\pi x)}{n^3\pi^3} \right) \tag{3.64}$$

First few terms of sum

> u(x,t):=sum(u[n](x,t),n=1..3):

ANIMATION

> animate(u(x,t),x=a..b,t=0..5,thickness=3);

Due to the additional heat loss from the lateral surface, we see that the temperature drops off more rapidly than in Example 3.6.1.

The preceding animation command illustrates the spatial-time-dependent solution for $u(x, t)$. The animation sequence shown in Figure 3.6 shows snapshots at times $t = 0, 1, 2, 3, 4$, and 5.

ANIMATION SEQUENCE

> u(x,0):=subs(t=0,u(x,t)):u(x,1):=subs(t=1,u(x,t)):
> u(x,2):=subs(t=2,u(x,t)):u(x,3):=subs(t=3,u(x,t)):
> u(x,4):=subs(t=4,u(x,t)):u(x,5):=subs(t=5,u(x,t)):
> plot({u(x,0),u(x,1),u(x,2),u(x,3),u(x,4),u(x,5),},x=a..b,thickness=10);

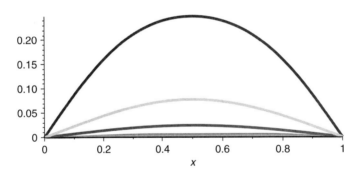

Figure 3.6

3.7 Verification of Solutions—Three-Step Verification Procedure

Earlier, we used the method of separation of variables to construct solutions to the initial value-boundary value problems. The natural question that arises is, does the solution satisfy all the conditions of the problem? We establish a three-step verification procedure:

1. Does the solution satisfy the partial differential equation?

2. Does the solution satisfy the boundary conditions?

3. Does the solution satisfy the initial conditions?

To answer all of these questions, we must first deal with some procedural problems that come about from having to take derivatives of infinite series. For a finite series, such problems do not exist because from ordinary calculus, we know that the derivative of a sum equals the sum of the derivatives. However, when the sum is an infinite sum, the preceding may or may not be true. We now list some theorems, without proof, that will play a role in the verification procedure for solutions of the preceding problems over the rectangular domain $D = \{(x, t) \mid a < x < b, t > 0\}$.

Theorem 3.7.1 (Convergence of Derivatives) Consider the following infinite series:

$$u(x) = \sum_{n=0}^{\infty} u_n(x)$$

If all the series terms $u_n(x)$ are differentiable on $I = [a, b]$ and if the series of differentiated terms

$$\sum_{n=0}^{\infty} \left(\frac{d}{dx} u_n(x) \right)$$

converges uniformly on I, then the series converges uniformly to the function $u(x)$ and the derivative of the series equals the series of the derivatives; that is,

$$\frac{d}{dx} u(x) = \sum_{n=0}^{\infty} \left(\frac{d}{dx} u_n(x) \right)$$

for all x in I.

Theorem 3.7.2 (The Weierstrass M-test for Uniform Convergence) If the terms of the preceding series satisfy the condition

$$|u_n(x)| \leq M_n$$

for all n and for all x in $I = [a, b]$, where the M_n are constants (independent of x), and if the series of constants

$$\sum_{n=0}^{\infty} M_n$$

converges, then the series

$$\sum_{n=0}^{\infty} u_n(x)$$

converges uniformly for all x in the interval I.

We now apply the three-step verification procedure to verify the solution to the illustration problem in Section 3.5.

The homogeneous diffusion equation reads

$$\frac{\partial}{\partial t}u(x,t) = k\left(\frac{\partial^2}{\partial x^2}u(x,t)\right)$$

The boundary conditions are type 1 at $x = 0$ and type 1 at $x = 1$

$$u(0,t) = 0 \quad \text{and} \quad u(1,t) = 0$$

The initial condition is

$$u_{x,0} = x(1-x)$$

From Section 3.5, the solution to the equation for $k = 1/10$ reads

$$u(x,t) = \sum_{n=1}^{\infty}\left(-\frac{4\big((-1)^n - 1\big)e^{-\frac{n^2\pi^2 t}{10}}\sin(n\pi x)}{n^3\pi^3}\right)$$

For the first step, we check to see if this solution satisfies the partial differential equation. Differentiating formally, once with respect to t and twice with respect to x, we get

$$\frac{\partial}{\partial t}u(x,t) = \sum_{n=1}^{\infty}\frac{2\big((-1)^n - 1\big)e^{-\frac{n^2\pi^2 t}{10}}\sin(n\pi x)}{5n\pi}$$

and

$$\frac{\partial^2}{\partial x^2}u(x,t) = \sum_{n=1}^{\infty}\frac{4\big((-1)^n - 1\big)e^{-\frac{n^2\pi^2 t}{10}}\sin(n\pi x)}{n\pi}$$

It is obvious that for $k = 1/10$, both sides of the partial differential equation are satisfied; that is,

$$\frac{\partial}{\partial t}u(x,t) = k\left(\frac{\partial^2}{\partial x^2}u(x,t)\right)$$

The preceding differentiations were done formally; that is, we wrote the derivative of the series as being the series of the derivatives. To verify the validity of such a move, we must use Theorems 3.7.1 and 3.7.2. The n-th term of both differentiated series given reads

$$\frac{2\big((-1)^n - 1\big)e^{-\frac{n^2\pi^2 t}{10}}\sin(n\pi x)}{5n\pi}$$

For x in the interval $I = [0, 1]$, the absolute value of the preceding term is less than or equal to the following term:

$$\frac{2e^{-\frac{n^2\pi^2 t}{10}}\left|\left((-1)^n - 1\right)\sin(n\pi x)\right|}{5n\pi} \leq \frac{e^{-\frac{n^2\pi^2 t}{10}}}{n}$$

Using the Weierstrass M-test on this inequality, in addition to using the ratio test on the following series

$$\sum_{n=1}^{\infty} \frac{e^{-\frac{n^2\pi^2 t}{10}}}{n}$$

indicates the series converges for $t > 0$. Thus, since the series converges absolutely for all x in I, then, from Theorem 3.7.2, both of the differentiated series converge uniformly, and this justifies the formal operation of differentiation.

The second step in the verification procedure is to confirm that the boundary conditions are satisfied. Since the solution is a generalized Fourier series expansion in terms of the eigenfunctons, which satisfy the same boundary conditions (this is always the case for homogeneous boundary conditions), then the boundary conditions on the solution are, indeed, satisfied. Obviously, substituting $x = 0$ and $x = 1$ into the solution yields

$$u(0, t) = 0 \quad \text{and} \quad u(1, t) = 0$$

The third and final step in the verification procedure is to check whether the initial condition is satisfied. If we substitute $t = 0$ into the preceding solution, we get

$$u(x, 0) = \sum_{n=1}^{\infty} \left(-\frac{4\left((-1)^n - 1\right)\sin(n\pi x)}{n^3\pi^3}\right)$$

Since the initial condition function $f(x)$ is required to be piecewise continuous over the interval I, we see that the given series is the generalized Fourier series expansion of $f(x) = x(1 - x)$ in terms of the "complete" set of orthonormalized eigenfunctions

$$X_n(x) = \sqrt{2}\sin(n\pi x)$$

A plot of both the initial condition function $f(x)$ and the series representation of the solution $u(x, 0)$ is shown in Figure 3.7. The accuracy of the series representation is obvious.

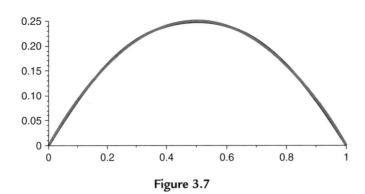

Figure 3.7

3.8 Diffusion Equation in the Cylindrical Coordinate System

The partial differential equation for diffusion or heat phenomena in the rectangular-cartesian coordinate system is presented in Section 3.2. The equivalent equation in the polar-cylindrical coordinate system for a circularly symmetric system, with spatially invariant thermal coefficients, is given as (see references for the conversion)

$$\frac{\partial}{\partial t}u(r,t) = \frac{k\left(\frac{\partial}{\partial r}u(r,t) + r\left(\frac{\partial^2}{\partial r^2}u(r,t)\right)\right)}{r} + h(r,t)$$

In this equation, r is the coordinate radius of the system. There is no angle dependence here because we have assumed circular symmetry. Further, there is no z dependence because we will be considering problems that have no extension along the z-axis (thin plates).

As for the rectangular coordinate system, we can write the preceding equation in terms of the linear operator for the diffusion equation in rectangular coordinates as

$$L(u) = h(r,t)$$

where the diffusion operator in cylindrical coordinates with cylindrical symmetry is

$$L(u) = \frac{\partial}{\partial t}u(r,t) - \frac{k\left(\frac{\partial}{\partial r}u(r,t) + r\left(\frac{\partial^2}{\partial r^2}u(r,t)\right)\right)}{r}$$

The homogeneous version of the diffusion equation can be written as

$$L(u) = 0$$

and this is generally written in the more familiar form

$$\frac{\partial}{\partial t}u(r,t) = \frac{k\left(\frac{\partial}{\partial r}u(r,t) + r\left(\frac{\partial^2}{\partial r^2}u(r,t)\right)\right)}{r}$$

We seek solutions to this partial differential equation over the finite interval $I = \{r \,|\, a < r < b\}$ subject to the nonregular homogeneous boundary conditions

$$|u(a, t)| < \infty$$

and

$$\kappa_1 u(b, t) + \kappa_2 u_r(b, t) = 0$$

and the initial condition

$$u(r, 0) = f(r)$$

We now attempt to solve this partial differential equation using the method of separation of variables. We set

$$u(r, t) = R(r)T(t)$$

Substituting this into the preceding homogeneous partial differential equation, we get

$$R(r)\left(\frac{d}{dt}T(t)\right) = \frac{k\left(\frac{d}{dr}R(r) + r\left(\frac{d^2}{dr^2}R(r)\right)\right)T(t)}{r}$$

Dividing both sides by the product solution yields

$$\frac{\frac{d}{dt}T(t)}{kT(t)} = \frac{\frac{d}{dr}R(r) + r\left(\frac{d^2}{dr^2}R(r)\right)}{R(r)r}$$

Since the left-hand side of the preceding is an exclusive function of t and the right-hand side an exclusive function of r, and r and t are independent, then the only way we can ensure equality for all r and t is to set each side equal to a constant.

Doing so, we arrive at the following two ordinary differential equations in terms of the separation constant λ^2:

$$\frac{d}{dt}T(t) + k\lambda^2 T(t) = 0$$

and

$$\frac{d^2}{dr^2}R(r) + \frac{\frac{d}{dr}R(r)}{r} + \lambda^2 R(r) = 0$$

The preceding differential equation in t is an ordinary first-order linear equation for which we already have the solution from Chapter 1.

The second differential equation in the variable r is recognized from Section 1.10 as being an ordinary Bessel differential equation of order zero. The solution of this equation is the Bessel function of the first kind of order zero.

It was noted in Section 2.6 that, for the Bessel differential equation, the point $r = 0$ is a regular singular point of the differential equation. With appropriate boundary conditions over an interval that includes the origin, we obtain a "nonregular" (singular) type Sturm-Liouville eigenvalue problem whose eigenfunctions form an orthogonal set.

Similar to regular Sturm-Liouville problems over finite intervals, there exists an infinite number of eigenvalues that can be indexed by the positive integers n. The indexed eigenvalues and corresponding eigenfunctions are given, respectively, as

$$\lambda_n, R_n(r)$$

for $n = 0, 1, 2, 3, \ldots$.

The eigenfunctions form a "complete" set with respect to any piecewise smooth function over the finite interval $I = \{r \mid a < r < b\}$. In Section 2.6, we examined the nature of the orthogonality of the Bessel functions, and we showed the eigenfunctions to be orthogonal with respect to the weight function $w(r) = r$ over a finite interval I. Further, the eigenfunctions can be normalized and the corresponding statement of orthonormality reads

$$\int_a^b R_n(r) R_m(r) r \, dr = \delta(n, m)$$

where the term on the right is the familiar Kronecker delta function.

Using arguments similar to those for the regular Sturm-Liouville problem, we can write our general solution to the partial differential equation as a superposition of the products of the solutions to each of the ordinary differential equations given earlier.

For the indexed values of λ, the solution to the preceding time-dependent equation is

$$T_n(t) = C(n)e^{-k\lambda_n^2 t}$$

where the coefficients $C(n)$ are unknown arbitrary constants.

By the method of separation of variables, we arrive at an infinite number of indexed solutions $u_n(r, t)$ $(n = 0, 1, 2, 3, \ldots)$ for the homogeneous diffusion partial differential equation, over a finite interval, given as

$$u_n(r, t) = R_n(r) C(n) e^{-k\lambda_n^2 t}$$

Because the differential operator is linear, then any superposition of solutions to the homogeneous equation is also a solution. Thus, the general solution can be written as the infinite sum

$$u(r, t) = \sum_{n=0}^{\infty} R_n(r) C(n) e^{-k\lambda_n^2 t}$$

We demonstrate the preceding concepts with an example diffusion problem in cylindrical coordinates.

DEMONSTRATION: We seek the temperature distribution $u(r, t)$ in a thin circularly symmetric plate over the finite interval $I = \{r \mid 0 < r < 1\}$ whose lateral surface is insulated, so there is no heat loss through the lateral surfaces. The periphery (edge) of the plate is at the fixed temperature of zero. The initial temperature distribution $f(r)$ is given below, and the diffusivity is $k = 1/20$.

SOLUTION: The homogeneous diffusion equation is

$$\frac{\partial}{\partial t} u(r, t) = \frac{\frac{\partial}{\partial r} u(r, t) + r\left(\frac{\partial^2}{\partial r^2} u(r, t)\right)}{20r}$$

The boundary conditions are type 1 at $r = 1$, and we require the solution to be finite at the origin

$$|u(0, t)| < \infty \quad \text{and} \quad u(1, t) = 0$$

The initial condition is

$$u(r, 0) = f(r)$$

From the method of separation of variables, we obtain the two ordinary differential equations:

$$\frac{d}{dt} T(t) + \frac{\lambda^2 T(t)}{20} = 0$$

and

$$\frac{d^2}{dr^2} R(r) + \frac{\frac{d}{dr} R(r)}{r} + \lambda^2 R(r) = 0$$

We first consider the spatial differential equation in r. This is a Bessel-type differential equation of the first kind of order zero with boundary conditions

$$|R(0)| < \infty \quad \text{and} \quad R(1) = 0$$

This same problem was considered in Example 2.6.2 in Chapter 2. The allowed eigenvalues are the roots of the eigenvalue equation

$$J(0, \lambda_n) = 0$$

for $n = 1, 2, 3, \ldots$, and the corresponding orthonormal eigenfunctions are

$$R_n(r) = \frac{\sqrt{2} J(0, \lambda_n r)}{J(1, \lambda_n)}$$

where $J(0, \lambda_n)$ and $J(1, \lambda_n)$ are the Bessel functions of the first kind of order zero and one, respectively. The corresponding statement of orthonormality with respect to the weight function $w(r) = r$ over the interval I is

$$\int_0^1 \frac{2J(0, \lambda_n r)J(0, \lambda_m r)r}{J(1, \lambda_n)J(1, \lambda_m)} dr = \delta(n, m)$$

We next consider the time-dependent differential equation. This is a first-order ordinary differential equation that we solved in Section 1.2. The solution for the allowed values of λ given earlier reads

$$T_n(t) = C(n)e^{-\frac{\lambda_n^2 t}{20}}$$

Thus, the eigenfunction expansion for the solution to the problem reads

$$u(r, t) = \sum_{n=1}^{\infty} \frac{C(n)e^{-\frac{\lambda_n^2 t}{20}} \sqrt{2}J(0, \lambda_n r)}{J(1, \lambda_n)}$$

The unknown coefficients $C(n)$, for $n = 1, 2, 3, \ldots$, are to be determined from the initial condition function imposed on the problem.

3.9 Initial Conditions for the Diffusion Equation in Cylindrical Coordinates

We now consider the initial conditions on the problem. If the initial condition temperature distribution is given as

$$u(r, 0) = f(r)$$

then substitution of this into the following general solution

$$u(r, t) = \sum_{n=0}^{\infty} R_n(r)C(n)e^{-k\lambda_n^2 t}$$

at time $t = 0$ yields

$$f(r) = \sum_{n=0}^{\infty} R_n(r)C(n)$$

This equation is the Fourier-Bessel series expansion of the function $f(r)$, and the coefficients $C(n)$ are the Fourier coefficients.

As we did before for the generalized Fourier series expansion of a piecewise smooth function over the finite interval I, we can evaluate the coefficients $C(n)$ by taking the inner product of both sides of the preceding equation with the orthonormalized eigenfunctions with respect to the weight function $w(r) = r$. Assuming validity of the interchange between the summation and integration operations, we get

$$\int_a^b f(r) R_m(r) r \, dr = \sum_{n=0}^{\infty} C(n) \left(\int_a^b R_n(r) R_m(r) r \, dr \right)$$

Taking advantage of the statement of orthonormality, this equation reduces to

$$\int_a^b f(r) R_m(r) r \, dr = \sum_{n=0}^{\infty} C(n) \delta(n, m)$$

Due to the mathematical character of the Kronecker delta function, only one term ($n = m$) in the sum survives, and we get

$$C(m) = \int_a^b f(r) R_m(r) r \, dr$$

Thus, we can write the final generalized solution to the diffusion equation in cylindrical coordinates in one dimension, subject to the given homogeneous boundary conditions and initial conditions, as

$$u(r, t) = \sum_{n=0}^{\infty} R_n(r) e^{-k\lambda_n^2 t} \left(\int_a^b f(s) R_n(s) s \, ds \right)$$

Again, all of the preceding operations are based on the assumption that the infinite series is uniformly convergent and the formal interchange between the operator and the summation is legitimate. It can be shown that if the initial condition function $f(r)$ is piecewise smooth and it satisfies the same boundary conditions as the eigenfunctions, then the preceding series is, indeed, uniformly convergent.

DEMONSTRATION: We now provide a demonstration of these concepts for the example problem given in Section 3.8 for the case where the initial temperature distribution is

$$f(r) = 1 - r^2$$

SOLUTION: The unknown Fourier coefficients are to be evaluated from the integral

$$C(n) = \int_0^1 \frac{(1 - r^2) \sqrt{2} J(0, \lambda_n r) r}{J(1, \lambda_n)} dr$$

Evaluation of this integral yields

$$C(n) = \frac{4\sqrt{2}}{\lambda_n^3}$$

for $n = 1, 2, 3, \ldots$. Thus, the final series solution to our problem reads

$$u(r, t) = \sum_{n=1}^{\infty} \frac{8e^{-\frac{\lambda_n^2 t}{20}} J(0, \lambda_n r)}{\lambda_n^3 J(1, \lambda_n)}$$

The detailed development of the solution of this problem along with the graphics are given in one of the Maple worksheet examples given later.

3.10 Example Diffusion Problems in Cylindrical Coordinates

We now consider several examples of partial differential equations for heat or diffusion phenomena under various homogeneous boundary conditions over finite intervals in the cylindrical coordinate system. We note that all the spatial ordinary differential equations in the cylindrical coordinate system are of the Bessel type and the solutions are Bessel functions of the first kind.

EXAMPLE 3.10.1: We seek the temperature distribution $u(r, t)$ in a thin circularly symmetric plate over the interval $I = \{r \mid 0 < r < 1\}$ whose lateral surface is insulated. The periphery (edge) of the plate is at the fixed temperature of zero. The initial temperature distribution $f(r)$ is given following, and the diffusivity is $k = 1/20$.

SOLUTION: The homogeneous diffusion equation is

$$\frac{\partial^2}{\partial t^2} u(r, t) = \frac{k\left(\frac{\partial}{\partial r} u(r, t) + r\left(\frac{\partial^2}{\partial r^2} u(r, t)\right)\right)}{r}$$

The boundary conditions are type 1 at $r = 1$, and we require a finite solution at $r = 0$.

$$|u(0, t)| < \infty \quad \text{and} \quad u(1, t) = 0$$

The initial condition is

$$u(r, 0) = 1 - r^2$$

Ordinary differential equations obtained from the method of separation of variables are

$$\frac{d}{dt} T(t) + k\lambda^2 T(t) = 0$$

and

$$\frac{d^2}{dr^2}R(r) + \frac{\frac{d}{dr}R(r)}{r} + \lambda^2 R(r) = 0$$

Boundary conditions on the spatial equation are

$$|R(0)| < \infty \quad \text{and} \quad R(1) = 0$$

Assignment of system parameters

> restart:with(plots):a:=0:b:=1:k:=1/20:

Allowed eigenvalues and orthonormal eigenfunctions are from Example 2.6.2. The eigenvalues are the roots of the eigenvalue equation

> BesselJ(0,lambda[n]*b)=0;

$$\text{BesselJ}(0, \lambda_n) = 0 \tag{3.65}$$

for $n = 1, 2, 3, \ldots$.

Orthonormal eigenfunctions

> R[n](r):=simplify(BesselJ(0,lambda[n]*r)/sqrt(int(BesselJ(0,lambda[n]*r)^2*r,r=a..b)));

$$R_n(r) := \frac{\text{BesselJ}(0, \lambda_n r)\sqrt{2}}{\sqrt{\text{BesselJ}(0, \lambda_n)^2 + \text{BesselJ}(1, \lambda_n)^2}} \tag{3.66}$$

Substitution of the eigenvalue equation simplifies the preceding equation

> R[n](r):=radsimp(subs(BesselJ(0,lambda[n])=0,R[n](r)));R[m](r):=subs(n=m,R[n](r)):

$$R_n(r) := \frac{\text{BesselJ}(0, \lambda_n r)\sqrt{2}}{\text{BesselJ}(1, \lambda_n)} \tag{3.67}$$

Statement of orthonormality with respect to the weight function $w(r) = r$

> w(r):=r:Int(R[n](r)*R[m](r)*w(r),r=a...b)=delta(n,m);

$$\int_0^1 \frac{2\,\text{BesselJ}(0, \lambda_n r)\,\text{BesselJ}(0, \lambda_m r)r}{\text{BesselJ}(1, \lambda_n)\,\text{BesselJ}(1, \lambda_m)}\,dr = \delta(n, m) \tag{3.68}$$

Time-dependent solution

> T[n](t):=C(n)*exp(-k*lambda[n]^2*t);u[n](r,t):=T[n](t)*R[n](r):

$$T_n(t) := C(n)e^{-\frac{1}{20}\lambda_n^2 t} \tag{3.69}$$

Generalized series terms

> u[n](r,t):=T[n](t)*R[n](r);

$$u_n(r, t) := \frac{C(n)e^{-\frac{1}{20}\lambda_n^2 t}\text{BesselJ}(0, \lambda_n r)\sqrt{2}}{\text{BesselJ}(1, \lambda_n)} \tag{3.70}$$

Eigenfunction expansion

> u(r,t):=Sum(u[n](r,t),n=1..infinity);

$$u(r, t) := \sum_{n=1}^{\infty} \frac{C(n)e^{-\frac{1}{20}\lambda_n^2 t}\text{BesselJ}(0, \lambda_n r)\sqrt{2}}{\text{BesselJ}(1, \lambda_n)} \tag{3.71}$$

Evaluation of Fourier coefficients for the specific initial condition

> f(r):=1−r^2;

$$f(r) := 1 - r^2 \tag{3.72}$$

> C(n):=Int(f(r)*R[n](r)*w(r),r=a..b);

$$C(n) := \int_0^1 \frac{(1-r^2)\text{BesselJ}(0, \lambda_n r)\sqrt{2}r}{\text{BesselJ}(1, \lambda_n)}dr \tag{3.73}$$

Substitution of the eigenvalue equation simplifies the preceding equation

> C(n):=simplify(subs(BesselJ(0,lambda[n])=0,value(%)));u[n](r,t):=eval(T[n](t)*R[n](r)):

$$C(n) := \frac{4\sqrt{2}}{\lambda_n^3} \tag{3.74}$$

Generalized series terms

> u[n](r,t):=eval(T[n](t)*R[n](r));

$$u_n(r, t) := \frac{8e^{-\frac{1}{20}\lambda_n^2 t}\text{BesselJ}(0, \lambda_n r)}{\lambda_n^3 \text{BesselJ}(1, \lambda_n)} \tag{3.75}$$

Series solution

> u(r,t):=Sum(u[n](r,t),n=1..infinity);

$$u(r, t) := \sum_{n=1}^{\infty} \frac{8e^{-\frac{1}{20}\lambda_n^2 t}\text{BesselJ}(0, \lambda_n r)}{\lambda_n^3 \text{BesselJ}(1, \lambda_n)} \tag{3.76}$$

Evaluation of the eigenvalues from the roots of the eigenvalue equation yields

> BesselJ(0,lambda[n]*b)=0;

$$\text{BesselJ}(0, \lambda_n) = 0 \tag{3.77}$$

> plot(BesselJ(0,v),v=0..20,thickness=10);

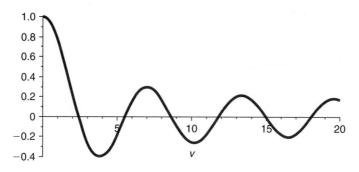

Figure 3.8

If we set $v = \lambda b$, then the eigenvalues are found from the intersection points of the curve with the v-axis shown in Figure 3.8. We evaluate a few of these eigenvalues using the Maple fsolve command:

> lambda[1]:=(1/b)*fsolve(BesselJ(0,v)=0,v=0..3);

$$\lambda_1 := 2.404825558 \tag{3.78}$$

> lambda[2]:=(1/b)*fsolve(BesselJ(0,v)=0,v=3..6);

$$\lambda_2 := 5.520078110 \tag{3.79}$$

> lambda[3]:=(1/b)*fsolve(BesselJ(0,v)=0,v=6..9);

$$\lambda_3 := 8.653727913 \tag{3.80}$$

First few terms in sum

> u(r,t):=eval(sum(u[n](r,t),n=1..3)):

ANIMATION

> animate(u(r,t),r=a..b,t=0..5,thickness=3);

The preceding animation command illustrates the spatial-time-dependent solution for $u(r, t)$. The animation sequence shown in Figure 3.9 shows snapshots at times $t = 0, 1, 2, 3, 4$, and 5.

ANIMATION SEQUENCE

```
> u(r,0):=subs(t=0,u(r,t)):u(r,1):=subs(t=1,u(r,t)):
> u(r,2):=subs(t=2,u(r,t)):u(r,3):=subs(t=3,u(r,t)):
> u(r,4):=subs(t=4,u(r,t)):u(r,5):=subs(t=5,u(r,t)):
> plot({u(r,0),u(r,1),u(r,2),u(r,3),u(r,4),u(r,5)},r=a..b,thickness=10);
```

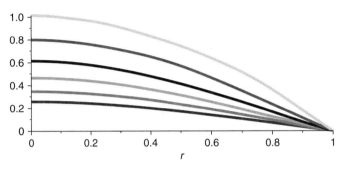

Figure 3.9

THREE-DIMENSIONAL ANIMATION

```
> u(x,y,t):=eval(subs(r=sqrt(x^2+y^2),u(r,t))):
> u(x,y,t):=(u(x,y,t))*Heaviside(1−sqrt(x^2+y^2)):
> animate3d(u(x,y,t),x=−b..b,y=−b..b,t=0..5,axes=framed,thickness=1);
```

EXAMPLE 3.10.2: We again seek the temperature distribution $u(r, t)$ in a thin circularly symmetric plate over the interval $I = \{r \mid 0 < r < 1\}$. The lateral surface and the periphery (edge) of the plate are insulated. The initial temperature distribution $f(r)$ is given below, and the diffusivity is $k = 1/50$.

SOLUTION: The homogeneous diffusion equation is

$$\frac{\partial}{\partial t}u(r, t) = \frac{k\left(\frac{\partial}{\partial r}u(r, t) + r\left(\frac{\partial^2}{\partial r^2}u(r, t)\right)\right)}{r}$$

The boundary conditions are type 2 at $r = 1$, and we require a finite solution at $r = 0$.

$$|u(0, t)| < \infty \quad \text{and} \quad u_r(1, t) = 0$$

The initial condition is

$$u(r, 0) = 1 - r^2$$

Ordinary differential equations obtained from the method of separation of variables are

$$\frac{d}{dt}T(t) + k\lambda^2 T(t) = 0$$

and

$$\frac{d^2}{dr^2}R(r) + \frac{\frac{d}{dr}R(r)}{r} + \lambda^2 R(r) = 0$$

Boundary conditions on the spatial equation

$$|R(0)| < \infty \quad \text{and} \quad R_r(1) = 0$$

Assignment of system parameters

> restart:with(plots):a:=0:b:=1:k:=1/50:

Allowed eigenvalues and orthonormal eigenfunctions are obtained from Example 2.6.3.

> lambda[0]:=0;

$$\lambda_0 := 0 \tag{3.81}$$

for $n = 0$.

Orthonormal eigenfunction

> R[0](r):=sqrt(2)/b;

$$R_0(r) := \sqrt{2} \tag{3.82}$$

For $n = 1, 2, 3, \ldots$, the eigenvalues are the roots of the eigenvalue equation

> subs(r=b,diff(BesselJ(0,lambda[n]*r),r))=0;

$$-\text{BesselJ}(1, \lambda_n)\lambda_n = 0 \tag{3.83}$$

Orthonormal eigenfunctions

> R[n](r):=simplify(BesselJ(0,lambda[n]*r)/sqrt(int(BesselJ(0,lambda[n]*r)^2*r,r=a..b)));

$$R_n(r) := \frac{\text{BesselJ}(0, \lambda_n r)\sqrt{2}}{\sqrt{\text{BesselJ}(0, \lambda_n)^2 + \text{BesselJ}(1, \lambda_n)^2}} \tag{3.84}$$

Substitution of the eigenvalue equation simplifies the preceding equation

> R[n](r):=radsimp(subs(BesselJ(1,lambda[n])=0,R[n](r)));R[m](r):=subs(n=m,R[n](r)):

$$R_n(r) := \frac{\text{BesselJ}(0, \lambda_n r)\sqrt{2}}{\text{BesselJ}(0, \lambda_n)} \tag{3.85}$$

Statement of orthonormality with respect to the weight function $w(r) = r$

> w(r):=r:Int(R[n](r)*R[m](r)*w(r),r=a...b)=delta(n,m);

$$\int_0^1 \frac{2\,\mathrm{BesselJ}(0, \lambda_n r)\mathrm{BesselJ}(0, \lambda_m r)r}{\mathrm{BesselJ}(0, \lambda_n)\mathrm{BesselJ}(0, \lambda_m)} dr = \delta(n, m) \tag{3.86}$$

Time-dependent solution for $n = 1, 2, 3, \ldots$ reads

> T[n](t):=C(n)*exp(−k*lambda[n]^2*t);u[n](r,t):=T[n](t)*R[n](r);

$$T_n(t) := C(n)e^{-\frac{1}{50}\lambda_n^2 t} \tag{3.87}$$

Generalized series terms

> u[n](r,t):=T[n](t)*R[n](r);

$$u_n(r, t) := \frac{C(n)e^{-\frac{1}{50}\lambda_n^2 t}\mathrm{BesselJ}(0, \lambda_n r)\sqrt{2}}{\mathrm{BesselJ}(0, \lambda_n)} \tag{3.88}$$

and for $n = 0$,

> T[0](t):=C(0);u[0](r,t):=T[0](t)*R[0](r):

$$T_0(t) := C(0) \tag{3.89}$$

Eigenfunction expansion

> u(r,t):=u[0](r,t)+Sum(u[n](r,t),n=1..infinity);

$$u(r, t) := C(0)\sqrt{2} + \sum_{n=1}^{\infty} \frac{C(n)e^{-\frac{1}{50}\lambda_n^2 t}\mathrm{BesselJ}(0, \lambda_n r)\sqrt{2}}{\mathrm{BesselJ}(0, \lambda_n)} \tag{3.90}$$

Evaluation of Fourier coefficients from the given specific initial condition

> f(r):=1−r^2;

$$f(r) := 1 - r^2 \tag{3.91}$$

yields, for $n = 0$,

> C(0):=eval(Int(f(r)*R[0](r)*r,r=a..b));

$$C(0) := \int_0^1 \left(1 - r^2\right)\sqrt{2}r\,dr \tag{3.92}$$

```
> C(0):=value(%);u[0](r,t):=eval(T[0](t)*R[0](r)):
```

$$C(0) := \frac{1}{4}\sqrt{2} \tag{3.93}$$

and for $n = 1, 2, 3, \ldots,$

```
> C(n):=Int(f(r)*R[n](r)*r,r=a..b);
```

$$C(n) := \int_0^1 \frac{\left(1 - r^2\right) \text{BesselJ}(0, \lambda_n r) \sqrt{2} r}{\text{BesselJ}(0, \lambda_n)} dr \tag{3.94}$$

Substitution of the eigenvalue equation simplifies the preceding equation

```
> C(n):=radsimp(subs(BesselJ(1,lambda[n]*b)=0,value(%)));
```

$$C(n) := -\frac{2\sqrt{2}}{\lambda_n^2} \tag{3.95}$$

Generalized series terms

```
> u[n](r,t):=eval(T[n](t)*R[n](r));
```

$$u_n(r, t) := -\frac{4 \, e^{-\frac{1}{50}\lambda_n^2 t} \text{BesselJ}(0, \lambda_n r)}{\lambda_n^2 \text{BesselJ}(0, \lambda_n)} \tag{3.96}$$

Series solution

```
> u(r,t):=u[0](r,t)+Sum(u[n](r,t),n=1..infinity);
```

$$u(r, t) := \frac{1}{2} + \sum_{n=1}^{\infty} \left(-\frac{4 \, e^{-\frac{1}{50}\lambda_n^2 t} \text{BesselJ}(0, \lambda_n r)}{\lambda_n^2 \text{BesselJ}(0, \lambda_n)} \right) \tag{3.97}$$

Evaluation of the eigenvalues from the roots of the eigenvalue equation yields

```
> BesselJ(1,lambda[n]*b)=0;
```

$$\text{BesselJ}(1, \lambda_n) = 0 \tag{3.98}$$

```
> plot(BesselJ(1,v),v=0..20,thickness=10);
```

If we set $v = \lambda b$, then the eigenvalues are found from the intersection points of the curve with the v-axis shown in Figure 3.10. We evaluate a few of these eigenvalues using the Maple fsolve command:

```
> lambda[1]:=(1/b)*fsolve(BesselJ(1,v)=0,v=1..4);
```

$$\lambda_1 := 3.831705970 \tag{3.99}$$

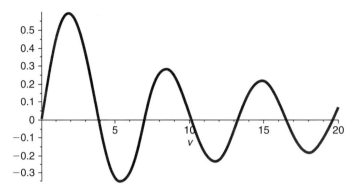

Figure 3.10

> lambda[2]:=(1/b)*fsolve(BesselJ(1,v)=0,v=4..8);

$$\lambda_2 := 7.015586670 \tag{3.100}$$

> lambda[3]:=(1/b)*fsolve(BesselJ(1,v)=0,v=8..12);

$$\lambda_3 := 10.17346814 \tag{3.101}$$

First few terms in the sum

> u(r,t):=u[0](r,t)+eval(sum(u[n](r,t),n=1..1)):

ANIMATION

> animate(u(r,t),r=a..b,t=0..5,thickness=3);

The preceding animation command illustrates the spatial-time-dependent solution for $u(r, t)$. The animation sequence shown in Figure 3.11 shows snapshots at times $t = 0, 1, 2, 3, 4,$ and 5.

ANIMATION SEQUENCE

> u(r,0):=subs(t=0,u(r,t)):u(r,1):=subs(t=1,u(r,t)):
> u(r,2):=subs(t=2,u(r,t)):u(r,3):=subs(t=3,u(r,t)):
> u(r,4):=subs(t=4,u(r,t)):u(r,5):=subs(t=5,u(r,t)):
> plot({u(r,0),u(r,1),u(r,2),u(r,3),u(r,4),u(r,5)},r=a..b,thickness=10);

THREE-DIMENSIONAL ANIMATION

> u(x,y,t):=eval(subs(r=sqrt(x^2+y^2),u(r,t))):
> u(x,y,t):=(u(x,y,t))*Heaviside(1−sqrt(x^2+y^2)):
> animate3d(u(x,y,t),x=−b..b,y=−b..b,t=0..5,axes=framed,thickness=1);

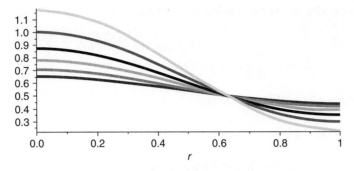

Figure 3.11

Chapter Summary

Nonhomogeneous one-dimensional diffusion equation in rectangular coordinates

$$\frac{\partial}{\partial t}u(x, t) = k\left(\frac{\partial^2}{\partial x^2}u(x, t)\right) + h(x, t)$$

Linear diffusion operator of one dimension in rectangular coordinates

$$L(u) = \frac{\partial}{\partial t}u(x, t) - k\left(\frac{\partial^2}{\partial x^2}u(x, t)\right)$$

Method of separation of variables solution

$$u(x, t) = X(x)T(t)$$

Eigenfunction expansion solution for rectangular coordinates

$$u(x, t) = \sum_{n=0}^{\infty} X_n(x)C(n)e^{-k\lambda_n t}$$

Initial condition Fourier coefficients for rectangular coordinates

$$C(n) = \int_{a}^{b} f(x)X_n(x)dx$$

Nonhomogeneous one-dimensional diffusion equation in cylindrical coordinates

$$\frac{\partial}{\partial t}u(r, t) = \frac{k\left(\frac{\partial}{\partial r}u(r, t) + r\left(\frac{\partial^2}{\partial r^2}u(r, t)\right)\right)}{r} + h(r, t)$$

Linear diffusion operator of one dimension in cylindrical coordinates

$$L(u) = \frac{\partial}{\partial t}u(r, t) - \frac{k\left(\frac{\partial}{\partial r}u(r, t) + r\left(\frac{\partial^2}{\partial r^2}u(r, t)\right)\right)}{r}$$

Method of separation of variables solution

$$u(r, t) = R(r)T(t)$$

Eigenfunction expansion solution in cylindrical coordinates

$$u(r, t) = \sum_{n=0}^{\infty} R_n(r)C(n)e^{-k\lambda_n^2 t}$$

Initial condition Fourier coefficients for cylindrical coordinates

$$C(n) = \int_a^b f(r)R_n(r)r\,dr$$

We have examined partial differential equations describing diffusion or heat diffusion phenomena in a single spatial dimension for both the rectangular and the cylindrical coordinate systems. We examine these same partial differential equations in steady-state and higher-dimensional systems later.

Exercises

We now consider exercise problems dealing with diffusion or heat equations with homogeneous boundary conditions in both the rectangular and the cylindrical coordinate systems. Use the method of separation of variables and eigenfunction expansions to evaluate the solutions.

3.1.   Consider the temperature distribution in a thin rod over the interval $I = \{x \mid 0 < x < 1\}$ whose lateral surface is insulated. The homogeneous partial differential equation reads

$$\frac{\partial}{\partial t}u(x, t) = k\left(\frac{\partial^2}{\partial x^2}u(x, t)\right)$$

with $k = 1/10$, initial condition $u(x, 0) = f(x)$ given following, and boundary conditions

$$u(0, t) = 0, u(1, t) = 0$$

that is, the left end of the rod is at the fixed temperature zero and the right end is at the fixed temperature zero. Evaluate the eigenvalues and corresponding orthonormalized eigenfunctions, and write the general solution for each of these three initial condition functions:

$$f1(x) = 1$$
$$f2(x) = x$$
$$f3(x) = x(1-x) \tag{3.102}$$

Generate the animated solution for each case, and plot the animated sequence for $0 < t < 5$. In the animated sequence, take note of the adherence of the solution to the boundary conditions.

3.2. Use the three-step verification procedure in Exercise 3.1 for the case $u(x, 0) = f3(x)$.

3.3. Consider the temperature distribution in a thin rod over the interval $I = \{x \mid 0 < x < 1\}$ whose lateral surface is insulated. The homogeneous partial differential equation reads

$$\frac{\partial}{\partial t} u(x, t) = k \left(\frac{\partial^2}{\partial x^2} u(x, t) \right)$$

with $k = 1/10$, initial condition $u(x, 0) = f(x)$ given later, and boundary conditions

$$u(0, t) = 0, \, u_x(1, t) = 0$$

that is, the left end of the rod is at the fixed temperature zero and the right end is insulated. Evaluate the eigenvalues and corresponding orthonormalized eigenfunctions, and write the general solution for each of these three initial condition functions:

$$f1(x) = x^2$$
$$f2(x) = x$$
$$f3(x) = x \left(1 - \frac{x}{2} \right)$$

Generate the animated solution for each case, and plot the animated sequence for $0 < t < 5$. In the animated sequence, take note of the adherence of the solution to the boundary conditions.

3.4. Use the three-step verification procedure in Exercise 3.3 for the case $u(x, 0) = f3(x)$.

3.5. Consider the temperature distribution in a thin rod over the interval $I = \{x \mid 0 < x < 1\}$ whose lateral surface is insulated. The homogeneous partial differential equation reads

$$\frac{\partial}{\partial t} u(x, t) = k \left(\frac{\partial^2}{\partial x^2} u(x, t) \right)$$

with $k = 1/10$, initial condition $u(x, 0) = f(x)$ given later, and boundary conditions

$$u_x(0, t) = 0, u(1, t) = 0$$

that is, the left end of the rod is insulated and the right end is at the fixed temperature zero. Evaluate the eigenvalues and the corresponding orthonormalized eigenfunctions, and write the general solution for each of these three initial condition functions:

$$f1(x) = x$$
$$f2(x) = 1$$
$$f3(x) = 1 - x^2$$

Generate the animated solution for each case, and plot the animated sequence for $0 < t < 5$. In the animated sequence, take note of the adherence of the solution to the boundary conditions.

3.6. Use the three-step verification procedure in Exercise 3.5 for the case $u(x, 0) = f3(x)$.

3.7. Consider the temperature distribution in a thin rod over the interval $I = \{x \mid 0 < x < 1\}$ whose lateral surface is insulated. The homogeneous partial differential equation reads

$$\frac{\partial}{\partial t} u(x, t) = k \left(\frac{\partial^2}{\partial x^2} u(x, t) \right)$$

with $k = 1/10$, initial condition $u(x, 0) = f(x)$ given later, and boundary conditions

$$u_x(0, t) = 0, u_x(1, t) = 0$$

that is, the left end of the rod is insulated and the right end is insulated. Evaluate the eigenvalues and corresponding orthonormalized eigenfunctions, and write the general solution for each of these three initial condition functions:

$$f1(x) = x$$
$$f2(x) = 1 - x^2$$
$$f3(x) = x^2 \left(1 - \frac{2x}{3} \right)$$

Generate the animated solution for each case, and plot the animated sequence for $0 < t < 5$. In the animated sequence, take note of the adherence of the solution to the boundary conditions.

3.8. Use the three-step verification procedure in Exercise 3.7 for the case $u(x, 0) = f3(x)$.

3.9.   Consider the temperature distribution in a thin rod over the interval $I = \{x \mid 0 < x < 1\}$ whose lateral surface is insulated. The homogeneous partial differential equation reads

$$\frac{\partial}{\partial t} u(x, t) = k\left(\frac{\partial^2}{\partial x^2} u(x, t)\right)$$

with $k = 1/10$, initial condition $u(x, 0) = f(x)$ given later, and boundary conditions

$$u(0, t) = 0, u(1, t) + u_x(1, t) = 0$$

that is, the left end of the rod is at a fixed temperature zero, and the right end is losing heat by convection into a zero temperature surrounding. Evaluate the eigenvalues and corresponding orthonormalized eigenfunctions, and write the general solution for each of these three initial condition functions:

$$f1(x) = 1$$
$$f2(x) = x$$
$$f3(x) = x\left(1 - \frac{2x}{3}\right)$$

Generate the animated solution for each case, and plot the animated sequence for $0 < t < 5$. In the animated sequence, take note of the adherence of the solution to the boundary conditions.

3.10.   Use the three-step verification procedure in Exercise 3.9 for the case $u(x, 0) = f3(x)$.

3.11.   Consider the temperature distribution in a thin rod over the interval $I = \{x \mid 0 < x < 1\}$ whose lateral surface is insulated. The homogeneous partial differential equation reads

$$\frac{\partial}{\partial t} u(x, t) = k\left(\frac{\partial^2}{\partial x^2} u(x, t)\right)$$

with $k = 1/10$, initial condition $u(x, 0) = f(x)$ given later, and boundary conditions

$$u(0, t) - u_x(0, t) = 0, u(1, t) = 0$$

that is, the left end of the rod is losing heat by convection into a zero temperature surrounding and the right end is at a fixed temperature zero. Evaluate the eigenvalues and corresponding orthonormalized eigenfunctions, and write the general solution for each of these three initial condition functions:

$$f1(x) = 1$$
$$f2(x) = 1 - x$$
$$f3(x) = -\frac{2x^2}{3} + \frac{x}{3} + \frac{1}{3}$$

Generate the animated solution for each case, and plot the animated sequence for $0 < t < 5$. In the animated sequence, take note of the adherence of the solution to the boundary conditions.

3.12.   Use the three-step verification procedure in Exercise 3.11 for the case $u(x, 0) = f3(x)$.

3.13.   Consider the temperature distribution in a thin rod over the interval $I = \{x \,|\, 0 < x < 1\}$ whose lateral surface is insulated. The homogeneous partial differential equation reads

$$\frac{\partial}{\partial t} u(x, t) = k\left(\frac{\partial^2}{\partial x^2} u(x, t)\right)$$

with $k = 1/10$, initial condition $u(x, 0) = f(x)$ given later, and boundary conditions

$$u_x(0, t) = 0, \, u(1, t) + u_x(1, t) = 0$$

that is, the left end of the rod is insulated and the right end is losing heat by convection into a zero temperature surrounding. Evaluate the eigenvalues and corresponding orthonormalized eigenfunctions, and write the general solution for each of these three initial condition functions:

$$f1(x) = x$$
$$f2(x) = 1$$
$$f3(x) = 1 - \frac{x^2}{3}$$

Generate the animated solution for each case, and plot the animated sequence for $0 < t < 5$. In the animated sequence, take note of the adherence of the solution to the boundary conditions.

3.14.   Use the three-step verification procedure in Exercise 3.13 for the case $u(x, 0) = f3(x)$.

3.15.   Consider the temperature distribution in a thin rod over the interval $I = \{x \,|\, 0 < x < 1\}$ whose lateral surface is insulated. The homogeneous partial differential equation reads

$$\frac{\partial}{\partial t} u(x, t) = k\left(\frac{\partial^2}{\partial x^2} u(x, t)\right)$$

with $k = 1/10$, initial condition $u(x, 0) = f(x)$ given later, and boundary conditions

$$u(0, t) - u_x(0, t) = 0, \, u_x(1, t) = 0$$

that is, the left end of the rod is losing heat by convection into a zero temperature surrounding and the right end is insulated. Evaluate the eigenvalues and corresponding orthonormalized eigenfunctions, and write the general solution for each of these three initial condition functions:

$$f1(x) = x$$
$$f2(x) = 1$$
$$f3(x) = 1 - \frac{(x-1)^2}{3}$$

Generate the animated solution for each case, and plot the animated sequence for $0 < t < 5$. In the animated sequence, take note of the adherence of the solution to the boundary conditions.

3.16. Use the three-step verification procedure in Exercise 3.15 for the case $u(x, 0) = f3(x)$.

3.17. Consider the temperature distribution in a thin rod over the interval $I = \{x \,|\, 0 < x < 1\}$ whose lateral surface is not insulated. The rod is experiencing a heat loss proportional to the difference between the rod temperature and the surrounding temperature at zero degrees. The homogeneous partial differential equation reads

$$\frac{\partial}{\partial t} u(x, t) = k \left(\frac{\partial^2}{\partial x^2} u(x, t) \right) - \beta u(x, t)$$

with $k = 1/10$, $\beta = 1/4$, initial condition $u(x, 0) = f(x)$ given later, and boundary conditions

$$u(0, t) = 0, u(1, t) = 0$$

that is, the left end of the rod is at the fixed temperature zero and the right end is at the fixed temperature zero. Evaluate the eigenvalues and corresponding orthonormalized eigenfunctions, and write the general solution for each of these three initial condition functions:

$$f1(x) = 1$$
$$f2(x) = x$$
$$f3(x) = x(1-x)$$

Generate the animated solution for each case, and plot the animated sequence for $0 < t < 5$. In the animated sequence, take note of the adherence of the solution to the boundary conditions.

3.18. Consider the temperature distribution in a thin rod over the interval $I = \{x \,|\, 0 < x < 1\}$ whose lateral surface is not insulated. The rod is experiencing a heat loss proportional to the difference between the rod temperature and the surrounding temperature at zero degrees. The homogeneous partial differential equation reads

$$\frac{\partial}{\partial t} u(x, t) = k\left(\frac{\partial^2}{\partial x^2} u(x, t)\right) - \beta u(x, t)$$

with $k = 1/10$, $\beta = 1/4$, initial condition $u(x, 0) = f(x)$ given later, and boundary conditions

$$u_x(0, t) = 0, \quad u(1, t) = 0$$

that is, the left end of the rod is insulated and the right end is at a fixed temperature zero. Evaluate the eigenvalues and corresponding orthonormalized eigenfunctions, and write the general solution for each of these three initial condition functions:

$$f1(x) = x$$
$$f2(x) = 1$$
$$f3(x) = 1 - x^2$$

Generate the animated solution for each case, and plot the animated sequence for $0 < t < 5$. In the animated sequence, take note of the adherence of the solution to the boundary conditions.

3.19. Consider the temperature distribution in a thin rod over the interval $I = \{x \,|\, 0 < x < 1\}$ whose lateral surface is not insulated. The rod is experiencing a heat loss proportional to the difference between the rod temperature and the surrounding temperature at zero degrees. The homogeneous partial differential equation reads

$$\frac{\partial}{\partial t} u(x, t) = k\left(\frac{\partial^2}{\partial x^2} u(x, t)\right) - \beta u(x, t)$$

with $k = 1/10$, $\beta = 1/4$, initial condition $u(x, 0) = f(x)$ given later, and boundary conditions

$$u_x(0, t) = 0, \quad u(1, t) + u_x(1, t) = 0$$

that is, the left end of the rod is insulated and the right end is losing heat by convection into a zero temperature surrounding. Evaluate the eigenvalues and corresponding orthonormalized eigenfunctions, and write the general solution for each of these three initial condition functions:

$$f1(x) = x$$
$$f2(x) = 1$$
$$f3(x) = 1 - \frac{x^2}{3}$$

Generate the animated solution for each case, and plot the animated sequence for $0 < t < 5$. In the animated sequence, take note of the adherence of the solution to the boundary conditions.

3.20. Consider the temperature distribution in a thin circularly symmetric plate over the interval $I = \{r \mid 0 < r < 1\}$. The lateral surface of the plate is insulated. The homogeneous partial differential equation reads

$$\frac{\partial}{\partial t} u(r, t) = \frac{k\left(\frac{\partial}{\partial r} u(r, t) + r\left(\frac{\partial^2}{\partial r^2} u(r, t)\right)\right)}{r}$$

with $k = 1/10$, initial condition $u(r, 0) = f(r)$ given later, and boundary conditions

$$|u(0, t)| < \infty, u(1, t) = 0$$

that is, the center of the plate has a finite temperature and the periphery is at a fixed temperature zero. Evaluate the eigenvalues and corresponding orthonormalized eigenfunctions, and write the general solution for each of these three initial condition functions:

$$f1(r) = r^2$$
$$f2(r) = 1$$
$$f3(r) = 1 - r^2$$

Generate the animated solution for each case, and plot the animated sequence for $0 < t < 5$. In the animated sequence, take note of the adherence of the solution to the boundary conditions.

3.21. Consider the temperature distribution in a thin circularly symmetric plate over the interval $I = \{r \mid 0 < r < 1\}$. The lateral surface of the plate is insulated. The homogeneous partial differential equation reads

$$\frac{\partial}{\partial t} u(r, t) = \frac{k\left(\frac{\partial}{\partial r} u(r, t) + r\left(\frac{\partial^2}{\partial r^2} u(r, t)\right)\right)}{r}$$

with $k = 1/10$, initial condition $u(r, 0) = f(r)$ given later, and boundary conditions

$$|u(0, t)| < \infty, u_r(1, t) = 0$$

that is, the center of the plate has a finite temperature and the periphery is insulated. Evaluate the eigenvalues and corresponding orthonormalized eigenfunctions, and write the general solution for each of these three initial condition functions:

$$f1(r) = r^2$$
$$f2(r) = 1$$
$$f3(r) = r^2 - \frac{r^4}{2}$$

Generate the animated solution for each case, and plot the animated sequence for $0 < t < 5$. In the animated sequence, take note of the adherence of the solution to the boundary conditions.

3.22. Consider the temperature distribution in a thin circularly symmetric plate over the interval $I = \{r \mid 0 < r < 1\}$. The lateral surface of the plate is insulated. The homogeneous partial differential equation reads

$$\frac{\partial}{\partial t} u(r, t) = \frac{k\left(\frac{\partial}{\partial r} u(r, t) + r\left(\frac{\partial^2}{\partial r^2} u(r, t)\right)\right)}{r}$$

with $k = 1/10$, initial condition $u(r, 0) = f(r)$ given later, and boundary conditions

$$|u(0, t)| < \infty, u_r(1, t) + u(1, t) = 0$$

that is, the center of the plate has a finite temperature and the periphery is losing heat by convection into a zero temperature surrounding. Evaluate the eigenvalues and corresponding orthonormalized eigenfunctions, and write the general solution for each of these three initial condition functions:

$$f1(r) = r^2$$
$$f2(r) = 1$$
$$f3(r) = 1 - \frac{r^2}{3}$$

Generate the animated solution for each case, and plot the animated sequence for $0 < t < 5$. In the animated sequence, take note of the adherence of the solution to the boundary conditions.

Significance of Thermal Diffusivity

The coefficient k in the heat equation is called the "thermal diffusivity" of the medium under consideration. This constant is equal to

$$k = \frac{K}{c\rho}$$

where c is the specific heat of the medium, ρ is the mass density, and K is the thermal conductivity of the medium. In a uniform, isotropic medium, these terms are all constants. By convention, we say that heat is a diffusion process whereby heat flows from high-temperature regions to low-temperature regions similar to how salt in a water solution diffuses from high-concentration regions to low-concentration regions. The magnitude of the thermal diffusivity k is an indication of the ability of the medium to conduct heat from one region to another. Thus, large values of k provide for rapid transfers and small values of k provide for slow transfers of heat within the medium.

In the following, we investigate the significance of the change in magnitude of the diffusivity k by noting its effect on the solution of a problem. In Exercises 3.23 through 3.28, multiply the diffusivity by the given factor, and develop the solution for the given initial condition function $f3(x)$. Develop the animated solution and take particular note of the change in the time dependence of the solution due to the increased magnitude of the diffusivity.

3.23.   In Exercise 3.1, multiply the diffusivity k by a factor of 5 and solve.

3.24.   In Exercise 3.5, multiply the diffusivity k by a factor of 10 and solve.

3.25.   In Exercise 3.9, multiply the diffusivity k by a factor of 5 and solve.

3.26.   In Exercise 3.17, multiply the diffusivity k by a factor of 10 and solve.

3.27.   In Exercise 3.20, multiply the diffusivity k by a factor of 5 and solve.

3.28.   In Exercise 3.22, multiply the diffusivity k by a factor of 10 and solve.

CHAPTER 4

The Wave Partial Differential Equation

4.1 Introduction

We begin by examining types of partial differential equations that exhibit wave phenomena. We see wave-type partial differential equations in many areas of engineering and physics, including acoustics, electromagnetic theory, quantum mechanics, and the study of the transmission of longitudinal and transverse disturbances in solids and liquids.

Similar to the partial differential equations that are descriptive of heat and diffusion phenomena, the wave partial differential equations that we examine here are also linear in that the partial differential operator L obeys the definition characteristics of a linear operator defined in Section 3.1.

4.2 One-Dimensional Wave Operator in Rectangular Coordinates

Wave phenomena in one dimension can be described by the following partial differential equation in the rectangular coordinate system:

$$\frac{\partial^2}{\partial t^2} u(x, t) = c^2 \left(\frac{\partial^2}{\partial x^2} u(x, t) \right) - \gamma \left(\frac{\partial}{\partial t} u(x, t) \right) + h(x, t)$$

In the preceding, $u(x, t)$ denotes the spatial-time-dependent wave amplitude, c denotes the wave speed, γ denotes the damping coefficient of the medium, and $h(x, t)$ denotes the presence of any external applied forces acting on the system. If the medium is uniform, then all the preceding coefficients are spatially invariant, and we can write them as constants. The wave speed c is generally proportional to the square root of the quotient of the tension in the system and the inertia of the system (see Exercises 4.19 through 4.25).

We can write the preceding nonhomogeneous equation in terms of the linear wave operator as

$$L(u) = h(x, t)$$

where the wave operator in rectangular coordinates is

$$L(u) = \frac{\partial^2}{\partial t^2} u(x, t) - c^2 \left(\frac{\partial^2}{\partial x^2} u(x, t) \right) + \gamma \left(\frac{\partial}{\partial t} u(x, t) \right)$$

If $h(x, t) = 0$, then there are no external applied forces acting on the system, and the preceding nonhomogeneous partial differential equation reduces to the corresponding homogeneous equation

$$L(u) = 0$$

For the case of no source terms and no damping in the system, the wave equation can be written in the more familiar form

$$\frac{\partial^2}{\partial t^2} u(x, t) = c^2 \left(\frac{\partial^2}{\partial x^2} u(x, t) \right)$$

This partial differential equation for wave phenomena is characterized by the fact that it relates the second partial derivative with respect to the time variable with the second partial derivative with respect to the spatial variable.

Similar to what we did for the diffusion equation, some motivation for the derivation of the wave equation can be provided in terms of some simple concepts from introductory calculus. Recall that the slope of a curve of a function is found from the first derivative, and the concavity of the curve is found from the second derivative with respect to the spatial variable x. In addition, the speed or time rate of change of a function is given as the first derivative, and its acceleration is given as the second derivative with respect to the time variable t.

We imagine the sideview profile of a taut string that is secured at both end points. We assume small-amplitude vibrations of the string so the tension can be assumed to be invariant with respect to time and position along the string. At those points along the string where the profile is most bent, as indicated by large values of concavity, we expect the string to want to resist further bending and to provide a large return force. From Newton's second law, large forces provide large accelerations; thus, at regions of large concavity, the string should experience large return accelerations. For a string whose local profile is concave down, we expect a negative return acceleration, which forces the string back to its unstretched position. Thus, the acceleration of the local profile is proportional to the spatial concavity of the profile: the exact behavior as described by the preceding partial differential equation.

A typical example wave problem follows. We consider the waves describing the longitudinal vibrations in a rigid bar over the finite interval $I = \{x \mid 0 < x < 1\}$. The left end of the bar is secure, and the right end is attached to an elastic hinge. The wave speed is c, and the damping term γ is very small. The initial displacement distribution is $f(x)$, and the initial speed distribution is $g(x)$.

The partial differential equation that describes the wave distribution $u(x, t)$ for longitudinal vibrations in the bar is

$$\frac{\partial^2}{\partial t^2} u(x, t) = c^2 \left(\frac{\partial^2}{\partial x^2} u(x, t) \right) - \gamma \left(\frac{\partial}{\partial t} u(x, t) \right)$$

Since the bar is secured at the left end and hinged at the right end, the boundary conditions are of type 1 at $x = 0$ and type 3 at $x = 1$.

$$u(0, t) = 0$$

and

$$u_x(1, t) + u(1, t) = 0$$

The initial displacement condition on the bar is

$$u(x, 0) = f(x)$$

and the initial speed distribution is

$$u_t(x, 0) = g(x)$$

We solve this particular problem later. For now, we develop generalized procedures for the solution to this homogeneous partial differential equation over a finite domain with homogeneous boundary conditions using the method of separation of variables.

4.3 Method of Separation of Variables for the Wave Equation

The homogeneous wave equation for a uniform system in one dimension in rectangular coordinates can be written as

$$\frac{\partial^2}{\partial t^2} u(x, t) - c^2 \left(\frac{\partial^2}{\partial x^2} u(x, t) \right) + \gamma \left(\frac{\partial}{\partial t} u(x, t) \right) = 0$$

This can be rewritten in the more familiar form as

$$\frac{\partial^2}{\partial t^2} u(x, t) + \gamma \left(\frac{\partial}{\partial t} u(x, t) \right) = c^2 \left(\frac{\partial^2}{\partial x^2} u(x, t) \right)$$

Generally, one seeks a solution to this problem over the finite one-dimensional domain $I = \{x \mid a < x < b\}$, subject to the regular homogeneous boundary conditions

$$\kappa_1 u(a, t) + \kappa_2 u_x(a, t) = 0$$

and

$$\kappa_3 u(b, t) + \kappa_4 u_x(b, t) = 0$$

and the two initial conditions

$$u(x, 0) = f(x)$$

and

$$u_t(x, 0) = g(x)$$

Here, $f(x)$ denotes the initial displacement of the wave, and $g(x)$ denotes the initial speed of the amplitude of the wave.

In the method of separation of variables, the character of the equation is such that we can assume a solution in the form of a product as follows:

$$u(x, t) = X(x)T(t)$$

As in the case of the diffusion equation, because the two independent variables in the partial differential equation are the spatial variable x and the time variable t, then the preceding assumed solution is written as a product of two functions; one is exclusively a function of x and the other an exclusive function of t.

Substituting this assumed solution into the partial differential equation yields

$$X(x)\left(\frac{d^2}{dt^2}T(t)\right) + \gamma X(x)\left(\frac{d}{dt}T(t)\right) = c^2\left(\frac{d^2}{dx^2}X(x)\right)T(t)$$

Dividing both sides of the preceding equation by the product solution gives us

$$\frac{\frac{d^2}{dt^2}T(t) + \gamma\left(\frac{d}{dt}T(t)\right)}{c^2 T(t)} = \frac{\frac{d^2}{dx^2}X(x)}{X(x)}$$

Since the left-hand side of the preceding is an exclusive function of t and the right-hand side an exclusive function of x, and t and x are independent, then the only way the preceding can hold for all values of t and x is to set each side equal to a constant. Doing so, we get the following two ordinary differential equations in terms of the separation constant λ:

$$\frac{d^2}{dt^2}T(t) + \gamma\left(\frac{d}{dt}T(t)\right) + c^2\lambda\, T(t) = 0$$

and

$$\frac{d^2}{dx^2}X(x) + \lambda X(x) = 0$$

This differential equation in t is an ordinary second-order linear homogeneous differential equation for which we already have the solution from Section 1.4. Similarly, the second differential equation in x is also an ordinary second-order linear homogeneous differential equation for which we already have the solution from Section 1.4. We easily recognize this differential equation as being of the Sturm-Liouville type.

4.4 Sturm-Liouville Problem for the Wave Equation

The second differential equation in the spatial variable x in the last section must be solved subject to the homogeneous spatial boundary conditions over the finite interval. The Sturm-Liouville problem for the wave equation consists of the ordinary differential equation

$$\frac{d^2}{dx^2}X(x) + \lambda X(x) = 0$$

along with the corresponding homogeneous boundary conditions

$$\kappa_1 X(a) + \kappa_2 X_x(a) = 0$$

and

$$\kappa_3 X(b) + \kappa_4 X_x(b) = 0$$

We see the preceding problem in the spatial variable x to be a regular Sturm-Liouville eigenvalue problem where the allowed values of λ are called the system "eigenvalues" and the corresponding solutions are called the "eigenfunctions." Note that the ordinary differential equation is of the Euler type and the weight function $w(x) = 1$.

It can be shown that for regular Sturm-Liouville problems over finite domains, an infinite number of eigenvalues exist that can be indexed by the integers n. The indexed eigenvalues and corresponding eigenfunctions are given, respectively, as

$$\lambda_n, X_n(x)$$

for $n = 0, 1, 2, 3, \ldots$.

The eigenfunctions form a "complete" set with respect to any piecewise smooth function over the finite interval $I = \{x \mid a < x < b\}$. Further, the eigenfunctions can be normalized and the corresponding statement of orthonormality reads

$$\int_a^b X_n(x) X_m(x)\, dx = \delta(n, m)$$

where the term on the right is the familiar Kronecker delta function defined in Section 2.2.

We now consider the time-dependent differential equation, which reads

$$\frac{d^2}{dt^2}T(t) + \gamma\left(\frac{d}{dt}T(t)\right) + c^2\lambda T(t) = 0$$

From Section 1.4 on second-order linear differential equations, the solution to the preceding time-dependent differential equation in terms of the allowed eigenvalues is

$$T_n(t) = A(n)e^{-\frac{\gamma t}{2}}\cos\left(\frac{\sqrt{4\lambda_n c^2 - \gamma^2}\,t}{2}\right) + B(n)e^{-\frac{\gamma t}{2}}\sin\left(\frac{\sqrt{4\lambda_n c^2 - \gamma^2}\,t}{2}\right)$$

The coefficients $A(n)$ and $B(n)$ are unknown arbitrary constants. Throughout most of our example problems, we make the assumption that the damping in the system is small—that is,

$$\gamma^2 < 4\lambda_n c^2$$

This assumption gives rise to time-dependent solutions that are oscillatory. The argument of the preceding sinusoidal sine and cosine expressions denotes ω_n, which is the damped angular frequency of oscillation of the n-th term in the system. The damped angular frequency of oscillation of the n-th term or mode is

$$\omega_n = \frac{\sqrt{4\lambda_n c^2 - \gamma^2}}{2}$$

Thus, by the method of separation of variables, we arrive at an infinite number of indexed solutions $u_n(x, t)$ $(n = 0, 1, 2, 3, \ldots)$ for the homogeneous wave partial differential equation, over a finite interval, given as

$$u_n(x, t) = X_n(x)e^{-\frac{\gamma t}{2}}(A(n)\cos(\omega_n t) + B(n)\sin(\omega_n t))$$

Because the differential operator is linear, any superposition of solutions to the homogeneous equation is also a solution to the problem. Thus, we can write the general solution to the homogeneous partial differential equation as the infinite sum

$$u(x, t) = \sum_{n=0}^{\infty} X_n(x)e^{-\frac{\gamma t}{2}}(A(n)\cos(\omega_n t) + B(n)\sin(\omega_n t))$$

This equation is the eigenfunction expansion form of the solution to the wave partial differential equation. The terms of the preceding sum form the "basis vectors" of the solution space of the partial differential equation. Thus, for the wave partial differential equation, there are an infinite number of basis vectors in the solution space, and we say the dimension of the solution space is infinite. This compares dramatically with an ordinary differential equation where the dimension of the solution space is finite and equal to the order of the equation. We demonstrate the preceding concepts with the solution of an illustrative problem given earlier in Section 4.2.

DEMONSTRATION: We seek the wave distribution $u(x, t)$ for the longitudinal vibrations in a rigid bar over the finite interval $I = \{x \mid 0 < x < 1\}$. The left end of the bar is secure, and the right end is attached to an elastic hinge. The wave speed is c and the damping term γ is very small. The initial displacement distribution is $f(x)$, and the initial speed distribution is $g(x)$.

SOLUTION: The partial differential equation that describes the longitudinal vibrations in the bar is

$$\frac{\partial^2}{\partial t^2} u(x, t) = c^2 \left(\frac{\partial^2}{\partial x^2} u(x, t) \right) - \gamma \left(\frac{\partial}{\partial t} u(x, t) \right)$$

Since the bar is secured at the left end and hinged at the right end, the boundary conditions are of type 1 at $x = 0$ and type 3 at $x = 1$.

$$u(0, t) = 0$$

and

$$u_x(1, t) + u(1, t) = 0$$

The initial conditions on the bar are

$$u(x, 0) = f(x)$$

and

$$u_t(x, 0) = g(x)$$

From the method of separation of variables, we get the two ordinary differential equations

$$\frac{d^2}{dt^2} T(t) + \gamma \left(\frac{d}{dt} T(t) \right) + c^2 \lambda T(t) = 0$$

and

$$\frac{d^2}{dx^2} X(x) + \lambda X(x) = 0$$

We consider the solution to the spatial equation first. Again, the reason for this is that the spatial differential equation and the spatial boundary conditions determine the system eigenvalues, and these eigenvalues must be determined before we consider the solution to the time-dependent differential equation. Here, we have an Euler-type differential equation with a type 1 condition at the left and a type 3 condition at the right—that is,

$$X(0) = 0$$

and

$$X_x(1) + X(1) = 0$$

We previously solved this problem in Example 2.5.4 in Chapter 2. The allowed eigenvalues are the roots of the eigenvalue equation

$$\tan\left(\sqrt{\lambda_n}\right) = -\sqrt{\lambda_n}$$

and the corresponding orthonormalized eigenfunctions are

$$X_n(x) = \frac{\sqrt{2}\sin\left(\sqrt{\lambda_n}x\right)}{\sqrt{\cos\left(\sqrt{\lambda_n}\right)^2 + 1}}$$

for $n = 1, 2, 3, \ldots$.

The statement of orthonormality for the spatial eigenfunctions, with respect to the weight function $w(x) = 1$, is

$$\int_0^1 2\,\frac{\sin\left(\sqrt{\lambda_n}\,x\right)\sin\left(\sqrt{\lambda_m}\,x\right)}{\sqrt{\cos\left(\sqrt{\lambda_n}\right)^2 + 1}\,\sqrt{\cos\left(\sqrt{\lambda_m}\right)^2 + 1}}\,dx = \delta(n, m)$$

for $n, m = 1, 2, 3, \ldots$.

The corresponding solution for the time-dependent differential equation is

$$T_n(t) = e^{-\frac{\gamma t}{2}}\left(A(n)\cos(\omega_n t) + B(n)\sin(\omega_n t)\right)$$

for $n = 1, 2, 3, \ldots$.

The final eigenfunction expansion solution to the problem is constructed from the superposition of the products of the preceding spatial and time-dependent solutions, and this reads

$$u(x, t) = \sum_{n=1}^{\infty} \frac{e^{-\frac{\gamma t}{2}}\left(A(n)\cos(\omega_n t) + B(n)\sin(\omega_n t)\right)\sqrt{2}\sin\left(\sqrt{\lambda_n}x\right)}{\sqrt{\cos\left(\sqrt{\lambda_n}\right)^2 + 1}}$$

The unknown coefficients $A(n)$ and $B(n)$ are to be determined from the initial conditions given in the problem.

4.5 Initial Conditions for the Wave Equation in Rectangular Coordinates

We now consider the initial conditions on the problem. At time $t = 0$, we have the initial displacement

$$u(x, 0) = f(x)$$

and the initial amplitude speed

$$u_t(x, 0) = g(x)$$

Substituting the first condition into the general solution

$$u(x, t) = \sum_{n=0}^{\infty} X_n(x) e^{-\frac{\gamma t}{2}} (A(n) \cos(\omega_n t) + B(n) \sin(\omega_n t))$$

at time $t = 0$ yields

$$f(x) = \sum_{n=0}^{\infty} X_n(x) A(n)$$

Substituting the second condition into the time derivative of the general solution at time $t = 0$ yields

$$g(x) = \sum_{n=0}^{\infty} X_n(x) \left(\omega_n B(n) - \frac{A(n)\gamma}{2} \right)$$

Recall that the eigenfunctions of the regular Sturm-Liouville problem form a complete set with respect to piecewise smooth functions over the finite interval $I = \{x \mid a < x < b\}$. Thus, both of the preceding equations are the generalized Fourier series expansions of the functions $f(x)$ and $g(x)$, respectively, in terms of the spatial eigenfunctions of the system. The terms $A(n)$ and $B(n)$ are the corresponding Fourier coefficients. We now evaluate these coefficients.

As we did in Section 2.3 for the generalized Fourier series expansion, we take the inner product of both sides of the preceding equations, with respect to the weight function $w(x) = 1$. Operating on the first equation, we get

$$\int_a^b f(x) X_m(x) \, dx = \sum_{n=0}^{\infty} A(n) \left(\int_a^b X_m(x) X_n(x) dx \right)$$

From the statement of orthonormality, this reduces to

$$\int_a^b f(x) X_m(x) \, dx = \sum_{n=0}^{\infty} A(n) \delta(n, m)$$

From the definition of the Kronecker delta function, only one term ($n = m$) survives in the sum, and we are able to extract the value

$$A(m) = \int_a^b f(x) X_m(x) \, dx$$

In a similar manner, operating on the second equation and taking the inner product of both sides, we get

$$\int_a^b g(x) X_m(x)\,dx = \sum_{n=0}^{\infty}\left(\omega_n B(n) - \frac{A(n)\gamma}{2}\right)\left(\int_a^b X_m(x) X_n(x)\,dx\right)$$

Again, taking advantage of the statement of orthonormality, the preceding equation reduces to

$$\int_a^b g(x) X_m(x)\,dx = \sum_{n=0}^{\infty}\left(\omega_n B(n) - \frac{A(n)\gamma}{2}\right)\delta(n,m)$$

Evaluating the sum (only the $n = m$ term survives) and inserting the known value for $A(m)$ yields

$$B(m) = \frac{\int_a^b \left(\frac{f(x)\gamma}{2} + g(x)\right) X_m(x)\,dx}{\omega_m}$$

Thus, we can write the final generalized solution to the wave equation in one dimension, subject to the earlier homogeneous boundary conditions and initial conditions, as

$$u(x,t) = \sum_{n=0}^{\infty} X_n(x) e^{-\frac{\gamma t}{2}}\left(\left(\int_a^b f(x) X_n(x)\,dx\right)\cos(\omega_n t) + \frac{\left(\int_a^b \left(\frac{f(x)\gamma}{2} + g(x)\right) X_n(x)\,dx\right)\sin(w_n t)}{\omega_n}\right)$$

 All of the operations given are based on the assumption that the infinite series is uniformly convergent and the formal interchange between the operator and the summation is legitimate. It can be shown that if the functions $f(x)$ and $g(x)$ satisfy the same boundary conditions as the eigenfunctions, then both of the preceding series are uniformly convergent. We demonstrate the evaluation of the coefficients $A(n)$ and $B(n)$ for the illustrative problem in Section 4.4.

DEMONSTRATION: Consider the case where the wave speed $c = 1/4$ and the damping factor $\gamma = 1/5$. The initial displacement distribution is

$$f(x) = x - \frac{2x^2}{3}$$

and the initial speed distribution is

$$g(x) = x$$

SOLUTION: The calculated value of ω_n is

$$\omega_n = \frac{\sqrt{25\lambda_n - 4}}{20}$$

To evaluate $A(n)$ and $B(n)$, we take the inner products of the initial condition functions $f(x)$ and $g(x)$, over the interval $I = \{x \mid 0 < x < 1\}$, with the corresponding eigenfunctions. Doing so, we get

$$A(n) = \int_0^1 \frac{\left(-\frac{2x^2}{3} + x\right)\sqrt{2}\sin\left(\sqrt{\lambda_n}x\right)}{\sqrt{\cos\left(\sqrt{\lambda_n}\right)^2 + 1}}\, dx$$

which evaluates to

$$A(n) = -\frac{4\sqrt{2}\left(\cos\left(\sqrt{\lambda_n}\right) - 1\right)}{3\lambda_n^{\left(\frac{3}{2}\right)}\sqrt{\cos\left(\sqrt{\lambda_n}\right)^2 + 1}}$$

and

$$B(n) = \frac{\int_0^1 \frac{\left(-\frac{x^2}{15} + \frac{11x}{10}\right)\sqrt{2}\sin\left(\sqrt{\lambda_n}x\right)}{\sqrt{\cos\left(\sqrt{\lambda_n}\right)^2 + 1}}\, dx}{\omega_n}$$

which evaluates to

$$B(n) = -\frac{8\sqrt{2}\left(15\cos\left(\sqrt{\lambda_n}\right)\lambda_n + \cos\left(\sqrt{\lambda_n}\right) - 1\right)}{3\sqrt{25\lambda_n - 4}\,\lambda_n^{\left(\frac{3}{2}\right)}\sqrt{\cos\left(\sqrt{\lambda_n}\right)^2 + 1}}$$

for $n = 1, 2, 3, \ldots$.

For $\lambda > 0$, the eigenvalues λ_n are determined from the roots of the eigenvalue equation

$$\tan\left(\sqrt{\lambda_n}\right) = -\sqrt{\lambda_n} \tag{4.1}$$

Figure 4.1 shows curves of the two functions $\tan(v)$ and $-v$ plotted on the same set of axes. If we set $v = \sqrt{\lambda}$, then the eigenvalues are the squares of the values of v at the intersection points of these curves.

The first three eigenvalues are

$$\lambda_1 = 4.115858365$$

$$\lambda_2 = 24.13934203$$

$$\lambda_3 = 63.65910654$$

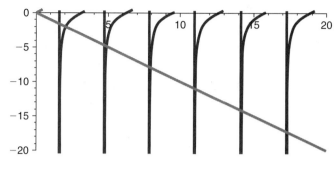

Figure 4.1

Combining all of the preceding, the final series solution to the problem reads

$$u(x, t) = \sum_{n=1}^{\infty} 2 \sin\left(\sqrt{\lambda_n}\, x\right) e^{-\frac{t}{10}} \left(-\frac{4\left(\cos\left(\sqrt{\lambda_n}\right) - 1\right) \cos\left(\frac{\sqrt{25\lambda_n - 4}\, t}{20}\right)}{3\lambda_n^{\left(\frac{3}{2}\right)} \left(\cos\left(\sqrt{\lambda_n}\right)^2 + 1\right)} \right.$$

$$\left. -\frac{8\left(15 \cos\left(\sqrt{\lambda_n}\right) \lambda_n + \cos\left(\sqrt{\lambda_n}\right) - 1\right) \sin\left(\frac{\sqrt{25\lambda_n - 4}\, t}{20}\right)}{3\sqrt{25\lambda_n - 4}\, \lambda_n^{\left(\frac{3}{2}\right)} \left(\cos\left(\sqrt{\lambda_n}\right)^2 + 1\right)} \right)$$

The detailed development of the solution of this problem and the graphics are given in one of the Maple worksheet examples.

4.6 Example Wave Equation Problems in Rectangular Coordinates

We now consider several examples of partial differential equations for wave phenomena under various homogeneous boundary conditions over finite intervals in the rectangular coordinate system. We note that all of the spatial ordinary differential equations in the rectangular coordinate system are of the Euler type; thus, the weight function $w(x) = 1$.

EXAMPLE 4.6.1: We seek the wave distribution $u(x, t)$ for transverse waves on a taut string, over the finite interval $I = \{x \,|\, 0 < x < 1\}$. The string is tied at both ends. It is vibrating in a viscous medium with small damping, and there are no external forces acting. The string has an initial displacement distribution $f(x)$ and an initial speed distribution $g(x)$ given following. The wave speed is $c = 1/4$, and the damping factor is $\gamma = 1/5$.

SOLUTION: The homogeneous wave equation is

$$\frac{\partial^2}{\partial t^2} u(x, t) = c^2 \left(\frac{\partial^2}{\partial x^2} u(x, t) \right) - \gamma \left(\frac{\partial}{\partial t} u(x, t) \right)$$

The boundary conditions are type 1 at $x = 0$ and type 1 at $x = 1$

$$u(0, t) = 0 \quad \text{and} \quad u(1, t) = 0$$

The initial conditions are

$$u(x, 0) = x(1 - x) \quad \text{and} \quad u_t(x, 0) = x(1 - x)$$

The ordinary differential equations from the method of separation of variables are

$$\frac{d^2}{dt^2} T(t) + \gamma \left(\frac{d}{dt} T(t) \right) + c^2 \lambda T(t) = 0$$

and

$$\frac{d^2}{dx^2} X(x) + \lambda X(x) = 0 \tag{4.2}$$

The boundary conditions on the spatial equation are

$$X(0) = 0 \quad \text{and} \quad X(1) = 0$$

Assignment of system parameters

> restart:with(plots):a:=0:b:=1:c:=1/4:unprotect(gamma):gamma:=1/5:

Allowed eigenvalues and orthonormal eigenfunctions are obtained from Example 2.5.1.

> lambda[n]:=(n*Pi/b)^2;

$$\lambda_n := n^2 \pi^2 \tag{4.3}$$

for $n = 1, 2, 3, \ldots$.

> X[n](x):=sqrt(2/b)*sin(n*Pi/b*x);X[m](x):=subs(n=m,X[n](x)):

$$X_n(x) := \sqrt{2} \sin(n\pi x) \tag{4.4}$$

Statement of orthonormality with respect to the weight function $w(x) = 1$

> w(x):=1:Int(X[n](x)*X[m](x)*w(x),x=a...b)=delta(n,m);

$$\int_0^1 2 \sin(n\pi x) \sin(m\pi x) \, dx = \delta(n, m) \tag{4.5}$$

Time-dependent solution

> T[n](t):=exp((−gamma/2)*t)*(A(n)*cos(omega[n]*t)+B(n)*sin(omega[n]*t));

$$T_n(t) := e^{-\frac{1}{10}t}(A(n)\cos(\omega_n t) + B(n)\sin(\omega_n t)) \tag{4.6}$$

Generalized series terms

> u[n](x,t):=T[n](t)*X[n](x):

Eigenfunction expansion

> u(x,t):=Sum(u[n](x,t),n=1..infinity);

$$u(x,t) := \sum_{n=1}^{\infty} e^{-\frac{1}{10}t}(A(n)\cos(\omega_n t) + B(n)\sin(\omega_n t))\sqrt{2}\sin(n\pi x) \tag{4.7}$$

> omega[n]:=sqrt(lambda[n]*c^2−gamma^2/4);

$$\omega_n := \frac{1}{20}\sqrt{25\,n^2\pi^2 - 4} \tag{4.8}$$

From Section 4.5, the coefficients $A(n)$ and $B(n)$ are to be determined from the inner product of the eigenfunctions and the given initial condition functions $u(x,0) = f(x)$ and $u_t(x,0) = g(x)$.

> f(x):=x*(1−x);

$$f(x) := x(1 - x) \tag{4.9}$$

> g(x):=x*(1−x);

$$g(x) := x(1 - x) \tag{4.10}$$

For $n = 1, 2, 3, \ldots$,

> A(n):=Int(f(x)*X[n](x)*w(x),x=a..b);A(n):=expand(value(%)):

$$A(n) := \int_0^1 x(1 - x)\sqrt{2}\sin(n\pi x)\,dx \tag{4.11}$$

> A(n):=simplify(subs({sin(n*Pi)=0,cos(n*Pi)=(−1)^n},A(n)));

$$A(n) := -\frac{2\sqrt{2}(-1 + (-1)^n)}{\pi^3 n^3} \tag{4.12}$$

> B(n):=Int((f(x)*gamma/2+g(x))*X[n](x)*w(x), x=a..b)/omega[n];B(n):=expand(value(%)):

$$B(n) := \frac{20\left(\int_0^1 \frac{11}{10}x(1-x)\sqrt{2}\sin(n\pi x)dx\right)}{\sqrt{25n^2\pi^2-4}} \tag{4.13}$$

> B(n):=radsimp(subs({sin(n*Pi)=0, cos(n*Pi)=(−1)^n},%));

$$B(n) := -\frac{44\sqrt{2}(-1+(-1)^n)}{\pi^3 n^3 \sqrt{25n^2\pi^2-4}} \tag{4.14}$$

> T[n](t):=exp((−gamma/2)*t)*(A(n)*cos(omega[n]*t)+B(n)*sin(omega[n]*t)):

Generalized series terms

> u[n](x,t):=eval(T[n](t)*X[n](x)):

Series solution

> u(x,t):=Sum(u[n](x,t),n=1..infinity);

$$u(x,t) := \sum_{n=1}^{\infty} e^{-\frac{1}{10}t}\left(-\frac{2\sqrt{2}(-1+(-1)^n)\cos\left(\frac{1}{20}\sqrt{25n^2\pi^2-4}\,t\right)}{\pi^3 n^3}\right.$$
$$\left.-\frac{44\sqrt{2}(-1+(-1)^n)\sin\left(\frac{1}{20}\sqrt{25n^2\pi^2-4}\,t\right)}{\pi^3 n^3 \sqrt{25n^2\pi^2-4}}\right)\sqrt{2}\sin(n\pi x) \tag{4.15}$$

First few terms of sum

> u(x,t):=sum(u[n](x,t),n=1..3):

ANIMATION

> animate(u(x,t),x=a..b,t=0..20,thickness=3);

The preceding animation command illustrates the spatial-time-dependent solution for $u(x,t)$. The following animation sequence in Figure 4.2 shows snapshots of the animation at times $t = 0, 1, 2, 3, 4$, and 5.

ANIMATION SEQUENCE

> u(x,0):=subs(t=0,u(x,t)):u(x,1):=subs(t=1,u(x,t)):
> u(x,2):=subs(t=2,u(x,t)):u(x,3):=subs(t=3,u(x,t)):
> u(x,4):=subs(t=4,u(x,t)):u(x,5):=subs(t=5,u(x,t)):
> plot({u(x,0),u(x,1),u(x,2),u(x,3),u(x,4),u(x,5)},x=a..b,thickness=10);

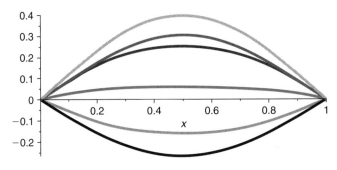

Figure 4.2

EXAMPLE 4.6.2: We seek the wave distribution $u(x, t)$ for longitudinal vibrations in a rigid bar over the finite interval $I = \{x \mid 0 < x < 1\}$. The left end of the bar is secured, and the right end is unsecured. The damping is very small, and the bar has an initial displacement distribution $f(x)$ and an initial speed distribution $g(x)$ given as follows. The wave speed is $c = 1/4$, and the damping factor is $\gamma = 1/5$.

SOLUTION: The homogeneous wave equation is

$$\frac{\partial^2}{\partial t^2} u(x, t) = c^2 \left(\frac{\partial^2}{\partial x^2} u(x, t) \right) - \gamma \left(\frac{\partial}{\partial t} u(x, t) \right)$$

The boundary conditions are type 1 at $x = 0$ and type 2 at $x = 1$

$$u(0, t) = 0 \quad \text{and} \quad u_x(1, t) = 0$$

The initial conditions are

$$u(x, 0) = x\left(1 - \frac{x}{2}\right) \quad \text{and} \quad u_t(x, 0) = x\left(1 - \frac{x}{2}\right)$$

The ordinary differential equations from the method of separation of variables are

$$\frac{d^2}{dt^2} T(t) + \gamma \left(\frac{d}{dt} T(t) \right) + c^2 \lambda T(t) = 0$$

and

$$\frac{d^2}{dx^2} X(x) + \lambda X(x) = 0 \tag{4.16}$$

The boundary conditions on the spatial equation are

$$X(0) = 0 \quad \text{and} \quad X_x(1) = 0$$

Assignment of system parameters

> restart:with(plots):a:=0:b:=1:c:=1/4:unprotect(gamma):gamma:=1/5:

Allowed eigenvalues and orthonormal eigenfunctions are obtained from Example 2.5.2.

> lambda[n]:=((2*n−1)*Pi/(2*b))^2;

$$\lambda_n := \frac{1}{4}(2n-1)^2\pi^2 \tag{4.17}$$

for $n = 1, 2, 3, \ldots$.

> X[n](x):=sqrt(2/b)*sin((2*n−1)*Pi/(2*b)*x);X[m](x):=subs(n=m,X[n](x)):

$$X_n(x) := \sqrt{2}\sin\left(\frac{1}{2}(2n-1)\pi x\right) \tag{4.18}$$

Statement of orthonormality with respect to the weight function $w(x) = 1$

> w(x):=1:Int(X[n](x)*X[m](x)*w(x),x=a..b)=delta(n,m);

$$\int_0^1 2\sin\left(\frac{1}{2}(2n-1)\pi x\right)\sin\left(\frac{1}{2}(2m-1)\pi x\right)dx = \delta(n,m) \tag{4.19}$$

Time-dependent solution

> T[n](t):=exp((−gamma/2)*t)*(A(n)*cos(omega[n]*t)+B(n)*sin(omega[n]*t));u[n](x,t):=
  T[n](t)*X[n](x):

$$T_n(t) := e^{-\frac{1}{10}t}(A(n)\cos(\omega_n t) + B(n)\sin(\omega_n t)) \tag{4.20}$$

Generalized series terms

> u[n](x,t):=T[n](t)*X[n](x):

Eigenfunction expansion

> u(x,t):=Sum(u[n](x,t),n=1..infinity);

$$u(x, t) := \sum_{n=1}^{\infty} e^{-\frac{1}{10}t}(A(n)\cos(\omega_n t) + B(n)\sin(\omega_n t))\sqrt{2}\sin\left(\frac{1}{2}(2n-1)\pi x\right) \tag{4.21}$$

> omega[n]:=sqrt(lambda[n]*c^2−gamma^2/4);

$$\omega_n := \frac{1}{40}\sqrt{25(2n-1)^2\pi^2 - 16} \tag{4.22}$$

From Section 4.5, the coefficients $A(n)$ and $B(n)$ are to be determined from the inner product of the eigenfunctions and the initial condition functions $u(x, 0) = f(x)$ and $u_t(x, 0) = g(x)$.

\> f(x):=x*(1−x/2);

$$f(x) := x\left(1 - \frac{1}{2}x\right) \tag{4.23}$$

\> g(x):=x*(1−x/2);

$$g(x) := x\left(1 - \frac{1}{2}x\right) \tag{4.24}$$

For $n = 1, 2, 3, \ldots,$

\> A(n):=Int(f(x)*X[n](x)*w(x),x=a..b);A(n):=(value(%)):

$$A(n) := \int_0^1 x\left(1 - \frac{1}{2}x\right)\sqrt{2}\sin\left(\frac{1}{2}(2n-1)\pi x\right) dx \tag{4.25}$$

\> A(n):=factor(subs({sin(n*Pi)=0,cos(n*Pi)=(−1)^n},A(n)));

$$A(n) := \frac{8\sqrt{2}}{(2n-1)^3\pi^3} \tag{4.26}$$

\> B(n):=Int((f(x)*gamma/2+g(x))*X[n](x)*w(x),x=a..b)/omega[n];B(n):=(value(%)):

$$B(n) := \frac{40\left(\int_0^1 \frac{11}{10}x\left(1 - \frac{1}{2}x\right)\sqrt{2}\sin\left(\frac{1}{2}(2n-1)\pi x\right) dx\right)}{\sqrt{25(2n-1)^2\pi^2 - 16}} \tag{4.27}$$

\> B(n):=subs({sin(n*Pi)=0,cos(n*Pi)=(−1)^n},B(n));

$$B(n) := \frac{352\sqrt{2}}{\pi^3(8n^3 - 12n^2 + 6n - 1)\sqrt{25(2n-1)^2\pi^2 - 16}} \tag{4.28}$$

Generalized series terms

\> u[n](x,t):=eval(T[n](t)*X[n](x)):

Series solution

\> u(x,t):=Sum(u[n](x,t),n=1..infinity);

$$u(x,t) := \sum_{n=1}^{\infty} e^{-\frac{1}{10}t}\left(\frac{8\sqrt{2}\cos\left(\frac{1}{40}\sqrt{25(2n-1)^2\pi^2 - 16}\,t\right)}{(2n-1)^3\pi^3}\right. \tag{4.29}$$

$$\left. + \frac{352\sqrt{2}\sin\left(\frac{1}{40}\sqrt{25(2n-1)^2\pi^2 - 16}\,t\right)}{\pi^3(8n^3 - 12n^2 + 6n - 1)\sqrt{25(2n-1)^2\pi^2 - 16}}\right)\sqrt{2}\sin\left(\frac{1}{2}(2n-1)\pi x\right)$$

First few terms of sum

> u(x,t):=sum(u[n](x,t),n=1..3):

ANIMATION

> animate(u(x,t),x=a..b,t=0..20,thickness=3);

The preceding animation command illustrates the spatial-time-dependent solution for $u(x, t)$. The following animation sequence in Figure 4.3 shows snapshots of the animation at times $t = 0, 1, 2, 3, 4$, and 5.

ANIMATION SEQUENCE

> u(x,0):=subs(t=0,u(x,t)):u(x,1):=subs(t=1,u(x,t)):
> u(x,2):=subs(t=2,u(x,t)):u(x,3):=subs(t=3,u(x,t)):
> u(x,4):=subs(t=4,u(x,t)):u(x,5):=subs(t=5,u(x,t)):
> plot({u(x,0),u(x,1),u(x,2),u(x,3),u(x,4),u(x,5)},x=a..b,thickness=10);

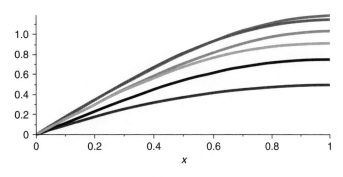

Figure 4.3

EXAMPLE 4.6.3: We again consider the waves describing the longitudinal vibrations in a rigid bar over the finite interval $I = \{x \mid 0 < x < 1\}$ with both ends unsecured. The damping is very small. The initial displacement distribution $f(x)$ and the initial speed distribution $g(x)$ are given following. The wave speed is $c = 1/4$, and the damping factor is $\gamma = 1/5$.

SOLUTION: The homogeneous wave equation is

$$\frac{\partial^2}{\partial t^2} u(x, t) = c^2 \left(\frac{\partial^2}{\partial x^2} u(x, t) \right) - \gamma \left(\frac{\partial}{\partial t} u(x, t) \right)$$

The boundary conditions are type 2 at $x = 0$ and type 2 at $x = 1$

$$u_x(0, t) = 0 \quad \text{and} \quad u_x(1, t) = 0$$

The initial conditions are

$$u(x, 0) = 4x^3 - 6x^2 + 1 \quad \text{and} \quad u_t(x, 0) = 0$$

The ordinary differential equations from the method of separation of variables are

$$\frac{d^2}{dt^2} T(t) + \gamma \left(\frac{d}{dt} T(t) \right) + c^2 \lambda T(t) = 0$$

and

$$\frac{d^2}{dx^2} X(x) + \lambda X(x) = 0 \tag{4.30}$$

The boundary conditions on the spatial equation are

$$X_x(0) = 0 \quad \text{and} \quad X_x(1) = 0$$

Assignment of system parameters

> restart:with(plots):a:=0:b:=1:c:=1/4:unprotect(gamma):gamma:=1/5:

Allowed eigenvalues and orthonormal eigenfunctions are obtained from Example 2.5.4.

For $n = 0$,

> lambda[0]:=0;

$$\lambda_0 := 0 \tag{4.31}$$

> X[0](x):=1/sqrt(b);

$$X_0(x) := 1 \tag{4.32}$$

For $n = 1, 2, 3, \ldots$,

> lambda[n]:=(n*Pi/b)^2;

$$\lambda_n := n^2 \pi^2 \tag{4.33}$$

> X[n](x):=sqrt(2/b)*cos(n*Pi/b*x);X[m](x):=subs(n=m,X[n](x)):

$$X_n(x) := \sqrt{2} \cos(n\pi x) \tag{4.34}$$

Statement of orthonormality with respect to the weight function $w(x) = 1$

> w(x):=1:Int(X[n](x)*X[m](x)*w(x),x=a..b)=delta(n,m);

$$\int_0^1 2\cos(n\pi x)\cos(m\pi x)\, dx = \delta(n, m) \tag{4.35}$$

Time-dependent solution

> T[0](t):=A(0)+B(0)*exp(−gamma*t);

$$T_0(t) := A(0) + B(0)e^{-\frac{1}{5}t} \tag{4.36}$$

For $n = 0$,

> T[n](t):=exp((−gamma/2)*t)*(A(n)*cos(omega[n]*t)+B(n)*sin(omega[n]*t));

$$T_n(t) := e^{-\frac{1}{10}t}(A(n)\cos(\omega_n t) + B(n)\sin(\omega_n t)) \tag{4.37}$$

For $n = 1, 2, 3, \ldots$,

Generalized series terms

> u[0](x,t):=T[0](t)*X[0](x):u[n](x,t):=T[n](t)*X[n](x):

Eigenfunction expansion

> u(x,t):=u[0](x,t)+Sum(u[n](x,t),n=1..infinity);

$$u(x, t) := A(0) + B(0)e^{-\frac{1}{5}t} + \sum_{n=1}^{\infty} e^{-\frac{1}{10}t}(A(n)\cos(\omega_n t) + B(n)\sin(\omega_n t))\sqrt{2}\cos(n\pi x) \tag{4.38}$$

> omega[n]:=sqrt(lambda[n]*c^2−gamma^2/4);

$$\omega_n := \frac{1}{20}\sqrt{25n^2\pi^2 - 4} \tag{4.39}$$

From Section 4.5, the coefficients $A(n)$ and $B(n)$ are to be determined from the inner product of the eigenfunctions and the initial condition functions $u(x, 0) = f(x)$ and $u_t(x, 0) = g(x)$.

> f(x):=4*x^3−6*x^2+1;

$$f(x) := 4x^3 - 6x^2 + 1 \tag{4.40}$$

> g(x):=0;

$$g(x) := 0 \tag{4.41}$$

For $n = 0$,

> A(0):=(Int((f(x)*gamma+g(x))/gamma*X[0](x)*w(x),x=a..b));

$$A(0) := \int_0^1 (4x^3 - 6x^2 + 1)\,dx \tag{4.42}$$

> A(0):=value(%);

$$A(0) := 0 \tag{4.43}$$

> B(0):=Int((−1/gamma)*g(x)*X[0](x)*w(x),x=a..b);

$$B(0) := \int_{0}^{1} 0 \, dx \tag{4.44}$$

> B(0):=value(%);

$$B(0) := 0 \tag{4.45}$$

For $n = 1, 2, 3, \dots,$

> A(n):=eval(Int(f(x)*X[n](x)*w(x),x=a..b));A(n):=expand(value(%)):

$$A(n) := \int_{0}^{1} (4x^3 - 6x^2 + 1)\sqrt{2} \cos(n\pi x) \, dx \tag{4.46}$$

> A(n):=simplify(subs({sin(n*Pi)=0,cos(n*Pi)=(−1)^n}, A(n)));

$$A(n) := -\frac{24\sqrt{2}\,(-1 + (-1)^n)}{n^4 \pi^4} \tag{4.47}$$

> B(n):=eval(Int((f(x)*gamma/2+g(x))*X[n](x)*w(x),x=a..b)/omega[n]);
  B(n):=expand(value(%)):

$$B(n) := \frac{20\left(\int_{0}^{1}\left(\frac{2}{5}x^3 - \frac{3}{5}x^2 + \frac{1}{10}\right)\sqrt{2}\cos(n\pi x)\,dx\right)}{\sqrt{25n^2\pi^2 - 4}} \tag{4.48}$$

> B(n):=simplify(subs({sin(n*Pi)=0,cos(n*Pi)=(−1)^n},B(n)));

$$B(n) := -\frac{48\sqrt{2}\,(-1 + (-1)^n)}{n^4 \pi^4 \sqrt{25n^2\pi^2 - 4}} \tag{4.49}$$

Generalized series terms

> u[0](x,t):=eval(T[0](t)*X[0](x)):u[n](x,t):=eval(T[n](t)*X[n](x)):

Series solution

> u(x,t):=u[0](x,t)+Sum(u[n](x,t),n=1..infinity);

$$u(x, t) := \sum_{n=1}^{\infty} e^{-\frac{1}{10}t} \left(-\frac{24\sqrt{2}(-1+(-1)^n)\cos\left(\frac{1}{20}\sqrt{25n^2\pi^2 - 4}\, t\right)}{n^4\pi^4} \right.$$

$$\left. -\frac{48\sqrt{2}(-1+(-1)^n)\sin\left(\frac{1}{20}\sqrt{25n^2\pi^2 - 4}\, t\right)}{n^4\pi^4\sqrt{25n^2\pi^2 - 4}} \right) \sqrt{2}\cos(n\pi x) \tag{4.50}$$

First few terms of sum

> u(x,t):=u[0](x,t)+sum(u[n](x,t),n=1..3):

ANIMATION

> animate(u(x,t),x=a..b,t=0..20,thickness=3);

The preceding animation command illustrates the spatial-time-dependent solution for $u(x, t)$. Of particular note is that since the bar is initially extended symmetrically and both ends of the bar are unsecured, the subsequent vibrations indicate a symmetric longitudinal oscillation of the bar about its center. The following animation sequence in Figure 4.4 shows snapshots of the animation at times $t = 0, 1, 2, 3, 4$, and 5.

ANIMATION SEQUENCE

> u(x,0):=subs(t=0,u(x,t)):u(x,1):=subs(t=1,u(x,t)):
> u(x,2):=subs(t=2,u(x,t)):u(x,3):=subs(t=3,u(x,t)):
> u(x,4):=subs(t=4,u(x,t)):u(x,5):=subs(t=5,u(x,t)):
> plot({u(x,0),u(x,1),u(x,2),u(x,3),u(x,4),u(x,5)},x=a..b,thickness=10);

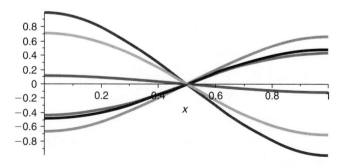

Figure 4.4

EXAMPLE 4.6.4: We seek the wave distribution $u(x, t)$ for longitudinal vibrations in a rigid bar over the finite interval $I = \{x \mid 0 < x < 1\}$. The left end of the bar is secure, and the right end is attached to an elastic hinge. The damping is small, and the bar has an initial displacement distribution $f(x)$ and an initial speed distribution $g(x)$ given as follows. The wave speed is $c = 1/4$, and the damping factor is $\gamma = 1/5$.

SOLUTION: The homogeneous wave equation is

$$\frac{\partial^2}{\partial t^2} u(x, t) = c^2 \left(\frac{\partial^2}{\partial x^2} u(x, t) \right) - \gamma \left(\frac{\partial}{\partial t} u(x, t) \right)$$

The boundary conditions are type 1 at $x = 0$ and type 3 at $x = 1$

$$u(0, t) = 0 \quad \text{and} \quad u_x(1, t) + u(1, t) = 0$$

The initial conditions are

$$u(x, 0) = x - \frac{2x^2}{3} \quad \text{and} \quad u_t(x, 0) = x$$

The ordinary differential equations from the method of separation of variables are

$$\frac{d^2}{dt^2} T(t) + \gamma \left(\frac{d}{dt} T(t) \right) + c^2 \lambda T(t) = 0$$

and

$$\frac{d^2}{dx^2} X(x) + \lambda X(x) = 0 \tag{4.51}$$

The boundary conditions on the spatial equation are

$$X(0) = 0 \quad \text{and} \quad X_x(1) + X(1) = 0$$

Assignment of system parameters

> restart:with(plots):a:=0:b:=1:c:=1/4:unprotect(gamma):gamma:=1/5:

Allowed eigenvalues and orthonormalized eigenfunctions are obtained from Example 2.5.4. The eigenvalues are the roots of the eigenvalue equation

> tan(sqrt(lambda[n]*b))=-sqrt(lambda[n]);

$$\tan\left(\sqrt{\lambda_n}\right) = -\sqrt{\lambda_n} \tag{4.52}$$

For $n = 1, 2, 3, \ldots,$

> X[n](x):=sqrt(2)*(1/sqrt((cos(sqrt(lambda[n]*b))^2+b)))*sin(sqrt(lambda[n])*x);
  X[m](x):=subs(n=m,X[n](x)):

$$X_n(x) := \frac{\sqrt{2} \sin\left(\sqrt{\lambda_n} x\right)}{\sqrt{\cos\left(\sqrt{\lambda_n}\right)^2 + 1}} \tag{4.53}$$

Statement of orthonormality with respect to the weight function $w(x) = 1$

> w(x):=1:Int(X[n](x)*X[m](x)*w(x),x=a..b)=delta(n,m);

$$\int_0^1 \frac{2\sin\left(\sqrt{\lambda_n}x\right)\sin\left(\sqrt{\lambda_m}x\right)}{\sqrt{\cos\left(\sqrt{\lambda_n}\right)^2+1}\sqrt{\cos\left(\sqrt{\lambda_m}\right)^2+1}}\,dx = \delta(n,m) \tag{4.54}$$

Time-dependent solution

> T[n](t):=exp((−gamma/2)*t)*(A(n)*cos(omega[n]*t)+B(n)*sin(omega[n]*t));

$$T_n(t) := e^{-\frac{1}{10}t}(A(n)\cos(\omega_n t) + B(n)\sin(\omega_n t)) \tag{4.55}$$

Generalized series terms

> u[n](x,t):=T[n](t)*X[n](x):

Eigenfunction expansion

> u(x,t):=Sum(u[n](x,t),n=1..infinity);

$$u(x,t) := \sum_{n=1}^{\infty} \frac{e^{-\frac{1}{10}t}(A(n)\cos(\omega_n t) + B(n)\sin(\omega_n t))\sqrt{2}\sin\left(\sqrt{\lambda_n}x\right)}{\sqrt{\cos\left(\sqrt{\lambda_n}\right)^2+1}} \tag{4.56}$$

> omega[n]:=sqrt(lambda[n]*c^2−gamma^2/4);

$$\omega_n := \frac{1}{20}\sqrt{25\lambda_n - 4} \tag{4.57}$$

From Section 4.5, the coefficients $A(n)$ and $B(n)$ are to be determined from the inner product of the eigenfunctions and the initial condition functions $u(x,0) = f(x)$ and $u_t(x,0) = g(x)$.

> f(x):=x−2/3*x^2;

$$f(x) := x - \frac{2}{3}x^2 \tag{4.58}$$

> g(x):=x;

$$g(x) := x \tag{4.59}$$

For $n = 1, 2, 3, \ldots,$

> A(n):=eval(Int(f(x)*X[n](x)*w(x),x=a..b));A(n):=expand(value(%)):

$$A(n) := \int_0^1 \frac{\left(x - \frac{2}{3}x^2\right)\sqrt{2}\sin\left(\sqrt{\lambda_n}x\right)}{\sqrt{\cos\left(\sqrt{\lambda_n}\right)^2+1}}\,dx \tag{4.60}$$

Substitution of the eigenvalue equation simplifies the preceding equation

> A(n):=simplify(subs(sin(sqrt(lambda[n])*b)=−sqrt(lambda[n])*cos(sqrt(lambda[n])*b), A(n)));

$$A(n) := -\frac{4}{3}\frac{\sqrt{2}\left(-1+\cos\left(\sqrt{\lambda_n}\right)\right)}{\lambda_n^{(3/2)}\sqrt{\cos\left(\sqrt{\lambda_n}\right)^2+1}} \tag{4.61}$$

> B(n):=eval(Int((f(x)*gamma/2+g(x))*X[n](x)*w(x),x=a..b)/omega[n]);
  B(n):=expand(value(%)):

$$B(n) := \frac{20\left(\displaystyle\int_0^1 \frac{\left(\frac{11}{10}x-\frac{1}{15}x^2\right)\sqrt{2}\sin\left(\sqrt{\lambda_n}x\right)}{\sqrt{\cos\left(\sqrt{\lambda_n}\right)^2+1}}\,dx\right)}{\sqrt{25\lambda_n-4}} \tag{4.62}$$

Substitution of the eigenvalue equation simplifies this equation

> B(n):=simplify(subs(sin(sqrt(lambda[n])*b)=−sqrt(lambda[n])*cos(sqrt(lambda[n])*b), B(n)));

$$B(n) := -\frac{8}{3}\frac{\sqrt{2}\left(-1+15\cos\left(\sqrt{\lambda_n}\right)\lambda_n+\cos\left(\sqrt{\lambda_n}\right)\right)}{\lambda_n^{(3/2)}\sqrt{\cos\left(\sqrt{\lambda_n}\right)^2+1}\sqrt{25\lambda_n-4}} \tag{4.63}$$

Generalized series terms

> u[n](x,t):=eval(T[n](t)*X[n](x)):

Series solution

> u(x,t):=Sum(u[n](x,t),n=1..infinity);

$$u(x,t) := \sum_{n=1}^{\infty}\frac{1}{\sqrt{\cos\left(\sqrt{\lambda_n}\right)^2+1}}\left(e^{-\frac{1}{10}t}\left(-\frac{4}{3}\frac{\sqrt{2}\left(-1+\cos\left(\sqrt{\lambda_n}\right)\right)\cos\left(\frac{1}{20}\sqrt{25\lambda_n-4}\,t\right)}{\lambda_n^{(3/2)}\sqrt{\cos\left(\sqrt{\lambda_n}\right)^2+1}}\right.\right.$$
$$\left.\left.-\frac{8}{3}\frac{\sqrt{2}\left(-1+15\cos\left(\sqrt{\lambda_n}\right)\lambda_n+\cos\left(\sqrt{\lambda_n}\right)\right)\sin\left(\frac{1}{20}\sqrt{25\lambda_n-4}\,t\right)}{\lambda_n^{(3/2)}\sqrt{\cos\left(\sqrt{\lambda_n}\right)^2+1}\sqrt{25\lambda_n-4}}\right)\sqrt{2}\sin\left(\sqrt{\lambda_n}x\right)\right) \tag{4.64}$$

Evaluation of the eigenvalues from the roots of the eigenvalue equation yields

> tan(sqrt(lambda[n])*b)=−sqrt(lambda[n]);

$$\tan\left(\sqrt{\lambda_n}\right) = -\sqrt{\lambda_n} \tag{4.65}$$

```
> plot({tan(v),−v},v=0..20,y=−20..0,thickness=10);
```

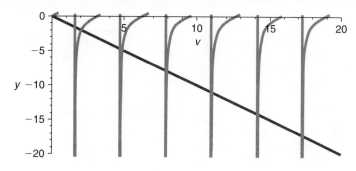

Figure 4.5

If we set $v = \sqrt{\lambda}b$, then the eigenvalues are determined from the squares of the values of v at the intersection points of the curves in Figure 4.5. Evaluation of the first few eigenvalues using the Maple fsolve command yields:

```
> lambda[1]:=(1/b^2)*(fsolve((tan(v)+v),v=1..3)^2);
```

$$\lambda_1 := 4.115858365 \tag{4.66}$$

```
> lambda[2]:=(1/b^2)*(fsolve((tan(v)+v),v=3..6)^2);
```

$$\lambda_2 := 24.13934203 \tag{4.67}$$

```
> lambda[3]:=(1/b^2)*(fsolve((tan(v)+v),v=6..9)^2);
```

$$\lambda_3 := 63.65910654 \tag{4.68}$$

First few terms of sum

```
> u(x,t):=eval(sum(u[n](x,t),n=1..3)):
```

ANIMATION

```
> animate(u(x,t),x=a..b,t=0..20,thickness=3);
```

The preceding animation command illustrates the spatial-time-dependent solution for $u(x, t)$. The following animation sequence in Figure 4.6 shows snapshots of the animation at times $t = 0, 1, 2, 3, 4,$ and 5.

ANIMATION SEQUENCE

```
> u(x,0):=subs(t=0,u(x,t)):u(x,1):=subs(t=1,u(x,t)):
> u(x,2):=subs(t=2,u(x,t)):u(x,3):=subs(t=3,u(x,t)):
```

> u(x,4):=subs(t=4,u(x,t)):u(x,5):=subs(t=5,u(x,t)):
> plot({u(x,0),u(x,1),u(x,2),u(x,3),u(x,4),u(x,5)},x=a..b,thickness=10);

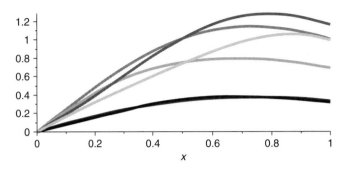

Figure 4.6

4.7 Wave Equation in the Cylindrical Coordinate System

The partial differential equation for wave phenomena in the cylindrical coordinate system for a circularly symmetric system is given below. Here, $u(r, t)$ is the spatial-time-dependent solution

$$\frac{\partial^2}{\partial t^2} u(r, t) = \frac{c^2 \left(\frac{\partial}{\partial r} u(r, t) + r \left(\frac{\partial^2}{\partial r^2} u(r, t) \right) \right)}{r} - \gamma \left(\frac{\partial}{\partial t} u(r, t) \right) + h(r, t)$$

and r is the coordinate radius of the system. See the references for the equivalency of the wave equation from the rectangular cartesian system to the cylindrical coordinate system. There is no angle dependence here because we have assumed circular symmetry, and there is no z dependence because we will be considering regions with no extension along the z-axis.

As for the rectangular coordinate system, we can write the preceding in terms of the linear operator for the wave equation in cylindrical coordinates as

$$L(u) = h(r, t)$$

where the wave operator in cylindrical coordinates reads as

$$L(u) = \frac{\partial^2}{\partial t^2} u(r, t) - \frac{c^2 \left(\frac{\partial}{\partial r} u(r, t) + r \left(\frac{\partial^2}{\partial r^2} u(r, t) \right) \right)}{r} + \gamma \left(\frac{\partial}{\partial t} u(r, t) \right)$$

We now consider the homogeneous version of the partial differential equation:

$$\frac{\partial^2}{\partial t^2} u(r, t) + \gamma \left(\frac{\partial}{\partial t} u(r, t) \right) = \frac{c^2 \left(\frac{\partial}{\partial r} u(r, t) + r \left(\frac{\partial^2}{\partial r^2} u(r, t) \right) \right)}{r}$$

We seek solutions to this partial differential equation over the finite interval $I = \{r \,|\, a < r < b\}$ subject to the nonregular homogeneous boundary conditions

$$|u(a, t)| < \infty$$

and

$$\kappa_1 u(b, t) + \kappa_2 u_r(b, t) = 0$$

with the initial conditions

$$u(r, 0) = f(r)$$

and

$$u_t(r, 0) = g(r)$$

Similar to the procedure for rectangular coordinates, we solve the partial differential equation using the method of separation of variables. We set

$$u(r, t) = R(r)T(t)$$

Substituting this into the preceding homogeneous partial differential equation yields

$$R(r)\left(\frac{d^2}{dt^2}T(t) + \gamma\left(\frac{d}{dt}T(t)\right)\right) = \frac{c^2 T(t)\left(\frac{d}{dr}R(r) + r\left(\frac{d^2}{dr^2}R(r)\right)\right)}{r}$$

Dividing both sides by the product solution, we get

$$\frac{\frac{d^2}{dt^2}T(t) + \gamma\left(\frac{d}{dt}T(t)\right)}{c^2 T(t)} = \frac{\frac{d}{dr}R(r) + r\left(\frac{d^2}{dr^2}R(r)\right)}{R(r)r}$$

Because the left-hand side of this equation is an exclusive function of t and the right-hand side an exclusive function of r, and t and r are independent, then the only way we can ensure equality of the preceding for all r and t is to set each side equal to a constant.

Doing so, we arrive at the following two ordinary differential equations in terms of the separation constant λ^2:

$$\frac{d^2}{dt^2}T(t) + \gamma\left(\frac{\partial}{\partial t}T(r, t)\right) + \lambda^2 c^2 T(t) = 0$$

and

$$\frac{d^2}{dr^2}R(r) + \frac{\frac{d}{dr}R(r)}{r} + \lambda^2 R(r) = 0$$

The preceding differential equation in t is an ordinary second-order linear differential equation for which we already have the solution from Chapter 1.

The second differential equation in the variable r is recognized from Section 1.11 as being an ordinary Bessel differential equation. The solution of this equation is the Bessel function of the first kind of order zero.

We noted in Section 2.6 that, for the Bessel differential equation, the point $r = 0$ is a regular singular point of the differential equation. With appropriate boundary conditions over an interval that includes the origin, we arrive at a "nonregular" (singular) type Sturm-Liouville eigenvalue problem whose eigenfunctions form an orthogonal set.

Similar to regular Sturm-Liouville problems over finite intervals, an infinite number of eigenvalues exist that can be indexed by the positive integers n. The indexed eigenvalues and corresponding eigenfunctions are given, respectively, as

$$\lambda_n, R_n(r)$$

for $n = 0, 1, 2, 3, \ldots$.

The eigenfunctions form a "complete" set with respect to any piecewise smooth function over the finite interval $I = \{r \mid a < r < b\}$. In Section 2.6, we examined the nature of the orthogonality of the Bessel functions, and we showed the eigenfunctions to be orthogonal with respect to the weight function $w(r) = r$ over the finite interval I. Further, the eigenfunctions can be normalized, and the corresponding statement of orthonormality reads

$$\int_a^b R_n(r) R_m(r) r \, dr = \delta(n, m)$$

where the term on the right is the familiar Kronecker delta function.

Using arguments similar to that used for the regular Sturm-Liouville problem, we can write our general solution to the partial differential equation as the superposition of the products of the solutions to each of the ordinary differential equations given earlier.

We now consider the time-dependent differential equation, which reads

$$\frac{d^2}{dt^2} T(t) + \gamma \left(\frac{\partial}{\partial t} T(r, t) \right) + \lambda^2 c^2 T(t) = 0$$

For the indexed values of λ, the solution to this equation is

$$T_n(t) = e^{-\frac{\gamma t}{2}} \left(A(n) \cos(\omega_n t) + B(n) \sin(\omega_n t) \right)$$

where the n-th term damped angular frequency ω_n is given in terms of the indexed λ_n as

$$\omega_n = \frac{\sqrt{4\lambda_n^2 c^2 - \gamma^2}}{2}$$

for $n = 0, 1, 2, 3, \ldots$. In the preceding equation, the coefficients $A(n)$ and $B(n)$ are unknown arbitrary constants and we have assumed small damping in the system—that is,

$$\gamma^2 < 4\lambda_n^2 c^2$$

Thus, by the method of separation of variables, we arrive at an infinite number of indexed solutions $u_n(r, t)$ $(n = 0, 1, 2, 3, \ldots)$ for the homogeneous wave partial differential equation, over a finite interval, given as

$$u_n(r, t) = R_n(r) e^{-\frac{\gamma t}{2}} (A(n) \cos(\omega_n t) + B(n) \sin(\omega_n t))$$

Because the differential operator is linear, then any superposition of solutions to the homogeneous equation is also a solution to the problem. Thus, we can write the general solution to the homogeneous partial differential equation as the infinite sum

$$u(r, t) = \sum_{n=0}^{\infty} R_n(r) e^{-\frac{\gamma t}{2}} (A(n) \cos(\omega_n t) + B(n) \sin(\omega_n t))$$

We demonstrate the preceding concepts with an example wave problem in cylindrical coordinates.

DEMONSTRATION: We seek the wave distribution $u(r, t)$ for transverse vibrations in a thin circularly symmetric membrane over the interval $I = \{r \mid 0 < r < 1\}$, which is secure at the periphery. The membrane vibrates in a medium with a small amount of damping. The membrane has an initial displacement distribution $f(r)$ and an initial speed distribution $g(r)$ given as follows. The wave speed is $c = 1/5$, and the damping factor is $\gamma = 2/5$.

SOLUTION: The homogeneous wave equation is

$$\frac{\partial^2}{\partial t^2} u(r, t) + \frac{2\left(\frac{\partial}{\partial t} u(r, t)\right)}{5} = \frac{\frac{\partial}{\partial r} u(r, t) + r\left(\frac{\partial^2}{\partial r^2} u(r, t)\right)}{25r}$$

The boundary conditions are type 1 at $r = 1$, and we require a finite solution at the origin

$$|u(0, t)| < \infty \quad \text{and} \quad u(1, t) = 0$$

The initial conditions are

$$u(r, 0) = f(r) \quad \text{and} \quad u_t(r, 0) = g(r)$$

From the method of separation of variables, the ordinary differential equations are

$$\frac{d^2}{dt^2} T(t) + \frac{2\left(\frac{d}{dt} T(t)\right)}{5} + \frac{\lambda^2 T(t)}{25} = 0$$

and

$$\frac{d^2}{dr^2} R(r) + \frac{\frac{d}{dr} R(r)}{r} + \lambda^2 R(r) = 0 \tag{4.69}$$

We first consider the spatial differential equation in r. This is a Bessel-type differential equation of the first kind of order zero with boundary conditions

$$|R(0)| < \infty \quad \text{and} \quad R(1) = 0$$

This same problem was considered in Example 2.6.2 in Chapter 2. The allowed eigenvalues are the roots of the eigenvalue equation

$$J(0, \lambda_n) = 0$$

and the corresponding orthonormal eigenfunctions are

$$R_n(r) = \frac{\sqrt{2} J(0, \lambda_n r)}{J(1, \lambda_n)}$$

for $n = 1, 2, 3, \ldots$. The functions $J(0, \lambda_n r)$ and $J(1, \lambda_n r)$ are Bessel functions of the first kind of order zero and one, respectively.

The corresponding statement of orthonormality with respect to the weight function $w(r) = r$ over the interval I is

$$\int_0^1 \frac{2 J(0, \lambda_n r) J(0, \lambda_m r) r}{J(1, \lambda_n) J(1, \lambda_m)} \, dr = \delta(n, m)$$

for $n = 1, 2, 3, \ldots$, and $m = 1, 2, 3, \ldots$.

We next consider the time-dependent differential equation. This is a second-order differential equation that we solved in Section 1.4. The solution, for the allowed values of λ given earlier, reads

$$T_n(t) = e^{-\frac{t}{5}} (A(n) \cos(\omega_n t) + B(n) \sin(\omega_n t))$$

where

$$\omega_n = \frac{\sqrt{\lambda_n^2 - 1}}{5}$$

Thus, the eigenfunction expansion solution to the problem reads

$$u(r, t) = \sum_{n=1}^{\infty} \frac{e^{-\frac{t}{5}} (A(n) \cos(\omega_n t) + B(n) \sin(\omega_n t)) \sqrt{2} J(0, \lambda_n r)}{J(1, \lambda_n)}$$

The unknown coefficients $A(n)$ and $B(n)$ are to be determined from the initial conditions imposed on the problem.

4.8 Initial Conditions for the Wave Equation in Cylindrical Coordinates

We now consider the initial conditions on the problem. At time $t = 0$ we have the initial displacement

$$u(r, 0) = f(r)$$

and the initial amplitude speed

$$u_t(r, 0) = g(r)$$

Substituting the first condition into the general solution

$$u(r, t) = \sum_{n=0}^{\infty} R_n(r) e^{-\frac{\gamma t}{2}} (A(n) \cos(\omega_n t) + B(n) \sin(\omega_n t))$$

at time $t = 0$ yields

$$f(r) = \sum_{n=0}^{\infty} R_n(r) A(n)$$

Substituting the second condition into the time derivative of the preceding at time $t = 0$ yields

$$g(r) = \sum_{n=0}^{\infty} R_n(r) \left(B(n) \omega_n - \frac{A(n) \gamma}{2} \right)$$

Because the eigenfunctions of this nonregular Sturm-Liouville problem form a complete set with respect to piecewise smooth functions over the finite interval $I = \{r \mid a < r < b\}$, then both of the preceding equations are the generalized Fourier series expansions of the functions $f(r)$ and $g(r)$, respectively, in terms of the spatial eigenfunctions of the system. We now evaluate the coefficients $A(n)$ and $B(n)$.

As we did in Section 2.3 for the generalized Fourier series expansion, we take the inner product of both sides of the preceding equations, with respect to the weight function $w(r) = r$. Operating on the first equation and using the statement of orthonormality, we get

$$\int_a^b f(r) R_m(r) r \, dr = \sum_{n=0}^{\infty} A(n) \delta(n, m)$$

From the mathematical character of the Kronecker delta function, only one term ($n = m$) survives in the preceding sum, and we are able to extract the value

$$A(m) = \int_a^b f(r) R_m(r) r \, dr$$

In a similar manner, operating on the second equation and taking the inner product of both sides and using the statement of orthonormality, we get

$$\int_a^b g(r) R_m(r) r \, dr = \sum_{n=0}^{\infty} \left(B(n)\omega_n - \frac{A(n)\gamma}{2} \right) \delta(n, m)$$

Evaluating the sum (only the $n = m$ term survives) and inserting the preceding known value for $A(m)$ yields

$$B(m) = \frac{\int_a^b \left(\frac{f(r)\gamma}{2} + g(r) \right) R_m(r) r \, dr}{\omega_m}$$

Thus, we can write the final generalized solution to the wave equation in one dimension, subject to the preceding homogeneous boundary and initial conditions, as

$$u(r, t) = \sum_{n=0}^{\infty} R_n(r) e^{-\frac{\gamma t}{2}} \left(\left(\int_a^b f(r) R_n(r) r dr \right) \cos(\omega_n t) + \frac{\left(\int_a^b \left(\frac{f(r)\gamma}{2} + g(r) \right) R_n(r) r dr \right) \sin(\omega_n t)}{\omega_n} \right)$$

$$(4.70)$$

All of the preceding operations are based on the assumption that the infinite series are uniformly convergent, and the formal interchange between the operator and the summation is legitimate. It can be shown that if the functions $f(r)$ and $g(r)$ satisfy the same boundary conditions as the eigenfunctions, then the preceding series are, indeed, uniformly convergent.

DEMONSTRATION: We now provide a demonstration of these concepts for the example problem given in Section 4.8 for the case where the initial conditions are

$$f(r) = 1 - r^2 \quad \text{and} \quad g(r) = 1 - r^2$$

SOLUTION: For the coefficient $A(n)$, we have

$$A(n) = \frac{\int_0^1 (1 - r^2)\sqrt{2} J(0, \lambda_n r) r \, dr}{J(1, \lambda_n)}$$

which evaluates to

$$A(n) = \frac{4\sqrt{2}}{\lambda_n^3}$$

For the coefficient $B(n)$, we have

$$B(n) = \frac{\int_0^1 6\sqrt{2}(1-r)^2 J(0, \lambda_n r) r \, dr}{J(1, \lambda_n)\sqrt{\lambda_n^2 - 1}}$$

which evaluates to

$$B(n) = \frac{24\sqrt{2}}{\lambda_n^3 \sqrt{\lambda_n^2 - 1}}$$

for $n = 1, 2, 3, \ldots$. Thus, the final series solution to our problem reads

$$u(r, t) = \sum_{n=1}^{\infty} \frac{8e^{-\frac{t}{5}}\left(\sqrt{\lambda_n^2 - 1}\cos\left(\frac{\sqrt{\lambda_n^2-1}\,t}{5}\right) + 6\sin\left(\frac{\sqrt{\lambda_n^2-1}\,t}{5}\right)\right) J(0, \lambda_n r)}{\lambda_n^3 \sqrt{\lambda_n^2 - 1}\, J(1, \lambda_n)}$$

The detailed development of the solution to this problem and its graphics are given in one of the Maple worksheet examples given later.

4.9 Example Wave Equation Problems in Cylindrical Coordinates

We now consider several examples of partial differential equations for wave phenomena under various homogeneous boundary conditions over finite intervals in the cylindrical coordinate system. We note that all the spatial ordinary differential equations are of the Bessel type of order zero and the weight function is $w(r) = r$.

EXAMPLE 4.9.1: We seek the wave distribution $u(r, t)$ for transverse vibrations in a thin circularly symmetric membrane over the interval $I = \{r \mid 0 < r < 1\}$ that is secure at the periphery. The membrane vibrates in a medium with a small amount of damping. The membrane has an initial displacement distribution $f(r)$ and an initial speed distribution $g(r)$ given as follows. The wave speed is $c = 1/5$, and the damping factor is $\gamma = 2/5$.

SOLUTION: The homogeneous wave equation is

$$\frac{\partial^2}{\partial t^2}u(r, t) + \gamma\left(\frac{\partial}{\partial t}u(r, t)\right) = \frac{c^2\left(\frac{\partial}{\partial r}u(r, t) + r\left(\frac{\partial^2}{\partial r^2}u(r, t)\right)\right)}{r}$$

The boundary conditions are type 1 at $r = 1$ and a finite solution at $r = 0$

$$|u(0, t)| < \infty \quad \text{and} \quad u(1, t) = 0$$

The initial conditions are

$$u(r, 0) = 1 - r^2 \quad \text{and} \quad u_t(r, 0) = 1 - r^2$$

The ordinary differential equations from the method of separation of variables are

$$\frac{d^2}{dt^2} T(t) + \gamma \left(\frac{d}{dt} T(t) \right) + c^2 \lambda^2 T(t) = 0$$

and

$$\frac{d^2}{dr^2} R(r) + \frac{\frac{d}{dr} R(r)}{r} + \lambda^2 R(r) = 0 \tag{4.71}$$

The boundary conditions on the spatial equation are

$$|R(0)| < \infty \quad \text{and} \quad R(1) = 0$$

Assignment of system parameters

> restart:with(plots):a:=0:b:=1:c:=1/5:unprotect(gamma):gamma:=2/5:

Allowed eigenvalues and orthonormalized eigenfunctions are obtained from Example 2.6.2. The eigenvalues are the roots of the eigenvalue equation

> BesselJ(0,lambda[n]*b)=0;

$$\text{BesselJ}(0, \lambda_n) = 0 \tag{4.72}$$

for $n = 1, 2, 3, \ldots$.

> R[n](r):=simplify(BesselJ(0,lambda[n]*r)/sqrt(int(BesselJ(0,lambda[n]*r)^2*r,r=a..b)));

$$R_n(r) := \frac{\text{BesselJ}(0, \lambda_n r)\sqrt{2}}{\sqrt{\text{BesselJ}(0, \lambda_n)^2 + \text{BesselJ}(1, \lambda_n)^2}} \tag{4.73}$$

Substitution of the eigenvalue equation simplifies the preceding equation

> R[n](r):=radsimp(subs(BesselJ(0,lambda[n]*b)=0,R[n](r)));R[m](r):=subs(n=m,R[n](r)):

$$R_n(r) := \frac{\text{BesselJ}(0, \lambda_n r)\sqrt{2}}{\text{BesselJ}(1, \lambda_n)} \tag{4.74}$$

Statement of orthonormality with respect to the weight function $w(r) = 1$

> w(r):=r:Int(R[n](r)*R[m](r)*w(r),r=a..b)=delta(n,m);

$$\int_0^1 \frac{2\,\text{BesselJ}(0, \lambda_n r)\text{BesselJ}(0, \lambda_m r)r}{\text{BesselJ}(1, \lambda_n)\text{BesselJ}(1, \lambda_m)}\,dr = \delta(n, m) \tag{4.75}$$

Time-dependent solution

> T[n](t):=exp((−gamma/2)*t)*(A(n)*cos(omega[n]*t)+B(n)*sin(omega[n]*t));

$$T_n(t) := e^{-\frac{1}{5}t}(A(n)\cos(\omega_n t) + B(n)\sin(\omega_n t)) \tag{4.76}$$

Generalized series terms

> u[n](r,t):=T[n](t)*R[n](r):

Eigenfunction expansion

> u(r,t):=Sum(u[n](r,t),n=1..infinity);

$$u(r,t) := \sum_{n=1}^{\infty} \frac{e^{-\frac{1}{5}t}(A(n)\cos(\omega_n t) + B(n)\sin(\omega_n t)) \; \text{BesselJ}(0, \lambda_n r)\sqrt{2}}{\text{BesselJ}(1, \lambda_n)} \tag{4.77}$$

> omega[n]:=sqrt(lambda[n]^2*c^2−gamma^2/4);

$$\omega_n := \frac{1}{5}\sqrt{\lambda_n^2 - 1} \tag{4.78}$$

From Section 4.8, the coefficients $A(n)$ and $B(n)$ are to be determined from the inner product of the eigenfunctions and the initial condition functions $u(r,0) = f(r)$ and $u_t(r,0) = g(r)$.

> f(r):=1−r^2;

$$f(r) := 1 - r^2 \tag{4.79}$$

> g(r):=1−r^2;

$$g(r) := 1 - r^2 \tag{4.80}$$

For $n = 1, 2, 3, \ldots,$

> A(n):=eval(Int(f(r)*R[n](r)*w(r),r=a..b));A(n):=value(%):

$$A(n) := \int_0^1 \frac{(1-r^2)\,\text{BesselJ}(0, \lambda_n r)\sqrt{2}\,r}{\text{BesselJ}(1, \lambda_n)}\,dr \tag{4.81}$$

Substitution of the eigenvalue equation simplifies the preceding equation

> A(n):=simplify(subs(BesselJ(0,lambda[n]*b)=0,A(n)));

$$A(n) := \frac{4\sqrt{2}}{\lambda_n^3} \tag{4.82}$$

> B(n):=eval(Int((f(r)*gamma/2+g(r))*R[n](r)*w(r),r=a..b))/omega[n];B(n):=value(%):

$$B(n) := \frac{5\left(\displaystyle\int_0^1 \frac{\left(\frac{6}{5}-\frac{6}{5}r^2\right)\,\mathrm{BesselJ}(0,\lambda_n r)\sqrt{2}\,r}{\mathrm{BesselJ}(1,\lambda_n)}\,\mathrm{d}r\right)}{\sqrt{\lambda_n^2-1}}$$

(4.83)

Substitution of the eigenvalue equation simplifies the preceding equation

> B(n):=simplify(subs(BesselJ(0,lambda[n]*b)=0,B(n)));

$$B(n) := \frac{24\sqrt{2}}{\lambda_n^3\sqrt{\lambda_n^2-1}}$$

(4.84)

Generalized series terms

> u[n](r,t):=simplify(eval(T[n](t)*R[n](r))):

Series solution

> u(r,t):=Sum(u[n](r,t),n=1..infinity);

$$u(r,t) := \sum_{n=1}^{\infty} \frac{8\,e^{-\frac{1}{5}t}\,\mathrm{BesselJ}(0,\lambda_n r)\left(\cos\left(\frac{1}{5}\sqrt{\lambda_n^2-1}\,t\right)\sqrt{\lambda_n^2-1}+6\sin\left(\frac{1}{5}\sqrt{\lambda_n^2-1}\,t\right)\right)}{\mathrm{BesselJ}(1,\lambda_n)\lambda_n^3\sqrt{\lambda_n^2-1}}$$

(4.85)

Evaluation of the eigenvalues from the roots of the eigenvalue equation yields

> BesselJ(0,lambda[n]*b)=0;

$$\mathrm{BesselJ}(0,\lambda_n)=0$$

(4.86)

> plot(BesselJ(0,v),v=0..20,thickness=10);

If we set $v=\lambda b$, then the eigenvalues are determined from the intersection points of the curve with the v-axis shown in Figure 4.7. Evaluation of a few of the eigenvalues using the Maple fsolve command yields

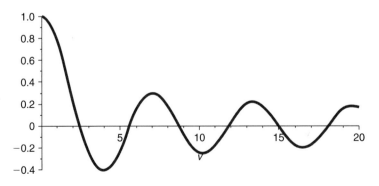

Figure 4.7

```
> lambda[1]:=(1/b)*fsolve(BesselJ(0,v)=0,v=0..3);
```

$$\lambda_1 := 2.404825558 \qquad\qquad (4.87)$$

```
> lambda[2]:=(1/b)*fsolve(BesselJ(0,v)=0,v=3..6);
```

$$\lambda_2 := 5.520078110 \qquad\qquad (4.88)$$

```
> lambda[3]:=(1/b)*fsolve(BesselJ(0,v)=0,v=6..9);
```

$$\lambda_3 := 8.653727913 \qquad\qquad (4.89)$$

First few terms in sum

```
> u(r,t):=eval(sum(u[n](r,t),n=1..3)):
```

ANIMATION

```
> animate(u(r,t),r=a..b,t=0..2,thickness=3);
```

The preceding animation command illustrates the spatial-time-dependent solution for $u(r, t)$. The following animation sequence in Figure 4.8 shows snapshots of the animation at times $t = 0, 1, 2, 3, 4$, and 5.

ANIMATION SEQUENCE

```
> u(r,0):=subs(t=0,u(r,t)):u(r,1):=subs(t=1,u(r,t)):
> u(r,2):=subs(t=2,u(r,t)):u(r,3):=subs(t=3,u(r,t)):
> u(r,4):=subs(t=4,u(r,t)):u(r,5):=subs(t=5,u(r,t)):
> plot({u(r,0),u(r,1),u(r,2),u(r,3),u(r,4),u(r,5)},r=a..b,thickness=10);
```

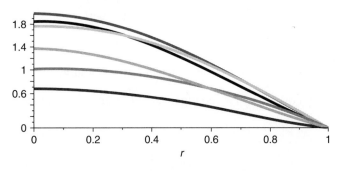

Figure 4.8

THREE-DIMENSIONAL ANIMATION

```
> u(x,y,t):=eval(subs(r=sqrt(x^2+y^2),u(r,t))):
> u(x,y,t):=(u(x,y,t))*Heaviside(1−sqrt(x^2+y^2)):
> animate3d(u(x,y,t),x=−b..b,y=−b..b,t=0..10,axes=framed,thickness=1);
```

EXAMPLE 4.9.2: We consider transverse vibrations in a thin circularly symmetric membrane over the interval $I = \{r \mid 0 < r < 1\}$, which is unsecured at the periphery. The membrane vibrates in a medium with very small damping. The membrane has an initial displacement distribution $f(r)$ and an initial speed distribution $g(r)$ given as follows. The wave speed is $c = 1/5$, and the damping factor is $\gamma = 2/5$.

SOLUTION: The homogeneous wave equation is

$$\frac{\partial^2}{\partial t^2} u(r, t) + \gamma \left(\frac{\partial}{\partial t} u(r, t) \right) = \frac{c^2 \left(\frac{\partial}{\partial r} u(r, t) + r \left(\frac{\partial^2}{\partial r^2} u(r, t) \right) \right)}{r}$$

The boundary conditions are type 2 at $r = 1$ and a finite solution at $r = 0$

$$|u(0, t)| < \infty \quad \text{and} \quad u_r(1, t) = 0$$

The initial conditions are

$$u(r, 0) = 1 - r^2 \quad \text{and} \quad u_t(r, 0) = 1 - r^2$$

The ordinary differential equations from the method of separation of variables are

$$\frac{d^2}{dt^2} T(t) + \gamma \left(\frac{d}{dt} T(t) \right) + c^2 \lambda^2 T(t) = 0$$

and

$$\frac{d^2}{dr^2} R(r) + \frac{\frac{d}{dr} R(r)}{r} + \lambda^2 R(r) = 0 \tag{4.90}$$

The boundary conditions on the spatial equation are

$$|R(0)| < \infty \quad \text{and} \quad R_r(1) = 0$$

Assignment of system parameters

> restart:with(plots):a:=0:b:=1:c:=1/5:unprotect(gamma):gamma:=2/5:

Allowed eigenvalues and orthonormalized eigenfunctions are obtained from Example 2.6.3.

> lambda[0]:=0;

$$\lambda_0 := 0 \tag{4.91}$$

for $n = 0$.

> R[0](r):=sqrt(2)/b;

$$R_0(r) := \sqrt{2} \tag{4.92}$$

For $n = 1, 2, 3, \ldots$, the eigenvalues are the roots of the eigenvalue equation

> subs(r=b,diff(BesselJ(0,lambda[n]*r),r)=0);

$$-\text{BesselJ}(1, \lambda_n)\lambda_n = 0 \tag{4.93}$$

> R[n](r):=simplify(BesselJ(0,lambda[n]*r)/sqrt(int(BesselJ(0,lambda[n]*r)^2*r,r=a..b)));

$$R_n(r) := \frac{\text{BesselJ}(0, \lambda_n r)\sqrt{2}}{\sqrt{\text{BesselJ}(0, \lambda_n)^2 + \text{BesselJ}(1, \lambda_n)^2}} \tag{4.94}$$

Substitution of the eigenvalue equation simplifies the preceding equation

> R[n](r):=radsimp(subs(BesselJ(1,lambda[n]*b)=0,R[n](r)));R[m](r):=subs(n=m,R[n](r)):

$$R_n(r) := \frac{\text{BesselJ}(0, \lambda_n r)\sqrt{2}}{\text{BesselJ}(0, \lambda_n)} \tag{4.95}$$

Statement of orthonormality with respect to the weight function $w(r) = r$

> w(r):=r:Int(R[n](r)*R[m](r)*w(r),r=a..b)=delta(n,m);

$$\int_0^1 \frac{2\,\text{BesselJ}(0, \lambda_n r)\,\text{BesselJ}(0, \lambda_m r)r}{\text{BesselJ}(0, \lambda_n)\,\text{BesselJ}(0, \lambda_m)}\,dr = \delta(n, m) \tag{4.96}$$

Time-dependent solution

> T[0](t):=A(0)+B(0)*exp(−gamma*t);

$$T_0(t) := A(0) + B(0)\mathrm{e}^{-\frac{2}{5}t} \tag{4.97}$$

for $n = 0$.

> T[n](t):=exp((−gamma/2)*t)*(A(n)*cos(omega[n]*t)+B(n)*sin(omega[n]*t));

$$T_n(t) := \mathrm{e}^{-\frac{1}{5}t}(A(n)\cos(\omega_n t) + B(n)\sin(\omega_n t)) \tag{4.98}$$

for $n = 1, 2, 3, \ldots$.

Generalized series terms

> u[0](r,t):=eval(T[0](t)*R[0](r)):u[n](r,t):=eval(T[n](t)*R[n](r)):

Eigenfunction expansion

> u(r,t):=u[0](r,t)+Sum(u[n](r,t),n=1..infinity);

$$u(r, t) := \left(A(0) + B(0)\mathrm{e}^{-\frac{2}{5}t} \right)\sqrt{2}$$
$$+ \sum_{n=1}^{\infty} \frac{\mathrm{e}^{-\frac{1}{5}t}(A(n)\cos(\omega_n t) + B(n)\sin(\omega_n t))\,\text{BesselJ}(0, \lambda_n r)\sqrt{2}}{\text{BesselJ}(0, \lambda_n)} \tag{4.99}$$

> omega[n]:=sqrt(lambda[n]^2*c^2−gamma^2/4);

$$\omega_n := \frac{1}{5}\sqrt{\lambda_n^2 - 1} \tag{4.100}$$

From Section 4.8, the coefficients $A(n)$ and $B(n)$ are to be determined from the inner product of the eigenfunctions and the initial condition functions $u(r, 0) = f(r)$ and $u_t(r, 0) = g(r)$.

> f(r):=1−r^2;

$$f(r) := 1 - r^2 \tag{4.101}$$

> g(r):=1−r^2;

$$g(r) := 1 - r^2 \tag{4.102}$$

For $n = 0$,

> A(0):=Int((f(r)*gamma+g(r))/gamma*R[0](r)*w(r),r=a..b);

$$A(0) := \int_0^1 \frac{5}{2}\left(\frac{7}{5} - \frac{7}{5}r^2\right)\sqrt{2}\,r\,dr \tag{4.103}$$

> A(0):=value(%);

$$A(0) := \frac{7}{8}\sqrt{2} \tag{4.104}$$

> B(0):=Int((−1/gamma)*g(r)*R[0](r)*w(r),r=a..b);

$$B(0) := \int_0^1 \left(-\frac{5}{2}(1 - r^2)\sqrt{2}\,r\right)dr \tag{4.105}$$

> B(0):=value(%);

$$B(0) := -\frac{5}{8}\sqrt{2} \tag{4.106}$$

For $n = 1, 2, 3, \ldots$,

> A(n):=Int(f(r)*R[n](r)*w(r),r=a..b);

$$A(n) := \int_0^1 \frac{(1 - r^2)\,\text{BesselJ}(0, \lambda_n r)\sqrt{2}\,r}{\text{BesselJ}(0, \lambda_n)}\,dr \tag{4.107}$$

Substitution of the eigenvalue equation simplifies the preceding equation

> A(n):=subs(BesselJ(1,lambda[n]*b)=0,value(%));

$$A(n) := -\frac{2\sqrt{2}}{\lambda_n^2} \tag{4.108}$$

> B(n):=(Int((f(r)*gamma/2+g(r))*R[n](r)*w(r),r=a..b))/omega[n];

$$B(n) := \frac{5\left(\displaystyle\int_0^1 \frac{\left(\frac{6}{5}-\frac{6}{5}r^2\right)\,\text{BesselJ}(0,\lambda_n r)\sqrt{2}\,r}{\text{BesselJ}(0,\lambda_n)}\,dr\right)}{\sqrt{\lambda_n^2-1}} \tag{4.109}$$

Substitution of the eigenvalue equation simplifies the preceding equation

> B(n):=subs(BesselJ(1,lambda[n]*b)=0,value(%));

$$B(n) := -\frac{12\sqrt{2}}{\lambda_n^2\sqrt{\lambda_n^2-1}} \tag{4.110}$$

Generalized series terms

> u[0](r,t):=simplify(eval(T[0](t)*R[0](r))):u[n](r,t):=simplify(eval(T[n](t)*R[n](r))):

Series solution

> u(r,t):=u[0](r,t)+Sum(u[n](r,t),n=1..infinity);

$$u(r,t) := \frac{7}{4} - \frac{5}{4}e^{-\frac{2}{5}t} \tag{4.111}$$

$$+ \sum_{n=1}^{\infty}\left(-\frac{4e^{-\frac{1}{5}t}\,\text{BesselJ}(0,\lambda_n r)\left(\cos\left(\frac{1}{5}\sqrt{\lambda_n^2-1}\,t\right)\sqrt{\lambda_n^2-1}+6\sin\left(\frac{1}{5}\sqrt{\lambda_n^2-1}\,t\right)\right)}{\text{BesselJ}(0,\lambda_n)\,\lambda_n^2\sqrt{\lambda_n^2-1}}\right)$$

Evaluation of the eigenvalues from the roots of the eigenvalue equation yields

> BesselJ(1,lambda[n]*b)=0;

$$\text{BesselJ}(1,\lambda_n) = 0 \tag{4.112}$$

> plot(BesselJ(1,v),v=0..20,thickness=10);

If we set $v = \lambda b$, then the eigenvalues are found from the intersection points of the curve with the v-axis shown in Figure 4.9. Evaluation of a few of these eigenvalues using the Maple fsolve command yields

> lambda[1]:=(1/b)*fsolve(BesselJ(1,v)=0,v=1..4);

$$\lambda_1 := 3.831705970 \tag{4.113}$$

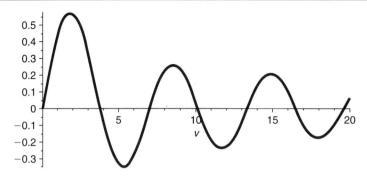

Figure 4.9

> lambda[2]:=(1/b)*fsolve(BesselJ(1,v)=0,v=4..8);

$$\lambda_2 := 7.015586670 \qquad (4.114)$$

> lambda[3]:=(1/b)*fsolve(BesselJ(1,v)=0,v=8..12);

$$\lambda_3 := 10.17346814 \qquad (4.115)$$

First few terms in sum

> u(r,t):=u[0](r,t)+eval(sum(u[n](r,t),n=1..1)):

ANIMATION

> animate(u(r,t),r=a..b,t=0..5,thickness=3);

The preceding animation command illustrates the spatial-time-dependent solution for $u(r, t)$. The following animation sequence in Figure 4.10 shows snapshots of the animation at times $t = 0, 1, 2, 3, 4,$ and 5.

ANIMATION SEQUENCE

> u(r,0):=subs(t=0,u(r,t)):u(r,1):=subs(t=1,u(r,t)):
> u(r,2):=subs(t=2,u(r,t)):u(r,3):=subs(t=3,u(r,t)):
> u(r,4):=subs(t=4,u(r,t)):u(r,5):=subs(t=5,u(r,t)):
> plot({u(r,0),u(r,1),u(r,2),u(r,3),u(r,4),u(r,5)},r=a..b,thickness=10);

THREE-DIMENSIONAL ANIMATION

> u(x,y,t):=eval(subs(r=sqrt(x^2+y^2),u(r,t))):
> u(x,y,t):=u(x,y,t)*Heaviside(1−sqrt(x^2+y^2)):
> animate3d(u(x,y,t),x=−b..b,y=−b..b,t=0..10,axes=framed,thickness=1);

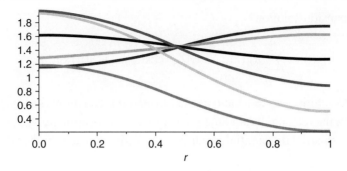

Figure 4.10

Chapter Summary

Nonhomogeneous wave equation of dimension one in rectangular coordinates

$$\frac{\partial^2}{\partial t^2}u(x,t) = c^2\left(\frac{\partial^2}{\partial x^2}u(x,t)\right) - \gamma\left(\frac{\partial}{\partial t}u(x,t)\right) + h(x,t)$$

Linear wave operator in rectangular coordinates of dimension one

$$L(u) = \frac{\partial^2}{\partial t^2}u(x,t) - c^2\left(\frac{\partial^2}{\partial x^2}u(x,t)\right) + \gamma\left(\frac{\partial}{\partial t}u(x,t)\right)$$

Method of separation of variables solution

$$u(x,t) = X(x)T(t)$$

Eigenfunction expansion solution

$$u(x,t) = \sum_{n=0}^{\infty} X_n(x)e^{-\frac{\gamma t}{2}}\left(A(n)\cos(\omega_n t) + B(n)\sin(\omega_n t)\right)$$

Nonhomogeneous wave equation of dimension one in cylindrical coordinates

$$\frac{\partial^2}{\partial t^2}u(r,t) = \frac{c^2\left(\frac{\partial}{\partial r}u(r,t) + r\left(\frac{\partial^2}{\partial r^2}u(r,t)\right)\right)}{r} - \gamma\left(\frac{\partial}{\partial t}u(r,t)\right) + h(r,t)$$

Linear wave operator in cylindrical coordinates of dimension one

$$L(u) = \frac{\partial^2}{\partial t^2}u(r,t) - \frac{c^2\left(\frac{\partial}{\partial r}u(r,t) + r\left(\frac{\partial^2}{\partial r^2}u(r,t)\right)\right)}{r} + \gamma\left(\frac{\partial}{\partial t}u(r,t)\right)$$

Method of separation of variables solution

$$u(r,t) = R(r)T(t)$$

Eigenfunction expansion solution

$$u(r, t) = \sum_{n=0}^{\infty} R_n(r) e^{-\frac{\gamma t}{2}} (A(n) \cos(\omega_n t) + B(n) \sin(\omega_n t))$$

We have examined solutions to the wave partial differential equation in a single spatial dimension for both the rectangular and the polar-cylindrical coordinate systems. We will examine these same partial differential equations in steady-state and higher-dimensional systems later.

Exercises

We now consider wave equations with homogeneous boundary conditions in both the rectangular and cylindrical coordinate systems. Use the method of separation of variables and eigenfunction expansions to evaluate the solutions to the following exercises.

Exercises in Rectangular Coordinates

4.1. Consider the transverse wave motion on a taut string over the finite interval $I = \{x \mid 0 < x < 1\}$ in a slightly damped medium. The homogeneous partial differential equation reads as

$$\frac{\partial^2}{\partial t^2} u(x, t) = c^2 \left(\frac{\partial^2}{\partial x^2} u(x, t) \right) - \gamma \left(\frac{\partial}{\partial t} u(x, t) \right)$$

with $c = 1/4$, $\gamma = 1/5$, and boundary conditions

$$u(0, t) = 0 \quad \text{and} \quad u(1, t) = 0$$

The left end and the right end are held fixed. The initial conditions are

$$u(x, 0) = f(x) \quad \text{and} \quad u_t(x, 0) = g(x)$$

Evaluate the eigenvalues and corresponding orthonormalized eigenfunctions, and write the general solution. Evaluate the solution for each of the four sets of initial conditions given:

$$f1(x) = 1 \quad \text{and} \quad g1(x) = 0$$
$$f2(x) = x \quad \text{and} \quad g2(x) = 1$$
$$f3(x) = x(1 - x) \quad \text{and} \quad g3(x) = 0$$

The plucked string. For $0 < x < 1/2$,

$$f4(x) = x \quad \text{and} \quad g4(x) = 0$$

For $1/2 < x < 1$,

$$f4(x) = 1 - x \quad \text{and} \quad g4(x) = 0$$

Generate the animated solution for each case, and plot the animated sequence for $0 < t < 5$.

4.2.   Consider longitudinal wave motion in a bar over the finite interval $I = \{x \,|\, 0 < x < 1\}$ with a slight amount of damping. The homogeneous partial differential equation reads as

$$\frac{\partial^2}{\partial t^2} u(x, t) = c^2 \left(\frac{\partial^2}{\partial x^2} u(x, t) \right) - \gamma \left(\frac{\partial}{\partial t} u(x, t) \right)$$

with $c = 1/5$, $\gamma = 1/5$, and boundary conditions

$$u(0, t) = 0 \quad \text{and} \quad u_x(1, t) = 0$$

The left end is secured and the right end is unsecured. The initial conditions are

$$u(x, 0) = f(x) \quad \text{and} \quad u_t(x, 0) = g(x)$$

Evaluate the eigenvalues and corresponding orthonormalized eigenfunctions, and write the general solution. Evaluate the solution for each of the three sets of initial conditions given:

$$f1(x) = \left(x - \frac{1}{2} \right)^2 \quad \text{and} \quad g1(x) = 0$$

$$f2(x) = x \quad \text{and} \quad g2(x) = 1$$

$$f3(x) = x \left(1 - \frac{x}{2} \right) \quad \text{and} \quad g3(x) = 0$$

Generate the animated solution for each case, and plot the animated sequence for $0 < t < 5$.

4.3.   Consider longitudinal wave motion in a bar over the finite interval $I = \{x \,|\, 0 < x < 1\}$ with a slight amount of damping. The homogeneous partial differential equation reads as

$$\frac{\partial^2}{\partial t^2} u(x, t) = c^2 \left(\frac{\partial^2}{\partial x^2} u(x, t) \right) - \gamma \left(\frac{\partial}{\partial t} u(x, t) \right)$$

with $c = 1/4$, $\gamma = 1/5$, and boundary conditions

$$u_x(0, t) = 0 \quad \text{and} \quad u(1, t) = 0$$

The left end is unsecured and the right end is secured. The initial conditions are

$$u(x, 0) = f(x) \quad \text{and} \quad u_t(x, 0) = g(x)$$

Evaluate the eigenvalues and corresponding orthonormalized eigenfunctions, and write the general solution. Evaluate the solution for each of the three sets of initial conditions given:

$$f1(x) = x \quad \text{and} \quad g1(x) = 0$$
$$f2(x) = 1 \quad \text{and} \quad g2(x) = 1$$
$$f3(x) = 1 - x^2 \quad \text{and} \quad g3(x) = 0$$

Generate the animated solution for each case, and plot the animated sequence for $0 < t < 5$.

4.4. Consider longitudinal wave motion in a bar over the finite interval $I = \{x \mid 0 < x < 1\}$ with a slight amount of damping. The homogeneous partial differential equation reads as

$$\frac{\partial^2}{\partial t^2} u(x, t) = c^2 \left(\frac{\partial^2}{\partial x^2} u(x, t) \right) - \gamma \left(\frac{\partial}{\partial t} u(x, t) \right)$$

with $c = 1/4$, $\gamma = 1/5$, and boundary conditions

$$u_x(0, t) = 0 \quad \text{and} \quad u_x(1, t) = 0$$

The left end and the right end are unsecured. The initial conditions are

$$u(x, 0) = f(x), u_t(x, 0) = g(x)$$

Evaluate the eigenvalues and corresponding orthonormalized eigenfunctions, and write the general solution. Evaluate the solution for each of the three sets of initial conditions given:

$$f1(x) = x \quad \text{and} \quad g1(x) = 0$$
$$f2(x) = 1 - x^2 \quad \text{and} \quad g2(x) = 1$$
$$f3(x) = x^2 \left(1 - \frac{2x}{3} \right) \quad \text{and} \quad g3(x) = 0$$

Generate the animated solution for each case, and plot the animated sequence for $0 < t < 5$.

4.5. Consider longitudinal wave motion in a bar over the finite interval $I = \{x \mid 0 < x < 1\}$ with a slight amount of damping. The homogeneous partial differential equation reads as

$$\frac{\partial^2}{\partial t^2} u(x, t) = c^2 \left(\frac{\partial^2}{\partial x^2} u(x, t) \right) - \gamma \left(\frac{\partial}{\partial t} u(x, t) \right)$$

with $c = 1/4$, $\gamma = 1/5$, and boundary conditions

$$u(0, t) = 0 \quad \text{and} \quad u_x(1, t) + u(1, t) = 0$$

The left end is secured and the right end is attached to an elastic hinge. The initial conditions are

$$u(x, 0) = f(x) \quad \text{and} \quad u_t(x, 0) = g(x)$$

Evaluate the eigenvalues and corresponding orthonormalized eigenfunctions, and write the general solution. Evaluate the solution for each of the three sets of initial conditions given:

$$f1(x) = 1 \quad \text{and} \quad g1(x) = 0$$
$$f2(x) = x \quad \text{and} \quad g2(x) = 1$$
$$f3(x) = x\left(1 - \frac{2x}{3}\right) \quad \text{and} \quad g3(x) = 0$$

Generate the animated solution for each case, and plot the animated sequence for $0 < t < 5$.

4.6. Consider longitudinal wave motion in a bar over the finite interval $I = \{x \mid 0 < x < 1\}$ with a slight amount of damping. The homogeneous partial differential equation reads as

$$\frac{\partial^2}{\partial t^2}u(x, t) = c^2\left(\frac{\partial^2}{\partial x^2}u(x, t)\right) - \gamma\left(\frac{\partial}{\partial t}u(x, t)\right)$$

with $c = 1/4$, $\gamma = 1/5$, and boundary conditions

$$u_x(0, t) - u(0, t) = 0 \quad \text{and} \quad u(1, t) = 0$$

The left end is attached to an elastic hinge and the right end is secured. The initial conditions are

$$u(x, 0) = f(x) \quad \text{and} \quad u_t(x, 0) = g(x)$$

Evaluate the eigenvalues and corresponding orthonormalized eigenfunctions, and write the general solution. Evaluate the solution for each of the three sets of initial conditions given:

$$f1(x) = 1 \quad \text{and} \quad g1(x) = 0$$
$$f2(x) = 1 - x \quad \text{and} \quad g2(x) = 1$$
$$f3(x) = -\frac{2x^2}{3} + \frac{x}{3} + \frac{1}{3} \quad \text{and} \quad g3(x) = 0$$

Generate the animated solution for each case, and plot the animated sequence for $0 < t < 5$.

4.7.   Consider longitudinal wave motion in a bar over the finite interval $I = \{x \mid 0 < x < 1\}$ with a slight amount of damping. The homogeneous partial differential equation reads as

$$\frac{\partial^2}{\partial t^2} u(x, t) = c^2 \left(\frac{\partial^2}{\partial x^2} u(x, t) \right) - \gamma \left(\frac{\partial}{\partial t} u(x, t) \right)$$

with $c = 1/4$, $\gamma = 1/5$, and boundary conditions

$$u_x(0, t) = 0 \quad \text{and} \quad u_x(1, t) + u(1, t) = 0$$

The left end is unsecured and the right end is attached to an elastic hinge. The initial conditions are

$$u(x, 0) = f(x) \quad \text{and} \quad u_t(x, 0) = g(x)$$

Evaluate the eigenvalues and corresponding orthonormalized eigenfunctions, and write the general solution. Evaluate the solution for each of the three sets of initial conditions given:

$$f1(x) = x \quad \text{and} \quad g1(x) = 0$$
$$f2(x) = 1 \quad \text{and} \quad g2(x) = 1$$
$$f3(x) = 1 - \frac{x^2}{3} \quad \text{and} \quad g3(x) = 0$$

Generate the animated solution for each case, and plot the animated sequence for $0 < t < 5$.

4.8.   Consider longitudinal wave motion in a bar over the finite interval $I = \{x \mid 0 < x < 1\}$ with a slight amount of damping. The homogeneous partial differential equation reads as

$$\frac{\partial^2}{\partial t^2} u(x, t) = c^2 \left(\frac{\partial^2}{\partial x^2} u(x, t) \right) - \gamma \left(\frac{\partial}{\partial t} u(x, t) \right)$$

with $c = 1/4$, $\gamma = 1/5$, and boundary conditions

$$u_x(0, t) - u(0, t) = 0 \quad \text{and} \quad u_x(1, t) = 0$$

The left end is attached to an elastic hinge and the right end is unsecured. The initial conditions are

$$u(x, 0) = f(x) \quad \text{and} \quad u_t(x, 0) = g(x)$$

Evaluate the eigenvalues and corresponding orthonormalized eigenfunctions, and write the general solution. Evaluate the solution for each of the three sets of initial conditions given:

$$f1(x) = x \quad \text{and} \quad g1(x) = 0$$

$$f2(x) = 1 \quad \text{and} \quad g2(x) = 1$$

$$f3(x) = 1 - \frac{(x-1)^2}{3} \quad \text{and} \quad g3(x) = 0$$

Generate the animated solution for each case, and plot the animated sequence for $0 < t < 5$.

4.9. We consider the transverse wave motion along a taut string over the finite interval $I = \{x \,|\, 0 < x < 1\}$. In addition to slight damping, the string is immersed in a medium that provides a restoring force that is proportional to the displacement of the string (Hooke's law–type of force). The resulting partial differential equation is called the "telegraph" equation:

$$\frac{\partial^2}{\partial t^2} u(x, t) = c^2 \left(\frac{\partial^2}{\partial x^2} u(x, t) \right) - \gamma \left(\frac{\partial}{\partial t} u(x, t) \right) - \zeta u(x, t)$$

with $c = 1/4$, $\gamma = 1/5$, $\zeta = 1/10$, and boundary conditions

$$u(0, t) = 0 \quad \text{and} \quad u(1, t) = 0$$

The left end of the string and the right end of the string are secured. The initial conditions are

$$u(x, 0) = f(x) \quad \text{and} \quad u_t(x, 0) = g(x)$$

Use the method of separation of variables to evaluate the eigenvalues and corresponding orthonormalized eigenfunctions, and write the general solution. Evaluate the solution for each of the three sets of initial conditions given:

$$f1(x) = 1 \quad \text{and} \quad g1(x) = 0$$

$$f2(x) = x \quad \text{and} \quad g2(x) = 1$$

$$f3(x) = x(x - 1) \quad \text{and} \quad g3(x) = 0$$

Generate the animated solution for each case, and plot the animated sequence for $0 < t < 5$.

4.10.   We again consider the telegraph partial differential equation over the finite interval
$I = \{x \mid 0 < x < 1\}$.

$$\frac{\partial^2}{\partial t^2} u(x, t) = c^2 \left(\frac{\partial^2}{\partial x^2} u(x, t) \right) - \gamma \left(\frac{\partial}{\partial t} u(x, t) \right) - \zeta u(x, t)$$

with $c = 1/4$, $\gamma = 1/5$, $\zeta = 1/10$, and boundary conditions

$$u(0, t) = 0 \quad \text{and} \quad u_x(1, t) = 0$$

The left end of the string is secured and the right end of the string is unsecured. The
initial conditions are

$$u(x, 0) = f(x) \quad \text{and} \quad u_t(x, 0) = g(x)$$

Use the method of separation of variables to evaluate the eigenvalues and
corresponding orthonormalized eigenfunctions, and write the general solution. Evaluate
the solution for each of the three sets of initial conditions given:

$$f1(x) = \left(x - \frac{1}{2} \right)^2 \quad \text{and} \quad g1(x) = 0$$

$$f2(x) = x \quad \text{and} \quad g2(x) = 1$$

$$f3(x) = x \left(1 - \frac{x}{2} \right) \quad \text{and} \quad g3(x) = 0$$

Generate the animated solution for each case, and plot the animated sequence for
$0 < t < 5$.

4.11.   We again consider the telegraph partial differential equation over the finite interval
$I = \{x \mid 0 < x < 1\}$.

$$\frac{\partial^2}{\partial t^2} u(x, t) = c^2 \left(\frac{\partial^2}{\partial x^2} u(x, t) \right) - \gamma \left(\frac{\partial}{\partial t} u(x, t) \right) - \zeta u(x, t)$$

with $c = 1/4$, $\gamma = 1/5$, $\zeta = 1/10$, and boundary conditions

$$u(0, t) = 0 \quad \text{and} \quad u_x(1, t) + u(1, t) = 0$$

The left end of the string is secured, and the right end of the string is attached to an
elastic hinge. The initial conditions are

$$u(x, 0) = f(x) \quad \text{and} \quad u_t(x, 0) = g(x)$$

Use the method of separation of variables to evaluate the eigenvalues and corresponding orthonormalized eigenfunctions, and write the general solution. Evaluate the solution for each of the three sets of initial conditions given:

$$f1(x) = 1 \quad \text{and} \quad g1(x) = 0$$
$$f2(x) = x \quad \text{and} \quad g2(x) = 1$$
$$f3(x) = x\left(1 - \frac{2x}{3}\right) \quad \text{and} \quad g3(x) = 0$$

Generate the animated solution for each case, and plot the animated sequence for $0 < t < 5$.

4.12. We again consider the telegraph partial differential equation over the finite interval $I = \{x \mid 0 < x < 1\}$. The telegraph equation reads as

$$\frac{\partial^2}{\partial t^2} u(x, t) = c^2 \left(\frac{\partial^2}{\partial x^2} u(x, t) \right) - \gamma \left(\frac{\partial}{\partial t} u(x, t) \right) - \zeta u(x, t)$$

with $c = 1/4$, $\gamma = 1/5$, $\zeta = 1/10$, and boundary conditions

$$u_x(0, t) - u(0, t) = 0 \quad \text{and} \quad u(1, t) = 0$$

The left end of the string is attached to an elastic hinge, and the right end of the string is secured. The initial conditions are

$$u(x, 0) = f(x) \quad \text{and} \quad u_t(x, 0) = g(x)$$

Use the method of separation of variables to evaluate the eigenvalues and corresponding orthonormalized eigenfunctions, and write the general solution. Evaluate the solution for each of the three sets of initial conditions given:

$$f1(x) = 1 \quad \text{and} \quad g1(x) = 0$$
$$f2(x) = 1 - x \quad \text{and} \quad g2(x) = 1$$
$$f3(x) = -\frac{2x^2}{3} + \frac{x}{3} + \frac{1}{3} \quad \text{and} \quad g3(x) = 0$$

Generate the animated solution for each case, and plot the animated sequence for $0 < t < 5$.

4.13. Here, we consider the transverse wave motion along a taut string over the finite interval $I = \{x \mid 0 < x < 1\}$. The damping on the string is negligible, and the string is immersed in a medium that provides a restoring force that is proportional to the displacement of the string (Hooke's law–type of force). The resulting partial differential equation is called the "Klein-Gordon" equation:

$$\frac{\partial^2}{\partial t^2} u(x, t) = c^2 \left(\frac{\partial^2}{\partial x^2} u(x, t) \right) - \zeta u(x, t)$$

with $c = 1/4$, $\zeta = 1/10$, and boundary conditions

$$u_x(0, t) = 0 \quad \text{and} \quad u(1, t) = 0$$

The left end of the string is unsecured and the right end of the string is secured. The initial conditions are

$$u(x, 0) = f(x) \quad \text{and} \quad u_t(x, 0) = g(x)$$

Use the method of separation of variables to evaluate the eigenvalues and corresponding orthonormalized eigenfunctions, and write the general solution. Evaluate the solution for each of the three sets of initial conditions given:

$$f1(x) = x \quad \text{and} \quad g1(x) = 0$$
$$f2(x) = 1 \quad \text{and} \quad g2(x) = 1$$
$$f3(x) = 1 - x^2 \quad \text{and} \quad g3(x) = 0$$

Generate the animated solution for each case, and plot the animated sequence for $0 < t < 5$.

4.14. We again consider the Klein-Gordon equation for transverse wave motion along a taut string over the finite interval $I = \{x \mid 0 < x < 1\}$. The Klein-Gordon equation reads

$$\frac{\partial^2}{\partial t^2} u(x, t) = c^2 \left(\frac{\partial^2}{\partial x^2} u(x, t) \right) - \zeta u(x, t)$$

with $c = 1/4$, $\zeta = 1/10$, and boundary conditions

$$u_x(0, t) = 0 \quad \text{and} \quad u_x(1, t) + u(1, t) = 0$$

The left end of the string is unsecured, and the right end of the string is attached to an elastic hinge. The initial conditions are

$$u(x, 0) = f(x) \quad \text{and} \quad u_t(x, 0) = g(x)$$

Use the method of separation of variables to evaluate the eigenvalues and corresponding orthonormalized eigenfunctions, and write the general solution. Evaluate the solution for each of the three sets of initial conditions given:

$$f1(x) = x \quad \text{and} \quad g1(x) = 0$$

$$f2(x) = 1 \quad \text{and} \quad g2(x) = 1$$

$$f3(x) = 1 - \frac{x^2}{3} \quad \text{and} \quad g3(x) = 0$$

Generate the animated solution for each case, and plot the animated sequence for $0 < t < 5$.

4.15. We again consider the Klein-Gordon equation for transverse wave motion along a taut string over the finite interval $I = \{x \,|\, 0 < x < 1\}$. The Klein-Gordon equation is

$$\frac{\partial^2}{\partial t^2} u(x, t) = c^2 \left(\frac{\partial^2}{\partial x^2} u(x, t) \right) - \zeta u(x, t)$$

with $c = 1/4$, $\zeta = 1/10$, and boundary conditions

$$u_x(0, t) - u(0, t) = 0 \quad \text{and} \quad u_x(1, t) = 0$$

The left end of the string is attached to an elastic hinge, and the right end of the string is unsecured. The initial conditions are

$$u(x, 0) = f(x) \quad \text{and} \quad u_t(x, 0) = g(x)$$

Use the method of separation of variables to evaluate the eigenvalues, and corresponding orthonormalized eigenfunctions, and write the general solution. Evaluate the solution for each of the three sets of initial conditions given:

$$f1(x) = x \quad \text{and} \quad g1(x) = 0$$

$$f2(x) = 1 \quad \text{and} \quad g2(x) = 1$$

$$f3(x) = 1 - \frac{(x - 1)^2}{3} \quad \text{and} \quad g3(x) = 0$$

Generate the animated solution for each case, and plot the animated sequence for $0 < t < 5$.

Exercises in Cylindrical Coordinates

4.16. Consider the transverse vibrations in a thin circularly symmetric membrane over the finite interval $I = \{r \,|\, 0 < r < 1$ with a slight amount of damping. The homogeneous partial differential equation reads as

$$\frac{\partial^2}{\partial t^2} u(r, t) = \frac{c^2 \left(\frac{\partial}{\partial r} u(r, t) + r \left(\frac{\partial^2}{\partial r^2} u(r, t) \right) \right)}{r} - \gamma \left(\frac{\partial}{\partial t} u(r, t) \right)$$

with $c = 1/4$, $\gamma = 1/5$, and boundary conditions

$$|u(0, t)| < \infty \quad \text{and} \quad u(1, t) = 0$$

The center of the membrane has a finite amplitude, and the periphery of the membrane is secure. The initial conditions are

$$u(r, 0) = f(r) \quad \text{and} \quad u_t(r, 0) = g(r)$$

Evaluate the eigenvalues and corresponding orthonormalized eigenfunctions, and write the general solution. Evaluate the solution for the three sets of initial conditions given:

$$f1(r) = r^2 \quad \text{and} \quad g1(r) = 0$$
$$f2(r) = 1 \quad \text{and} \quad g2(r) = 1$$
$$f3(r) = 1 - r^2 \quad \text{and} \quad g3(r) = 0$$

Generate the animated solution for each case, and plot the animated sequence for $0 < t < 5$.

4.17. Consider the transverse vibrations in a thin circularly symmetric membrane over the finite interval $I = \{r \mid 0 < r < 1\}$ with a slight amount of damping. The homogeneous partial differential equation reads as

$$\frac{\partial^2}{\partial t^2} u(r, t) = \frac{c^2 \left(\frac{\partial}{\partial r} u(r, t) + r \left(\frac{\partial^2}{\partial r^2} u(r, t) \right) \right)}{r} - \gamma \left(\frac{\partial}{\partial t} u(r, t) \right)$$

with $c = 1/4$, $\gamma = 1/5$, and boundary conditions

$$|u(0, t)| < \infty \quad \text{and} \quad u_r(1, t) = 0$$

The center of the membrane has a finite amplitude, and the periphery of the membrane is unsecured. The initial conditions are

$$u(r, 0) = f(r) \quad \text{and} \quad u_t(r, 0) = g(r)$$

Evaluate the eigenvalues and corresponding orthonormalized eigenfunctions, and write the general solution. Evaluate the solution for the three sets of initial conditions given:

$$f1(r) = r^2 \quad \text{and} \quad g1(r) = 0$$
$$f2(r) = 1 \quad \text{and} \quad g2(r) = 1$$
$$f3(r) = r^2 - \frac{r^4}{2} \quad \text{and} \quad g3(r) = 0$$

Generate the animated solution for each case, and plot the animated sequence for $0 < t < 5$.

4.18. Consider the transverse vibrations in a thin circularly symmetric membrane over the finite interval $I = \{r \,|\, 0 < r < 1\}$ with a slight amount of damping. The homogeneous partial differential equation reads as

$$\frac{\partial^2}{\partial t^2}u(r,t) = \frac{c^2\left(\frac{\partial}{\partial r}u(r,t) + r\left(\frac{\partial^2}{\partial r^2}u(r,t)\right)\right)}{r} - \gamma\left(\frac{\partial}{\partial t}u(r,t)\right)$$

with $c = 1/4$, $\gamma = 1/5$, and boundary conditions

$$|u(0,t)| < \infty \quad \text{and} \quad u_r(1,t) + u(1,t) = 0$$

The center of the membrane has a finite amplitude, and the periphery of the membrane is attached to an elastic hinge. The initial conditions are

$$u(r,0) = f(r) \quad \text{and} \quad u_t(r,0) = g(r)$$

Evaluate the eigenvalues and corresponding orthonormalized eigenfunctions, and write out the general solution. Evaluate the solution for the three sets of initial conditions given:

$$f1(r) = r^2 \quad \text{and} \quad g1(r) = 0$$
$$f2(r) = 1 \quad \text{and} \quad g2(r) = 1$$
$$f3(r) = 1 - \frac{r^2}{3} \quad \text{and} \quad g3(r) = 0$$

Generate the animated solution for each case, and plot the animated sequence for $0 < t < 5$.

Significance of Wave Speed

The speed of propagation of the wave in the preceding exercises was indicated by the factor c. (This speed is not to be confused with the speed of motion of the particles that make up the medium and whose collective motion constitutes the wave.) In the preceding exercises, we assumed the medium to be homogeneous and isotropic so that all terms that evaluate c are treated as constants.

In the case of transverse waves on a taut string, the constant c is given as

$$c = \sqrt{\frac{T_0}{\rho}}$$

where To is the tension in the string and ρ is the linear density of the string. In the case of longitudinal waves in a bar, the constant c is given as

$$c = \sqrt{\frac{Eo}{\rho}}$$

where Eo is the Young's modulus of elasticity and ρ is the mass density of the bar. In the case of transverse waves in a membrane, the constant c is given as

$$c = \sqrt{\frac{To}{\rho}}$$

where To is the tension per unit length in the membrane and ρ is the areal density (mass per unit area) of the membrane. In the MKS system, wave speed has dimensions of meters per second.

In the following exercises, you are asked to multiply the speed term by a given factor and to develop the solution for the initial conditions set $\{f3(x), g3(x)\}$. Develop the animated solution, and take particular note of how the increase in wave speed affects the vibration frequency of the various modes of vibration of the medium.

4.19.   In Exercise 4.2, multiply the speed term c by a factor of 2 and solve.

4.20.   In Exercise 4.4, multiply the speed term c by a factor of 4 and solve.

4.21.   In Exercise 4.7, multiply the speed term c by a factor of 2 and solve.

4.22.   In Exercise 4.9, multiply the speed term c by a factor of 4 and solve.

4.23.   In Exercise 4.13, multiply the speed term c by a factor of 2 and solve.

4.24.   In Exercise 4.16, multiply the speed term c by a factor of 4 and solve.

4.25.   In Exercise 4.17, multiply the speed term c by a factor of 2 and solve.

CHAPTER 5

The Laplace Partial Differential Equation

5.1 Introduction

The Laplace partial differential equation can be shown to be the steady-state equivalent of either the wave equation or the diffusion equation. In electrostatics, the Laplace equation describes the potential in a charge-free region, and, for this reason, it is sometimes called the "potential" equation.

In Chapter 3, we discussed the diffusion or heat partial differential equation in rectangular coordinates in one spatial dimension with no internal sources. If we view the spatial derivative operator as the first term of the Laplacian operator, then an extension of this diffusion equation from one to two spatial dimensions reads

$$\frac{\partial}{\partial t} u(x, y, t) = k \left(\frac{\partial^2}{\partial x^2} u(x, y, t) + \frac{\partial^2}{\partial y^2} u(x, y, t) \right)$$

If we consider the steady-state or time-invariant version of this equation, we get

$$\frac{\partial^2}{\partial x^2} u(x, y) + \frac{\partial^2}{\partial y^2} u(x, y) = 0$$

The preceding partial differential equation is the familiar homogeneous Laplace partial differential equation in the two-dimensional rectangular coordinate system.

In Chapter 4, we discussed the wave partial differential equation in rectangular coordinates in one spatial dimension with no damping in the system and no external applied forces. Again, if we view the spatial derivative operator as the first term of the Laplacian operator, then an extension of the wave equation from one to two spatial dimensions reads

$$\frac{\partial^2}{\partial t^2} u(x, y, t) = c^2 \left(\frac{\partial^2}{\partial x^2} u(x, y, t) + \frac{\partial^2}{\partial y^2} u(x, y, t) \right)$$

275

Again, if we consider the steady-state or time-invariant version of this equation, we get

$$\frac{\partial^2}{\partial x^2}u(x, y) + \frac{\partial^2}{\partial y^2}u(x, y) = 0$$

Thus, we see that in the steady-state situation, both the diffusion and the wave partial differential equation reduce to the familiar Laplace equation.

A typical example steady-state problem is as follows. We seek the steady-state temperature distribution $u(x, y)$ in a thin plate over the rectangular domain $D = \{(x, y) \mid 0 < x < 1, 0 < y < 1\}$. The lateral surfaces of the plate are insulated so that no heat escapes from the lateral surfaces. The sides $y = 0$ and $y = 1$ are insulated, the side $x = 1$ has a fixed temperature of zero, and the side $x = 0$ has a temperature distribution $f(y)$ given as follows.

The Laplace partial differential equation that describes the steady-state temperature in the plate is

$$\frac{\partial^2}{\partial x^2}u(x, y) + \frac{\partial^2}{\partial y^2}u(x, y) = 0$$

Because the sides $y = 0$ and $y = 1$ are insulated, the side $x = 0$ is at a fixed temperature zero, and the side $x = 0$ has a given temperature distribution, the boundary conditions on the problem are

$$u(0, y) = f(y) \quad \text{and} \quad u(1, y) = 0$$

and

$$u_y(x, 0) = 0 \quad \text{and} \quad u_y(x, 1) = 0$$

We develop the solution for this problem in Section 5.3 later.

5.2 Laplace Equation in the Rectangular Coordinate System

We now focus on the solution to the Laplace equation over the finite two-dimensional domain $D = \{(x, y) \mid 0 < x < a, 0 < y < b\}$ in rectangular coordinates. From the preceding section, the relevant partial differential equation can be rewritten to read

$$\frac{\partial^2}{\partial x^2}u(x, y) = -\left(\frac{\partial^2}{\partial y^2}u(x, y)\right)$$

Based on our success in solving the diffusion equation and the wave equation in Chapters 3 and 4, we use the method of separation of variables to solve this equation. We set

$$u(x, y) = X(x)Y(y)$$

Substituting this assumed solution into the partial differential equation, we get

$$\left(\frac{d^2}{dx^2}X(x)\right)Y(y) = -X(x)\left(\frac{d^2}{dy^2}Y(y)\right)$$

Dividing the preceding by the product solution yields

$$\frac{\frac{d^2}{dx^2}X(x)}{X(x)} = -\frac{\frac{d^2}{dy^2}Y(y)}{Y(y)}$$

Because the left-hand side of the equation is an exclusive function of x and the right-hand side is an exclusive function of y, and x and y are independent, then in order for the equation to hold for all x and y, we must set each side equal to a constant λ. Doing so gives us the following two ordinary differential equations:

$$\frac{d^2}{dx^2}X(x) + \lambda X(x) = 0$$

and

$$\frac{d^2}{dy^2}Y(y) - \lambda Y(y) = 0$$

Both of these ordinary differential equations are of the Euler type. To proceed further, we now consider the boundary conditions on the problem. Generally, one seeks a solution to this problem over the finite two-dimensional domain $D = \{(x,y) \mid 0 < x < a, 0 < y < b\}$ subject to the regular homogeneous boundary conditions on one of the variables, x or y. We seek that variable whose boundary conditions are homogeneous.

For example, consider the case where the boundary conditions on $u(x,y)$ are homogeneous with respect to the x variable:

$$\kappa_1 u(0, y) + \kappa_2 u_x(0, y) = 0$$

and

$$\kappa_3 u(a, y) + \kappa_4 u_x(a, y) = 0 \tag{5.1}$$

The earlier ordinary differential equation in x, combined with the equivalent homogeneous boundary conditions, constitutes a Sturm-Liouville problem in the variable x, for which we already have solutions.

5.3 Sturm-Liouville Problem for the Laplace Equation in Rectangular Coordinates

The differential equation in the spatial variable x in the preceding equation must be solved subject to the equivalent homogeneous spatial boundary conditions over the finite interval. The Sturm-Liouville problem for the x-dependent equation consists of the ordinary differential equation

$$\frac{d^2}{dx^2} X(x) + \lambda X(x) = 0$$

along with the equivalent corresponding homogeneous boundary conditions on the x variable as shown here:

$$\kappa_1 X(0) + \kappa_2 X_x(0) = 0$$

and

$$\kappa_3 X(a) + \kappa_4 X_x(a) = 0$$

We recognize this problem in the spatial variable x to be a regular Sturm-Liouville eigenvalue problem whereby the allowed values of λ are called the system "eigenvalues" and the corresponding solutions are called the "eigenfunctions." Note that the preceding ordinary differential equation is of the Euler type with the weight function $w(x) = 1$.

As before, it can be shown that for regular Sturm-Liouville problems over finite domains, an infinite number of eigenvalues exists that can be indexed by the integer numbers n. The indexed eigenvalues and corresponding eigenfunctions are given, respectively, as

$$\lambda_n, X_n(x)$$

for $n = 0, 1, 2, 3, \ldots$.

From the stated properties of regular Sturm-Liouville eigenvalue problems in Section 2.1, the eigenfunctions form a "complete" set with respect to any piecewise smooth function over the finite x-dependent interval $I = \{x \mid 0 < x < a\}$.

From Section 1.4 on second-order linear differential equations, the solution to the corresponding y-dependent differential equation in terms of the allowed eigenvalues is

$$Y_n(y) = A(n) \cosh\left(\sqrt{\lambda_n} y\right) + B(n) \sinh\left(\sqrt{\lambda_n} y\right)$$

Thus, for each allowed value of the index n, we have a solution

$$u_n(x, y) = X_n(x)\left(A(n) \cosh\left(\sqrt{\lambda_n} y\right) + B(n) \sinh\left(\sqrt{\lambda_n} y\right)\right)$$

for $n = 0, 1, 2, 3, \ldots$.

These functions form a set of basis vectors for the solution space of the partial differential equation. Using the same concepts of the superposition principle as in Chapters 3 and 4, the general solution to the homogeneous partial differential equation can now be written as an infinite sum over all basis vectors as follows:

$$u(x, y) = \sum_{n=0}^{\infty} X_n(x) \left(A(n) \cosh\left(\sqrt{\lambda_n} y\right) + B(n) \sinh\left(\sqrt{\lambda_n} y\right) \right)$$

The arbitrary constants $A(n)$ and $B(n)$ are evaluated by taking advantage of the orthonormality of the x-dependent eigenfunctions over the finite interval $I = \{x \mid 0 < x < a\}$, with respect to the weight function $w(x) = 1$. The statement of orthonormality reads

$$\int_0^a X_n(x) X_m(x) dx = \delta(n, m)$$

for $n, m = 0, 1, 2, 3, \ldots$.

We mentioned earlier that in using the method of separation of variables successfully, we must seek out that variable, x or y, that experiences the homogeneous boundary conditions. In the preceding, conditions on x forced it to be that variable.

We now consider the alternative possibility. If the homogeneous boundary conditions on $u(x, y)$ are on the variable y instead of x, then we rewrite the two earlier ordinary differential equations as

$$\frac{d^2}{dx^2} X(x) - \lambda X(x) = 0$$

and

$$\frac{d^2}{dy^2} Y(y) + \lambda Y(y) = 0$$

The homogeneous boundary conditions on $u(x, y)$ with respect to the variable y are as shown:

$$\kappa_1 u(x, 0) + \kappa_2 u_y(x, 0) = 0$$

and

$$\kappa_3 u(x, b) + \kappa_4 u_y(x, b) = 0$$

The given ordinary differential equation on y, combined with the preceding equivalent regular homogeneous boundary conditions, again constitutes a Sturm-Liouville problem, in terms of the variable y, for which we already have solutions.

The differential equation in the spatial variable y must be solved subject to the homogeneous spatial boundary conditions over the finite interval. The Sturm-Liouville problem for the y-dependent equation consists of the ordinary differential equation

$$\frac{d^2}{dy^2} Y(y) + \lambda Y(y) = 0$$

along with the equivalent corresponding homogeneous boundary conditions on y shown:

$$\kappa_1 Y(0) + \kappa_2 Y_y(0) = 0$$

and

$$\kappa_3 Y(b) + \kappa_4 Y_y(b) = 0$$

Again, we see this problem in the spatial variable y to be a regular Sturm-Liouville eigenvalue problem whereby the allowed values of λ are called the system "eigenvalues" and the corresponding solutions are called the "eigenfunctions." Note that the preceding ordinary differential equation is of the Euler type and the weight function is $w(y) = 1$.

As before, we can show that for regular Sturm-Liouville problems over finite domains, an infinite number of eigenvalues exist that can be indexed by the positive integers n. The indexed eigenvalues and corresponding eigenfunctions are given, respectively, as

$$\lambda_n, Y_n(y)$$

for $n = 0, 1, 2, 3, \ldots$.

Again, from the stated properties of regular Sturm-Liouville eigenvalue problems in Section 2.1, the eigenfunctions form a "complete" set with respect to any piecewise smooth function over the finite y-dependent interval $I = \{y \mid 0 < y < b\}$.

From Section 1.4 on second-order linear differential equations, the solution to the corresponding x-dependent differential equation in terms of the allowed eigenvalues is

$$X_n(x) = A(n) \cosh\left(\sqrt{\lambda_n} x\right) + B(n) \sinh\left(\sqrt{\lambda_n} x\right) \tag{5.2}$$

Thus, for each allowed value of the index n, we have a solution

$$u_n(x, y) = Y_n(y)\left(A(n) \cosh\left(\sqrt{\lambda} x\right) + B(n) \sinh\left(\sqrt{\lambda} x\right)\right)$$

for $n = 0, 1, 2, 3, \ldots$.

Using the concept of the superposition principle as in Chapters 3 and 4, the general solution to the homogeneous partial differential equation can again be written as the infinite sum of the preceding basis vectors in terms of the eigenfunctions with respect to the variable y as follows:

$$u(x, y) = \sum_{n=0}^{\infty} Y_n(y)\left(A(n) \cosh\left(\sqrt{\lambda_n} x\right) + B(n) \sinh\left(\sqrt{\lambda_n} x\right)\right)$$

Again, the arbitrary constants $A(n)$ and $B(n)$ can be evaluated by taking advantage of the orthonormality of the y-dependent eigenfunctions, with respect to the weight function $w(y) = 1$ over the interval $I = \{y \,|\, 0 < y < b\}$. The statement of orthonormality reads

$$\int_0^b Y_n(y) Y_m(y) \mathrm{d}y = \delta(n, m) \tag{5.3}$$

for $n, m = 0, 1, 2, 3, \ldots$.

The boundary conditions on the problem determine whether the solution is written in terms of the eigenfunctions with respect to x or y. Remember that the decision comes from our seeking that particular variable that experiences the homogeneous boundary conditions.

DEMONSTRATION: We seek the steady-state temperature distribution in a thin rectangular plate over the rectangular domain $D = \{(x, y) \,|\, 0 < x < 1, 0 < y < 1\}$. The lateral surfaces of the plate are insulated. The sides $y = 0$ and $y = 1$ are insulated. The side $x = 1$ has a fixed temperature of zero, and the side $x = 0$ has a temperature distribution $f(y)$ given as follows.

SOLUTION: The Laplace partial differential equation is

$$\frac{\partial^2}{\partial x^2} u(x, y) + \frac{\partial^2}{\partial y^2} u(x, y) = 0$$

The boundary conditions are

$$u(0, y) = f(y) \quad \text{and} \quad u(1, y) = 0$$

and

$$u_y(x, 0) = 0 \quad \text{and} \quad u_y(x, 1) = 0$$

We seek that variable whose boundary conditions are homogeneous. Here, we see that the homogeneous boundary conditions are with respect to the y variable. Thus, by the method of separation of variables, we write the two ordinary differential equations as

$$\frac{\mathrm{d}^2}{\mathrm{d}x^2} X(x) - \lambda X(x) = 0$$

and

$$\frac{\mathrm{d}^2}{\mathrm{d}y^2} Y(y) + \lambda Y(y) = 0$$

Because both ends along the y-axis are insulated, the corresponding homogeneous boundary conditions on the y equation are type 2 at $y = 0$ and type 2 at $y = 1$.

$$Y_y(0) = 0 \quad \text{and} \quad Y_y(1) = 0$$

The allowed eigenvalues and corresponding orthonormal eigenfunctions are obtained from Example 2.5.3 in Chapter 2. For $\lambda_0 = 0$, we got

$$Y_0(y) = 1$$

and for $\lambda_n = n^2\pi^2$, we got

$$Y_n(y) = \sqrt{2}\,\cos(n\pi y)$$

for $n = 1, 2, 3, \ldots$.

The statement of orthonormality with respect to the weight function $w(y) = 1$ reads

$$\int_0^1 2\cos(n\pi y)\cos(m\pi y)\,dy = \delta(n, m)$$

for $n, m = 1, 2, 3, \ldots$.

A set of basis vectors for the x-dependent equation for $\lambda_n = n^2\pi^2$ reads

$$x1(x) = \sinh(n\pi(1 - x))$$

and

$$x2(x) = \cosh(n\pi(1 - x))$$

and the general solution to the x equation is

$$X_n(x) = A(n)\cosh(n\pi(1 - x)) + B(n)\sinh(n\pi(1 - x))$$

Substitution of the boundary condition at $x = 1$ yields $A(n) = 0$. Thus, the solution for the x-dependent equation is

$$X_n(x) = B(n)\sinh(n\pi(1 - x))$$

for $n = 1, 2, 3, \ldots$.

Similarly, for $\lambda_0 = 0$, a set of basis vectors is

$$x1(x) = 1$$

and

$$x2(x) = x$$

and a solution that satisfies the boundary condition at $x = 1$ reads

$$X_0(x) = B(0)(1 - x)$$

The generalized eigenfunction expansion of the solution is constructed from the sum of the product of the x-dependent solutions and the preceding orthonormal y-dependent eigenfunctions. This yields

$$u(x, y) = B(0)(1 - x) + \sum_{n=1}^{\infty} B(n) \sinh(n\pi(1 - x))\sqrt{2}\cos(n\pi y)$$

The Fourier coefficients $B(n)$ are determined from the remaining boundary condition. Substitution of the boundary condition $u(0, y) = f(y)$ at $x = 0$ yields

$$f(y) = B(0) + \sum_{n=1}^{\infty} B(n) \sinh(n\pi)\sqrt{2}\cos(n\pi y)$$

This equation is the Fourier series expansion of $f(y)$ in terms of the orthonormal y-dependent Sturm-Liouville eigenfunctions found earlier. To evaluate the Fourier coefficients $B(n)$, we take the inner product of both sides with respect to the orthonormal eigenfunctions, and taking advantage of the statement of orthonormality, we get

$$B(n) = \frac{\int_0^1 f(y)\sqrt{2}\cos(n\pi y)\,dy}{\sinh(n\pi)}$$

for $n = 1, 2, 3, \ldots$, and

$$B(0) = \int_0^1 f(y)\,dy$$

for $n = 0$.

We consider the special case for the boundary condition along $x = 0$ to be $f(y) = y(1 - y)$. The integral for $B(n)$ reads

$$B(n) = \frac{\int_0^1 y(1 - y)\sqrt{2}\cos(n\pi y)\,dy}{\sinh(n\pi)}$$

Evaluation of this integral yields

$$B(n) = -\frac{\sqrt{2}((-1)^n + 1)}{\sinh(n\pi)n^2\pi^2}$$

for $n = 1, 2, 3, \ldots$.

Similarly, for $n = 0$, the integral for $B(0)$ reads

$$B(0) = \int_0^1 y(1 - y)\,dy$$

Evaluation of this integral yields

$$B(0) = \frac{1}{6}$$

Finally, the solution for the steady-state temperature of the system is

$$u(x, y) = \frac{1-x}{6} + \sum_{n=1}^{\infty} \left(-\frac{2\left((-1)^n + 1\right) \sinh(n\pi(1-x)) \cos(n\pi y)}{\sinh(n\pi)n^2\pi^2} \right)$$

The detailed development of this solution and the graphics are given later in one of the Maple worksheet examples.

5.4 Example Laplace Problems in the Rectangular Coordinate System

We look at steady-state solutions to wave and diffusion phenomena in the two-dimensional, rectangular coordinate (x, y) plane. We assume conditions such that there is no dependence of the solution on the z coordinate.

EXAMPLE 5.4.1: We seek the electrostatic potential in a charge-free rectangular domain $D = \{(x, y) \mid 0 < x < 1, 0 < y < 1\}$. The sides $x = 0$, $x = 1$, and $y = 1$ are held at a fixed potential zero, and the side $y = 0$ has a potential distribution $u(x, 0) = f(x)$ given below.

SOLUTION: The homogeneous Laplace equation is

$$\frac{\partial^2}{\partial x^2} u(x, y) + \frac{\partial^2}{\partial y^2} u(x, y) = 0$$

The boundary conditions are

$$u(0, y) = 0 \quad \text{and} \quad u(1, y) = 0$$

and

$$u(x, 0) = x(1-x) \quad \text{and} \quad u(x, 1) = 0$$

Here, we see that the boundary conditions are homogeneous with respect to the x variable.

Ordinary differential equations from the method of separation of variables are

$$\frac{d^2}{dx^2} X(x) + \lambda X(x) = 0$$

and

$$\frac{d^2}{dy^2} Y(y) - \lambda Y(y) = 0$$

Homogeneous boundary conditions on the x equation are type 1 at $x = 0$ and type 1 at $x = 1$

$$X(0) = 0 \quad \text{and} \quad X(1) = 0$$

Assignment of system parameters

> restart:with(plots):a:=1:b:=1:

Allowed eigenvalues and corresponding orthonormal eigenfunctions for the x equation are obtained from Example 2.5.1.

> lambda[n]:=(n*Pi/a)^2;

$$\lambda_n := n^2 \pi^2 \tag{5.4}$$

for $n = 1, 2, 3, \ldots$.

Orthonormal eigenfunctions

> X[n](x):=sqrt(2/a)*sin(n*Pi/a*x);X[n](s):=subs(x=s,X[n](x)):X[m](x):=subs(n=m,X[n](x)):

$$X_n(x) := \sqrt{2}\sin(n\pi x) \tag{5.5}$$

Statement of orthonormality with respect to the weight function $w(x) = 1$

> w(x):=1:Int(X[n](x)*X[m](x)*w(x),x=0..a)=delta(n,m);

$$\int_0^1 2\sin(n\pi x)\sin(m\pi x)\,dx = \delta(n, m) \tag{5.6}$$

Basis vectors for the y equation

> y1(y):=sinh(sqrt(lambda)*(b−y));y2(y):=cosh(sqrt(lambda)*(b−y));

$$y1(y) := \sinh\left(\sqrt{\lambda}(1 - y)\right)$$
$$y2(y) := \cosh\left(\sqrt{\lambda}(1 - y)\right) \tag{5.7}$$

General solution to the y equation

> Y[n](y):=A(n)*cosh(n*Pi/a*(b−y))+B(n)*sinh(n*Pi/a*(b−y));

$$Y_n(y) := A(n)\cosh(n\pi(1 - y)) + B(n)\sinh(n\pi(1 - y)) \tag{5.8}$$

Substituting into the y-dependent boundary condition yields

> eval(subs(y=b,Y[n](y)))=0;

$$A(n) = 0 \tag{5.9}$$

The *y*-dependent solution

> Y[n](y):=B(n)*sinh(n*Pi/a*(b−y));

$$Y_n(y) := B(n) \sinh(n\pi(1-y)) \tag{5.10}$$

Generalized series terms

> u[n](x,y):=X[n](x)*Y[n](y);

$$u_n(x, y) := \sqrt{2} \sin(n\pi x) B(n) \sinh(n\pi(1-y)) \tag{5.11}$$

Eigenfunction expansion

> u(x,y):=Sum(u[n](x,y),n=1..infinity);

$$u(x, y) := \sum_{n=1}^{\infty} \sqrt{2} \sin(n\pi x) B(n) \sinh(n\pi(1-y)) \tag{5.12}$$

Evaluation of Fourier coefficients from the specific remaining boundary condition $u(x, 0) = f(x)$ where

> f(x):=x*(1−x);

$$f(x) := x(1-x) \tag{5.13}$$

Substituting this condition into the earlier solution yields

> f(x)=subs(y=0,u(x,y));

$$x(1-x) = \sum_{n=1}^{\infty} \sqrt{2} \sin(n\pi x) B(n) \sinh(n\pi) \tag{5.14}$$

The preceding equation is the Fourier series expansion of $f(x)$ in terms of the orthonormal Sturm-Liouville eigenfunctions found earlier. To evaluate the Fourier cofficients, we take the inner product of both sides with respect to the orthonormal eigenfunctions, and, taking advantage of the previous statement of orthonormality, we get

> B(n):=(1/sinh(n*Pi*b/a))*Int(f(x)*X[n](x)*w(x),x=0..a);B(n):=expand(value(%)):

$$B(n) := \frac{\displaystyle\int_0^1 x(1-x)\sqrt{2}\sin(n\pi x)\,dx}{\sinh(n\pi)} \tag{5.15}$$

Evaluation of this integral yields

> B(n):=simplify(subs({sin(n*Pi)=0,cos(n*Pi)=(−1)^n},B(n)));

$$B(n) := -\frac{2\sqrt{2}\left(-1+(-1)^n\right)}{\sinh(n\pi)n^3\pi^3} \tag{5.16}$$

Generalized series terms

> u[n](x,y):=eval(X[n](x)*Y[n](y));

$$u_n(x, y) := -\frac{4\sin(n\pi x)\left(-1+(-1)^n\right)\sinh(n\pi(1-y))}{\sinh(n\pi)n^3\pi^3} \tag{5.17}$$

Series solution

> u(x,y):=Sum(u[n](x,y),n=1..infinity);

$$u(x, y) := \sum_{n=1}^{\infty}\left(-\frac{4\sin(n\pi x)(-1+(-1)^n)\sinh(n\pi(1-y))}{\sinh(n\pi)n^3\pi^3}\right) \tag{5.18}$$

First few terms of expansion

> u(x,y):=sum(u[n](x,y),n=1..5):

> plot3d(u(x,y),x=0..a,y=0..b,axes=framed,thickness=1);

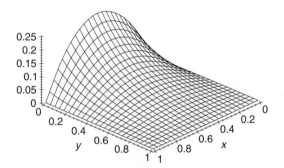

Figure 5.1

The three-dimensional surface shown in Figure 5.1 depicts the electrostatic potential distribution $u(x, y)$ over the rectangular region. Note how the edges of the surface adhere to the given boundary conditions. The equipotential lines can be obtained from Maple by clicking on the figure and then choosing the special option "Render the plot using the polygon patch and contour style" in the graphics bar and then clicking the "redraw" button.

EXAMPLE 5.4.2: We seek the steady-state temperature distribution in a thin rectangular plate over the domain $D = \{(x, y) \mid 0 < x < 1, 0 < y < 1\}$ whose lateral surface is insulated. The sides $y = 0$ and $y = 1$ are insulated, the side $x = 1$ has a fixed temperature of zero, and the side $x = 0$ follows the temperature distribution $u(0, y) = f(y)$ given as follows.

SOLUTION: The homogeneous Laplace equation is

$$\frac{\partial^2}{\partial x^2}u(x, y) + \frac{\partial^2}{\partial y^2}u(x, y) = 0$$

The boundary conditions are

$$u(0, y) = y(1 - y) \quad \text{and} \quad u(1, y) = 0$$

and

$$u_y(x, 0) = 0 \quad \text{and} \quad u_y(x, 1) = 0$$

Here, we see that the boundary conditions are homogeneous with respect to the y variable.

Ordinary differential equations from the method of separation of variables are

$$\frac{d^2}{dx^2}X(x) - \lambda X(x) = 0$$

and

$$\frac{d^2}{dy^2}Y(y) + \lambda Y(y) = 0$$

Homogeneous boundary conditions on the y equation are type 2 at $y = 0$ and type 2 at $y = 1$

$$Y_y(0) = 0 \quad \text{and} \quad Y_y(1) = 0$$

Assignment of system parameters

> restart:with(plots):a:=1:b:=1:

Allowed eigenvalues and corresponding orthonormal eigenfunctions for the y equation are obtained from Example 2.5.3. For $n = 0$,

> lambda[0]:=0;

$$\lambda_0 := 0 \tag{5.19}$$

Orthonormal eigenfunction

> Y[0](y):=1/sqrt(b);

$$Y_0(y) := 1 \tag{5.20}$$

For $n = 1, 2, 3, \ldots,$

> lambda[n]:=(n*Pi/b)^2;

$$\lambda_n := n^2\pi^2 \tag{5.21}$$

Orthonormal eigenfunctions

> Y[n](y):=sqrt(2/b)*cos(n*Pi/b*y);Y[n](s):=subs(y=s,Y[n](y)):Y[m](y):=subs(n=m,Y[n](y)):

$$Y_n(y) := \sqrt{2}\,\cos(n\pi y) \tag{5.22}$$

Statement of orthonormality with respect to the weight function $w(y) = 1$

> w(y):=1:Int(Y[n](y)*Y[m](y)*w(y),y=0..b)=delta(n,m);

$$\int_0^1 2\cos(n\pi y)\cos(m\pi y)\,\mathrm{d}y = \delta(n,m) \qquad (5.23)$$

Basis vectors for the x equation

> x1(x):=sinh(sqrt(lambda)*(a−x));x2(x):=cosh(sqrt(lambda)*(a−x));

$$x1(x) := \sinh\left(\sqrt{\lambda}(1-x)\right)$$
$$x2(x) := \cosh\left(\sqrt{\lambda}(1-x)\right) \qquad (5.24)$$

General solution to the x equation for $n = 1, 2, 3, \ldots$

> X[n](x):=A(n)*cosh(n*Pi/b*(a−x))+B(n)*sinh(n*Pi/b*(a−x));

$$X_n(x) := A(n)\cosh(n\pi(1-x)) + B(n)\sinh(n\pi(1-x)) \qquad (5.25)$$

Substituting into the x-dependent boundary condition yields

> eval(subs(x=a,X[n](x)))=0;

$$A(n) = 0 \qquad (5.26)$$

For $n = 1, 2, 3, \ldots$, we have for the x-dependent solution

> X[n](x):=B(n)*sinh(n*Pi/b*(a−x));

$$X_n(x) := B(n)\sinh(n\pi(1-x)) \qquad (5.27)$$

Similarly, for $n = 0$, the x-dependent solution is

> X[0](x):=B(0)*(a−x);

$$X_0(x) := B(0)(1-x) \qquad (5.28)$$

Generalized series terms

> u[0](x,y):=X[0](x)*Y[0](y);u[n](x,y):=X[n](x)*Y[n](y);

$$u_0(x, y) := B(0)(1 - x)$$
$$u_n(x, y) := B(n)\sinh(n\pi(1-x))\sqrt{2}\cos(n\pi y) \qquad (5.29)$$

Eigenfunction expansion

> u(x,y):=u[0](x,y)+Sum(u[n](x,y),n=1..infinity);

$$u(x, y) := B(0)(1 - x) + \sum_{n=1}^{\infty} B(n) \sinh(n\pi(1 - x))\sqrt{2}\cos(n\pi y) \tag{5.30}$$

Evaluation of Fourier coefficients from the specific remaining boundary condition $u(0, y) = f(y)$

> f(y):=y*(1−y);

$$f(y) := y(1 - y) \tag{5.31}$$

Substituting this condition into the preceding solution yields

> f(y)=subs(x=0,u(x,y));

$$y(1 - y) = B(0) + \sum_{n=1}^{\infty} B(n) \sinh(n\pi)\sqrt{2}\cos(n\pi y) \tag{5.32}$$

This equation is the Fourier series expansion of $f(y)$ in terms of the orthonormal Sturm-Liouville eigenfunctions found earlier. To evaluate the Fourier cofficients, we take the inner product of both sides with respect to the orthonormal eigenfunctions, and, taking advantage of the previous statement of orthonormality, we get

> B(n):=(1/sinh(n*Pi*a/b))*Int(f(y)*Y[n](y)*w(y),y=0..b);B(n):=expand(value(%)):

$$B(n) := \frac{\int_{0}^{1} y(1 - y)\sqrt{2}\cos(n\pi y)\,dy}{\sinh(n\pi)} \tag{5.33}$$

> B(n):=simplify(subs({sin(n*Pi)=0,cos(n*Pi)=(−1)^n},B(n)));

$$B(n) := -\frac{\sqrt{2}(1 + (-1)^n)}{\sinh(n\pi)n^2\pi^2} \tag{5.34}$$

for $n = 1, 2, 3, \ldots$. Similarly, for $n = 0$, we get

> B(0):=eval((1/a)*Int(f(y)*Y[0](y)*w(y),y=0..b));B(0):=value(%):

$$B(0) := \int_{0}^{1} y(1 - y)\,dy \tag{5.35}$$

> B(0):=subs({sin(n*Pi)=0,cos(n*Pi)=(−1)^n},B(0));

$$B(0) := \frac{1}{6} \tag{5.36}$$

Generalized series terms

> u[0](x,y):=eval(X[0](x)*Y[0](y));u[n](x,y):=eval(X[n](x)*Y[n](y));

$$u_0(x, y) := \frac{1}{6} - \frac{1}{6}x$$

$$u_n(x, y) := -\frac{2(1 + (-1)^n) \sinh(n\pi(1 - x)) \cos(n\pi y)}{\sinh(n\pi)n^2\pi^2} \tag{5.37}$$

Series solution

> u(x,y):=u[0](x,y)+Sum(u[n](x,y),n=1..infinity);

$$u(x, y) := \frac{1}{6} - \frac{1}{6}x + \sum_{n=1}^{\infty}\left(-\frac{2(1 + (-1)^n) \sinh(n\pi(1 - x)) \cos(n\pi y)}{\sinh(n\pi)n^2\pi^2}\right) \tag{5.38}$$

First few terms of expansion

> u(x,y):=sum(u[0](x,y)+u[n](x,y),n=1..5):

> plot3d(u(x,y),x=0..a,y=0..b,axes=framed,thickness=1);

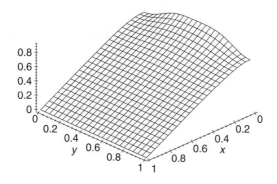

Figure 5.2

The three-dimensional surface shown in Figure 5.2 depicts the steady-state temperature distribution $u(x, y)$ over the rectangular region. Note how the edges of the surface adhere to the given boundary conditions. The temperature isotherms can be obtained from Maple by clicking on the figure, choosing the special option "Render the plot using the polygon patch and contour style" in the graphics bar, and then clicking the "redraw" button.

EXAMPLE 5.4.3: We seek the steady-state temperature distribution in a thin rectangular plate over the domain $D = \{(x, y) \mid 0 < x < 1, 0 < y < 1\}$ whose lateral surface is insulated. The side $y = 1$ is insulated, the sides $x = 1$ and $y = 0$ are held at a fixed temperature of zero, and the side $x = 0$ follows the temperature distribution $u(0, y) = f(y)$ given as follows.

SOLUTION: The homogeneous Laplace equation is

$$\frac{\partial^2}{\partial x^2}u(x, y) + \frac{\partial^2}{\partial y^2}u(x, y) = 0$$

The boundary conditions are

$$u(0, y) = y(1 - y) \quad \text{and} \quad u(1, y) = 0$$

and

$$u(x, 0) = 0 \quad \text{and} \quad u_y(x, 1) = 0$$

Here, we see that the boundary conditions are homogeneous with respect to the y variable.

Ordinary differential equations from the method of separation of variables are

$$\frac{d^2}{dx^2}X(x) - \lambda\, X(x) = 0$$

and

$$\frac{d^2}{dy^2}Y(y) + \lambda Y(y) = 0$$

Homogeneous boundary conditions on the y equation are type 1 at $y = 0$ and type 2 at $y = 1$

$$Y(0) = 0 \quad \text{and} \quad Y_y(1) = 0$$

Assignment of system parameters

> restart:with(plots):a:=1:b:=1:

Allowed eigenvalues and corresponding orthonormal eigenfunctions for the y equation are obtained from Example 2.5.2.

> lambda[n]:=((2*n−1)*Pi/(2*b))^2;

$$\lambda_n := \frac{1}{4}(2n - 1)^2 \pi^2 \tag{5.39}$$

for $n = 1, 2, 3, \ldots$.

Orthonormal eigenfunctions

> Y[n](y):=sqrt(2/b)*sin((2*n−1)*Pi/(2*b)*y);Y[m](y):=subs(n=m,Y[n](y)):

$$Y_n(y) := \sqrt{2}\sin\left(\frac{1}{2}(2n - 1)\pi y\right) \tag{5.40}$$

Statement of orthonormality with respect to the weight function $w(y) = 1$

> w(y):=1:Int(Y[n](y)*Y[m](y)*w(y),y=0..b)=delta(n,m);

$$\int_0^1 2\sin\left(\frac{1}{2}(2n-1)\pi y\right)\sin\left(\frac{1}{2}(2m-1)\pi y\right)dy = \delta(n,m) \qquad (5.41)$$

Basis vectors for the x equation

> x1(x):=sinh(sqrt(lambda)*(a−x));x2(x):=cosh(sqrt(lambda)*(a−x));

$$x1(x) := \sinh\left(\sqrt{\lambda}(1-x)\right)$$

$$x2(x) := \cosh\left(\sqrt{\lambda}(1-x)\right) \qquad (5.42)$$

General solution to the x equation

> X[n](x):=A(n)*cosh((2*n−1)*Pi/(2*b)*(a−x))+B(n)*sinh((2*n−1)*Pi/(2*b)*(a−x));

$$X_n(x) := A(n)\cosh\left(\frac{1}{2}(2n-1)\pi(1-x)\right) + B(n)\sinh\left(\frac{1}{2}(2n-1)\pi(1-x)\right) \qquad (5.43)$$

Substituting into the x-dependent boundary condition yields

> eval(subs(x=a,X[n](x)))=0;

$$A(n) = 0 \qquad (5.44)$$

The x-dependent solution

> X[n](x):=B(n)*sinh((2*n−1)*Pi/(2*b)*(a−x));

$$X_n(x) := B(n)\sinh\left(\frac{1}{2}(2n-1)\pi(1-x)\right) \qquad (5.45)$$

Generalized series terms

> u[n](x,y):=X[n](x)*Y[n](y);

$$u_n(x, y) := B(n)\sinh\left(\frac{1}{2}(2n-1)\pi(1-x)\right)\sqrt{2}\sin\left(\frac{1}{2}(2n-1)\pi y\right) \qquad (5.46)$$

Eigenfunction expansion

> u(x,y):=Sum(u[n](x,y),n=1..infinity);

$$u(x, y) := \sum_{n=1}^{\infty} B(n)\sinh\left(\frac{1}{2}(2n-1)\pi(1-x)\right)\sqrt{2}\sin\left(\frac{1}{2}(2n-1)\pi y\right) \qquad (5.47)$$

Evaluation of Fourier coefficients from the specific remaining boundary condition
$u(0, y) = f(y)$

```
> f(y):=y*(1−y);
```

$$f(y) := y(1 - y) \qquad (5.48)$$

Substituting this condition into the preceding solution yields

```
> f(y)=subs(x=0,u(x,y));
```

$$y(1 - y) = \sum_{n=1}^{\infty} B(n) \sinh\left(\frac{1}{2}(2n - 1)\pi\right) \sqrt{2} \sin\left(\frac{1}{2}(2n - 1)\pi y\right) \qquad (5.49)$$

The preceding equation is the Fourier series expansion of $f(y)$ in terms of the orthonormalized Sturm-Liouville eigenfunctions found earlier. To evaluate the Fourier cofficients, we take the inner product of both sides with respect to the orthonormalized eigenfunctions, and, taking advantage of the previous statement of orthonormality, we get, for $n = 1, 2, 3, \ldots,$

```
> B(n):=(1/sinh((2*n−1)*Pi*a/(2*b)))*Int(f(y)*Y[n](y)*w(y),y=0..b);B(n):=value(%):
```

$$B(n) := \frac{\int_0^1 y(1 - y)\sqrt{2} \sin\left(\frac{1}{2}(2n - 1)\pi y\right) dy}{\sinh\left(\frac{1}{2}(2n - 1)\pi\right)} \qquad (5.50)$$

```
> B(n):=simplify(subs({sin(n*Pi)=0,cos(n*Pi)=(−1)^n,sin((2*n+1)*Pi/2)=(−1)^n,
  cos((2*n+1)*Pi/2)=0},B(n)));
```

$$B(n) := \frac{4\sqrt{2}(4 + 2\pi n(-1)^n - \pi(-1)^n)}{\sinh\left(\frac{1}{2}(2n - 1)\pi\right)\pi^3(8n^3 - 12n^2 + 6n - 1)} \qquad (5.51)$$

Generalized series terms

```
> u[n](x,y):=eval(X[n](x)*Y[n](y));
```

$$u_n(x, y) := \frac{8(4 + 2\pi n(-1)^n - \pi(-1)^n) \sinh\left(\frac{1}{2}(2n - 1)\pi(1 - x)\right) \sin\left(\frac{1}{2}(2n - 1)\pi y\right)}{\sinh\left(\frac{1}{2}(2n - 1)\pi\right)\pi^3(8n^3 - 12n^2 + 6n - 1)} \qquad (5.52)$$

Series solution

```
> u(x,y):=Sum(u[n](x,y),n=1..infinity);
```

$$u(x, y) := \sum_{n=1}^{\infty} \frac{8(4 + 2\pi n(-1)^n - \pi(-1)^n) \sinh\left(\frac{1}{2}(2n - 1)\pi(1 - x)\right) \sin\left(\frac{1}{2}(2n - 1)\pi y\right)}{\sinh\left(\frac{1}{2}(2n - 1)\pi\right)\pi^3(8n^3 - 12n^2 + 6n - 1)}$$

$$\qquad (5.53)$$

First few terms of expansion

```
> u(x,y):=sum(u[n](x,y),n=1..5):
> plot3d(u(x,y),x=0..a,y=0..b,axes=framed,thickness=1);
```

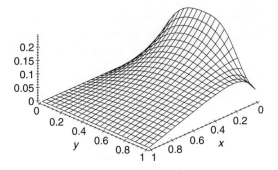

Figure 5.3

The three-dimensional surface shown in Figure 5.3 depicts the steady-state temperature distribution $u(x, y)$ over the rectangular region. Note how the edges of the surface adhere to the given boundary conditions. The temperature isotherms can be obtained from Maple by clicking on the figure, choosing the special option "Render the plot using the polygon patch and contour style" in the graphics bar, and then clicking the "redraw" button.

EXAMPLE 5.4.4: We seek the steady-state temperature distribution in a thin rectangular plate over the domain $D = \{(x, y) \mid 0 < x < 1, 0 < y < 1\}$ whose lateral surface is insulated. The side $x = 0$ is insulated, the side $y = 1$ has a fixed temperature of zero, the side $y = 0$ follows the temperature distribution $f(x)$, and the side $x = 1$ is experiencing a convection heat loss into a zero temperature surrounding.

SOLUTION: The homogeneous Laplace equation is

$$\frac{\partial^2}{\partial x^2} u(x, y) + \frac{\partial^2}{\partial y^2} u(x, y) = 0$$

The boundary conditions are

$$u_x(0, y) = 0 \quad \text{and} \quad u(1, y) + u_x(1, y) = 0$$

and

$$u(x, 0) = 1 - x^2 \quad \text{and} \quad u(x, 1) = 0$$

Here, we see that the boundary conditions are homogeneous with respect to the x variable.

Ordinary differential equations from the method of separation of variables are

$$\frac{d^2}{dx^2} X(x) + \lambda X(x) = 0$$

and

$$\frac{d^2}{dy^2} Y(y) - \lambda Y(y) = 0$$

Homogeneous boundary conditions on the x equation are type 2 at $x = 0$ and type 3 at $x = 1$:

$$X_x(0) = 0 \quad \text{and} \quad X_x(1) + X(1) = 0$$

Assignment of system parameters

> restart:with(plots):a:=1:b:=1:

Allowed eigenvalues and corresponding orthonormal eigenfunctions for the x equation are obtained from Example 2.5.5. The eigenvalues are the roots of the eigenvalue equation

> tan(sqrt(lambda[n])*a)=1/sqrt(lambda[n]);

$$\tan\left(\sqrt{\lambda_n}\right) = \frac{1}{\sqrt{\lambda_n}} \tag{5.54}$$

for $n = 1, 2, 3, \ldots$.

Orthonormal eigenfunctions

> X[n](x):=sqrt(2)*cos(sqrt(lambda[n])*x)/sqrt(((sin(sqrt(lambda[n])*a))^2+a));X[m](x):=
  subs(n=m,X[n](x)):

$$X_n(x) := \frac{\sqrt{2}\cos\left(\sqrt{\lambda_n}\,x\right)}{\sqrt{\sin\left(\sqrt{\lambda_n}\right)^2 + 1}} \tag{5.55}$$

Statement of orthonormality with respect to the weight function $w(x) = 1$

> w(x):=1:Int(X[n](x)*X[m](x)*w(x),x=0..a)=delta(n,m);

$$\int_0^1 \frac{2\cos\left(\sqrt{\lambda_n}\,x\right)\cos\left(\sqrt{\lambda_m}\,x\right)}{\sqrt{\sin\left(\sqrt{\lambda_n}\right)^2 + 1}\sqrt{\sin\left(\sqrt{\lambda_m}\right)^2 + 1}}\,dx = \delta(n, m) \tag{5.56}$$

Basis vectors for the y equation

> y1(y):=sinh(sqrt(lambda)*(b−y));y2(y):=cosh(sqrt(lambda)*(b−y));

$$y1(y) := \sinh\left(\sqrt{\lambda}(1 - y)\right)$$

$$y2(y) := \cosh\left(\sqrt{\lambda}(1 - y)\right) \tag{5.57}$$

General solution to the y equation

> Y[n](y):=A(n)*cosh(sqrt(lambda[n])*(b−y))+B(n)*sinh(sqrt(lambda[n])*(b−y));

$$Y_n(y) := A(n)\cosh\left(\sqrt{\lambda_n}(1 - y)\right) + B(n)\sinh\left(\sqrt{\lambda_n}(1 - y)\right) \tag{5.58}$$

Substituting into the y-dependent boundary condition yields

> eval(subs(y=b,Y[n](y)))=0;

$$A(n) = 0 \tag{5.59}$$

The y-dependent solution

> Y[n](y):=B(n)*sinh(sqrt(lambda[n])*(b−y));

$$Y_n(y) := B(n) \sinh\left(\sqrt{\lambda_n}(1-y)\right) \tag{5.60}$$

> u[n](x,y):=X(n)(x)*Y[n](y):

Eigenfunction expansion

> u(x,y):=Sum(u[n](x,y),n=1..infinity);

$$u(x, y) := \sum_{n=1}^{\infty} \frac{\sqrt{2}\cos\left(\sqrt{\lambda_n}x\right) B(n) \sinh\left(\sqrt{\lambda_n}(1-y)\right)}{\sqrt{\sin\left(\sqrt{\lambda_n}\right)^2 + 1}} \tag{5.61}$$

Evaluation of Fourier coefficients from the specific remaining boundary condition $u(x, 0) = f(x)$

> f(x):=1−x^2;

$$f(x) := 1 - x^2 \tag{5.62}$$

Substituting this condition into the preceding solution yields

> f(x)=subs(y=0,u(x,y));

$$1 - x^2 = \sum_{n=1}^{\infty} \frac{\sqrt{2}\cos\left(\sqrt{\lambda_n}x\right) B(n) \sinh\left(\sqrt{\lambda_n}\right)}{\sqrt{\sin\left(\sqrt{\lambda_n}\right)^2 + 1}} \tag{5.63}$$

This equation is the Fourier series expansion of $f(x)$ in terms of the orthonormal Sturm-Liouville eigenfunctions found earlier. To evaluate the Fourier coefficients, we take the inner product of both sides with respect to the orthonormal eigenfunctions, and, taking advantage of the previous statement of orthonormality, we get, for $n = 1, 2, 3, \ldots$,

> B(n):=(1/sinh(sqrt(lambda[n])*b))*Int(f(x)*X[n](x)*w(x),x=0..a);B(n):=value(%):

$$B(n) := \frac{\int_0^1 \frac{(1-x^2)\sqrt{2}\cos\left(\sqrt{\lambda_n}x\right)}{\sqrt{\sin\left(\sqrt{\lambda_n}\right)^2 + 1}}\,dx}{\sinh\left(\sqrt{\lambda_n}\right)} \tag{5.64}$$

Substitution of the eigenvalue equation simplifies the preceding equation

> B(n):=radsimp(subs(sin(sqrt(lambda[n])*a)=(1/sqrt(lambda[n]))*cos(sqrt(lambda[n])*a), B(n)));

$$B(n) := -\frac{2\sqrt{2}\cos\left(\sqrt{\lambda_n}\right)(\lambda_n - 1)}{\sqrt{2 - \cos\left(\sqrt{\lambda_n}\right)^2}\,\lambda_n^2\,\sinh\left(\sqrt{\lambda_n}\right)} \tag{5.65}$$

Generalized series terms

> u[n](x,y):=eval(X[n](x)*Y[n](y));

$$u_n(x, y) := -\frac{4\cos\left(\sqrt{\lambda_n}\,x\right)\cos\left(\sqrt{\lambda_n}\right)(\lambda_n - 1)\sinh\left(\sqrt{\lambda_n}(1 - y)\right)}{\sqrt{\sin\left(\sqrt{\lambda_n}\right)^2 + 1}\sqrt{2 - \cos\left(\sqrt{\lambda_n}\right)^2}\,\lambda_n^2\,\sinh\left(\sqrt{\lambda_n}\right)} \tag{5.66}$$

Series solution

> u(x,y):=Sum(u[n](x,y),n=1..infinity);

$$u(x, y) := \sum_{n=1}^{\infty}\left(-\frac{4\cos\left(\sqrt{\lambda_n}x\right)\cos\left(\sqrt{\lambda_n}\right)(\lambda_n - 1)\sinh\left(\sqrt{\lambda_n}(1 - y)\right)}{\sqrt{\sin\left(\sqrt{\lambda_n}\right)^2 + 1}\sqrt{2 - \cos\left(\sqrt{\lambda_n}\right)^2}\,\lambda_n^2\,\sinh\left(\sqrt{\lambda_n}\right)}\right) \tag{5.67}$$

Evaluation of the eigenvalues from the roots of the eigenvalue equation gives

> tan(sqrt(lambda[n])*a)=(1/sqrt(lambda[n]));

$$\tan\left(\sqrt{\lambda_n}\right) = \frac{1}{\sqrt{\lambda_n}} \tag{5.68}$$

> plot({tan(v),1/v},v=0..20,y=0..3/2,thickness=10);

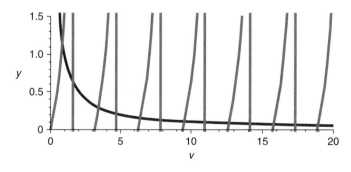

Figure 5.4

If we set $v = \sqrt{\lambda}a$, then the eigenvalues are found from the intersection points of the curves shown in Figure 5.4. An evaluation of the first few eigenvalues using the Maple fsolve command yields

> lambda[1]:=(1/a^2)*(fsolve((tan(v)−1/v),v=0..1))^2;

$$\lambda_1 := 0.7401738844 \tag{5.69}$$

> lambda[2]:=(1/a^2)*(fsolve((tan(v)−1/v),v=1..4))^2;

$$\lambda_2 := 11.73486183 \tag{5.70}$$

> lambda[3]:=(1/a^2)*(fsolve((tan(v)−1/v),v=4..7))^2;

$$\lambda_3 := 41.43880785 \tag{5.71}$$

First few terms of expansion

> u(x,y):=sum(u[n](x,y),n=1..3):

> plot3d(u(x,y),x=0..a,y=0..b,axes=framed,thickness=1);

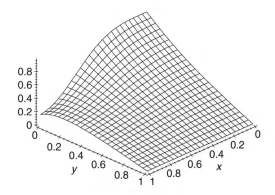

Figure 5.5

The three-dimensional surface shown in Figure 5.5 depicts the steady-state temperature distribution $u(x, y)$ over the rectangular region. Note how the edges of the surface adhere to the given boundary conditions. The temperature isotherms can be obtained from Maple by clicking on the figure, choosing the special option "Render the plot using the polygon patch and contour style" in the graphics bar, and then clicking the "redraw" button.

5.5 Laplace Equation in Cylindrical Coordinates

We now focus on the solution to the Laplace equation over the finite two-dimensional domain $D = \{(r, \theta) \,|\, 0 < r < a, 0 < \theta < b\}$ in cylindrical coordinates. We will be considering problems in regions that have no extension in z; thus, the following laplacian is z independent.

In the cylindrical coordinate system, the Laplace equation reads

$$\frac{\frac{\partial}{\partial r}u(r, \theta) + r\left(\frac{\partial^2}{\partial r^2}u(r, \theta)\right)}{r} + \frac{\frac{\partial^2}{\partial \theta^2}u(r, \theta)}{r^2} = 0$$

We can rewrite this as

$$\frac{\frac{\partial}{\partial r}u(r,\theta)+r\left(\frac{\partial^2}{\partial r^2}u(r,\theta)\right)}{r}=-\frac{\frac{\partial^2}{\partial \theta^2}u(r,\theta)}{r^2}$$

We use the method of separation of variables to solve the preceding equation. We set

$$u(r,\theta)=R(r)\Theta(\theta)$$

Substituting this assumed solution into the partial differential equation, we get

$$\frac{\left(\frac{d}{dr}R(r)\right)\Theta(\theta)+r\left(\frac{d^2}{dr^2}R(r)\right)\Theta(\theta)}{r}=-\frac{R(r)\left(\frac{d^2}{d\theta^2}\Theta(\theta)\right)}{r^2}$$

Dividing the preceding by the product solution yields

$$\frac{r\left(\frac{d}{dr}R(r)+r\left(\frac{d^2}{dr^2}R(r)\right)\right)}{R(r)}=-\frac{\frac{d^2}{d\theta^2}\Theta(\theta)}{\Theta(\theta)} \tag{5.72}$$

Setting each side of the preceding equal to a constant λ, we arrive at the two ordinary differential equations

$$r^2\left(\frac{d^2}{dr^2}R(r)\right)+r\left(\frac{d}{dr}R(r)\right)-\lambda R(r)=0$$

and

$$\frac{d^2}{d\theta^2}\Theta(\theta)+\lambda\Theta(\theta)=0$$

The first differential equation shown is of the Cauchy-Euler type and the second is of the Euler type. To proceed further, we now consider the boundary conditions on the problem. Generally, one seeks a solution to this problem over the finite two-dimensional domain $D=\{(r,\theta)\,|\,0<r<a,0<\theta<b\}$ subject to the regular homogeneous boundary conditions on one of the variables, r or θ. We seek out that variable whose boundary conditions are homogeneous.

For example, consider the case where the boundary conditions on $u(r,\theta)$ are homogeneous with respect to the variable θ; that is,

$$\kappa_1 u(r,0)+\kappa_2 u_\theta(r,0)=0$$

and

$$\kappa_3 u(r,b)+\kappa_4 u_\theta(r,b)=0$$

The earlier ordinary differential equation on θ combined with the preceding equivalent homogeneous boundary conditions constitute a Sturm-Liouville problem in the variable θ. We already have solutions for this type of problem.

5.6 Sturm-Liouville Problem for the Laplace Equation in Cylindrical Coordinates

The differential equation in the spatial variable θ in the preceding section must be solved subject to the equivalent homogeneous spatial boundary conditions over the finite interval $I = \{\theta \mid 0 < \theta < b\}$. The Sturm-Liouville problem for the θ-dependent equation consists of the ordinary differential equation

$$\frac{d^2}{d\theta^2}\Theta(\theta) + \lambda\Theta(\theta) = 0$$

along with the corresponding equivalent homogeneous boundary conditions on the θ variable, which read as

$$\kappa_1\Theta(0) + \kappa_2\Theta_\theta(0) = 0$$

and

$$\kappa_3\Theta(b) + \kappa_4\Theta_\theta(b) = 0$$

We see that the problem in the variable θ is a regular Sturm-Liouville eigenvalue problem where the allowed values of λ are called the system "eigenvalues" and the corresponding solutions are called the "eigenfunctions." Note that the preceding ordinary differential equation is of the Euler type with the weight function $w(\theta) = 1$.

As before, we can show that for regular Sturm-Liouville problems over finite domains, an infinite number of eigenvalues exist that can be indexed by the positive integer n. The indexed eigenvalues and corresponding eigenfunctions are given, respectively, as

$$\lambda_n, \Theta_n(\theta)$$

for $n = 0, 1, 2, 3, \ldots$.

What we stated in Section 2.1 about regular Sturm-Liouville eigenvalue problems holds here; that is, the eigenfunctions form a "complete" set with respect to any piecewise smooth function over the finite interval $I = \{\theta \mid 0 < \theta < b\}$.

The remaining r-dependent differential equation is a Cauchy-Euler-type differential equation and, from Section 1.5, the solution to the corresponding r-dependent differential equation in terms of the allowed eigenvalues is

$$R_n(r) = C(n)r^{\sqrt{\lambda_n}} + D(n)r^{-\sqrt{\lambda_n}}$$

Thus, for each allowed value of the index n, we have a solution

$$u_n(r, \theta) = \left(C(n)r^{\sqrt{\lambda_n}} + D(n)r^{-\sqrt{\lambda_n}}\right)\Theta_n(\theta) \tag{5.73}$$

for $n = 0, 1, 2, 3, \ldots$.

These functions form a set of basis vectors for the solution space of the partial differential equation, and using the same concepts of the superposition principle as in Chapters 3 and 4, the general solution to the homogeneous partial differential equation can now be written as an infinite sum over all basis vectors as follows:

$$u(r, \theta) = \sum_{n=0}^{\infty} \left(C(n) r^{\sqrt{\lambda_n}} + D(n) r^{-\sqrt{\lambda_n}} \right) \Theta_n(\theta)$$

The arbitrary constants $C(n)$ and $D(n)$ are evaluated by taking advantage of the orthonormality of the θ-dependent eigenfunctions over the finite interval $I = \{\theta \mid 0 < \theta < b\}$ with respect to the weight function $w(\theta) = 1$. The statement of orthonormality reads

$$\int_0^b \Theta_n(\theta) \Theta_m(\theta) \, d\theta = \delta(n, m)$$

for $n, m = 0, 1, 2, 3, \ldots$.

We now consider the alternative possibility. If the homogeneous boundary conditions on $u(r, \theta)$ are with respect to the variable r instead of θ, then we rewrite the two earlier ordinary differential equations as

$$r^2 \left(\frac{d^2}{dr^2} R(r) \right) + r \left(\frac{d}{dr} R(r) \right) + \lambda R(r) = 0$$

and

$$\frac{d^2}{d\theta^2} \Theta(\theta) - \lambda \Theta(\theta) = 0$$

We consider the solution to this problem over the interval $I = \{r \mid 1 < r < a\}$; we are avoiding the origin because the differential equation in r is not normal (it is singular) at $r = 0$. Let the equivalent homogeneous boundary conditions with respect to the variable r read as

$$\kappa_1 u(1, \theta) + \kappa_2 u_r(1, \theta) = 0$$

and

$$\kappa_3 u(a, \theta) + \kappa_4 u_r(a, \theta) = 0$$

The preceding ordinary differential equation in r combined with the preceding homogeneous boundary conditions again constitutes a Sturm-Liouville problem, for which we already have solutions from Chapter 2.

The differential equation in the preceding spatial variable r must be solved subject to the homogeneous spatial boundary conditions over the finite interval. The Sturm-Liouville problem for the r-dependent equation consists of the ordinary differential equation

$$r^2\left(\frac{d^2}{dr^2}R(r)\right) + r\left(\frac{d}{dr}R(r)\right) + \lambda R(r) = 0 \tag{5.74}$$

along with the equivalent corresponding regular homogeneous boundary conditions on r, which read

$$\kappa_1 R(1) + \kappa_2 R_r(1) = 0$$

and

$$\kappa_3 R(a) + \kappa_4 R_r(a) = 0$$

We see the preceding problem in the spatial variable r over the interval $I = \{r \mid 1 < r < a\}$ to be a regular Sturm-Liouville eigenvalue problem where the allowed values of λ are called the system "eigenvalues" and the corresponding solutions are called the "eigenfunctions." Note that the preceding ordinary differential equation is of the Cauchy-Euler type, and, from Section 2.5, the weight function is $w(r) = 1/r$.

As before, we can show that for regular Sturm-Liouville problems over finite domains, an infinite number of eigenvalues exist that can be indexed by a positive integer n. The indexed eigenvalues and corresponding eigenfunctions are given, respectively, as

$$\lambda_n, R_n(r)$$

for $n = 0, 1, 2, 3, \ldots$.

What we stated in Section 2.1 about regular Sturm-Liouville eigenvalue problems holds here; that is, the eigenfunctions form a "complete" set with respect to any piecewise smooth function over the finite r interval $I = \{r \mid 1 < r < a\}$.

From Section 1.4 on second-order linear differential equations, the solution to the remaining θ-dependent differential equation in terms of the allowed eigenvalues is

$$\Theta_n(\theta) = A(n)\cosh\left(\sqrt{\lambda_n}\,\theta\right) + B(n)\sinh\left(\sqrt{\lambda_n}\,\theta\right)$$

Thus, for each allowed value of the index n, we have a solution

$$u_n(r, \theta) = R_n(r)\left(A(n)\cosh\left(\sqrt{\lambda_n}\,\theta\right) + B(n)\sinh\left(\sqrt{\lambda_n}\,\theta\right)\right)$$

for $n = 0, 1, 2, 3, \ldots$.

Using the concept of the superposition principle as in Chapters 3 and 4, the general solution to the homogeneous partial differential equation can again be written as the infinite sum of the given basis vectors, in terms of the eigenfunctions with respect to the variable r, as follows:

$$u(r, \theta) = \sum_{n=0}^{\infty} R_n(r)\left(A(n)\cosh\left(\sqrt{\lambda_n}\,\theta\right) + B(n)\sinh\left(\sqrt{\lambda_n}\,\theta\right)\right)$$

Again, the arbitrary constants $A(n)$ and $B(n)$ can be evaluated by taking advantage of the orthonormality of the r-dependent eigenfunctions, over the finite interval $I = \{r \mid 1 < r < a\}$, with respect to the weight function $w(r) = 1/r$. The statement of orthonormality reads

$$\int_1^a \frac{R_n(r)R_m(r)}{r}\,dr = \delta(n, m)$$

for $n, m = 0, 1, 2, 3, \ldots$.

The original statement of the boundary conditions on the problem determines whether the solution is written in terms of the eigenfunctions with respect to r or θ. Remember that the decision comes from our seeking that particular variable that is experiencing the homogeneous boundary conditions.

DEMONSTRATION: We seek the steady-state temperature distribution in a thin cylindrical plate over the domain $D = \{(r, \theta) \mid 0 < r < 1, 0 < \theta < \pi/3\}$ whose lateral surface is insulated. The sides $\theta = 0$ and $\theta = \pi/3$ are fixed at zero temperature, and the side $r = 1$ has a temperature distribution $f(\theta)$ given as follows. The solution is required to be finite at the origin.

SOLUTION: The homogeneous Laplace equation is

$$\frac{\frac{\partial}{\partial r}u(r, \theta) + r\left(\frac{\partial^2}{\partial r^2}u(r, \theta)\right)}{r} + \frac{\frac{\partial^2}{\partial \theta^2}u(r, \theta)}{r^2} = 0$$

The boundary conditions are

$$|u(0, \theta)| < \infty \quad \text{and} \quad u(1, \theta) = f(\theta)$$

and

$$u(r, 0) = 0 \quad \text{and} \quad u\left(r, \frac{\pi}{3}\right) = 0$$

We seek that variable whose boundary conditions are homogeneous. Since the boundary conditions are homogeneous with respect to the θ variable, we write the ordinary differential equations from the method of separation of variables as follows:

$$\frac{d^2}{d\theta^2}\Theta(\theta) + \lambda\Theta(\theta) = 0$$

and

$$r^2 \left(\frac{d^2}{dr^2} R(r) \right) + r \left(\frac{d}{dr} R(r) \right) - \lambda R(r) = 0$$

We first consider the θ-dependent differential equation, since the boundary conditions on this equation are homogeneous. The boundary conditions on the θ equation are type 1 at $\theta = 0$ and type 1 at $\theta = \pi/3$:

$$\Theta(0) = 0 \quad \text{and} \quad \Theta\left(\frac{\pi}{3} \right) = 0$$

This same boundary value problem was considered in Example 2.5.1 in Chapter 2. The allowed eigenvalues are

$$\lambda_n = 9n^2$$

for $n = 1, 2, 3, \ldots$, and the corresponding orthonormal eigenfunctions are

$$\Theta_n(\theta) = \sqrt{\frac{6}{\pi}} \sin(3n\theta)$$

The statement of orthonormality with respect to the weight function $w(\theta) = 1$ over the interval $I = \{\theta \mid 0 < \theta < \pi/3\}$ reads

$$\int_0^{\frac{\pi}{3}} \frac{6 \sin(3n\theta) \sin(3m\theta)}{\pi} \, d\theta = \delta(n, m)$$

Focusing on the r-dependent equation, we see that we have a Cauchy-Euler-type differential equation and a set of basis vectors for this equation for $\lambda_n = 9n^2$ is

$$r1 = r^{3n}$$

and

$$r2 = r^{-3n}$$

Thus, the general solution to the r equation reads

$$R_n(r) = C(n)r^{3n} + D(n)r^{-3n}$$

Since the r-dependent solution must be finite at $r = 0$, we must set the coefficient $D(n) = 0$, and the final solution to the r-dependent equation reads

$$R_n(r) = C(n)r^{3n}$$

From the superposition of the products of the θ- and r-dependent solutions, the final eigenfunction expansion for the solution reads

$$u(r, \theta) = \sum_{n=1}^{\infty} C(n) r^{3n} \sqrt{\frac{6}{\pi}} \sin(3n\theta)$$

The Fourier coefficients $C(n)$ are to be determined from the specific remaining boundary condition $u(1, \theta) = f(\theta)$. We consider the special case for $f(\theta) = \theta \left(\frac{\pi}{3} - \theta\right)$. Substitution of this condition into the preceding solution yields

$$\theta\left(\frac{\pi}{3} - \theta\right) = \sum_{n=1}^{\infty} C(n) \sqrt{\frac{6}{\pi}} \sin(3n\theta) \tag{5.75}$$

This equation is the Fourier series expansion of the boundary function $f(\theta)$ in terms of the orthonormal Sturm-Liouville eigenfunctions found earlier. To evaluate the Fourier cofficients, we take the inner product of both sides with respect to the orthonormal eigenfunctions over the interval $I = \{\theta \mid 0 < \theta < \pi/3\}$ with respect to the weight function $w(\theta) = 1$. Doing so yields

$$\int_0^{\frac{\pi}{3}} \theta\left(\frac{\pi}{3} - \theta\right) \sqrt{\frac{6}{\pi}} \sin(3m\theta)\, d\theta = \sum_{n=1}^{\infty} C(n) \left(\int_0^{\frac{\pi}{3}} \sqrt{\frac{6}{\pi}} \sin(3n\theta) \sqrt{\frac{6}{\pi}} \sin(3m\theta)\, d\theta \right)$$

Taking advantage of the preceding statement of orthonormality, the Fourier coefficients are

$$C(n) = \int_0^{\frac{\pi}{3}} \theta\left(\frac{\pi}{3} - \theta\right) \sqrt{\frac{6}{\pi}} \sin(3n\theta)\, d\theta$$

Evaluation of this integral yields

$$C(n) = \frac{\sqrt{\frac{24}{\pi}} (1 - (-1)^n)}{(3n)^3}$$

for $n = 1, 2, 3, \ldots$. Thus, the final series solution to our problem reads

$$u(r, \theta) = \sum_{n=1}^{\infty} \frac{4(1 - (-1)^n) r^{3n} \sin(3n\theta)}{9n^3 \pi}$$

The details for the development of the solution and the graphics are given later in one of the Maple worksheet examples.

5.7 Example Laplace Problems in the Cylindrical Coordinate System

We now consider Laplace equation examples in the cylindrical coordinate system. The examples here have no extension along the z-axis; thus, the Laplacian here is z-independent. Note that all the r-dependent differential equations are of the Cauchy-Euler type and the θ-dependent equations are of the Euler type.

EXAMPLE 5.7.1: We seek the steady-state temperature distribution in a thin cylindrical plate over the domain $D = \{(r, \theta) \mid 0 < r < 1, 0 < \theta < \pi/3\}$ whose lateral surface is insulated. The sides $\theta = 0$ and $\theta = \pi/3$ are fixed at temperature zero, and the side $r = 1$ has a temperature distribution $f(\theta)$ given as follows. The solution is required to be finite at the origin.

SOLUTION: The homogeneous Laplace equation is

$$\frac{\frac{\partial}{\partial r}u(r,\theta) + r\left(\frac{\partial^2}{\partial r^2}u(r,\theta)\right)}{r} + \frac{\frac{\partial^2}{\partial\theta^2}u(r,\theta)}{r^2} = 0$$

The boundary conditions are

$$|u(0,\theta)| < \infty \quad \text{and} \quad u(1,\theta) = \theta\left(\frac{\pi}{3} - \theta\right)$$

and

$$u(r,0) = 0 \quad \text{and} \quad u\left(r, \frac{\pi}{3}\right) = 0$$

Here we see that the boundary conditions are homogeneous with respect to the θ variable.

Ordinary differential equations from the method of separation of variables are

$$\frac{d^2}{d\theta^2}\Theta(\theta) + \lambda\Theta(\theta) = 0$$

and

$$r^2\left(\frac{d^2}{dr^2}R(r)\right) + r\left(\frac{d}{dr}R(r)\right) - \lambda R(r) = 0$$

Homogeneous boundary conditions on the θ equation are type 1 at $\theta = 0$ and type 1 at $\theta = \pi/3$

$$\Theta(0) = 0 \quad \text{and} \quad \Theta\left(\frac{\pi}{3}\right) = 0$$

Assignment of system parameters

> restart:with(plots):a:=1:b:=Pi/3:

The allowed eigenvalues and corresponding orthonormal eigenfunctions are obtained from Example 2.5.1.

> lambda[n]:=(n*Pi/b)^2;

$$\lambda_n := 9n^2 \tag{5.76}$$

for $n = 1, 2, 3, \ldots$.

Orthonormal eigenfunctions

> Theta[n](theta):=sqrt(2/b)*sin(n*Pi/b*theta);Theta[m](theta):=subs(n=m,Theta[n](theta)):

$$\Theta_n(\theta) := \frac{\sqrt{6}\sin(3n\theta)}{\sqrt{\pi}} \tag{5.77}$$

Statement of orthonormality with respect to the weight function $w(\theta) = 1$

> w(theta):=1:Int(Theta[n](theta)*Theta[m](theta)*w(theta),theta=0..b)=delta(n,m);

$$\int_0^{\frac{1}{3}\pi} \frac{6\sin(3n\theta)\sin(3m\theta)}{\pi}\, d\theta = \delta(n, m) \tag{5.78}$$

Basis vectors for the r equation

> r1(r):=r^(sqrt(lambda));r2(r):=r^(−sqrt(lambda));

$$r1(r) := r^{\sqrt{\lambda}}$$
$$r2(r) := r^{-\sqrt{\lambda}} \tag{5.79}$$

General solution to the r equation

> R[n](r):=C(n)*r^(sqrt(lambda[n]))+D(n)*r^(−sqrt(lambda[n]));

$$R_n(r) := C(n)r^{3\sqrt{n^2}} + D(n)r^{-3\sqrt{n^2}} \tag{5.80}$$

Substituting into the r-dependent boundary condition requiring a finite solution at $r = 0$ yields

> D(n):=0;

$$D(n) := 0 \tag{5.81}$$

The r-dependent solution for the given eigenvalues reads

> R[n](r):=C(n)*r^(n*Pi/b);

$$R_n(r) := C(n)r^{3n} \tag{5.82}$$

Generalized series terms

> u[n](r,theta):=R[n](r)*Theta[n](theta);

$$u_n(r, \theta) := \frac{C(n)r^{3n}\sqrt{6}\sin(3n\theta)}{\sqrt{\pi}} \tag{5.83}$$

Eigenfunction expansion

> u(r,theta):=Sum(u[n](r,theta),n=1..infinity);

$$u(r, \theta) := \sum_{n=1}^{\infty} \frac{C(n)r^{3n}\sqrt{6}\sin(3n\theta)}{\sqrt{\pi}} \tag{5.84}$$

Evaluation of Fourier coefficients from the specific remaining boundary condition $u(1, \theta) = f(\theta)$

> f(theta):=theta*(Pi/3−theta);

$$f(\theta) := \theta\left(\frac{1}{3}\pi - \theta\right) \tag{5.85}$$

Substituting this condition into the preceding solution yields

> f(theta)=subs(r=a,u(r,theta));

$$\theta\left(\frac{1}{3}\pi - \theta\right) = \sum_{n=1}^{\infty} \frac{C(n)\sqrt{6}\sin(3n\theta)}{\sqrt{\pi}} \tag{5.86}$$

This equation is the Fourier series expansion of $f(\theta)$ in terms of the orthonormal Sturm-Liouville eigenfunctions found earlier. To evaluate the Fourier cofficients, we take the inner product of both sides with respect to the orthonormal eigenfunctions, and, taking advantage of the previous statement of orthonormality, we get, for $n = 1, 2, 3, \ldots,$

> C(n):=(1/(a^(n*Pi/b)))*Int(f(theta)*Theta[n](theta)*w(theta),theta=0..b);
  C(n):=expand(value(%)):

$$C(n) := \int_{0}^{\frac{1}{3}\pi} \frac{\theta\left(\frac{1}{3}\pi - \theta\right)\sqrt{6}\sin(3n\theta)}{\sqrt{\pi}} \, d\theta \tag{5.87}$$

> C(n):=simplify(subs({sin(n*Pi)=0,cos(n*Pi)=(−1)^n},C(n)));

$$C(n) := -\frac{2}{27}\frac{\sqrt{6}(-1+(-1)^n)}{\sqrt{\pi}n^3} \tag{5.88}$$

Generalized series terms

> u[n](r,theta):=eval(R[n](r)*Theta[n](theta));

$$u_n(r, \theta) := -\frac{4}{9}\frac{(-1+(-1)^n)r^{3n}\sin(3n\theta)}{\pi n^3} \tag{5.89}$$

Series solution

> u(r,theta):=Sum(u[n](r,theta),n=1..infinity);

$$u(r, \theta) := \sum_{n=1}^{\infty}\left(-\frac{4}{9}\frac{(-1+(-1)^n)r^{3n}\sin(3n\theta)}{\pi n^3}\right) \tag{5.90}$$

First few terms of expansion

> u(r,theta):=sum(u[n](r,theta),n=1..5):

> cylinderplot([r,theta,u(r,theta)],r=0..a,theta=0..b,axes=framed,orientation=[−31,61],
thickness=1);

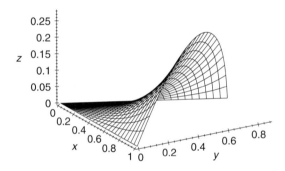

Figure 5.6

The three-dimensional surface shown in Figure 5.6 depicts the steady-state temperature distribution $u(r, \theta)$ over the cylindrical region. Note how the edges of the surface adhere to the given boundary conditions. The temperature isotherms can be obtained from Maple by clicking on the figure, choosing the special option "Render the plot using the polygon patch and contour style" in the graphics bar, and then clicking the "redraw" button.

EXAMPLE 5.7.2: We seek the steady-state temperature distribution in a thin cylindrical plate over the domain $D = \{(r, \theta) \mid 0 < r < 1, 0 < \theta < \pi/3\}$ whose lateral surface is insulated. The sides $\theta = 0$ and $\theta = \pi/3$ are insulated, the solution is finite at $r = 0$, and the side $r = 1$ has a temperature distribution $f(\theta)$ given as follows.

SOLUTION: The homogeneous Laplace equation is

$$\frac{\frac{\partial}{\partial r}u(r, \theta) + r\left(\frac{\partial^2}{\partial r^2}u(r, \theta)\right)}{r} + \frac{\frac{\partial^2}{\partial \theta^2}u(r, \theta)}{r^2} = 0$$

The boundary conditions are

$$|u(0,\theta)| < \infty \quad \text{and} \quad u(1,\theta) = \theta\left(\frac{\pi}{3} - \theta\right)$$

and

$$u_\theta(r,0) = 0 \quad \text{and} \quad u_\theta\left(r, \frac{\pi}{3}\right) = 0$$

Here we see that the boundary conditions are homogeneous with respect to the θ variable.

Ordinary differential equations from the method of separation of variables are

$$\frac{d^2}{d\theta^2}\Theta(\theta) + \lambda\Theta(\theta) = 0$$

and

$$r^2\left(\frac{d^2}{dr^2}R(r)\right) + r\left(\frac{d}{dr}R(r)\right) - \lambda R(r) = 0$$

Homogeneous boundary conditions on the θ equation are type 2 at $\theta = 0$ and type 2 at $\theta = \pi/3$

$$\Theta_\theta(0) = 0 \quad \text{and} \quad \Theta_\theta\left(\frac{\pi}{3}\right) = 0$$

Assignment of system parameters

> restart:with(plots):a:=1:b:=Pi/3:

Allowed eigenvalues and corresponding orthonormal eigenfunctions for θ equation are obtained from Example 2.5.3. For $n = 0$,

> lambda[0]:=0;

$$\lambda_0 := 0 \tag{5.91}$$

Orthonormal eigenfunctions

> Theta[0](theta):=1/sqrt(b);

$$\Theta_0(\theta) := \frac{\sqrt{3}}{\sqrt{\pi}} \tag{5.92}$$

For $n = 1, 2, 3, \ldots,$

> lambda[n]:=(n*Pi/b)^2;

$$\lambda_n := 9n^2 \tag{5.93}$$

Orthonormal eigenfunctions

> Theta[n](theta):=sqrt(2/b)*cos(n*Pi/b*theta);Theta[m](theta):=subs(n=m,Theta[n](theta)):

$$\Theta_n(\theta) := \frac{\sqrt{6}\cos(3n\theta)}{\sqrt{\pi}} \tag{5.94}$$

Statement of orthonormality with respect to the weight function $w(\theta) = 1$

> w(theta):=1:Int(Theta[n](theta)*Theta[m](theta)*w(theta),theta=0..b)=delta(n,m);

$$\int_0^{\frac{1}{3}\pi} \frac{6\cos(3n\theta)\cos(3m\theta)}{\pi}\, d\theta = \delta(n,m) \tag{5.95}$$

Basis vectors for the r equation for $\lambda > 0$

> r1(r):=r^(sqrt(lambda));r2(r):=r^(-sqrt(lambda));

$$r1(r) := r^{\sqrt{\lambda}}$$

$$r2(r) := r^{-\sqrt{\lambda}} \tag{5.96}$$

General solution to the r equation for $\lambda > 0$ is

> R[n](r):=A(n)*r^(sqrt(lambda))+B(n)*r^(-sqrt(lambda));

$$R_n(r) := A(n)r^{\sqrt{\lambda}} + B(n)r^{-\sqrt{\lambda}} \tag{5.97}$$

Substituting into the r-dependent boundary condition requiring a finite solution at $r = 0$ yields

> B(n):=0;

$$B(n) := 0 \tag{5.98}$$

The r-dependent solution for $n = 1, 2, 3, \ldots$, for the given eigenvalues, reads

> R[n](r):=A(n)*r^(n*Pi/b);

$$R_n(r) := A(n)r^{3n} \tag{5.99}$$

Basis vectors for the r equation for $\lambda = 0$

> r1(r):=1;r2(r):=ln(r);

$$r1(r) := 1$$

$$r2(r) := \ln(r) \tag{5.100}$$

General solution to the r equation for $\lambda = 0$

> R[0](r):=A(0)+B(0)*ln(r);

$$R_0(r) := A(0) + B(0)\ln(r) \tag{5.101}$$

Substituting into the r-dependent boundary condition at $r = 0$ yields

> B(0):=0;

$$B(0) := 0 \tag{5.102}$$

The r-dependent solution for $n = 0$ reads

> R[0](r):=A(0);

$$R_0(r) := A(0) \tag{5.103}$$

Generalized series terms

> u[0](r,theta):=R[0](r)*Theta[0](theta);u[n](r,theta):=R[n](r)*Theta[n](theta);

$$u_0(r, \theta) := \frac{A(0)\sqrt{3}}{\sqrt{\pi}}$$

$$u_n(r, \theta) := \frac{A(n)r^{3n}\sqrt{6}\cos(3n\theta)}{\sqrt{\pi}} \tag{5.104}$$

Eigenfunction expansion

> u(r,theta):=u[0](r,theta)+Sum(u[n](r,theta),n=1..infinity);

$$u(r, \theta) := \frac{A(0)\sqrt{3}}{\sqrt{\pi}} + \sum_{n=1}^{\infty} \frac{A(n)r^{3n}\sqrt{6}\cos(3n\theta)}{\sqrt{\pi}} \tag{5.105}$$

Evaluation of Fourier coefficients from the specific remaining boundary condition $u(1, \theta) = f(\theta)$

> f(theta):=theta*(Pi/3−theta);

$$f(\theta) := \theta\left(\frac{1}{3}\pi - \theta\right) \tag{5.106}$$

Substituting this condition into the preceding solution yields

> f(theta)=subs(r=a,u(r,theta));

$$\theta\left(\frac{1}{3}\pi - \theta\right) = \frac{A(0)\sqrt{3}}{\sqrt{\pi}} + \sum_{n=1}^{\infty} \frac{A(n)\sqrt{6}\cos(3n\theta)}{\sqrt{\pi}} \tag{5.107}$$

This equation is the Fourier series expansion of $f(\theta)$ in terms of the orthonormal Sturm-Liouville eigenfunctions found earlier. To evaluate the Fourier cofficients, we take the inner product of both sides with respect to the orthonormal eigenfunctions, and, taking advantage of the preceding orthonormality statement, we get, for $n = 1, 2, 3, \ldots$,

```
> A(n):=eval((1/(a^(n*Pi/b)))*Int(f(theta)*Theta[n](theta)*w(theta),theta=0..b));
  A(n):=expand(value(%)):
```

$$A(n) := \int_0^{\frac{1}{3}\pi} \frac{\theta\left(\frac{1}{3}\pi - \theta\right)\sqrt{6}\cos(3n\theta)}{\sqrt{\pi}} \, d\theta \tag{5.108}$$

```
> A(n):=simplify(subs({sin(n*Pi)=0,cos(n*Pi)=(-1)^n},A(n)));
```

$$A(n) := -\frac{1}{27}\frac{\sqrt{\pi}\sqrt{6}(1 + (-1)^n)}{n^2} \tag{5.109}$$

For $n = 0$,

```
> A(0):=Int(f(theta)*Theta[0](theta)*w(theta),theta=0..b);
```

$$A(0) := \int_0^{\frac{1}{3}\pi} \frac{\theta\left(\frac{1}{3}\pi - \theta\right)\sqrt{3}}{\sqrt{\pi}} \, d\theta \tag{5.110}$$

```
> A(0):=value(%);
```

$$A(0) := \frac{1}{162}\sqrt{3}\pi^{5/2} \tag{5.111}$$

Generalized series terms

```
> u[0](r,theta):=eval(R[0](r)*Theta[0](theta));u[n](r,theta):=eval(R[n](r)*Theta[n](theta));
```

$$u_0(r, \theta) := \frac{1}{54}\pi^2$$

$$u_n(r, \theta) := -\frac{2}{9}\frac{(1 + (-1)^n)r^{3n}\cos(3n\theta)}{n^2} \tag{5.112}$$

Series solution

```
> u(r,theta):=u[0](r,theta)+Sum(u[n](r,theta),n=1..infinity);
```

$$u(r, \theta) := \frac{1}{54}\pi^2 + \sum_{n=1}^{\infty}\left(-\frac{2}{9}\frac{(1 + (-1)^n)r^{3n}\cos(3n\theta)}{n^2}\right) \tag{5.113}$$

First few terms of expansion

```
> u(r,theta):=u[0](r,theta)+sum(u[n](r,theta),n=1..5):
```

> cylinderplot([r,theta,u(r,theta)],r=0..a,theta=0..b,axes=framed,orientation=[−31,61],
  thickness=1);

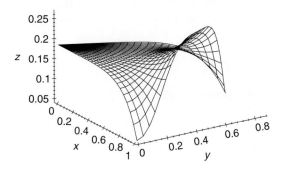

Figure 5.7

The three-dimensional surface shown in Figure 5.7 depicts the steady-state temperature distribution $u(r, \theta)$ over the cylindrical region. Note how the edges of the surface adhere to the given boundary conditions. The temperature isotherms can be obtained from Maple by clicking on the figure, choosing the special option "Render the plot using the polygon patch and contour style" in the graphics bar, and then clicking the "redraw" button.

EXAMPLE 5.7.3: We seek the electrostatic potential in a cylindrical region over the domain $D = \{(r, \theta) \,|\, 1 < r < 2, 0 < \theta < \pi/2\}$. The sides $r = 1, r = 2$, and $\theta = \pi/2$ are held at fixed potential zero. The side $\theta = 0$ has an electrostatic potential distribution $f(r)$ given as follows.

SOLUTION: The homogeneous Laplace equation is

$$\frac{\frac{\partial}{\partial r}u(r, \theta) + r\left(\frac{\partial^2}{\partial r^2}u(r, \theta)\right)}{r} + \frac{\frac{\partial^2}{\partial \theta^2}u(r, \theta)}{r^2} = 0$$

The boundary conditions are

$$u(1, \theta) = 0 \quad \text{and} \quad u(2, \theta) = 0$$

and

$$u(r, 0) = (r - 1)(2 - r) \quad \text{and} \quad u\left(r, \frac{\pi}{2}\right) = 0$$

Here we see that the boundary conditions are homogeneous with respect to the r variable.

Ordinary differential equations from the method of separation of variables are

$$\frac{d^2}{d\theta^2}\Theta(\theta) - \lambda\Theta(\theta) = 0$$

and

$$r^2\left(\frac{d^2}{dr^2}R(r)\right) + r\left(\frac{d}{dr}R(r)\right) + \lambda R(r) = 0$$

Homogeneous boundary conditions on the r equation are type 1 at $r = 1$ and type 1 at $r = 2$

$$R(1) = 0 \quad \text{and} \quad R(2) = 0$$

Assignment of system parameters

> restart:with(plots):a:=2:b:=Pi/2:

Allowed eigenvalues and corresponding orthonormal eigenfunctions for the r equation are obtained from Example 2.5.7.

> lambda[n]:=(n*Pi/(ln(a)))^2;

$$\lambda_n := \frac{n^2 \pi^2}{\ln(2)^2} \tag{5.114}$$

for $n = 1, 2, 3, \ldots$.

Orthonormal eigenfunctions

> R[n](r):=sqrt(2/ln(a))*sin(n*Pi*ln(r)/(ln(a)));R[n](s):=subs(r=s,R[n](r)):R[m](r):=
subs(n=m,R[n](r)):

$$R_n(r) := \frac{\sqrt{2} \sin\left(\dfrac{n\pi \ln(r)}{\ln(2)}\right)}{\sqrt{\ln(2)}} \tag{5.115}$$

Statement of orthonormality with respect to the weight function $w(r) = \frac{1}{r}$

> w(r):=1/r:Int(R[n](r)*R[m](r)/r,r=1..a)=delta(n,m);

$$\int_1^2 \frac{2 \sin\left(\dfrac{n\pi \ln(r)}{\ln(2)}\right) \sin\left(\dfrac{m\pi \ln(r)}{\ln(2)}\right)}{\ln(2)r} \, dr = \delta(n, m) \tag{5.116}$$

Basis vectors for the θ equation

> theta1(theta):=sinh(sqrt(lambda)*(b−theta));theta2(theta):=cosh(sqrt(lambda)*(b−theta));

$$\theta1(\theta) := \sinh\left(\sqrt{\lambda}\left(\frac{1}{2}\pi - \theta\right)\right)$$

$$\theta2(\theta) := \cosh\left(\sqrt{\lambda}\left(\frac{1}{2}\pi - \theta\right)\right) \tag{5.117}$$

General solution to the θ equation

> Theta[n](theta):=A(n)*cosh(n*Pi/(ln(a))*(b−theta))+B(n)*sinh(n*Pi/(ln(a))*(b−theta));

$$\Theta_n(\theta) := A(n)\cosh\left(\frac{n\pi\left(\frac{1}{2}\pi-\theta\right)}{\ln(2)}\right) + B(n)\sinh\left(\frac{n\pi\left(\frac{1}{2}\pi-\theta\right)}{\ln(2)}\right) \qquad (5.118)$$

Substituting into the θ-dependent boundary condition yields

> eval(subs(theta=b,Theta[n](theta)))=0;

$$A(n) = 0 \qquad (5.119)$$

The θ-dependent solution

> Theta[n](theta):=B(n)*sinh(n*Pi*(b−theta)/ln(a));

$$\Theta_n(\theta) := B(n)\sinh\left(\frac{n\pi\left(\frac{1}{2}\pi-\theta\right)}{\ln(2)}\right) \qquad (5.120)$$

Generalized series terms

> u[n](r,theta):=R[n](r)*Theta[n](theta);

$$u_n(r,\theta) := \frac{\sqrt{2}\sin\left(\frac{n\pi\ln(r)}{\ln(2)}\right)B(n)\sinh\left(\frac{n\pi\left(\frac{1}{2}\pi-\theta\right)}{\ln(2)}\right)}{\sqrt{\ln(2)}} \qquad (5.121)$$

Eigenfunction expansion

> u(r,theta):=Sum(u[n](r,theta),n=1..infinity);

$$u(r,\theta) := \sum_{n=1}^{\infty} \frac{\sqrt{2}\sin\left(\frac{n\pi\ln(r)}{\ln(2)}\right)B(n)\sinh\left(\frac{n\pi\left(\frac{1}{2}\pi-\theta\right)}{\ln(2)}\right)}{\sqrt{\ln(2)}} \qquad (5.122)$$

Evaluation of Fourier coefficients from the specific remaining boundary condition $u(r,0) = f(r)$

> f(r):=(r−1)*(2−r);

$$f(r) := (r-1)(2-r) \qquad (5.123)$$

Substituting this equation into the preceding solution yields

> f(r)=subs(theta=0,u(r,theta));

$$(r-1)(2-r) = \sum_{n=1}^{\infty} \frac{\sqrt{2}\sin\left(\dfrac{n\pi \ln(r)}{\ln(2)}\right) B(n) \sinh\left(\dfrac{1}{2}\dfrac{n\pi^2}{\ln(2)}\right)}{\sqrt{\ln(2)}} \tag{5.124}$$

This equation is the Fourier series expansion of $f(r)$ in terms of the orthonormal Sturm-Liouville eigenfunctions found earlier. To evaluate the Fourier cofficients, we take the inner product of both sides with respect to the orthonormal eigenfunctions, and, taking advantage of the previous statement of orthonormality, we get, for $n = 1, 2, 3, \ldots$,

> B(n):=(1/sinh(n*Pi*b/ln(a)))*Int(f(r)*R[n](r)*w(r),r=1..a);B(n):=simplify(value(%)):

$$B(n) := \frac{\displaystyle\int_1^2 \frac{(r-1)(2-r)\sqrt{2}\sin\left(\dfrac{n\pi \ln(r)}{\ln(2)}\right)}{\sqrt{\ln(2)}\,r}\,dr}{\sinh\left(\dfrac{1}{2}\dfrac{n\pi^2}{\ln(2)}\right)} \tag{5.125}$$

> B(n):=simplify(subs({sin(n*Pi/2)=(exp(i*n*Pi/2)−exp(−i*n*Pi/2))/(2*i),cos(n*Pi/2)=
  (exp(i*n*Pi/2)+exp(−i*n*Pi/2))/2},B(n))):B(n):=simplify(subs({exp(i*n*Pi)=(−1)^n,
  exp(−i*n*Pi)=(−1)^n},B(n)));

$$B(n) := \frac{\sqrt{2}\ln(2)^{5/2}(n^2\pi^2 + 10(-1)^{1+n}n^2\pi^2 - 8\ln(2)^2 + 8\ln(2)^2(-1)^n)}{n\pi \sinh\left(\dfrac{1}{2}\dfrac{n\pi^2}{\ln(2)}\right)(n^4\pi^4 + 5n^2\pi^2 \ln(2)^2 + 4\ln(2)^4)} \tag{5.126}$$

Generalized series terms

> u[n](r,theta):=eval(R[n](r)*Theta[n](theta));

$$u_n(r,\theta) := \left(2\ln(2)^2 \sin\left(\frac{n\pi \ln(r)}{\ln(2)}\right)\left(n^2\pi^2 + 10(-1)^{1+n}n^2\pi^2 - 8\ln(2)^2 + 8\ln(2)^2(-1)^n\right)\right.$$

$$\left.\times \sinh\left(\frac{n\pi\left(\frac{1}{2}\pi - \theta\right)}{\ln(2)}\right)\right) \Bigg/ \left(n\pi \sinh\left(\frac{1}{2}\frac{n\pi^2}{\ln(2)}\right)\left(n^4\pi^4 + 5n^2\pi^2 \ln(2)^2 + 4\ln(2)^4\right)\right)$$

$$\tag{5.127}$$

Series solution

> u(r,theta):=Sum(u[n](r,theta),n=1..infinity);

$$u(r, \theta) := \sum_{n=1}^{\infty} \left(2 \ln(2)^2 \sin\left(\frac{n\pi \ln(r)}{\ln(2)} \right) \left(n^2 \pi^2 + 10(-1)^{1+n} n^2 \pi^2 - 8 \ln(2)^2 + 8 \ln(2)^2 (-1)^n \right) \right.$$

$$\left. \times \sinh\left(\frac{n\pi \left(\frac{1}{2}\pi - \theta \right)}{\ln(2)} \right) \right) \bigg/ \left(\pi n \sinh\left(\frac{1}{2} \frac{n\pi^2}{\ln(2)} \right) \left(n^4 \pi^4 + 5n^2 \pi^2 \ln(2)^2 + 4 \ln(2)^4 \right) \right)$$

$$(5.128)$$

First few terms of expansion

> u(r,theta):=sum(u[n](r,theta),n=1..5):
> cylinderplot([r,theta,u(r,theta)],r=1..a,theta=0..b,axes=framed,thickness=1);

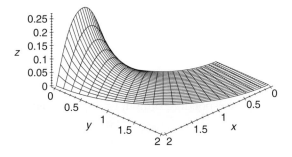

Figure 5.8

The three-dimensional surface shown in Figure 5.8 depicts the electrostatic potential distribution $u(r, \theta)$ over the cylindrical region. Note how the edges of the surface adhere to the given boundary conditions. The equipotential lines can be obtained from Maple by clicking on the figure, choosing the special option "Render the plot using the polygon patch and contour style" in the graphics bar, and then clicking the "redraw" button.

EXAMPLE 5.7.4: We seek the steady-state temperature distribution in a thin cylindrical plate over the domain $D = \{(r, \theta) \,|\, 0 < r < 1, -\pi < \theta < \pi\}$ whose lateral surface is insulated. The side $r = 1$ has a temperature distribution $f(\theta)$ given as follows, the solution is finite at $r = 0$, and we force periodic boundary conditions on the angle θ.

SOLUTION: The homogeneous Laplace equation is

$$\frac{\frac{\partial}{\partial r} u(r, \theta) + r \left(\frac{\partial^2}{\partial r^2} u(r, \theta) \right)}{r} + \frac{\frac{\partial^2}{\partial \theta^2} u(r, \theta)}{r^2} = 0$$

The boundary conditions are

$$|u(0, \theta)| < \infty \quad \text{and} \quad u(1, \theta) = \sin(3\theta)$$

and

$$u(r, -\pi) = u(r, \pi) \quad \text{and} \quad u_\theta(r, -\pi) = u_\theta(r, \pi)$$

Here we see that the boundary conditions are periodic with respect to the θ variable.

Ordinary differential equations from the method of separation of variables are

$$\frac{d^2}{d\theta^2}\Theta(\theta) + \lambda\Theta(\theta) = 0$$

and

$$r^2\left(\frac{d^2}{dr^2}R(r)\right) + r\left(\frac{d}{dr}R(r)\right) - \lambda R(r) = 0$$

Periodic conditions on the θ equation

$$\Theta(-\pi) = \Theta(\pi) \quad \text{and} \quad \Theta_\theta(-\pi) = \Theta_\theta(\pi)$$

Assignment of system parameters

> restart:with(plots):a:=1:

The allowed eigenvalues and corresponding orthonormal eigenfunctions are obtained from Example 2.6.1. For $n = 0$,

> lambda[0]:=0;

$$\lambda_0 := 0 \tag{5.129}$$

For $n = 1, 2, 3, \ldots$,

> lambda[n]:=n^2;

$$\lambda_n := n^2 \tag{5.130}$$

Orthonormal eigenfunctions

For $n = 1, 2, 3, \ldots$,

> phi[n](theta):=sqrt(1/Pi)*cos(n*theta);psi[n](theta):=sqrt(1/Pi)*sin(n*theta);phi[m](theta):= subs(n=m,phi[n](theta)):psi[m](theta):=subs(n=m,psi[n](theta)):

$$\phi_n(\theta) := \frac{\cos(n\theta)}{\sqrt{\pi}}$$

$$\psi_n(\theta) := \frac{\sin(n\theta)}{\sqrt{\pi}} \tag{5.131}$$

For $n = 0$,

> phi[0](theta):=1/sqrt(2*Pi);

$$\phi_0(\theta) := \frac{1}{2}\frac{\sqrt{2}}{\sqrt{\pi}} \tag{5.132}$$

Statement of orthonormality with respect to the weight function $w(\theta) = 1$

> w(theta):=1:Int(phi[n](theta)*phi[m](theta)*w(theta),theta=−Pi..Pi)=delta(n,m);

$$\int_{-\pi}^{\pi} \frac{\cos(n\theta)\cos(m\theta)}{\pi} d\theta = \delta(n,m) \tag{5.133}$$

> Int(psi[n](theta)*psi[m](theta)*w(theta),theta=−Pi..Pi)=delta(n,m);

$$\int_{-\pi}^{\pi} \frac{\sin(n\theta)\sin(m\theta)}{\pi} d\theta = \delta(n,m) \tag{5.134}$$

Basis vectors for the r equation for $\lambda > 0$

> r1(r):=r^(sqrt(lambda));r2(r):=r^(−sqrt(lambda));

$$r1(r) := r^{\sqrt{\lambda}}$$
$$r2(r) := r^{-\sqrt{\lambda}} \tag{5.135}$$

General solution to the r equation for $\lambda > 0$

> R[n](r):=C(n)*r^(sqrt(lambda))+D(n)*r^(−sqrt(lambda));

$$R_n(r) := C(n)r^{\sqrt{\lambda}} + D(n)r^{-\sqrt{\lambda}} \tag{5.136}$$

Substituting into the r-dependent boundary condition and requiring a finite solution at $r = 0$ yields

> D(n):=0;

$$D(n) := 0 \tag{5.137}$$

The r-dependent solution for $n = 1, 2, 3, \ldots$, for the given eigenvalues, reads

> R[n](r):=r^(n);

$$R_n(r) := r^n \tag{5.138}$$

Basis vectors for the r equation for $\lambda = 0$

> r1(r):=1;r2(r):=ln(r);

$$r1(r) := 1$$
$$r2(r) := \ln(r) \qquad (5.139)$$

General solution to the r equation for $\lambda = 0$ is

> R[0](r):=A(0)+B(0)*ln(r);

$$R_0(r) := A(0) + B(0) \ln(r) \qquad (5.140)$$

Substituting into the r-dependent boundary condition at $r = 0$ yields

> B(0):=0;

$$B(0) := 0 \qquad (5.141)$$

The r-dependent solution for $n = 0$ reads

> R[0](r):=A(0);

$$R_0(r) := A(0) \qquad (5.142)$$

Generalized series terms

> u[0](r,theta):=R[0](r)*phi[0](theta);u[n](r,theta):=R[n](r)*(A(n)*phi[n](theta)+B(n)*psi[n](theta));

$$u_0(r, \theta) := \frac{1}{2} \frac{A(0)\sqrt{2}}{\sqrt{\pi}}$$
$$u_n(r, \theta) := r^n \left(\frac{A(n)\cos(n\theta)}{\sqrt{\pi}} + \frac{B(n)\sin(n\theta)}{\sqrt{\pi}} \right) \qquad (5.143)$$

Eigenfunction expansion

> u(r,theta):=u[0](r,theta)+Sum(u[n](r,theta),n=1..infinity);

$$u(r, \theta) := \frac{1}{2} \frac{A(0)\sqrt{2}}{\sqrt{\pi}} + \sum_{n=1}^{\infty} r^n \left(\frac{A(n)\cos(n\theta)}{\sqrt{\pi}} + \frac{B(n)\sin(n\theta)}{\sqrt{\pi}} \right) \qquad (5.144)$$

Evaluation of Fourier coefficients from the specific remaining boundary condition $u(1, \theta) = f(\theta)$

> f(theta):=sin(3*theta);

$$f(\theta) := \sin(3\theta) \qquad (5.145)$$

Substituting this condition into the preceding solution yields

> f(theta)=subs(r=a,u(r,theta));

$$\sin(3\theta) = \frac{1}{2}\frac{A(0)\sqrt{2}}{\sqrt{\pi}} + \sum_{n=1}^{\infty}\left(\frac{A(n)\cos(n\theta)}{\sqrt{\pi}} + \frac{B(n)\sin(n\theta)}{\sqrt{\pi}}\right) \tag{5.146}$$

This equation is the Fourier series expansion of $f(\theta)$ in terms of the orthonormalized Sturm-Liouville eigenfunctions found earlier. To evaluate the Fourier cofficients, we take the inner product of both sides with respect to the orthonormalized eigenfunctions, and, taking advantage of the previous statement of orthonormality, we get, for $n = 1, 2, 3, \ldots,$

> A(n):=(1/(a^(n)))*Int(f(theta)*phi[n](theta)*w(theta),theta=−Pi..Pi);A(n):=
  expand(value(%)):

$$A(n) := \int_{-\pi}^{\pi} \frac{\sin(3\theta)\cos(n\theta)}{\sqrt{\pi}}\, d\theta \tag{5.147}$$

> A(n):=simplify(subs({sin(n*Pi)=0,cos(n*Pi)=(−1)^n},A(n)));

$$A(n) := 0 \tag{5.148}$$

> B(n):=(1/a^n)*Int(f(theta)*psi[n](theta)*w(theta),theta=−Pi..Pi);

$$B(n) := \int_{-\pi}^{\pi} \frac{\sin(3\theta)\sin(n\theta)}{\sqrt{\pi}}\, d\theta \tag{5.149}$$

> B(n):=expand(value(%));

$$B(n) := -\frac{6\sin(\pi n)}{\sqrt{\pi}(-9+n^2)} \tag{5.150}$$

The only term that survives here is the $n = 3$ term, which is evaluated to be the limit of the preceding equation:

> B(3):=limit(B(n),n=3);

$$B(3) := \sqrt{\pi} \tag{5.151}$$

Note that, because of the previous statement of orthonormality, only the $n = 3$ term survives. We use the Kronecker delta to write the $B(n)$ term as

> B(n):=B(3)*delta(n,3);

$$B(n) := \sqrt{\pi}\delta(n, 3) \tag{5.152}$$

For $n = 0,$

> A(0):=Int(f(theta)*phi[0](theta)*w(theta),theta=−Pi..Pi);A(0):=value(%):

$$A(0) := \int_{-\pi}^{\pi} \frac{1}{2} \frac{\sin(3\theta)\sqrt{2}}{\sqrt{\pi}} d\theta \qquad (5.153)$$

> A(0):=value(%);

$$A(0) := 0 \qquad (5.154)$$

Generalized series terms

> u[0](r,theta):=eval(R[0](r)*phi[0](theta));u[n](r,theta):=R[n](r)*(A(n)*phi[n](theta)
+B(n)*psi[n](theta));

$$u_0(r, \theta) := 0$$

$$u_n(r, \theta) := r^n \delta(n, 3) \sin(n\theta) \qquad (5.155)$$

Series solution

> u(r,theta):=u[0](r,theta)+Sum(u[n](r,theta),n=1..infinity);

$$u(r, \theta) := \sum_{n=1}^{\infty} r^n \delta(n, 3) \sin(n\theta) \qquad (5.156)$$

First few terms of expansion

> u(r,theta):=r^3*sin(3*theta);

$$u(r, \theta) := r^3 \sin(3\theta) \qquad (5.157)$$

> cylinderplot([r,theta,u(r,theta)],r=0..a,theta=−Pi..Pi,axes=framed,orientation=[21,69],
thickness=1);

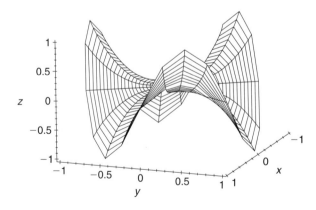

Figure 5.9

The three-dimensional surface shown in Figure 5.9 depicts the steady-state temperature
distribution $u(r, \theta)$ over the cylindrical region. Note how the edges of the surface adhere to the

given boundary conditions. The temperature isotherms can be obtained from Maple by clicking on the figure, choosing the special option "Render the plot using the polygon patch and contour style" in the graphics bar, and then clicking the "redraw" button.

Chapter Summary

Laplace equation in rectangular coordinates

$$\frac{\partial^2}{\partial x^2} u(x, y) + \frac{\partial^2}{\partial y^2} u(x, y) = 0$$

Sturm-Liouville problem with respect to the x variable

$$\frac{d^2}{dx^2} X(x) + \lambda X(x) = 0$$

Homogeneous boundary conditions

$$\kappa_1 X(0) + \kappa_2 X_x(0) = 0$$

and

$$\kappa_3 X(a) + \kappa_4 X_x(a) = 0$$

Solution in terms of eigenfunctions with respect to the x variable

$$u(x, y) = \sum_{n=0}^{\infty} X_n(x)\left(A(n)\cosh\left(\sqrt{\lambda_n}\, y\right) + B(n)\sinh\left(\sqrt{\lambda_n}\, y\right)\right)$$

Sturm-Liouville problem with respect to the y variable

$$\frac{d^2}{dy^2} Y(y) + \lambda Y(y) = 0$$

Homogeneous boundary conditions

$$\kappa_1 Y(0) + \kappa_2 Y_y(0) = 0$$

and

$$\kappa_3 Y(b) + \kappa_4 Y_y(b) = 0$$

Solution in terms of eigenfunctions with respect to the y variable

$$u(x, y) = \sum_{n=0}^{\infty} Y_n(y)\left(A(n)\cosh\left(\sqrt{\lambda_n}x\right) + B(n)\sinh\left(\sqrt{\lambda_n}x\right)\right)$$

Laplace equation in cylindrical coordinates

$$\frac{\frac{\partial}{\partial r}u(r,\theta)+r\left(\frac{\partial^2}{\partial r^2}u(r,\theta)\right)}{r}+\frac{\frac{\partial^2}{\partial\theta^2}u(r,\theta)}{r^2}=0$$

Sturm-Liouville problem with respect to the θ variable

$$\frac{d^2}{d\theta^2}\Theta(\theta)+\lambda\Theta(\theta)=0$$

Homogeneous boundary conditions

$$\kappa_1\Theta(0)+\kappa_2\Theta_\theta(0)=0$$

and

$$\kappa_3\Theta(b)+\kappa_4\Theta_\theta(b)=0$$

Solution in terms of eigenfunctions with respect to the θ variable

$$u(r,\theta)=\sum_{n=0}^{\infty}\Theta_n(\theta)\left(A(n)r^{\sqrt{\lambda_n}}+B(n)r^{-\sqrt{\lambda_n}}\right)$$

Sturm-Liouville problem in terms of the r variable

$$r^2\left(\frac{d^2}{dr^2}R(r)\right)+r\left(\frac{d}{dr}R(r)\right)+\lambda R(r)=0$$

Homogeneous boundary conditions

$$\kappa_1 R(1)+\kappa_2 R_r(1)=0$$

and

$$\kappa_3 R(b)+\kappa_4 R_r(b)=0$$

Solution in terms of eigenfunctions with respect to the r variable

$$u(r,\theta)=\sum_{n=0}^{\infty}R_n(r)\left(A(n)\cosh\left(\sqrt{\lambda_n}\theta\right)+B(n)\sinh\left(\sqrt{\lambda_n}\theta\right)\right)$$

We have considered solutions to the steady-state or time-invariant wave and diffusion equation. This partial differential equation is called the "Laplace" equation. In electrostatics, this equation is referred to as the "potential" equation. Solutions to the Laplace equation are said to be "harmonic."

Exercises

In the following exercises, we are asked to develop the graphics showing temperature isotherms and electrostatic equipotential lines. These graphics can be obtained from Maple by clicking on the figure, choosing the special option "Render the plot using the polygon patch and contour style" in the graphics bar, and then clicking the "redraw" button.

Exercises in Rectangular Coordinates

5.1. We seek the steady-state temperature distribution in a thin plate over the rectangular domain $D = \{(x, y) \mid 0 < x < 1, 0 < y < 1\}$ whose lateral surface is insulated. The sides $x = 0$ and $y = 0$ are at the fixed temperature of zero, the side $x = 1$ is insulated, and the side $y = 1$ has a given temperature distribution. The boundary conditions are

$$u(0, y) = 0 \quad \text{and} \quad u_x(1, y) = 0$$

$$u(x, 0) = 0 \quad \text{and} \quad u(x, 1) = x\left(1 - \frac{x}{2}\right)$$

Develop the graphics for the three-dimensional temperature surface showing the isotherms.

5.2. We seek the steady-state temperature distribution in a thin plate over the rectangular domain $D = \{(x, y) \mid 0 < x < 1, 0 < y < 1\}$ whose lateral surface is insulated. The sides $y = 0$ and $y = 1$ are at a fixed temperature of zero, the side $x = 1$ is insulated, and the side $x = 0$ has a given temperature distribution. The boundary conditions are

$$u(0, y) = y(1 - y) \quad \text{and} \quad u_x(1, y) = 0$$

$$u(x, 0) = 0 \quad \text{and} \quad u(x, 1) = 0$$

Develop the graphics for the three-dimensional temperature surface showing the isotherms.

5.3. We seek the electrostatic potential in a charge-free rectangular domain $D = \{(x, y) \mid 0 < x < 1, 0 < y < 1\}$. The sides $x = 0$, $x = 1$, and $y = 1$ are held at a fixed potential of zero, and the side $y = 0$ has a given potential distribution. The boundary conditions are

$$u(0, y) = 0 \quad \text{and} \quad u(1, y) = 0$$

$$u(x, 0) = x(1 - x) \quad \text{and} \quad u(x, 1) = 0$$

Develop the graphics for the three-dimensional equipotential surfaces showing the equipotential lines.

5.4. We seek the steady-state temperature distribution in a thin plate over the rectangular domain $D = \{(x, y) \mid 0 < x < 1, 0 < y < 1\}$ whose lateral surface is insulated. The sides

$x = 0$ and $y = 1$ are at a fixed temperature of zero, the side $y = 0$ is insulated, and the side $x = 1$ has a given temperature distribution.

$$u(0, y) = 0 \quad \text{and} \quad u(1, y) = 1 - y^2$$
$$u_y(x, 0) = 0 \quad \text{and} \quad u(x, 1) = 0$$

Develop the graphics for the three-dimensional temperature surface showing the isotherms.

5.5. We seek the electrostatic potential in a charge-free rectangular domain $D = \{(x, y) \mid 0 < x < 1, 0 < y < 1\}$. The sides $x = 0$, $y = 0$, and $y = 1$ are held at a fixed potential of zero, and the side $x = 1$ has a given potential distribution. The boundary conditions are

$$u(0, y) = 0 \quad \text{and} \quad u(1, y) = y(1 - y)$$
$$u(x, 0) = 0 \quad \text{and} \quad u(x, 1) = 0$$

Develop the graphics for the three-dimensional equipotential surfaces showing the equipotential lines.

5.6. We seek the steady-state temperature distribution in a thin plate over the rectangular domain $D = \{(x, y) \mid 0 < x < 1, 0 < y < 1\}$ whose lateral surface is insulated. The sides $x = 0$ and $x = 1$ are at the fixed temperature of zero, the side $y = 0$ is insulated, and the side $y = 1$ has a given temperature distribution. The boundary conditions are

$$u(0, y) = 0 \quad \text{and} \quad u(1, y) = 0$$
$$u_y(x, 0) = 0 \quad \text{and} \quad u(x, 1) = x(1 - x)$$

Develop the graphics for the three-dimensional temperature surface showing the isotherms.

5.7. We seek the electrostatic potential in a charge-free rectangular domain $D = \{(x, y) \mid 0 < x < 1, 0 < y < 1\}$. The sides $x = 1$, $y = 0$, and $y = 1$ are held at a fixed potential of zero, and the side $x = 0$ has a given potential distribution. The boundary conditions are

$$u(0, y) = y(1 - y) \quad \text{and} \quad u(1, y) = 0$$
$$u(x, 0) = 0 \quad \text{and} \quad u(x, 1) = 0$$

Develop the graphics for the three-dimensional equipotential surfaces showing the equipotential lines.

5.8. We seek the steady-state temperature distribution in a thin plate over the rectangular domain $D = \{(x, y) \mid 0 < x < 1, 0 < y < 1\}$ whose lateral surface is insulated. The sides

$x = 0$ and $y = 1$ are held at the fixed temperature of zero, the side $x = 1$ is losing heat by convection into a surrounding temperature of zero, and the side $y = 0$ has a given temperature distribution. The boundary conditions are

$$u(0, y) = 0 \quad \text{and} \quad u(1, y) + u_x(1, y) = 0$$

$$u(x, 0) = x\left(1 - \frac{2x}{3}\right) \quad \text{and} \quad u(x, 1) = 0$$

Develop the graphics for the three-dimensional temperature surface showing the isotherms.

5.9.   We seek the steady-state temperature distribution in a thin plate over the rectangular domain $D = \{(x, y) \mid 0 < x < 1, 0 < y < 1\}$ whose lateral surface is insulated. The sides $y = 0$ and $y = 1$ are at a fixed temperature of zero, the side $x = 1$ is losing heat by convection into a surrounding temperature of zero, and the side $x = 0$ has a given temperature distribution. The boundary conditions are

$$u(0, y) = y(1 - y) \quad \text{and} \quad u(1, y) + u_x(1, y) = 0$$

$$u(x, 0) = 0 \quad \text{and} \quad u(x, 1) = 0$$

Develop the graphics for the three-dimensional temperature surface showing the isotherms.

5.10.   We seek the electrostatic potential in a charge-free rectangular domain $D = \{(x, y) \mid 0 < x < 1, 0 < y < 1\}$. The sides $x = 0$, $x = 1$, and $y = 0$ are held at a fixed potential of zero, and the side $y = 1$ has a given potential distribution. The boundary conditions are

$$u(0, y) = 0 \quad \text{and} \quad u(1, y) = 0$$

$$u(x, 0) = 0 \quad \text{and} \quad u(x, 1) = x(1 - x)$$

Develop the graphics for the three-dimensional equipotential surfaces showing the equipotential lines.

5.11.   We seek the steady-state temperature distribution in a thin plate over the rectangular domain $D = \{(x, y) \mid 0 < x < 1, 0 < y < 1\}$ whose lateral surface is insulated. The side $y = 0$ is at a fixed temperature of zero, the side $y = 1$ is insulated, the side $x = 0$ is losing heat by convection into a surrounding medium of zero, and the side $x = 1$ has a given temperature distribution. The boundary conditions are

$$u(0, y) - u_x(0, y) = 0 \quad \text{and} \quad u(1, y) = y\left(1 - \frac{y}{2}\right)$$

$$u(x, 0) = 0 \quad \text{and} \quad u_y(x, 1) = 0$$

Develop the graphics for the three-dimensional temperature surface showing the isotherms.

5.12. We seek the steady-state temperature distribution in a thin plate over the rectangular domain $D = \{(x,y) \mid 0 < x < 1, 0 < y < 1\}$ whose lateral surface is insulated. The sides $x = 0$ and $y = 0$ are at a fixed temperature of zero, the side $y = 1$ is losing heat by convection into a surrounding temperature of zero, and the side $x = 1$ has a given temperature distribution. The boundary conditions are

$$u(0, y) = 0 \quad \text{and} \quad u(1, y) = y\left(1 - \frac{2y}{3}\right)$$

$$u(x, 0) = 0 \quad \text{and} \quad u(x, 1) + u_y(x, 1) = 0$$

Develop the graphics for the three-dimensional temperature surface showing the isotherms.

5.13. We seek the steady-state temperature distribution in a thin plate over the rectangular domain $D = \{(x,y) \mid 0 < x < 1, 0 < y < 1\}$ whose lateral surface is insulated. The side $y = 1$ is insulated, the side $x = 0$ is losing heat by convection into a surrounding temperature of zero, the side $x = 1$ is at a fixed temperature of zero, and the side $y = 0$ has a given temperature distribution. The boundary conditions are

$$u_x(0, y) - u(0, y) = 0 \quad \text{and} \quad u(1, y) = 0$$

$$u(x, 0) = -\frac{2x^2}{3} + \frac{x}{3} + \frac{1}{3} \quad \text{and} \quad u_y(x, 1) = 0$$

Develop the graphics for the three-dimensional temperature surface showing the isotherms.

5.14. We seek the steady-state temperature distribution in a thin plate over the rectangular domain $D = \{(x,y) \mid 0 < x < 1, 0 < y < 1\}$ whose lateral surface is insulated. The sides $x = 0$ and $x = 1$ are at a fixed temperature of zero, the side $y = 1$ is losing heat by convection into a surrounding temperature of zero, and the side $y = 0$ has a given temperature distribution. The boundary conditions are

$$u(0, y) = 0 \quad \text{and} \quad u(1, y) = 0$$

$$u(x, 0) = x(1 - x) \quad \text{and} \quad u(x, 1) + u_y(x, 1) = 0$$

Develop the graphics for the three-dimensional temperature surface showing the isotherms.

5.15. We seek the steady-state temperature distribution in a thin plate over the rectangular domain $D = \{(x,y) \mid 0 < x < 1, 0 < y < 1\}$ whose lateral surface is insulated. The side $x = 0$ is at a fixed temperature of zero, the side $y = 0$ is losing heat by convection into a surrounding temperature of zero, the side $x = 1$ is insulated, and the side $y = 1$ has a given temperature distribution. The boundary conditions are

$$u(0, y) = 0 \quad \text{and} \quad u_x(1, y) = 0$$

$$u_y(x, 0) - u(x, 0) = 0 \quad \text{and} \quad u(x, 1) = x\left(1 - \frac{x}{2}\right)$$

Develop the graphics for the three-dimensional temperature surface showing the isotherms.

5.16. We seek the steady-state temperature distribution in a thin plate over the rectangular domain $D = \{(x, y) \mid 0 < x < 1, 0 < y < 1\}$ whose lateral surface is insulated. The side $x = 1$ is insulated, the side $y = 0$ is losing heat by convection into a surrounding temperature of zero, the side $y = 1$ is at a fixed temperature of zero, and the side $x = 0$ has a given temperature distribution. The boundary conditions are

$$u(0, y) = -\frac{2y^2}{3} + \frac{y}{3} + \frac{1}{3} \quad \text{and} \quad u_x(1, y) = 0$$

$$u_y(x, 0) - u(x, 0) = 0 \quad \text{and} \quad u(x, 1) = 0$$

Develop the graphics for the three-dimensional temperature surface showing the isotherms.

Superposition of Solutions

In all the preceding exercises, homogeneous boundary conditions occurred with respect to either the x or y coordinate. We now consider problems whereby we do not have a set of homogeneous boundary conditions. To treat these problems, we must write the solution as a superposition of two solutions; that is, we set $u(x, y) = u1(x, y) + u2(x, y)$, and we partition the two subsequent problems in a manner that gives homogeneous boundary conditions.

For example, consider the problem of finding the steady-state temperature in a two-dimensional domain $D = \{(x, y) \mid 0 < x < 1, 0 < y < 1\}$ that satisfies the boundary conditions

$$u(0, y) = g(y) \quad \text{and} \quad u(1, y) = 0$$

and

$$u(x, 0) = 0 \quad \text{and} \quad u(x, 1) = f(x)$$

Note that we lack a set of homogeneous boundary conditions with respect to both the x and y coordinates. We write the solution as the superposition of two solutions:

$$u(x, y) = u1(x, y) + u2(x, y)$$

We partition the boundary conditions in the following manner so as to give a set of homogeneous boundary conditions for each problem

$$u1(0, y) = g(y) \quad \text{and} \quad u1(1, y) = 0 \quad \text{and} \quad u1(x, 0) = 0 \quad \text{and} \quad u1(x, 1) = 0$$

and

$$u2(0, y) = 0 \quad \text{and} \quad u2(1, y) = 0 \quad \text{and} \quad u2(x, 0) = 0 \quad \text{and} \quad u2(x, 1) = f(x)$$

The solution for $u1(x, y)$, for a specific $g(y)$, has been evaluated in Exercise 5.7, and the solution for $u2(x, y)$, for a specific $f(x)$, has been evaluated in Exercise 5.10. Thus, the sum of these two solutions gives the answer to our original problem.

The following exercises require the solution to be partitioned as a superposition of two solutions $u1(x, y)$ and $u2(x, y)$. The problems are such that they can be resolved into two problems, each of which has already been solved in Exercises 5.1 through 5.16. Seek those preceding solutions, and write the general solution as a superposition.

5.17. We seek the steady-state temperature distribution in a thin plate over the rectangular domain $D = \{(x, y) \mid 0 < x < 1, 0 < y < 1\}$ whose lateral surface is insulated. The side $x = 1$ is insulated, the side $y = 0$ is at a fixed temperature of zero, and the sides $x = 0$ and $y = 1$ have given temperature distributions. The boundary conditions are

$$u(0, y) = y(1 - y) \quad \text{and} \quad u_x(1, y) = 0$$
$$u(x, 0) = 0 \quad \text{and} \quad u(x, 1) = x\left(1 - \frac{x}{2}\right)$$

Develop the graphics for the three-dimensional temperature surface showing the isotherms.

5.18. We seek the steady-state temperature distribution in a thin plate over the rectangular domain $D = \{(x, y) \mid 0 < x < 1, 0 < y < 1\}$ whose lateral surface is insulated. The side $y = 0$ is insulated, the side $x = 0$ is at a fixed temperature of zero, and the sides $x = 1$ and $y = 1$ have given temperature distributions. The boundary conditions are

$$u(0, y) = 0 \quad \text{and} \quad u(1, y) = 1 - y^2$$
$$u_y(x, 0) = 0 \quad \text{and} \quad u(x, 1) = x(1 - x)$$

Develop the graphics for the three-dimensional temperature surface showing the isotherms.

5.19. We seek the steady-state temperature distribution in a thin plate over the rectangular domain $D = \{(x, y) \mid 0 < x < 1, 0 < y < 1\}$ whose lateral surface is insulated. The side $y = 1$ is at a fixed temperature of zero, the side $x = 1$ is losing heat by convection into a surrounding temperature of zero, and the sides $x = 0$ and $y = 0$ have given temperature distributions. The boundary conditions are

$$u(0, y) = y(1 - y) \quad \text{and} \quad u(1, y) + u_x(1, y) = 0$$

$$u(x, 0) = x\left(1 - \frac{2x}{3}\right) \quad \text{and} \quad u(x, 1) = 0$$

Develop the graphics for the three-dimensional temperature surface showing the isotherms.

5.20. We seek the steady-state temperature distribution in a thin plate over the rectangular domain $D = \{(x, y) \,|\, 0 < x < 1, 0 < y < 1\}$ whose lateral surface is insulated. The side $x = 0$ is at a fixed temperature of zero, the side $y = 1$ is losing heat by convection into a surrounding temperature of zero, and the sides $x = 1$ and $y = 0$ have given temperature distributions. The boundary conditions are

$$u(0, y) = 0 \quad \text{and} \quad u(1, y) = y\left(1 - \frac{2y}{3}\right)$$

$$u(x, 0) = x(1 - x) \quad \text{and} \quad u(x, 1) + u_y(x, 1) = 0$$

Develop the graphics for the three-dimensional temperature surface showing the isotherms.

5.21 We seek the steady-state temperature distribution in a thin plate over the rectangular domain $D = \{(x, y) \,|\, 0 < x < 1, 0 < y < 1\}$ whose lateral surface is insulated. The side $y = 1$ is insulated, the side $x = 0$ is losing heat by convection into a surrounding temperature of zero, and the sides $x = 1$ and $y = 0$ have given temperature distributions. The boundary conditions are

$$u(0, y) - u_x(0, y) = 0 \quad \text{and} \quad u(1, y) = y\left(1 - \frac{y}{2}\right)$$

$$u(x, 0) = -\frac{2x^2}{3} + \frac{x}{3} + \frac{1}{3} \quad \text{and} \quad u_y(x, 1) = 0$$

Develop the graphics for the three-dimensional temperature surface showing the isotherms.

5.22. We seek the steady-state temperature distribution in a thin plate over the rectangular domain $D = \{(x, y) \,|\, 0 < x < 1, 0 < y < 1\}$ whose lateral surface is insulated. The side $x = 1$ is insulated, the side $y = 0$ is losing heat by convection into a surrounding temperature of zero, and the sides $x = 0$ and $y = 1$ have given temperature distributions.

The boundary conditions are

$$u(0, y) = -\frac{2y^2}{3} + \frac{y}{3} + \frac{1}{3} \quad \text{and} \quad u_x(1, y) = 0$$

$$u(x, 0) - u_y(x, 0) = 0 \quad \text{and} \quad u(x, 1) = x\left(1 - \frac{x}{2}\right)$$

Develop the graphics for the three-dimensional temperature surface showing the isotherms.

Exercises in Cylindrical Coordinates

5.23. We seek the steady-state temperature distribution in a thin plate over the cylindrical domain $D = \{(r, \theta) \mid 0 < r < 1, 0 < \theta < \pi/3\}$ whose lateral surface is insulated. The temperature at the origin is finite. The side $\theta = 0$ is at a fixed temperature of zero, the side $\theta = \pi/3$ is insulated, and the side $r = 1$ has a given temperature distribution. The boundary conditions are

$$|u(0, \theta)| < \infty, \ u(1, \theta) = \theta\left(1 - \frac{3\theta}{2\pi}\right)$$

$$u(r, 0) = 0 \quad \text{and} \quad u_\theta\left(r, \frac{\pi}{3}\right) = 0$$

Develop the graphics for the three-dimensional temperature surface showing the isotherms.

5.24. We seek the steady-state temperature distribution in a thin plate over the cylindrical domain $D = \{(r, \theta) \mid 0 < r < 1, 0 < \theta < \pi/3\}$ whose lateral surface is insulated. The temperature at the origin is finite. The sides $\theta = 0$ and $\theta = \pi/3$ are at a fixed temperature of zero, and the side $r = 1$ has a given temperature distribution. The boundary conditions are

$$|u(0, \theta)| < \infty \quad \text{and} \quad u(1, \theta) = \theta\left(1 - \frac{3\theta}{\pi}\right)$$

$$u(r, 0) = 0 \quad \text{and} \quad u\left(r, \frac{\pi}{3}\right) = 0$$

Develop the graphics for the three-dimensional temperature surface showing the isotherms.

5.25 We seek the steady-state temperature distribution in a thin plate over the cylindrical domain $D = \{(r, \theta) \mid 0 < r < 1, 0 < \theta < \pi/3\}$ whose lateral surface is insulated. The temperature at the origin is finite. The side $\theta = 0$ is insulated, the side $\theta = \pi/3$ is at a fixed temperature of zero, and the side $r = 1$ has a given temperature distribution. The boundary conditions are

$$|u(0, \theta)| < \infty \quad \text{and} \quad u(1, \theta) = 1 - \frac{9\theta^2}{\pi^2}$$

$$u_\theta(r, 0) = 0 \quad \text{and} \quad u\left(r, \frac{\pi}{3}\right) = 0$$

Develop the graphics for the three-dimensional temperature surface showing the isotherms.

5.26.   We seek the steady-state temperature distribution in a thin plate over the cylindrical domain $D = \{(r,\theta) \mid 1 < r < 2, 0 < \theta < \pi/2\}$ whose lateral surface is insulated. The sides $r = 1, r = 2$, and $\theta = 0$ are at a fixed temperature of zero, and the side $\theta = \pi/2$ has a given temperature distribution. The boundary conditions are

$$u(1, \theta) = 0 \quad \text{and} \quad u(2, \theta) = 0$$

$$u(r, 0) = 0 \quad \text{and} \quad u\left(r, \frac{\pi}{2}\right) = (r - 1)(2 - r)$$

Develop the graphics for the three-dimensional temperature surface showing the isotherms.

5.27.   We seek the steady-state temperature distribution in a thin plate over the cylindrical domain $D = \{(r,\theta) \mid 1 < r < 2, 0 < \theta < \pi/2\}$ whose lateral surface is insulated. The sides $r = 1$ and $\theta = 0$ are at a fixed temperature of zero, the side $r = 2$ is insulated, and the side $\theta = \pi/2$ has a given temperature distribution. The boundary conditions are

$$u(1, \theta) = 0 \quad \text{and} \quad u_r(2, \theta) = 0$$

$$u(r, 0) = 0 \quad \text{and} \quad u\left(r, \frac{\pi}{2}\right) = -r^2 + 4r - 3$$

Develop the graphics for the three-dimensional temperature surface showing the isotherms.

5.28   We seek the steady-state temperature distribution in a thin plate over the cylindrical domain $D = \{(r,\theta) \mid 1 < r < 2, 0 < \theta < \pi/2\}$ whose lateral surface is insulated. The sides $r = 1, r = 2$, and $\theta = \pi/2$ are at a fixed temperature of zero, and the side $\theta = 0$ has a given temperature distribution. The boundary conditions are

$$u(1, \theta) = 0 \quad \text{and} \quad u(2, \theta) = 0$$

$$u(r, 0) = (r - 1)(2 - r) \quad \text{and} \quad u\left(r, \frac{\pi}{2}\right) = 0$$

Develop the graphics for the three-dimensional temperature surface showing the isotherms.

5.29.   We seek the steady-state temperature distribution in a thin plate over the cylindrical
domain $D = \{(r,\theta) \mid 1 < r < 2, 0 < \theta < \pi/2\}$ whose lateral surface is insulated. The sides
$r = 1$ and $\theta = \pi/2$ are at a fixed temperature of zero, the side $r = 2$ is insulated, and the
side $\theta = 0$ has a given temperature distribution. The boundary conditions are

$$u(1, \theta) = 0 \quad \text{and} \quad u_r(2, \theta) = 0$$

$$u(r, 0) = -r^2 + 4r - 3 \quad \text{and} \quad u\left(r, \frac{\pi}{2}\right) = 0$$

Develop the graphics for the three-dimensional temperature surface showing the
isotherms.

5.30.   We seek the steady-state temperature distribution in a thin plate over the cylindrical
domain $D = \{(r,\theta) \mid 1 < r < 2, 0 < \theta < \pi/2\}$ whose lateral surface is insulated. The sides
$r = 2$ and $\theta = 0$ are at a fixed temperature of zero, the side $r = 1$ is insulated, and the
side $\theta = \pi/2$ has a given temperature distribution. The boundary conditions are

$$u_r(1, \theta) = 0 \quad \text{and} \quad u(2, \theta) = 0$$

$$u(r, 0) = 0 \quad \text{and} \quad u\left(r, \frac{\pi}{2}\right) = -r^2 + 2r$$

Develop the graphics for the three-dimensional temperature surface showing the
isotherms.

Superposition of Solutions in Cylindrical Coordinates

5.31.   We seek the steady-state temperature distribution in a thin plate over the cylindrical
domain $D = \{(r,\theta) \mid 1 < r < 2, 0 < \theta < \pi/2\}$ whose lateral surface is insulated. The sides
$r = 1$ and $r = 2$ are at a fixed temperature of zero, and the sides $\theta = 0$ and $\theta = \pi/2$ have
given temperature distributions. The boundary conditions are

$$u(1, \theta) = 0 \quad \text{and} \quad u(2, \theta) = 0$$

$$u(r, 0) = (r - 1)(2 - r) \quad \text{and} \quad u\left(r, \frac{\pi}{2}\right) = 1$$

Develop the graphics for the three-dimensional temperature surface showing the
isotherms.

5.32.   We seek the steady-state temperature distribution in a thin plate over the cylindrical
domain $D = \{(r,\theta) \mid 1 < r < 2, 0 < \theta < \pi/2\}$ whose lateral surface is insulated. The side
$r = 1$ is at a fixed temperature of zero, the side $r = 2$ is insulated, and the sides $\theta = 0$ and

$\theta = \pi/2$ have given temperature distributions. The boundary conditions are

$$u(1, \theta) = 0 \quad \text{and} \quad u_r(2, \theta) = 0$$

$$u(r, 0) = -r^2 + 4r - 3 \quad \text{and} \quad u\left(r, \frac{\pi}{2}\right) = 1$$

Develop the graphics for the three-dimensional temperature surface showing the isotherms.

CHAPTER 6

The Diffusion Equation in Two Spatial Dimensions

6.1 Introduction

In Chapter 3, we examined the heat or diffusion partial differential equation in only one spatial dimension. Here, we examine the diffusion equation in two spatial dimensions. Thus, we will be considering problems with a total of three independent variables: the two spatial variables and time. The method of separation of variables will be used, and we shall see the development of two independent Sturm-Liouville problems—one for each of the two spatial variables. The final solutions will have the form of a double Fourier series. We will consider problems in both the rectangular and the cylindrical coordinate systems.

6.2 Two-Dimensional Diffusion Operator in Rectangular Coordinates

In the one-dimensional diffusion equation, as discussed in Chapter 3, the second-order partial derivative with respect to the spatial variable x can be viewed as the first term of the Laplacian operator. Thus, an extension of this Laplacian operator to higher dimensions is natural, and it can be shown that heat or diffusion phenomena in two dimensions can be described by the following nonhomogeneous partial differential equation in the rectangular coordinate system:

$$\frac{\partial}{\partial t} u(x, y, t) = k\left(\frac{\partial^2}{\partial x^2} u(x, y, t) + \frac{\partial^2}{\partial y^2} u(x, y, t)\right) + h(x, y, t)$$

where, for the situation of heat phenomena, $u(x, y, t)$ denotes the spatial-time-dependent temperature and the constant k, called "thermal diffusivity," incorporates the thermal characteristics of the medium such as thermal conductivity, specific heat, and mass density. The term $h(x, y, t)$ accounts for the time-rate of input of any external heat source into the medium. We have assumed the medium to be isotropic and uniform so that the preceding thermal coefficients are spatially invariant, and we treat them as constants.

Most of the example problems that we will be considering deal with thin plates with no extension along the z-axis and whose lateral surfaces are insulated; thus, we have no z-dependent terms in the Laplacian.

The partial differential equation for diffusion or heat phenomena is characterized by the fact that it relates the first partial derivative with respect to the time variable with the second partial derivatives with respect to the spatial variables.

We can write this nonhomogeneous equation in terms of the linear heat or diffusion operator as

$$L(u) = h(x, y, t)$$

The diffusion operator in rectangular coordinates in two spatial dimensions reads

$$L(u) = \frac{\partial}{\partial t} u(x, y, t) - k\left(\frac{\partial^2}{\partial x^2} u(x, y, t) + \frac{\partial^2}{\partial y^2} u(x, y, t) \right)$$

If there are no external heat sources in the system, then $h(x, y, t) = 0$, and the preceding nonhomogeneous partial differential equation reduces to its corresponding homogeneous equivalent

$$L(u) = 0$$

A typical homogeneous diffusion problem in two spatial dimensions reads as follows. We seek the temperature distribution $u(x, y, t)$ in a thin rectangular plate over the finite two-dimensional domain $D = \{(x, y) \mid 0 < x < 1, 0 < y < 1\}$. The lateral surfaces of the plate are insulated, so there is no heat loss through the lateral surface. The boundaries $y = 0$ and $y = 1$ are fixed at temperature zero, the boundary $x = 0$ is insulated, and the boundary $x = 1$ is losing heat by convection into a surrounding medium at temperature zero. There are no heat sources in the system, the initial temperature distribution $f(x, y)$ is given following, and the thermal diffusivity of the plate material is $k = 1/10$.

The diffusion homogeneous partial differential equation reads

$$\frac{\partial}{\partial t} u(x, y, t) = \frac{\frac{\partial^2}{\partial x^2} u(x, y, t) + \frac{\partial^2}{\partial y^2} u(x, y, t)}{10}$$

The given boundary conditions on the problem are type 2 at $x = 0$, type 3 at $x = 1$, type 1 at $y = 0$, and type 1 at $y = 1$.

$$u_x(0, y, t) = 0 \quad \text{and} \quad u_x(1, y, t) + u(1, y, t) = 0$$

and

$$u(x, 0, t) = 0 \quad \text{and} \quad u(x, 1, t) = 0$$

The initial condition function is

$$u(x, y, 0) = f(x, y)$$

We now examine the general procedures for the solution to this homogeneous partial differential equation over finite domains, with homogeneous boundary conditions, using the method of separation of variables.

6.3 Method of Separation of Variables for the Diffusion Equation in Two Dimensions

The homogeneous diffusion equation for a uniform system in two dimensions in rectangular coordinates can be written

$$\frac{\partial}{\partial t} u(x, y, t) - k \left(\frac{\partial^2}{\partial x^2} u(x, y, t) + \frac{\partial^2}{\partial y^2} u(x, y, t) \right) = 0$$

This can be rewritten in the more familiar form

$$\frac{\partial}{\partial t} u(x, y, t) = k \left(\frac{\partial^2}{\partial x^2} u(x, y, t) + \frac{\partial^2}{\partial y^2} u(x, y, t) \right)$$

Generally, one seeks a solution to this problem over the finite two-dimensional rectangular domain $D = \{(x, y) \mid 0 < x < a, 0 < y < b\}$ subject to the homogeneous boundary conditions

$$\kappa_1 u(0, y, t) + \kappa_2 u_x(0, y, t) = 0$$

$$\kappa_3 u(a, y, t) + \kappa_4 u_x(a, y, t) = 0$$

$$\kappa_5 u(x, 0, t) + \kappa_6 u_y(x, 0, t) = 0$$

$$\kappa_7 u(x, b, t) + \kappa_8 u_y(x, b, t) = 0$$

and the initial condition

$$u(x, y, 0) = f(x, y)$$

In the method of separation of variables, as done in Chapter 3, the character of the equation is such that we can assume a solution in the form of a product as

$$u(x, y, t) = X(x)Y(y)T(t)$$

Because the three independent variables in the partial differential equation are the spatial variables x and y, and the time variable t, then the preceding assumed solution is written as a product of three functions; one is an exclusive function of x, one is an exclusive function of y, and the last an exclusive function of t.

Substituting this assumed solution into the partial differential equation gives

$$X(x)\,Y(y)\left(\frac{d}{dt}T(t)\right) = k\left(\left(\frac{d^2}{dx^2}X(x)\right)Y(y)\,T(t) + X(x)\left(\frac{d^2}{dy^2}Y(y)\right)T(t)\right)$$

Dividing both sides of the preceding by the product solution yields

$$\frac{\frac{d}{dt}T(t)}{kT(t)} = \frac{\left(\frac{d^2}{dx^2}X(x)\right)Y(y) + X(x)\left(\frac{d^2}{dy^2}Y(y)\right)}{X(x)Y(y)}$$

Because the left-hand side of this equation is an exclusive function of t and the right-hand side an exclusive function of both x and y, and the variables x, y, and t are independent, then the only way the preceding equation can hold for all values of x, y, and t is to set each side equal to a constant. Doing so, we get the following three ordinary differential equations in terms of the separation constants λ, α, and β:

$$\frac{d}{dt}T(t) + k\lambda\,T(t) = 0$$

$$\frac{d^2}{dx^2}X(x) + \alpha\,X(x) = 0$$

$$\frac{d^2}{dy^2}Y(y) + \beta\,Y(y) = 0$$

with the coupling equation

$$\lambda = \alpha + \beta$$

The preceding differential equation in t is an ordinary first-order linear homogeneous differential equation for which we already have the solution from Section 1.2. The second and third differential equations in x and y are ordinary second-order linear homogeneous differential equations of the Euler type for which we already have the solutions from Section 1.4. We easily recognize each of the two spatial differential equations as being Sturm-Liouville types.

6.4 Sturm-Liouville Problem for the Diffusion Equation in Two Dimensions

The second and third differential equations given in the previous section, in the spatial variables x and y, each must be solved subject to their respective homogeneous spatial boundary conditions.

We first consider the x-dependent differential equation over the finite interval $I = \{x \mid 0 < x < a\}$. The Sturm-Liouville problem for the x-dependent portion of the diffusion equation consists of the ordinary differential equation

$$\frac{d^2}{dx^2}X(x) + \alpha\, X(x) = 0$$

along with the respective corresponding regular homogeneous boundary conditions, which read as

$$\kappa_1 X(0) + \kappa_2 X_x(0) = 0$$

and

$$\kappa_3 X(a) + \kappa_4 X_x(a) = 0$$

We recognize this problem in the spatial variable x to be a regular Sturm-Liouville eigenvalue problem where the allowed values of α are called the system "eigenvalues" and the corresponding solutions are called the "eigenfunctions." Note that this ordinary differential equation is of the Euler type and the weight function is $w(x) = 1$. The indexed eigenvalues and corresponding eigenfunctions are given, respectively, as

$$\alpha_m,\, X_m(x)$$

for $m = 0, 1, 2, 3, \ldots$.

These eigenfunctions can be normalized and the corresponding statement of orthonormality, with respect to the weight function $w(x) = 1$, reads

$$\int_0^a X_m(x) X_r(x)\, dx = \delta(m, r)$$

Here, $\delta(m, r)$ is the Kronecker delta function defined earlier. We now focus on the third differential equation in the spatial variable y subject to the homogeneous spatial boundary conditions over the finite interval $I = \{y \,|\, 0 < y < b\}$. The Sturm-Liouville problem for the y-dependent portion of the diffusion equation consists of the ordinary differential equation

$$\frac{d^2}{dy^2}Y(y) + \beta\, Y(y) = 0$$

along with the respective corresponding regular homogeneous boundary conditions, which read

$$\kappa_5 Y(0) + \kappa_6\, Y_y(0) = 0$$

and

$$\kappa_7 Y(b) + \kappa_8\, Y_y(b) = 0$$

Again, we recognize this problem in the spatial variable y to be a regular Sturm-Liouville eigenvalue problem where the allowed values of β are called the system "eigenvalues" and the corresponding solutions are called the "eigenfunctions." Note that this ordinary differential equation is of the Euler type and the weight function is $w(y) = 1$. The indexed eigenvalues and corresponding eigenfunctions are given, respectively, as

$$\beta_n, Y_n(y)$$

for $n = 0, 1, 2, 3, \ldots$.

Similar to the x-dependent eigenfunctions, the preceding eigenfunctions can be normalized and the corresponding statement of orthonormality, with respect to the weight function $w(y) = 1$, reads

$$\int_0^b Y_n(y)Y_s(y)\mathrm{d}y = \delta(n, s)$$

Here, $\delta(n, s)$ is the Kronecker delta function defined earlier. The statements made in Section 2.2 about regular Sturm-Liouville eigenvalue problems in one dimension can be extended to two dimensions here; that is, the sum of the products of the eigenfunctions $X_m(x)$ and $Y_n(y)$ form a "complete" set with respect to any piecewise smooth function $f(x, y)$ over the finite two-dimensional domain $D = \{(x, y) \mid 0 < x < a, 0 < y < b\}$.

Finally, we focus on the solution to the time-dependent differential equation, which reads

$$\frac{\mathrm{d}}{\mathrm{d}t}T(t) + k\lambda T(t) = 0$$

From Section 1.2 on first-order linear differential equations, the solution to this time-dependent differential equation in terms of the allowed eigenvalues $\lambda_{m,n}$ is

$$T_{m,n}(t) = C(m, n)\,\mathrm{e}^{-k\lambda_{m,n}t}$$

where, from the coupling equation, we have

$$\lambda_{m,n} = \alpha_m + \beta_n$$

for $m, n = 0, 1, 2, 3, \ldots$. The coefficients $C(m, n)$ are unknown arbitrary constants.

Thus, by the method of separation of variables, we arrive at an infinite number of indexed solutions for the homogeneous diffusion partial differential equation, over a finite interval, given as the product

$$u_{m,n}(x, y, t) = X_m(x)Y_n(y)C(m, n)\,\mathrm{e}^{-k\lambda_{m,n}t}$$

for $m = 0, 1, 2, 3, \ldots$, and $n = 0, 1, 2, 3, \ldots$.

Because the partial differential operator is linear, any superposition of solutions to the homogeneous equation is also a solution to the problem. Thus, we can write the general solution to the homogeneous partial differential equation as the double-infinite sum

$$u(x, y, t) = \sum_{n=0}^{\infty} \left(\sum_{m=0}^{\infty} X_m(x) Y_n(y) C(m, n) e^{-k\lambda_{m,n}t} \right)$$

This is the double Fourier series eigenfunction expansion form of the solution to the partial differential equation. The terms of the preceding double series form the "basis vectors" of the solution space of the partial differential equation. Similar to a partial differential equation in only one dimension, there are an infinite number of basis vectors in the solution space, and we say the dimension of the solution space is infinite.

DEMONSTRATION: We seek the temperature distribution in a thin rectangular plate over the finite two-dimensional domain $D = \{(x, y) \mid 0 < x < 1, 0 < y < 1\}$. The lateral surfaces of the plate are insulated. The boundaries $y = 0$ and $y = 1$ are fixed at temperature 0, the boundary $x = 0$ is insulated, and the boundary $x = 1$ is losing heat by convection into a surrounding medium at temperature 0. The initial temperature distribution $f(x, y)$ is given below, and the thermal diffusivity of the plate material is $k = 1/10$.

SOLUTION: The two-dimensional diffusion partial differential equation reads

$$\frac{\partial}{\partial t} u(x, y, t) = \frac{\frac{\partial^2}{\partial x^2} u(x, y, t) + \frac{\partial^2}{\partial y^2} u(x, y, t)}{10}$$

The given boundary conditions on the problem are type 2 at $x = 0$, type 3 at $x = 1$, type 1 at $y = 0$, and type 1 at $y = 1$.

$$u_x(0, y, t) = 0 \quad \text{and} \quad u_x(1, y, t) + u(1, y, t) = 0$$

and

$$u(x, 0, t) = 0 \quad \text{and} \quad u(x, 1, t) = 0$$

The initial condition function is given as

$$u(x, y, 0) = f(x, y)$$

From the method of separation of variables, we obtain the three ordinary differential equations

$$\frac{d}{dt}T(t) + \frac{\lambda\,T(t)}{10} = 0$$

$$\frac{d^2}{dx^2}X(x) + \alpha\,X(x) = 0$$

$$\frac{d^2}{dy^2}Y(y) + \beta\,Y(y) = 0$$

with the coupling equation

$$\lambda = \alpha + \beta$$

We first focus on the solution for the two spatial differential equations. For the x equation, we have a differential equation of the Euler type, and the corresponding boundary conditions are type 2 at the left and type 3 at the right

$$X_x(0) = 0 \quad \text{and} \quad X_x(1) + X(1) = 0$$

This same boundary value problem was considered in Example 2.5.5 in Chapter 2. The allowed eigenvalues are determined from the roots of the equation

$$\tan\left(\sqrt{\alpha_m}\right) = \frac{1}{\sqrt{\alpha_m}}$$

and the corresponding orthonormal eigenfunctions are

$$X_m(x) = \frac{\sqrt{2}\cos\left(\sqrt{\alpha_m}x\right)}{\sqrt{\sin\left(\sqrt{\alpha_m}\right)^2 + 1}}$$

for $m = 1, 2, 3, \ldots$.

Similarly, for the y equation, we have a differential equation of the Euler type, and the corresponding boundary conditions are type 1 at the left and type 1 at the right

$$Y(0) = 0 \quad \text{and} \quad Y(1) = 0$$

The allowed eigenvalues and corresponding orthonormalized eigenfunctions, obtained from Example 2.5.1 in Chapter 2, are

$$\beta_n = n^2\pi^2$$

and

$$Y_n(y) = \sqrt{2}\sin\left(n\pi y\right)$$

for $n = 1, 2, 3, \ldots$.

For each of the two spatial equations, the statements of orthonormality, with their respective weight functions, are

$$\int_0^1 \frac{2\cos\left(\sqrt{\alpha_m}\,x\right)\cos\left(\sqrt{\alpha_r}\,x\right)}{\sqrt{\sin\left(\sqrt{\alpha_m}\right)^2+1}\sqrt{\sin\left(\sqrt{\alpha_r}\right)^2+1}}\,dx = \delta(m,r)$$

and

$$\int_0^1 2\sin(n\pi y)\sin(s\pi y)\,dy = \delta(n,s)$$

Finally, the time-dependent differential equation has the solution

$$T_{m,n}(t) = C(m,n)\,e^{-\frac{\lambda_{m,n}t}{10}}$$

where

$$\lambda_{m,n} = \alpha_m + n^2\pi^2$$

for $m = 1, 2, 3, \ldots$ and $n = 1, 2, 3, \ldots$. The coefficients $C(m,n)$ are arbitrary constants.

The final eigenfunction series expansion form of the solution is constructed from the sum of the products of the time-dependent solution and the preceding x- and y-dependent eigenfunctions. Forming this sum, we get

$$u(x,y,t) = \sum_{n=1}^{\infty}\left(\sum_{m=1}^{\infty}\frac{2C(m,n)e^{-\frac{\lambda_{m,n}t}{10}}\cos\left(\sqrt{\alpha_m}\,x\right)\sin(n\pi y)}{\sqrt{\sin\left(\sqrt{\alpha_m}\right)^2+1}}\right)$$

The coefficients $C(m,n)$ are yet to be determined from the initial condition function $f(x,y)$.

6.5 Initial Conditions for the Diffusion Equation in Rectangular Coordinates

We now consider evaluation of the coefficients from the initial conditions on the problem. At time $t = 0$ we have

$$u(x,y,0) = f(x,y)$$

Substituting this into the following general form of the solution

$$u(x,y,t) = \sum_{n=0}^{\infty}\left(\sum_{m=0}^{\infty}X_m(x)Y_n(y)C(m,n)e^{-k\lambda_{m,n}t}\right)$$

at time $t = 0$ yields

$$f(x, y) = \sum_{n=0}^{\infty} \left(\sum_{m=0}^{\infty} C(m, n) X_m(x) Y_n(y) \right)$$

Because the eigenfunctions of the regular Sturm-Liouville problem form a complete set with respect to piecewise smooth functions over the finite two-dimensional domain, the preceding is the generalized double Fourier series expansion of the function $f(x, y)$ in terms of the allowed eigenfunctions and the double Fourier coefficients $C(m, n)$. We now evaluate these coefficients.

As we did in Chapter 2, Section 2.4 for the generalized Fourier series expansion, we take the double inner products of both sides of the preceding equation, with respect to the x- and y-dependent eigenfunctions, and this yields

$$\int_0^b \int_0^a f(x, y) X_r(x) Y_s(y) \, dx \, dy = \int_0^b \int_0^a \left(\sum_{n=0}^{\infty} \left(\sum_{m=0}^{\infty} C(m, n) X_m(x) Y_n(y) \right) \right) X_r(x) \, Y_s(y) \, dx \, dy$$

If we assume the necessary conditions of uniform convergence of the series to ensure the validity of the interchange between the integration operators and the summation operators, then the double inner products of both sides of this equation can be written as

$$\int_0^b \int_0^a f(x, y) X_r(x) \, Y_s(y) \, dx \, dy = \sum_{n=0}^{\infty} \left(\sum_{m=0}^{\infty} C(m, n) \left(\int_0^a X_m(x) X_r(x) \, dx \right) \left(\int_0^b Y_n(y) \, Y_s(y) \, dy \right) \right)$$

Taking advantage of the statements of orthonormality for the two integrals on the right yields

$$\int_0^b \int_0^a f(x, y) \, X_r(x) Y_s(y) \, dx \, dy = \sum_{n=0}^{\infty} \left(\sum_{m=0}^{\infty} C(m, n) \delta(m, r) \delta(n, s) \right)$$

Due to the mathematical nature of the Kronecker delta functions, the double sum on the right is easy to determine (only the $m = r, n = s$ term survives), and the double Fourier coefficients $C(m, n)$ are evaluated to be

$$C(m, n) = \int_0^b \int_0^a f(x, y) X_m(x) Y_n(y) \, dx \, dy \tag{6.1}$$

We now demonstrate an example for evaluating these coefficients.

DEMONSTRATION: We seek the solution to the example problem in Section 6.4 for the special case where the initial temperature distribution is $u(x, y, 0) = f(x, y)$ where

$$f(x, y) = \left(1 - \frac{x^2}{3} \right) y(1 - y)$$

SOLUTION: Substituting $t = 0$ in the solution for $u(x, y, t)$ yields

$$\left(1 - \frac{x^2}{3}\right) y(1 - y) = \sum_{n=0}^{\infty} \left(\sum_{m=0}^{\infty} C(m, n) X_m(x) Y_n(y)\right)$$

where, from the preceding example, the orthonormalized eigenfunctions are

$$X_m(x) = \frac{\sqrt{2} \cos\left(\sqrt{\alpha_m} x\right)}{\sqrt{\sin\left(\sqrt{\alpha_m}\right)^2 + 1}}$$

and

$$Y_n(y) = \sqrt{2} \sin(n\pi y)$$

for $m = 1, 2, 3, \ldots$, and $n = 1, 2, 3, \ldots$. Insertion of these eigenfunctions into the above series yields

$$\left(1 - \frac{x^2}{3}\right) y(1 - y) = \sum_{n=0}^{\infty} \left(\sum_{m=0}^{\infty} \frac{2C(m, n) \cos\left(\sqrt{\alpha_m} x\right) \sin(n\pi y)}{\sqrt{\sin\left(\sqrt{\alpha_m}\right)^2 + 1}}\right)$$

This is the double Fourier series expansion of $f(x, y)$ in terms of the evaluated eigenfunctions from Section 6.4. The Fourier coefficients $C(m, n)$ were evaluated by taking the double inner product of the initial condition function $f(x, y)$ with respect to these corresponding orthonormalized eigenfunctions. Doing so yielded the integral

$$C(m, n) = \int_0^1 \int_0^1 \left(1 - \frac{x^2}{3}\right) y(1 - y) X_m(x) Y_n(y) \, dx \, dy$$

which, for the preceding eigenfunctions, reduces to

$$C(m, n) = \int_0^1 \int_0^1 \frac{2\left(1 - \frac{x^2}{3}\right) y(1 - y) \cos\left(\sqrt{\alpha_m} x\right) \sin(n\pi y)}{\sqrt{\sin\left(\sqrt{\alpha_m}\right)^2 + 1}} \, dx \, dy$$

for $n = 1, 2, 3, \ldots$, and $m = 1, 2, 3, \ldots$. Evaluation of this integral yields

$$C(m, n) = \frac{8 \cos\left(\sqrt{\alpha_m}\right) \left(1 - (-1)^n\right)}{3 n^3 \pi^3 \alpha_m^{\left(\frac{3}{2}\right)} \sqrt{\cos\left(\sqrt{\alpha_m}\right)^2 + \alpha_m}}$$

Thus, the time-dependent solution becomes

$$T_{m,n}(t) = \frac{8\cos\left(\sqrt{\alpha_m}\right)\left(1-(-1)^n\right)e^{-\frac{\left(\alpha_m+n^2\pi^2\right)t}{10}}}{3n^3\pi^3\alpha_m^{\left(\frac{3}{2}\right)}\sqrt{\cos\left(\sqrt{\alpha_m}\right)^2+\alpha_m}}$$

The eigenvalues α_m are determined from the roots of the equation

$$\tan\left(\sqrt{\alpha_m}\right) = \frac{1}{\sqrt{\alpha_m}} \tag{6.2}$$

If we set $v = \sqrt{\alpha}$, then the eigenvalues α_m are given as the squares of the values at the intersection points of the curves $\tan(v)$ and $1/v$ shown in Figure 6.1.

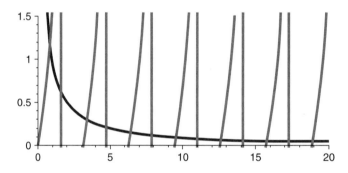

Figure 6.1

The first three eigenvalues are

$$\alpha_1 = 0.74017$$
$$\alpha_2 = 11.7348$$
$$\alpha_3 = 41.4388$$

With knowledge of the eigenvalues α_m, the final series solution to the problem is

$$u(x,y,t) = \sum_{n=1}^{\infty}\left(\sum_{n=1}^{\infty}\frac{16\cos\left(\sqrt{\alpha_m}\right)\left(1-(-1)^n\right)e^{-\frac{\left(\alpha_m+n^2\pi^2\right)t}{10}}\cos\left(\sqrt{\alpha_m}\,x\right)\sin(n\pi y)}{3n^3\pi^3\alpha_m^{\left(\frac{3}{2}\right)}\sqrt{\cos\left(\sqrt{\alpha_m}\right)^2+\alpha_m}\sqrt{\sin\left(\sqrt{\alpha_m}\right)^2+1}}\right)$$

The details for the development of this solution and the graphics are given later in one of the Maple worksheet examples.

6.6 Example Diffusion Problems in Rectangular Coordinates

We now consider several examples of partial differential equations for heat or diffusion phenomena under various homogeneous boundary conditions over finite two-dimensional domains in the rectangular coordinate system. We note that all the spatial ordinary differential equations are of the Euler type.

EXAMPLE 6.6.1: We seek the temperature distribution in a thin rectangular plate over the finite two-dimensional domain $D = \{(x, y) \mid 0 < x < 1, 0 < y < 1\}$. The lateral surfaces of the plate are insulated. The boundaries are all fixed at temperature 0, the plate has an initial temperature distribution $f(x, y)$ given as follows, and the thermal diffusivity is $k = 1/80$.

SOLUTION: The two-dimensional homogeneous diffusion equation is

$$\frac{\partial}{\partial t} u(x, y, t) = k \left(\frac{\partial^2}{\partial x^2} u(x, y, t) + \frac{\partial^2}{\partial y^2} u(x, y, t) \right)$$

The boundary conditions are type 1 at $x = 0$, type 1 at $x = 1$, type 1 at $y = 0$, and type 1 at $y = 1$:

$$u(0, y, t) = 0 \quad \text{and} \quad u(1, y, t) = 0$$

and

$$u(x, 0, t) = 0 \quad \text{and} \quad u(x, 1, t) = 0$$

The initial condition is

$$u(x, y, 0) = x(1 - x)y(1 - y)$$

The ordinary differential equations from the method of separation of variables are

$$\frac{d}{dt} T(t) + k\lambda T(t) = 0$$

$$\frac{d^2}{dx^2} X(x) + \alpha X(x) = 0$$

$$\frac{d^2}{dy^2} Y(y) + \beta Y(y) = 0$$

The coupling equation reads

$$\lambda = \alpha + \beta$$

The boundary conditions on the spatial variables are

$$X(0) = 0, \ X(1) = 0$$

and

$$Y(0) = 0, \ Y(1) = 0$$

Assignment of system parameters

> restart:with(plots):a:=1:b:=1:k:=1/80:

Allowed eigenvalues and corresponding orthonormal eigenfunctions are obtained from Example 2.5.1.

> alpha[m]:=(m*Pi/a)^2;beta[n]:=(n*Pi/b)^2;lambda[m,n]:=alpha[m]+beta[n];

$$\alpha_m := m^2\pi^2$$

$$\beta_n := n^2\pi^2$$

$$\lambda_{m,n} := m^2\pi^2 + n^2\pi^2 \tag{6.3}$$

for $m = 1, 2, 3, \ldots, n = 1, 2, 3, \ldots$.

Orthonormal eigenfunctions

> X[m](x):=sqrt(2/a)*sin(m*Pi/a*x);X[r](x):=subs(m=r,X[m](x)):

$$X_m(x) := \sqrt{2}\sin(m\pi x) \tag{6.4}$$

> Y[n](y):=sqrt(2/b)*sin(n*Pi/b*y);Y[s](y):=subs(n=s,Y[n](y)):

$$Y_n(y) := \sqrt{2}\sin(n\pi y) \tag{6.5}$$

Statements of orthonormality with their respective weight functions

> w(x):=1:Int(X[m](x)*X[r](x)*w(x),x=0...a)=delta(m,r);

$$\int_0^1 2\sin(m\pi x)\sin(r\pi x)\,dx = \delta(m,r) \tag{6.6}$$

> w(y):=1:Int(Y[n](y)*Y[s](y)*w(y),y=0...b)=delta(n,s);

$$\int_0^1 2\sin(n\pi y)\sin(s\pi y)\,dy = \delta(n,s) \tag{6.7}$$

Time-dependent solution

> T[m,n](t):=C(m,n)*exp(-k*lambda[m,n]*t);u[m,n](x,y,t):=T[m,n](t)*X[m](x)*Y[n](y):

$$T_{m,n}(t) := C(m,n)e^{-\frac{1}{80}(m^2\pi^2+n^2\pi^2)t} \tag{6.8}$$

Generalized series terms

> u[m,n](x,y,t):=T[m,n](t)*X[m](x)*Y[n](y);

$$u_{m,n}(x, y, t) := 2C(m, n)e^{-\frac{1}{80}(m^2\pi^2+n^2\pi^2)t}\sin(m\pi x)\sin(n\pi y) \tag{6.9}$$

Eigenfunction expansion

> u(x,y,t):=Sum(Sum(u[m,n](x,y,t),m=1..infinity),n=1..infinity);

$$u(x, y, t) := \sum_{n=1}^{\infty}\left(\sum_{m=1}^{\infty} 2C(m, n)e^{-\frac{1}{80}(m^2\pi^2+n^2\pi^2)t}\sin(m\pi x)\sin(n\pi y)\right) \tag{6.10}$$

The Fourier coefficients $C(m, n)$ are to be determined from the initial condition function $u(x, y, 0) = f(x, y)$. We consider the special case where

> f(x,y):=x*(1−x)*y*(1−y);

$$f(x, y) := x(1 - x)y(1 - y) \tag{6.11}$$

At time $t = 0$ we have

> f(x,y)=eval(subs(t=0,u(x,y,t)));

$$x(1 - x)y(1 - y) = \sum_{n=1}^{\infty}\left(\sum_{m=1}^{\infty} 2C(m, n)\sin(m\pi x)\sin(n\pi y)\right) \tag{6.12}$$

This is the double Fourier series expansion of $f(x, y)$. From Section 6.5, the Fourier coefficients $C(m, n)$ were evaluated by taking the double inner product of the initial condition function $f(x, y)$ with respect to the preceding corresponding orthonormalized eigenfunctions. This yielded the double integral

> C(m,n):=eval(Int(Int(f(x,y)*X[m](x)*w(x)*Y[n](y)*w(y),x=0..a),y=0..b));C(m,n):=
  expand(value(%)):

$$C(m, n) := \int_0^1\int_0^1 2x(1 - x)y(1 - y)\sin(m\pi x)\sin(n\pi y)dx\,dy \tag{6.13}$$

> C(m,n):=factor(subs({sin(m*Pi)=0,cos(m*Pi)=(−1)^m,sin(n*Pi)=0,cos(n*Pi)=(−1)^n},
  C(m,n)));

$$C(m, n) := \frac{8\left(-1+(-1)^n\right)\left(-1+(-1)^m\right)}{m^3\pi^6 n^3} \tag{6.14}$$

> T[m,n](t):=eval(T[m,n](t));

$$T_{m,n}(t) := \frac{8\left(-1+(-1)^n\right)\left(-1+(-1)^m\right)e^{-\frac{1}{80}(m^2\pi^2+n^2\pi^2)t}}{m^3\pi^6 n^3} \tag{6.15}$$

Generalized series terms

> u[m,n](x,y,t):=eval(T[m,n](t)*X[m](x)*Y[n](y));

$$u_{m,n}(x, y, t) := \frac{16\left(-1+(-1)^n\right)\left(-1+(-1)^m\right)e^{-\frac{1}{80}(m^2\pi^2+n^2\pi^2)t}\sin(m\pi x)\sin(m\pi y)}{m^3\pi^6 n^3}$$

(6.16)

Series solution

> u(x,y,t):=Sum(Sum(u[m,n](x,y,t),m=1..infinity),n=1..infinity);

$$u(x, y, t) := \sum_{n=1}^{\infty}\left(\sum_{n=1}^{\infty}\frac{16\left(-1+(-1)^n\right)\left(-1+(-1)^m\right)e^{-\frac{1}{80}(m^2\pi^2+n^2\pi^2)t}\sin(m\pi x)\sin(m\pi y)}{m^3\pi^6 n^3}\right)$$

(6.17)

First few terms of sum

> u(x,y,t):=sum(sum(u[m,n](x,y,t),m=1..3),n=1..3):

ANIMATION

> animate3d(u(x,y,t),x=0..a,y=0..b,t=0..10,axes=framed,thickness=1);

The preceding animation command illustrates the spatial-time-dependent solution for $u(x, y, t)$. The animation sequence in Figures 6.2 and 6.3 shows snapshots of the animation at the two different times $t = 0$ and $t = 2$.

ANIMATION SEQUENCE

> u(x,y,0):=subs(t=0,u(x,y,t)):plot3d(u(x,y,0),x=0..a,y=0..b,axes=framed,thickness=1);

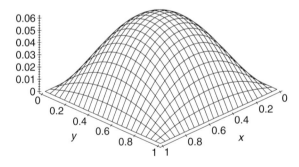

Figure 6.2

> u(x,y,3):=subs(t=3,u(x,y,t)):plot3d(u(x,y,3),x=0..a,y=0..b,axes=framed,thickness=1);

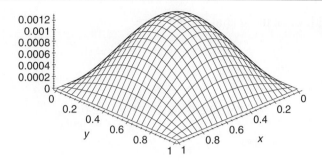

Figure 6.3

EXAMPLE 6.6.2: We seek the temperature distribution in a thin rectangular plate over the finite two-dimensional domain $D = \{(x, y) \mid 0 < x < 1, 0 < y < 1\}$. The lateral surfaces of the plate are insulated. The boundaries $x = 0$ and $x = 1$ are fixed at temperature 0 and the boundaries $y = 0$ and $y = 1$ are insulated. The initial temperature distribution $f(x, y)$ is given below and the thermal diffusivity $k = 1/50$.

SOLUTION: The two-dimensional homogeneous diffusion equation is

$$\frac{\partial}{\partial t} u(x, y, t) = k \left(\frac{\partial^2}{\partial x^2} u(x, y, t) + \frac{\partial^2}{\partial y^2} u(x, y, t) \right)$$

The boundary conditions are type 1 at $x = 0$, type 1 at $x = 1$, type 2 at $y = 0$, and type 2 at $y = 1$:

$$u(0, y, t) = 0 \quad \text{and} \quad u(1, y, t) = 0$$

and

$$u_y(x, 0, t) = 0 \quad \text{and} \quad u_y(x, 1, t) = 0$$

The initial condition is

$$u(x, y, 0) = x(1 - x)y$$

The ordinary differential equations from the method of separation of variables are

$$\frac{d}{dt} T(t) + k\lambda T(t) = 0$$

$$\frac{d^2}{dx^2} X(x) + \alpha X(x) = 0$$

$$\frac{d^2}{dy^2} Y(y) + \beta Y(y) = 0$$

The coupling equation reads

$$\lambda = \alpha + \beta$$

The boundary conditions on the spatial variables are

$$X(0) = 0 \quad \text{and} \quad X(1) = 0$$

and

$$Y_y(0) = 0 \quad \text{and} \quad Y_y(1) = 0$$

Assignment of system parameters

> restart:with(plots):a:=1:b:=1:k:=1/50:

Allowed eigenvalues and corresponding orthonormal eigenfunctions are obtained from Examples 2.5.1 and 2.5.3.

> alpha[m]:=(m*Pi/a)^2;beta[n]:=(n*Pi/b)^2;lambda[m,n]:=alpha[m]+beta[n];

$$\alpha_m := m^2 \pi^2$$
$$\beta_n := n^2 \pi^2$$
$$\lambda_{m,n} := m^2 \pi^2 + n^2 \pi^2 \tag{6.18}$$

for $m = 1, 2, 3, \ldots, n = 1, 2, 3, \ldots,$

> lambda[m,0]:=alpha[m];

$$\lambda_{m,0} := m^2 \pi^2 \tag{6.19}$$

for $n = 0$.

Orthonormal eigenfunctions

> X[m](x):=sqrt(2/a)*sin(m*Pi/a*x);X[r](x):=subs(m=r,X[m](x)):

$$X_m(x) := \sqrt{2} \sin(m\pi x) \tag{6.20}$$

> Y[n](y):=sqrt(2/b)*cos(n*Pi/b*y);Y[s](y):=subs(n=s,Y[n](y)):

$$Y_n(y) := \sqrt{2} \cos(n\pi y) \tag{6.21}$$

> Y[0](y):=1/sqrt(b);

$$Y_0(y) := 1 \tag{6.22}$$

Statements of orthonormality with their respective weight functions

> w(x):=1:Int(X[m](x)*X[r](x)*w(x),x=0...a)=delta(m,r);

$$\int_0^1 2 \sin(m\pi x) \sin(r\pi x) dx = \delta(m, r) \tag{6.23}$$

```
> w(y):=1:Int(Y[n](y)*Y[s](y)*w(y),y=0...b)=delta(n,s);
```

$$\int_0^1 2\cos(n\pi y)\cos(s\pi y)\mathrm{d}y = \delta(n, s) \tag{6.24}$$

Time-dependent solution

For $n = 1, 2, 3, \ldots, m = 1, 2, 3, \ldots,$

```
> T[m,n](t):=C(m,n)*exp(−k*lambda[m,n]*t);
```

$$T_{m,n}(t) := C(m, n)\mathrm{e}^{-\frac{1}{50}(m^2\pi^2+n^2\pi^2)t} \tag{6.25}$$

Generalized series terms

```
> u[m,n](x,y,t):=T[m,n](t)*X[m](x)*Y[n](y);
```

$$u_{m,n}(x, y, t) := 2C(m, n)\mathrm{e}^{-\frac{1}{50}(m^2\pi^2+n^2\pi^2)t}\sin(m\pi x)\cos(n\pi y) \tag{6.26}$$

For $n = 0, m = 1, 2, 3, \ldots,$

```
> T[m,0](t):=C(m,0)*exp(−k*lambda[m,0]*t);u[m,0](x,y,t):=T[m,0](t)*X[m](x)*Y[0](y):
```

$$T_{m,0}(t) := C(m, 0)\mathrm{e}^{-\frac{1}{50}m^2\pi^2 t} \tag{6.27}$$

Eigenfunction expansion

```
> u(x,y,t):=Sum(u[m,0](x,y,t),m=1..infinity)+Sum(Sum(u[m,n](x,y,t),m=1..infinity),n=
  1..infinity);
```

$$u(x, y, t) := \sum_{m=1}^{\infty} C(m, 0)\mathrm{e}^{-\frac{1}{50}m^2\pi^2 t}\sqrt{2}\sin(m\pi x)$$

$$+ \sum_{n=1}^{\infty}\left(\sum_{m=1}^{\infty} 2C(m, n)\mathrm{e}^{-\frac{1}{50}(m^2\pi^2+n^2\pi^2)t}\sin(m\pi x)\cos(n\pi y)\right) \tag{6.28}$$

The Fourier coefficients $C(m, n)$ are to be determined from the initial condition function $u(x, y, 0) = f(x, y)$. We consider the special case where

```
> f(x,y):=x*(1−x)*y;
```

$$f(x, y) := x(1 - x)y \tag{6.29}$$

At time $t = 0$ we have

```
> f(x,y)=eval(subs(t=0,u(x,y,t)));
```

$$x(1 - x)y = \sum_{m=1}^{\infty} C(m, 0)\sqrt{2}\sin(m\pi x) + \sum_{n=1}^{\infty}\left(\sum_{m=1}^{\infty} 2C(m, n)\sin(m\pi x)\cos(n\pi y)\right) \tag{6.30}$$

This is the double Fourier series expansion of $f(x, y)$. From Section 6.5, the Fourier coefficients $C(m, n)$ were evaluated by taking the double inner product of the initial condition function $f(x, y)$ with respect to the preceding corresponding orthonormalized eigenfunctions. This yielded the double integral

> C(m,n):=eval(Int(Int(f(x,y)*X[m](x)*w(x)*Y[n](y)*w(y),x=0..a),y=0..b));C(m,0):=
  expand(value(%)):

$$C(m, n) := \int_0^1 \int_0^1 2x(1 - x)y\sin(m\pi x)\cos(n\pi y)dx\,dy \tag{6.31}$$

> C(m,n):=factor(subs({sin(m*Pi)=0,cos(m*Pi)=(−1)^m,sin(n*Pi)=0,cos(n*Pi)=(−1)^n},
  C(m,n)));

$$C(m, n) := -\frac{4(-1 + (-1)^n)(-1 + (-1)^m)}{m^3\pi^5 n^2} \tag{6.32}$$

> C(m,0):=eval(Int(Int(f(x,y)*X[m](x)*w(x)*Y[0](y)*w(y),x=0..a),y=0..b));C(m,0):=
  expand(value(%)):

$$C(m, 0) := \int_0^1 \int_0^1 x(1 - x)y\sqrt{2}\sin(m\pi x)dx\,dy \tag{6.33}$$

> C(m,0):=simplify(subs({sin(m*Pi)=0,cos(m*Pi)=(−1)^m,sin(n*Pi)=0,cos(n*Pi)=(−1)^n},
  C(m,0)));

$$C(m, 0) := -\frac{\sqrt{2}(-1 + (-1)^m)}{m^3\pi^3} \tag{6.34}$$

> T[m,n](t):=eval(T[m,n](t));

$$T_{m,n}(t) := -\frac{4(-1 + (-1)^n)(-1 + (-1)^m)e^{-\frac{1}{50}(m^2\pi^2 + n^2\pi^2)t}}{m^3\pi^5 n^2} \tag{6.35}$$

> T[m,0](t):=eval(T[m,0](t));

$$T_{m,0}(t) := -\frac{\sqrt{2}(-1 + (-1)^m)e^{-\frac{1}{50}m^2\pi^2 t}}{m^3\pi^3} \tag{6.36}$$

Generalized series terms for $n = 0, m = 1, 2, 3, \ldots,$

> u[m,0](x,y,t):=eval(T[m,0](t)*X[m](x)*Y[0](y));

$$u_{m,0}(x, y, t) := -\frac{2(-1 + (-1)^m)e^{-\frac{1}{50}m^2\pi^2 t}\sin(m\pi x)}{m^3\pi^3} \tag{6.37}$$

Generalized series terms for $n = 1, 2, 3, \ldots, m = 1, 2, 3, \ldots,$

```
> u[m,n](x,y,t):=eval(T[m,n](t)*X[m](x)*Y[n](y));
```

$$u_{m,n}(x, y, t) := -\frac{8(-1+(-1)^n)(-1+(-1)^m)e^{-\frac{1}{50}(m^2\pi^2+n^2\pi^2)t}\sin(m\pi x)\cos(n\pi y)}{m^3\pi^5 n^2} \quad (6.38)$$

Series solution

```
> u(x,y,t):=Sum(u[m,0](x,y,t),m=1..infinity)+Sum(Sum(u[m,n](x,y,t),m=1..infinity),
  n=1..infinity);
```

$$u(x, y, t) := \sum_{m=1}^{\infty}\left(-\frac{2(-1+(-1)^m)e^{-\frac{1}{50}m^2\pi^2 t}\sin(m\pi x)}{m^3\pi^3}\right) \quad (6.39)$$

$$+\sum_{n=1}^{\infty}\left(\sum_{m=1}^{\infty}\left(-\frac{8(-1+(-1)^n)(-1+(-1)^m)e^{-\frac{1}{50}(m^2\pi^2+n^2\pi^2)t}\sin(m\pi x)\cos(n\pi y)}{m^3\pi^5 n^2}\right)\right)$$

First few terms of sum

```
> u(x,y,t):=sum(u[m,0](x,y,t),m=1..3)+sum(sum(u[m,n](x,y,t),m=1..3),n=1..3):
```

ANIMATION

```
> animate3d(u(x,y,t),x=0..a,y=0..b,t=0..5,axes=framed,thickness=1);
```

The preceding animation command illustrates the spatial-time-dependent solution for $u(x, y, t)$. The animation sequence in Figures 6.4 and 6.5 shows snapshots of the animation at two different times $t = 0$ and $t = 5$.

ANIMATION SEQUENCE

```
> u(x,y,0):=subs(t=0,u(x,y,t)):plot3d(u(x,y,0),x=0..a,y=0..b,axes=framed,thickness=1);
```

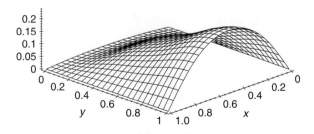

Figure 6.4

> u(x,y,5):=subs(t=5,u(x,y,t)):plot3d(u(x,y,5),x=0..a,y=0..b,axes=framed,thickness=1);

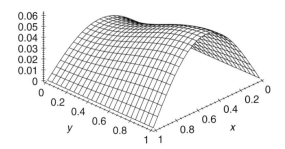

Figure 6.5

EXAMPLE 6.6.3: We seek the temperature distribution in a thin rectangular plate over the finite two-dimensional domain $D = \{(x, y) \mid 0 < x < 1, 0 < y < 1\}$. The lateral surfaces of the plate are insulated. The boundaries $y = 0$ and $y = 1$ are fixed at temperature 0, the boundary $x = 0$ is insulated, and the boundary $x = 1$ is losing heat by convection into a surrounding medium at temperature 0. The initial temperature distribution $f(x, y)$ is given as follows, and the thermal diffusivity is $k = 1/50$.

SOLUTION: The two-dimensional homogeneous diffusion equation is

$$\frac{\partial}{\partial t}u(x, y, t) = k\left(\frac{\partial^2}{\partial x^2}u(x, y, t) + \frac{\partial^2}{\partial y^2}u(x, y, t)\right)$$

The boundary conditions are type 2 at $x = 0$, type 3 at $x = 1$, type 1 at $y = 0$, and type 1 at $y = 1$:

$$u_x(0, y, t) = 0 \quad \text{and} \quad u_x(1, y, t) + u(1, y, t) = 0$$

and

$$u(x, 0, t) = 0 \quad \text{and} \quad u(x, 1, t) = 0$$

The initial condition is

$$u(x, y, 0) = \left(1 - \frac{x^2}{3}\right)y(1 - y)$$

The ordinary differential equations from the method of separation of variables are

$$\frac{d}{dt}T(t) + k\lambda T(t) = 0$$

$$\frac{d^2}{dx^2}X(x) + \alpha X(x) = 0$$

$$\frac{d^2}{dy^2}Y(y) + \beta Y(y) = 0$$

The coupling equation reads

$$\lambda = \alpha + \beta$$

The boundary conditions on the spatial variables are

$$X_x(0) = 0 \quad \text{and} \quad X_x(1) + X(1) = 0$$

and

$$Y(0) = 0 \quad \text{and} \quad Y(1) = 0$$

Assignment of system parameters

> restart:with(plots):a:=1:b:=1:k:=1/50:

Allowed eigenvalues and corresponding orthonormal eigenfunctions are obtained from Examples 2.5.1 and 2.5.5. Eigenvalues α_m are the roots of the eigenvalue equation

> tan(sqrt(alpha[m])*a)=1/sqrt(alpha[m]);

$$\tan\left(\sqrt{\alpha_m}\right) = \frac{1}{\alpha_m} \tag{6.40}$$

> beta[n]:=(n*Pi/b)^2;lambda[m,n]:=alpha[m]+beta[n];

$$\beta_n := n^2\pi^2$$

$$\lambda_{m,n} := \alpha_m + n^2\pi^2 \tag{6.41}$$

for $m = 1, 2, 3, \ldots, n = 1, 2, 3, \ldots$.

Orthonormal eigenfunctions

> X[m](x):=sqrt(2)*(1/sqrt((sin(sqrt(alpha[m])*a)^2+a)))*cos(sqrt(alpha[m])*x);X[r](x):= subs(m=r,X[m](x)):

$$X_m(x) := \frac{\sqrt{2}\cos\left(\sqrt{\alpha_m}\,x\right)}{\sqrt{\sin\left(\sqrt{\alpha_m}\right)^2 + 1}} \tag{6.42}$$

> Y[n](y):=sqrt(2/b)*sin(n*Pi/b*y);Y[s](y):=subs(n=s,Y[n](y)):

$$Y_n(y) := \sqrt{2}\sin(n\pi y) \tag{6.43}$$

Statements of orthonormality with their respective weight functions

> w(x):=1:Int(X[m](x)*X[r](x)*w(x),x=0..a)=delta(m,r);

$$\int_0^1 \frac{2\cos\left(\sqrt{\alpha_m}\,x\right)\cos\left(\sqrt{\alpha_r}\,x\right)}{\sqrt{\sin\left(\sqrt{\alpha_m}\right)^2 + 1}\sqrt{\sin\left(\sqrt{\alpha_r}\right)^2 + 1}}\,dx = \delta(m, r) \tag{6.44}$$

> w(y):=1:Int(Y[n](y)*Y[s](y)*w(y),y=0..b)=delta(n,s);

$$\int_0^1 2 \sin(n\pi y) \sin(s\pi y) \mathrm{d}y = \delta(n, s) \tag{6.45}$$

Time-dependent solution

> T[m,n](t):=C(m,n)*exp(−k*lambda[m,n]*t);

$$T_{m,n}(t) := C(m, n)\mathrm{e}^{-\frac{1}{50}\left(\alpha_m + n^2\pi^2\right)t} \tag{6.46}$$

Generalized series terms

> u[m,n](x,y,t):=T[m,n](t)*X[m](x)*Y[n](y);

$$u_{m,n}(x, y, t) := \frac{2C(m, n)\mathrm{e}^{-\frac{1}{50}\left(\alpha_m + n^2\pi^2\right)t} \cos\left(\sqrt{\alpha_m}x\right) \sin(n\pi y)}{\sqrt{\sin\left(\sqrt{\alpha_m}\right)^2 + 1}} \tag{6.47}$$

Eigenfunction expansion

> u(x,y,t):=Sum(Sum(u[m,n](x,y,t),m=1..infinity),n=1..infinity);

$$u(x, y, t) := \sum_{n=1}^{\infty}\left(\sum_{n=1}^{\infty} \frac{2C(m, n)\mathrm{e}^{-\frac{1}{50}\left(\alpha_m + n^2\pi^2\right)t} \cos\left(\sqrt{\alpha_m}x\right) \sin(n\pi y)}{\sqrt{\sin\left(\sqrt{\alpha_m}\right)^2 + 1}}\right) \tag{6.48}$$

The Fourier coefficients $C(m, n)$ are to be determined from the initial condition function $u(x, y, 0) = f(x, y)$. We consider the special case

> f(x,y):=(1−x^2/3)*y*(1−y);

$$f(x, y) := \left(1 - \frac{1}{3}x^2\right) y (1 - y) \tag{6.49}$$

At time $t = 0$, we have

> f(x,y)=eval(subs(t=0,u(x,y,t)));

$$\left(1 - \frac{1}{3}x^2\right) y (1 - y) = \sum_{n=1}^{\infty}\left(\sum_{m=1}^{\infty} \frac{2C(m, n) \cos\left(\sqrt{\alpha_m}\, x\right) \sin(n\pi y)}{\sqrt{\sin\left(\sqrt{\alpha_m}\right)^2 + 1}}\right) \tag{6.50}$$

This is the double Fourier series expansion of $f(x, y)$. From Section 6.5, the Fourier coefficients $C(m, n)$ were evaluated by taking the double inner product of the initial condition function $f(x, y)$ with respect to the preceding corresponding orthonormalized eigenfunctions. This yielded the double integral

> C(m,n):=eval(Int(Int(f(x,y)*X[m](x)*w(x)*Y[n](y)*w(y),x=0..a),y=0..b));C(m,n):=expand
  (value(%)):

$$C(m, n) := \int_0^1 \int_0^1 \frac{2\left(1 - \frac{1}{3}x^2\right) y\,(1-y)\cos\left(\sqrt{\alpha_m}\,x\right)\sin(n\pi y)}{\sqrt{\sin\left(\sqrt{\alpha_m}\right)^2 + 1}}\,dx\,dy \qquad (6.51)$$

> C(m,n):=radsimp(subs({sin(sqrt(alpha[m])*a)=1/sqrt(alpha[m])*cos(sqrt(alpha[m])*a),
  sin(n*Pi)=0,cos(n*Pi)=(−1)^n},C(m,n)));

$$C(m, n) := -\frac{8}{3}\frac{\cos\left(\sqrt{\alpha_m}\right)\left(-1 + (-1)^n\right)}{\alpha_m^2 n^3 \pi^3 \sqrt{2 - \cos\left(\sqrt{\alpha_m}\right)^2}} \qquad (6.52)$$

> T[m,n](t):=eval(T[m,n](t));

$$T_{m,n}(t) := -\frac{8}{3}\frac{\cos\left(\sqrt{\alpha_m}\right)\left(-1 + (-1)^n\right)e^{-\frac{1}{50}\left(\alpha_m + n^2\pi^2\right)t}}{\alpha_m^2\, n^3\, \pi^3 \sqrt{2 - \cos\left(\sqrt{\alpha_m}\right)^2}} \qquad (6.53)$$

Generalized series terms

> u[m,n](x,y,t):=eval(T[m,n](t)*X[m](x)*Y[n](y));

$$u_{m,n}(x, y, t) := -\frac{16}{3}\frac{\cos\left(\sqrt{\alpha_m}\right)\left(-1 + (-1)^n\right)e^{-\frac{1}{50}\left(\alpha_m + n^2\pi^2\right)t}\cos\left(\sqrt{\alpha_m}\,x\right)\sin(n\pi y)}{\alpha_m^2\, n^3\, \pi^3 \sqrt{2 - \cos\left(\sqrt{\alpha_m}\right)^2}\,\sqrt{\sin\left(\sqrt{\alpha_m}\right)^2 + 1}} \qquad (6.54)$$

Series solution

> u(x,y,t):=Sum(Sum(u[m,n](x,y,t),m=1..infinity),n=1..infinity);

$$u(x, y, t) := \sum_{n=1}^{\infty}\left(\sum_{n=1}^{\infty}\left(-\frac{16}{3}\frac{\cos\left(\sqrt{\alpha_m}\right)\left(-1 + (-1)^n\right)e^{-\frac{1}{50}\left(\alpha_m + n^2\pi^2\right)t}\cos\left(\sqrt{\alpha_m}x\right)\sin(n\pi y)}{\alpha_m^2\, n^3\pi^3\sqrt{2 - \cos\left(\sqrt{\alpha_m}\right)^2}\,\sqrt{\sin\left(\sqrt{\alpha_m}\right)^2 + 1}}\right)\right) \qquad (6.55)$$

Evaluation of the eigenvalues α_m from the roots of the eigenvalue equation yields

> tan(sqrt(alpha[m])*a)=1/sqrt(alpha[m]);

$$\tan\left(\sqrt{\alpha_m}\right) = \frac{1}{\sqrt{\alpha_m}} \qquad (6.56)$$

> plot({tab(v),1/v},v=0..20,y=0..3/2,thickness=10);

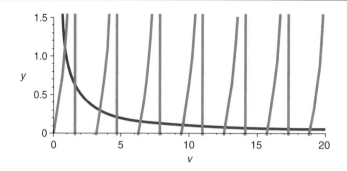

Figure 6.6

If we set $v = a\sqrt{\alpha}$, then the eigenvalues α_m are found from the squares of the values of v at the intersection points of the curves shown in Figure 6.6. Using the Maple fsolve command to evaluate the eigenvalues yields

> alpha[1]:=(1/a^2)*(fsolve((tan(v)−1/v),v=0..1)^2);

$$\alpha_1 := 0.7401738844 \tag{6.57}$$

> alpha[2]:=(1/a^2)*(fsolve((tan(v)−1/v),v=0..4)^2);

$$\alpha_2 := 11.73486183 \tag{6.58}$$

> alpha[3]:=(1/a^2)*(fsolve((tan(v)−1/v),v=4..7)^2);

$$\alpha_3 := 41.43880785 \tag{6.59}$$

First few terms of sum

> u(x,y,t):=eval(sum(sum(u[m,n](x,y,t),m=1..3),n=1..3)):

ANIMATION

> animate3d(u(x,y,t),x=0..a,y=0..b,t=0..5,axes=framed,thickness=1);

The preceding animation command illustrates the spatial-time-dependent solution for $u(x, y, t)$. The animation sequence in Figures 6.7 and 6.8 shows snapshots of the animation at the two different times $t = 0$ and $t = 3$.

ANIMATION SEQUENCE

> u(x,y,0):=subs(t=0,u(x,y,t)):plot3d(u(x,y,0),x=0..a,y=0..b,axes=framed,thickness=1);

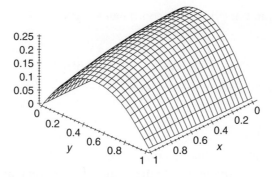

Figure 6.7

> u(x,y,3):=subs(t=3,u(x,y,t)):plot3d(u(x,y,3),x=0..a,y=0..b,axes=framed,thickness=1);

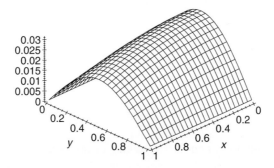

Figure 6.8

6.7 Diffusion Equation in the Cylindrical Coordinate System

If we replace the preceding Laplacian operator in two dimensions from rectangular to cylindrical coordinates, then the nonhomogeneous partial differential equation for diffusion or heat phenomena in the cylindrical coordinate system in two dimensions is given as

$$\frac{\partial}{\partial t} u(r, \theta, t) = k \left(\frac{\left(\frac{\partial}{\partial r} u(r, \theta, t) + r \left(\frac{\partial^2}{\partial r^2} u(r, \theta, t) \right) \right)}{r} + \frac{\frac{\partial^2}{\partial \theta^2} u(r, \theta, t)}{r^2} \right) + h(r, \theta, t)$$

where r is the coordinate radius of the system and θ is the polar angle. There is no z dependence because we will only be considering very thin circular plates with no extension along the z-axis.

Similar to the rectangular coordinate system, we can write the preceding equation in terms of the linear operator for the diffusion equation in cylindrical coordinates as

$$L(u) = h(r, \theta, t)$$

where the diffusion operator in cylindrical coordinates reads

$$L(u) = \frac{\partial}{\partial t} u(r, \theta, t) - k \left(\frac{\frac{\partial}{\partial r} u(r, \theta, t) + \left(r \frac{\partial^2}{\partial r^2} u(r, \theta, t) \right)}{r} + \frac{\frac{\partial^2}{\partial \theta^2} u(r, \theta, t)}{r^2} \right)$$

Generally, one seeks a solution to this equation over the finite two-dimensional domain $D = \{(r, \theta) \mid 0 < r < a, 0 < \theta < b\}$ subject to the homogeneous boundary conditions

$$\kappa_1 u(a, \theta, t) + \kappa_2 u_r(a, \theta, t) = 0$$

$$\kappa_3 u(r, 0, t) + \kappa_4 u_\theta(r, 0, t) = 0$$

$$\kappa_5 u(r, b, t) + \kappa_6 u_\theta(r, b, t) = 0$$

$$|u(0, \theta, t)| < \infty$$

along with the initial condition

$$u(r, \theta, 0) = f(r, \theta)$$

We now attempt to solve the homogeneous version [for no internal heat sources we set $h(r, \theta, t) = 0$] of this partial differential equation using the method of separation of variables. We set

$$u(r, \theta, t) = R(r) \, \Theta(\theta) \, T(t)$$

Substituting this into the preceding homogeneous partial differential equation, we get

$$R(r)\Theta(\theta)\left(\frac{d}{dt} T(t) \right) =$$

$$k \left(\frac{\left(\frac{d}{dr} R(r) \right) \Theta(\theta) T(t) + r \left(\frac{d^2}{dr^2} R(r) \right) \Theta(\theta) T(t)}{r} + \frac{R(r) \left(\frac{d^2}{d\theta^2} \Theta(\theta) \right) T(t)}{r^2} \right)$$

Dividing both sides by the product solution yields

$$\frac{\frac{d}{dt} T(t)}{kT(t)} = \frac{\Theta(\theta) r \left(\frac{d}{dr} R(r) \right) + \Theta(\theta) r^2 \left(\frac{d^2}{dr^2} R(r) \right) + R(r) \left(\frac{d^2}{d\theta^2} \Theta(\theta) \right)}{R(r) \Theta(\theta) r^2}$$

Because the left-hand side of the preceding is an exclusive function of t and the right-hand side an exclusive function of r and θ, and r, θ, and t are independent, the only way we can ensure equality of the preceding for all values of r, θ, and t is to set each side equal to a constant.

Doing so, we arrive at the following three ordinary differential equations in terms of the separation constants (for convenience here, we square the constants to ensure positivity).

$$\frac{d}{dt} T(t) + k\lambda^2 T(t) = 0$$

$$\frac{d^2}{d\theta^2} \Theta(\theta) + \mu^2 \Theta(\theta) = 0$$

$$r^2 \left(\frac{d^2}{dr^2} R(r) \right) + r \left(\frac{d}{dr} R(r) \right) + \left(\lambda^2 r^2 - \mu^2 \right) R(r) = 0$$

The preceding differential equation in t is an ordinary first-order linear equation for which we already have the solution from Chapter 1. The second differential equation in the variable θ is recognized as being an Euler-type differential equation for which we generated solutions in Section 1.4. The third differential equation is a Bessel-type differential equation whose solutions are Bessel functions of the first kind of order μ as discussed in Section 1.11. We recognize each of the two spatial differential equations in θ and r as being Sturm-Liouville-type differential equations.

We first consider the second differential equation with respect to the variable θ subject to the homogeneous boundary conditions over the finite interval $I = \{\theta \,|\, 0 < \theta < b\}$. The Sturm-Liouville problem for the θ-dependent portion of the diffusion equation consists of the ordinary differential equation

$$\frac{d^2}{d\theta^2} \Theta(\theta) + \mu^2 \, \Theta(\theta) = 0$$

along with the corresponding regular homogeneous boundary conditions, which read

$$\kappa_3 \Theta(0) + \kappa_4 \Theta_\theta(0) = 0$$

and

$$\kappa_5 \Theta(b) + \kappa_6 \Theta_\theta(b) = 0$$

We see the preceding problem in the variable θ to be a regular Sturm-Liouville eigenvalue problem where the allowed values of μ are called the system "eigenvalues" and the corresponding solutions are called the "eigenfunctions." Note that this ordinary differential equation is of the Euler type and the weight function is $w(\theta) = 1$. The indexed eigenvalues and corresponding eigenfunctions are given, respectively, as

$$\mu_m, \ \Theta_m(\theta)$$

for $m = 0, 1, 2, 3, \ldots$.

These eigenfunctions can be normalized and the corresponding statement of orthonormality with respect to the weight function $w(\theta) = 1$ reads

$$\int_0^b \Theta_m(\theta)\Theta_q(\theta)\, d\theta = \delta(m, q)$$

We now consider the r-dependent differential equation over the finite interval $I = \{r \,|\, 0 < r < a\}$. The Sturm-Liouville problem for the r-dependent portion of the diffusion equation consists of the ordinary differential equation

$$r^2\left(\frac{d^2}{dr^2}R(r)\right) + r\left(\frac{d}{dr}R(r)\right) + (\lambda^2 r^2 - \mu^2)R(r) = 0$$

along with the respective corresponding nonregular homogeneous boundary conditions

$$|R(0)| < \infty$$

and

$$\kappa_1 R(a) + \kappa_2 R_r(a) = 0$$

We see the preceding problem in the spatial variable r to be a nonregular Sturm-Liouville eigenvalue problem. (The problem is nonregular because we seek a solution that is valid over an interval that includes the origin $r = 0$, which is a singular point of the differential equation. Also, the boundary condition at the origin is not of the standard form.) We seek a solution of the system where the allowed values of λ are called the system "eigenvalues" and the corresponding solutions are called the "eigenfunctions." We note that the allowed values of λ are also dependent on the indexed values of μ. The preceding ordinary differential equation is of the Bessel type, and, from Section 2.6, the weight function is $w(r) = r$. The indexed eigenvalues and corresponding eigenfunctions are given, respectively, as

$$\lambda_{m,n}, \ R_{m,n}(r)$$

for $m = 0, 1, 2, 3, \ldots, n = 0, 1, 2, 3, \ldots$.

These eigenfunctions can be normalized, and the corresponding statement of orthonormality with respect to the weight function $w(r) = r$ reads

$$\int_0^a R_{m,n}(r)\, R_{m,p}(r)r\, dr = \delta(n, p)$$

Here, $\delta(n, p)$ is the Kronecker delta term defined earlier. What was stated in Section 2.2 about regular Sturm-Liouville eigenvalue problems in one dimension can be extended to two dimensions here; that is, the product of the preceding eigenfunctions in r and θ forms a "complete" set with respect to any piecewise smooth function $f(r, \theta)$ over the finite two-dimensional domain $D = \{(r, \theta) \,|\, 0 < r < a, 0 < \theta < b\}$.

Finally, we focus on the solution to the first-order linear time-dependent differential equation, which reads

$$\frac{d}{dt}T(t) + k\lambda T(t) = 0$$

For the indexed values of $\lambda_{m,n}$ given earlier, the solution to this equation, from Section 1.2, is

$$T_{m,n}(t) = C(m,n)e^{-k\lambda_{m,n}^2 t} \tag{6.60}$$

where the $C(m,n)$ are arbitrary constants.

Using similar arguments as we did for the regular Sturm-Liouville problem in rectangular coordinates given earlier, we can write our general solution to the partial differential equation as the superposition of the products of the solutions to each of the preceding ordinary differential equations.

Thus, the general solution to the homogeneous partial differential equation can be written as the double-infinite sum

$$u(r, \theta, t) = \sum_{n=0}^{\infty}\left(\sum_{m=0}^{\infty} R_{m,n}(r)\Theta_m(\theta)C(m,n)e^{-k\lambda_{m,n}^2 t}\right)$$

As before, the unknown Fourier coefficients $C(m,n)$ are evaluated from the initial conditions on the problem. We now demonstrate an example problem.

DEMONSTRATION: We seek the temperature distribution in a thin circular plate over the two-dimensional domain $D = \{(r, \theta) \,|\, 0 < r < 1, 0 < \theta < \pi\}$. The lateral surfaces of the plate are insulated, so no heat can escape from the lateral surfaces. The sides $\theta = 0$ and $\theta = \pi$ are at a fixed temperature of 0, and the edge $r = 1$ is insulated. The initial temperature distribution $u(r, \theta, 0) = f(r, \theta)$ is given as follows, and the thermal diffusivity is $k = 1/50$.

SOLUTION: The two-dimensional homogeneous diffusion equation for the problem reads

$$50\left(\frac{\partial}{\partial t}u(r, \theta, t)\right) = \frac{\frac{\partial}{\partial r}u(r, \theta, t) + r\frac{\partial^2}{\partial r^2}u(r, \theta, t)}{r} + \frac{\frac{\partial^2}{\partial \theta^2}u(r, \theta, t)}{r^2}$$

The boundary conditions are type 2 at $r = 1$, type 1 at $\theta = 0$, type 1 at $\theta = \pi$, and we require a finite solution at the origin

$$|u(0, \theta, t)| < \infty \quad \text{and} \quad u_r(1, \theta, t) = 0$$

and

$$u(r, 0, t) = 0 \quad \text{and} \quad u(r, \pi, t) = 0$$

The initial condition is

$$u(r, \theta, 0) = f(r, \theta)$$

The ordinary differential equations from the method of separation of variables are

$$\frac{d}{dt} T(t) + \frac{\lambda^2 T(t)}{50} = 0$$

$$r^2 \left(\frac{d^2}{dr^2} R(r) \right) + r \left(\frac{d}{dr} R(r) \right) + (\lambda^2 r^2 - \mu^2) R(r) = 0$$

$$\frac{d^2}{d\theta^2} \Theta(\theta) + \mu^2 \Theta(\theta) = 0$$

The boundary conditions on the spatial variables are

$$|R(0)| < \infty \quad \text{and} \quad R_r(1) = 0$$

and

$$\Theta(0) = 0 \quad \text{and} \quad \Theta(\pi) = 0$$

We consider the θ-dependent equation first. Here, we have a Sturm-Liouville differential equation of the Euler type with weight function $w(\theta) = 1$. From Example 2.5.1, for the given boundary conditions on the θ equation, the eigenvalues and corresponding orthonormal eigenfunctions are

$$\mu_m = m^2$$

and

$$\Theta_m = \sqrt{\frac{2}{\pi}} \sin(m\theta)$$

for $m = 1, 2, 3, \ldots$.

We next consider the r-dependent equation. Here, we have a Sturm-Liouville differential equation of the Bessel type with weight function $w(r) = r$. From Example 2.6.3, for the given boundary conditions on the r-equation, the eigenvalues $\lambda_{m,n}$ are the roots of the eigenvalue equation

$$-J(m+1, \lambda_{m,n}) + \frac{m J (m, \lambda_{m,n})}{\lambda_{m,n}} = 0$$

for $m = 1, 2, 3, \ldots$, and $n = 1, 2, 3, \ldots$. The corresponding normalized r-dependent eigenfunctions are

$$R_{m,n}(r) = \frac{J(m, \lambda_{m,n} r)}{\int_0^1 J(m, \lambda_{m,n} r)^2 r \, dr}$$

where $J(m, \lambda_{m,n}r)$ denotes the Bessel function of the first kind of order m. The corresponding statements of orthonormality, with respective weight functions, are

$$\int_0^1 \frac{J(m, \lambda_{m,n}r)\, J(m, \lambda_{m,p}r)\, r}{\sqrt{\int_0^1 J\left(m, \lambda_{m,n}r\right)^2 r\, dr}\sqrt{\int_0^1 J\left(m, \lambda_{m,p}r\right)^2 r\, dr}}\, dr = \delta(n, p)$$

and

$$\int_0^\pi \frac{2\sin(m\theta)\sin(q\theta)}{\pi}\, d\theta = \delta(m, q)$$

for $m, q = 1, 2, 3, \ldots$, and $n, p = 1, 2, 3, \ldots$.

We now focus on the t-dependent differential equation. This is a simple first-order differential equation, which we examined in Section 1.2 and whose solution is

$$T_{m,n}(t) = C(m, n)\mathrm{e}^{-\frac{\lambda_{m,n}^2 t}{50}}$$

From the superposition of the product of the solutions to the three preceding ordinary differential equations, the general eigenfunction expansion solution of the partial differential equation reads

$$u(r, \theta, t) = \sum_{n=1}^\infty \left(\sum_{m=1}^\infty \frac{C(m, n)\mathrm{e}^{-\frac{\lambda_{m,n}^2 t}{50}} J(m, \lambda_{m,n}r)\sqrt{\frac{2}{\pi}}\sin(m\theta)}{\sqrt{\int_0^1 J\left(m, \lambda_{m,n}r\right)^2 r\, dr}} \right)$$

The Fourier coefficients $C(m, n)$ are to be determined from the initial condition function $u(r, \theta, 0) = f(r, \theta)$.

6.8 Initial Conditions for the Diffusion Equation in Cylindrical Coordinates

We now consider evaluation of the Fourier coefficients. If the initial temperature distribution condition is

$$u(r, \theta, 0) = f(r, \theta)$$

then substitution of this into the following general solution

$$u(r, \theta, t) = \sum_{n=0}^\infty \left(\sum_{m=0}^\infty R_{m,n}(r)\Theta_m(\theta)\, C(m, n)\mathrm{e}^{-k\lambda_{m,n}^2 t} \right)$$

at time $t = 0$ yields

$$f(r, \theta) = \sum_{n=0}^{\infty} \left(\sum_{m=0}^{\infty} C(m, n) \, R_{m,n}(r) \, \Theta_m(\theta) \right)$$

This equation is the generalized double Fourier series expansion of the function $f(r, \theta)$ in terms of the allowed eigenfunctions and the double Fourier coefficients $C(m, n)$.

Thus, as we did in Section 2.4 for the generalized Fourier series expansion, we take the double inner products of both sides of the preceding equation, with respect to the r- and θ-dependent eigenfunctions. This yields

$$\int_0^b \int_0^a f(r, \theta) R_{q,p}(r) \Theta_q(\theta) r \, dr \, d\theta = \int_0^b \int_0^a \left(\sum_{n=0}^{\infty} \left(\sum_{m=0}^{\infty} C(m, n) R_{m,n}(r) \Theta_m(\theta) \right) \right) R_{q,p}(r) \Theta_q(\theta) r \, dr \, d\theta$$

Assuming the validity of the interchange between the summation and integration operations, we get

$$\int_0^b \int_0^a f(r, \theta) R_{q,p}(r) \Theta_q(\theta) r \, dr \, d\theta = \sum_{n=0}^{\infty} \left(\sum_{m=0}^{\infty} C(m, n) \left(\int_0^a R_{q,p}(r) R_{m,n}(r) r \, dr \right) \left(\int_0^b \Theta_m(\theta) \Theta_q(\theta) \, d\theta \right) \right)$$

Taking advantage of the statements of orthonormality for the two integrals on the right yields

$$\int_0^b \int_0^a f(r, \theta) R_{q,p}(r) \, \Theta_q(\theta) r \, dr \, d\theta = \sum_{n=0}^{\infty} \left(\sum_{m=0}^{\infty} C(m, n) \, \delta(n, p) \, \delta(m, q) \right) \tag{6.61}$$

With the Kronecker delta functions in the sum, only the $n = p, m = q$ term survives, and we are able to evaluate the double Fourier coefficients $C(m, n)$ to be

$$C(m, n) = \int_0^b \int_0^a f(r, \theta) R_{m,n}(r) \Theta_m(\theta) r \, dr \, d\theta$$

We now demonstrate the evaluation of the Fourier coefficients $C(m, n)$ for the example problem given in Section 6.7.

DEMONSTRATION: We consider the special case where the initial temperature distribution function is $u(r, \theta, 0) = f(r, \theta)$ where

$$f(r, \theta) = \left(r - \frac{r^3}{3} \right) \sin(\theta)$$

SOLUTION: At time $t = 0$, the solution $u(r, \theta, t)$ reads

$$\left(r - \frac{r^3}{3}\right) \sin(\theta) = \sum_{n=1}^{\infty} \left(\sum_{m=1}^{\infty} C(m, n) R_{m,n}(r) \Theta_m(\theta)\right)$$

where, from the preceding example, the orthonormalized eigenfunctions are

$$\Theta_m = \sqrt{\frac{2}{\pi}} \sin(m\theta)$$

and

$$R_{m,n}(r) = \frac{J(m, \lambda_{m,n} r)}{\int_0^1 J\left(m, \lambda_{m,n} r\right)^2 r \, dr}$$

for $m = 1, 2, 3, \ldots$, and $n = 1, 2, 3, \ldots$. Insertion of these eigenfunctions into the series yields

$$\left(r - \frac{r^3}{3}\right) \sin(\theta) = \sum_{n=1}^{\infty} \left(\sum_{m=1}^{\infty} C(m, n) R_{m,n}(r) \sqrt{\frac{2}{\pi}} \sin(m\theta)\right)$$

This is the double Fourier series expansion of $f(r, \theta)$. From the preceding, the Fourier coefficients $C(m, n)$ were evaluated by taking the double inner product of the initial condition function $f(r, \theta)$ with respect to the corresponding orthonormal eigenfunctions. This yielded the integral

$$C(m, n) = \int_0^1 \int_0^\pi \left(r - \frac{r^3}{3}\right) \sin(\theta) R_{m,n}(r) r \Theta_m(\theta) \, d\theta \, dr$$

which, for the given eigenfunctions, reduces to

$$C(m, n) = \int_0^1 \int_0^\pi \left(r - \frac{r^3}{3}\right) \sin(\theta) R_{m,n}(r) r \sqrt{\frac{2}{\pi}} \sin(m\theta) \, d\theta \, dr$$

for $m, n = 1, 2, 3, \ldots$. We integrate first with respect to θ. Since $f(r, \theta)$ depends on the term $\sin(m\theta)$ for the case $m = 1$, then from the orthogonality condition on the θ equation, we see that only the $m = 1$ solution survives here. Thus, we get

$$C(1, n) = \int_0^1 \int_0^\pi \left(r - \frac{r^3}{3}\right) \sin(\theta)^2 R_{1,n}(r) r \sqrt{\frac{2}{\pi}} \, d\theta \, dr$$

where, for $m = 1$,

$$R_{1,n} = \frac{J(1, \lambda_{1,n} r)}{\sqrt{\int_0^1 J(1, \lambda_{1,n}^2 r) r \, dr}}$$

Evaluation of the above for $R_{1,n}$ yields

$$R_{1,n} = \frac{J(1, \lambda_{1,n}r)\sqrt{2}\lambda_{1,n}}{\sqrt{\lambda_{1,n}^2 - 1}\, J(1, \lambda_{1,n})}$$

Inserting this into $C(1, n)$ and integrating yields

$$C(1, n) = \frac{8\sqrt{\pi}}{3\lambda_{1,n}^3 \sqrt{\lambda_{1,n}^2 - 1}}$$

Thus, the final solution to the problem reads

$$u(r, \theta, t) = \sum_{n=1}^{\infty} \frac{16 e^{-\frac{\lambda_{1,n}^2 t}{50}}\, J(1, \lambda_{1,n}r)\sin(\theta)}{3\lambda_{1,n}^2 \left(\lambda_{1,n}^2 - 1\right) J(1, \lambda_{1,n})}$$

This solution is completed upon evaluation of the eigenvalues $\lambda_{1,n}$ from the eigenvalue equation

$$J(1, \lambda_{1,n}) - \lambda_{1,n}\, J(2, \lambda_{1,n}) = 0$$

The details for the development of the solution and the graphics are given later in one of the Maple worksheets.

6.9 Example Diffusion Problems in Cylindrical Coordinates

We now consider several examples of partial differential equations for heat or diffusion phenomena under various homogeneous boundary conditions over finite intervals in the cylindrical coordinate system. We note that the r-dependent ordinary differential equations in the cylindrical coordinate system are of the Bessel type and the θ-dependent ones are of the Euler type.

EXAMPLE 6.9.1: We seek the temperature distribution $u(r, \theta, t)$ in a thin circular plate over the two-dimensional domain $D = \{(r, \theta) \,|\, 0 < r < 1, 0 < \theta < \pi/2\}$. The lateral surfaces of the plate are insulated. The edges $r = 1$ and $\theta = 0$ are at a fixed temperature of 0, and the edge $\theta = \pi/2$ is insulated. The initial temperature distribution $u(r, \theta, 0) = f(r, \theta)$ is given following. The thermal diffusivity is $k = 1/50$.

SOLUTION: The two-dimensional homogeneous diffusion equation is

$$\frac{\partial}{\partial t} u(r, \theta, t) = k \left(\frac{\frac{\partial}{\partial r} u(r, \theta, t) + r\left(\frac{\partial^2}{\partial r^2} u(r, \theta, t)\right)}{r} + \frac{\frac{\partial^2}{\partial \theta^2} u(r, \theta, t)}{r^2} \right)$$

The boundary conditions are type 1 at $r = 1$, type 1 at $\theta = 0$, and type 2 at $\theta = \pi/2$ and a finite solution at the origin:

$$|u(0, \theta, t)| < \infty \quad \text{and} \quad u(1, \theta, t) = 0$$

and

$$u(r, 0, t) = 0 \quad \text{and} \quad u_\theta\left(r, \frac{\pi}{2}, t\right) = 0$$

The initial condition is

$$u(r, \theta, 0) = (r - r^3)\sin(\theta)$$

The ordinary differential equations from the method of separation of variables are

$$\frac{d}{dt} T(t) + k\lambda^2 T(t) = 0$$

$$r^2\left(\frac{d^2}{dr^2} R(r)\right) + r\left(\frac{d}{dr} R(r)\right) + (\lambda^2 r^2 - \mu^2) R(r) = 0$$

$$\frac{d^2}{d\theta^2}\Theta(\theta) + \mu^2\Theta(\theta) = 0$$

The boundary conditions on the spatial variables are

$$|R(0)| < \infty \quad \text{and} \quad R(1) = 0$$

and

$$\Theta(0) = 0 \quad \text{and} \quad \Theta_\theta\left(\frac{\pi}{2}\right) = 0$$

Assignment of system parameters

> restart:with(plots):a:=1:b:=Pi/2:k:=1/50:

Allowed eigenvalues and corresponding orthonormal eigenfunctions are obtained from Examples 2.5.2 and 2.6.2.

> mu[m]:=(2*m−1)*Pi/(2*b);

$$\mu_m := 2m - 1 \tag{6.62}$$

for $m = 1, 2, 3, \ldots$.

Eigenvalues $\lambda_{m,n}$ are the roots of the eigenvalue equation

> BesselJ(mu[m],lambda[m,n]*a)=0;

$$\text{BesselJ}(2m - 1, \lambda_{m,n}) = 0 \tag{6.63}$$

for $m = 1, 2, 3, \ldots, n = 1, 2, 3, \ldots$.

Orthonormal eigenfunctions

> R[m,n](r):=BesselJ(mu[m],lambda[m,n]*r)/sqrt(Int(BesselJ(mu[m],lambda[m,n]*r)^2*r,
  r=0..a));R[m,p](r):=subs(n=p,R[m,n](r)):

$$R_{m,n}(r) := \frac{\text{BesselJ}(2m-1, \lambda_{m,n}\, r)}{\sqrt{\int_0^1 \text{BesselJ}(2m-1, \lambda_{m,n}\, r)^2\, r\, dr}} \tag{6.64}$$

> Theta[m](theta):=sqrt(2/b)*sin((2*m−1)*Pi/(2*b)*theta);
  Theta[q](theta):=subs(m=q,Theta[m](theta)):

$$\Theta_m(\theta) := \frac{2\,\sin((2m-1)\theta)}{\sqrt{\pi}} \tag{6.65}$$

Statements of orthonormality with their respective weight functions

> w(r):=r:Int(R[m,n](r)*R[m,p](r)*w(r),r=0...a)=delta(n,p);

$$\int_0^1 \frac{\text{BesselJ}(2m-1, \lambda_{m,n}r)\,\text{BesselJ}(2m-1, \lambda_{m,p}r)r}{\sqrt{\int_0^1 \text{BesselJ}(2m-1, \lambda_{m,n}r)^2\, r\, dr}\sqrt{\int_0^1 \text{BesselJ}(2m-1, \lambda_{m,p}r)^2\, r\, dr}}\, dr = \delta(n,\, p) \tag{6.66}$$

> w(theta):=1:Int(Theta[m](theta)*Theta[q](theta)*w(theta),theta=0...b)=delta(m,q);

$$\int_0^{\frac{1}{2}\pi} \frac{4\sin((2m-1)\theta)\,\sin((2q-1)\theta)}{\pi}\, d\theta = \delta(m,\, q) \tag{6.67}$$

Time-dependent solution

> T[m,n](t):=C(m,n)*exp(−k*lambda[m,n]^2*t);

$$T_{m,n}(t) := C(m,n)\, e^{-\frac{1}{50}\lambda_{m,n}^2\, t} \tag{6.68}$$

Generalized series terms

> u[m,n](r,theta,t):=T[m,n](t)*R[m,n](r)*Theta[m](theta);

$$u_{m,n}(r,\theta,t) := \frac{2C(m,n)e^{-\frac{1}{50}\lambda_{m,n}^2\, t}\,\text{BesselJ}(2m-1, \lambda_{m,n}r)\,\sin((2m-1)\theta)}{\sqrt{\int_0^1 \text{BesselJ}(2m-1, \lambda_{m,n}r)^2 r\, dr}\,\sqrt{\pi}} \tag{6.69}$$

Eigenfunction expansion

> u(r,theta,t):=Sum(Sum(u[m,n](r,theta,t),m=1..infinity),n=1..infinity);

$$u(r,\theta,t) := \sum_{n=1}^{\infty}\left(\sum_{m=1}^{\infty} \frac{2C(m,n)e^{-\frac{1}{50}\lambda_{m,n}^2\, t}\,\text{BesselJ}(2m-1, \lambda_{m,n}r)\,\sin((2m-1)\theta)}{\sqrt{\int_0^1 \text{BesselJ}(2m-1, \lambda_{m,n}r)^2\, r\, dr}\,\sqrt{\pi}}\right) \tag{6.70}$$

The Fourier coefficients $C(m, n)$ are to be determined from the initial condition function $u(r, \theta, 0) = f(r, \theta)$. We consider the special case where

> f(r,theta):=(r−r^3)*sin(theta);

$$f(r, \theta) := (r - r^3) \sin(\theta) \tag{6.71}$$

At time $t = 0$, we have

> f(r,theta)=eval(subs(t=0,u(r,theta,t)));

$$(r - r^3) \sin(\theta) = \sum_{n=1}^{\infty} \left(\sum_{m=1}^{\infty} \frac{2\, C(m, n)\, \text{BesselJ}(2m-1, \lambda_{m,n}\, r) \sin((2m-1)\theta)}{\sqrt{\int_0^1 \text{BesselJ}(2m-1, \lambda_{m,n}\, r))^2 r\, dr}\, \sqrt{\pi}} \right) \tag{6.72}$$

This is the double Fourier series expansion of $f(r, \theta)$. From Section 6.8, the Fourier coefficients $C(m, n)$ were evaluated by taking the double inner product of the initial condition function $f(r, \theta)$ with respect to the corresponding orthonormalized eigenfunctions given earlier.

Because $f(r, \theta)$ depends on $\sin((2m-1)\theta)$ for the case for $m = 1$, then, from the preceding orthogonality statement for θ, only the $m = 1$ solution survives here, and we set

> Theta[1](theta):=subs(m=1,eval(Theta[m](theta)));

$$\Theta_1(\theta) := \frac{2 \sin(\theta)}{\sqrt{\pi}} \tag{6.73}$$

Eigenvalues $\lambda_{1,n}$ are the roots of the eigenvalue equation

> subs({m=1,r=a},BesselJ(mu[m],lambda[m,n]*r))=0;

$$\text{BesselJ}(1, \lambda_{1,n}) = 0 \tag{6.74}$$

Evaluation of the orthonormalized eigenfunctions for $m = 1$ yields

> R[1,n](r):=eval(subs(m=1,BesselJ(mu[m],lambda[m,n]*r)/sqrt(expand(int(subs(m=1, BesselJ(mu[m],lambda[m,n]*r))^2*r,r=0..a))))));

$$R_{1,n}(r) := \frac{2\, \text{BesselJ}(1, \lambda_{1,n}\, r)}{\sqrt{2\, \text{BesselJ}(0, \lambda_{1,n})^2 - \dfrac{4\, \text{BesselJ}(0, \lambda_{1,n})\, \text{BesselJ}(1, \lambda_{1,n})}{\lambda_{1,n}} + 2\, \text{BesselJ}(1, \lambda_{1,n})^2}} \tag{6.75}$$

Substitution of the preceding eigenvalue equation simplifies this

> R[1,n](r):=radsimp(subs(BesselJ(1,lambda[1,n]*a)=0,R[1,n](r)));

$$R_{1,n}(r) := \frac{\text{BesselJ}(1, \lambda_{1,n}\, r)\, \sqrt{2}}{\text{BesselJ}(0, \lambda_{1,n})} \tag{6.76}$$

Evaluation of the Fourier coefficients yields

```
> C(1,n):=Int(Int(f(r,theta)*R[1,n](r)*w(r)*Theta[1](theta)*w(theta),r=0..a),theta=0..b);
```

$$C(1, n) := \int_0^{\frac{1}{2}\pi} \int_0^1 \frac{2(r - r^3)\sin(\theta)^2 \, \text{BesselJ}(1, \lambda_{1,n}r)\sqrt{2}\,r}{\text{BesselJ}(0, \lambda_{1,n})\sqrt{\pi}} \, dr \, d\theta \qquad (6.77)$$

```
> C(1,n):=radsimp(subs(BesselJ(1,lambda[1,n]*a)=0,value(%)));
```

$$C(1, n) := -\frac{4\sqrt{2}\sqrt{\pi}}{\lambda_{1,n}^3} \qquad (6.78)$$

```
> T[1,n](t):=eval(subs(m=1,T[m,n](t)));
```

$$T_{1,n}(t) := -\frac{4\sqrt{2}\sqrt{\pi}\,e^{-\frac{1}{50}\lambda_{1,n}^2 t}}{\lambda_{1,n}^3} \qquad (6.79)$$

Generalized series terms

```
> u[1,n](r,theta,t):=(T[1,n](t)*R[1,n](r)*Theta[1](theta));
```

$$u_{1,n}(r, \theta, t) := -\frac{16\,e^{-\frac{1}{50}\lambda_{1,n}^2 t}\,\text{BesselJ}(1, \lambda_{1,n}\,r)\,\sin(\theta)}{\lambda_{1,n}^3\,\text{BesselJ}(0, \lambda_{1,n})} \qquad (6.80)$$

Series solution

```
> u(r,theta,t):=Sum(u[1,n](r,theta,t),n=1..infinity);
```

$$u(r, \theta, t) := \sum_{n=1}^{\infty} \left(-\frac{16\,e^{-\frac{1}{50}\lambda_{1,n}^2 t}\,\text{BesselJ}(1, \lambda_{1,n}\,r)\,\sin(\theta)}{\lambda_{1,n}^3\,\text{BesselJ}(0, \lambda_{1,n})} \right) \qquad (6.81)$$

Evaluation of the eigenvalues $\lambda_{1,n}$ from the roots of the eigenvalue equation yields

```
> subs({m=1,r=a},BesselJ(mu[m],lambda[m,n]*r))=0;
```

$$\text{BesselJ}(1, \lambda_{1,n}) = 0 \qquad (6.82)$$

```
> plot(BesselJ(1,v),v=0..20,thickness=10);
```

If we set $v = \lambda a$, then the eigenvalues $\lambda_{1,n}$ are found from the intersection points of the curve of $J(1, v)$ with the v-axis shown in Figure 6.9. Using the Maple fsolve command to evaluate the first few eigenvalues yields

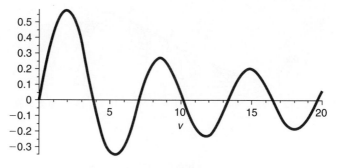

Figure 6.9

> lambda[1,1]:=(1/a)*fsolve(BesselJ(1,v)=0,v=1..6);

$$\lambda_{1,1} := 3.831705970 \tag{6.83}$$

> lambda[1,2]:=(1/a)*fsolve(BesselJ(1,v)=0,v=6..9);

$$\lambda_{1,2} := 7.015586670 \tag{6.84}$$

> lambda[1,3]:=(1/a)*fsolve(BesselJ(1,v)=0,v=9..12);

$$\lambda_{1,3} := 10.17346814 \tag{6.85}$$

First few terms in sum

> u(r,theta,t):=sum(u[1,n](r,theta,t),n=1..3):

ANIMATION

> u(x,y,t):=subs({r=sqrt(x^2+y^2),theta=arctan(y/x)},u(r,theta,t))*Heaviside(1−sqrt(x^2+y^2)):
> animate3d(u(x,y,t),x=0..a,y=0..a,t=0..8,axes=framed,thickness=1);

The preceding animation command illustrates the spatial-time-dependent solution for $u(r, \theta, t)$. The animation sequence in Figures 6.10 and 6.11 shows snapshots of the animation at the two different times $t = 0$ and $t = 3$.

ANIMATION SEQUENCE

> u(r,theta,0):=subs(t=0,u(r,theta,t)):cylinderplot([r,theta,u(r,theta,0)],r=0..a,theta=0..b,
  axes=framed,thickness=1);

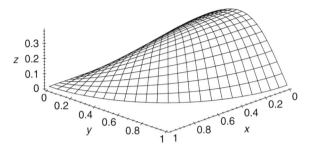

Figure 6.10

```
> u(r,theta,3):=subs(t=3,u(r,theta,t)):cylinderplot([r,theta,u(r,theta,3)],r=0..a,theta=0..b,
  axes=framed,thickness=1);
```

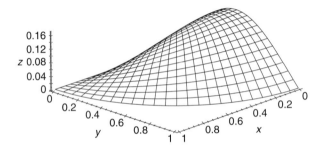

Figure 6.11

EXAMPLE 6.9.2: We seek the temperature distribution in a thin circular plate over the two-dimensional domain $D = \{(r, \theta) \mid 0 < r < 1, 0 < \theta < \pi\}$. The lateral surfaces of the plate are insulated. The sides $\theta = 0$ and $\theta = \pi$ are at a fixed temperature of 0, and the edge $r = 1$ is insulated. The initial temperature distribution $u(r, \theta, 0) = f(r, \theta)$ is given as follows, and the thermal diffusivity is $k = 1/25$.

SOLUTION: The two-dimensional homogeneous diffusion equation is

$$\frac{\partial}{\partial t} u(r, \theta, t) = k \left(\frac{\frac{\partial}{\partial r} u(r, \theta, t) + r \left(\frac{\partial^2}{\partial r^2} u(r, \theta, t) \right)}{r} + \frac{\frac{\partial^2}{\partial \theta^2} u(r, \theta, t)}{r^2} \right)$$

The boundary conditions are type 2 at $r = 1$, type 1 at $\theta = 0$, and type 1 at $\theta = \pi$ and a finite solution at the origin:

$$|u(0, \theta, t)| < \infty \quad \text{and} \quad u_r(1, \theta, t) = 0$$

and

$$u(r, 0, t) = 0 \quad \text{and} \quad u(r, \pi, t) = 0$$

The initial condition is

$$u(r, \theta, 0) = \left(r - \frac{r3}{3}\right) \sin(\theta)$$

The ordinary differential equations from the method of separation of variables are

$$\frac{d}{dt}T(t) + k\lambda^2 T(t) = 0$$

$$r^2 \left(\frac{d^2}{dr^2}R(r)\right) + r\left(\frac{d}{dr}R(r)\right) + \left(\lambda^2 r^2 - \mu^2\right) R(r) = 0$$

$$\frac{d^2}{d\theta^2}\Theta(\theta) + \mu^2\Theta(\theta) = 0$$

The boundary conditions on the spatial variables are

$$|R(0)| < \infty \quad \text{and} \quad R_r(1) = 0$$

and

$$\Theta(0) = 0 \quad \text{and} \quad \Theta(\pi) = 0$$

Assignment of system parameters

> restart:with(plots):a:=1:b:=Pi:k:=1/25:

Allowed eigenvalues and corresponding orthonormal eigenfunctions are obtained from Examples 2.5.1 and 2.6.3.

> mu[m]:=m*Pi/b;

$$\mu_m := m \tag{6.86}$$

for $m = 1, 2, 3, \ldots$.

Eigenvalues $\lambda_{m,n}$ are the roots of the eigenvalue equation

> subs(r=a,diff(BesselJ(mu[m],lambda[m,n]*r),r))=0;

$$\left(-\text{BesselJ}(m+1, \lambda_{m,n}) + \frac{m\,\text{BesselJ}(m, \lambda_{m,n})}{\lambda_{m,n}}\right)\lambda_{m,n} = 0 \tag{6.87}$$

for $m = 1, 2, 3, \ldots, n = 1, 2, 3, \ldots$.

Orthonormal eigenfunctions

> R[m,n](r):=BesselJ(mu[m],lambda[m,n]*r)/sqrt(Int(BesselJ(mu[m],lambda[m,n]*r)^2*r,
  r=0..a));R[m,p](r):=subs(n=p,R[m,n](r)):

$$R_{m,n}(r) := \frac{\text{BesselJ}(m, \lambda_{m,n}r)}{\sqrt{\int\limits_0^1 \text{BesselJ}(m, \lambda_{m,n}r)^2 r\, dr}} \tag{6.88}$$

> Theta[m](theta):=sqrt(2/b)*sin(m*Pi/b*theta);Theta[q](theta):=subs(m=q,Theta[m](theta)):

$$\Theta_m(\theta) := \frac{\sqrt{2}\sin(m\theta)}{\sqrt{\pi}} \tag{6.89}$$

Statements of orthonormality with their respective weight functions

> w(r):=r:Int(R[m,n](r)*R[m,p](r)*w(r),r=0...a)=delta(n,p);

$$\int\limits_0^1 \frac{\text{BesselJ}(m, \lambda_{m,n}r)\text{BesselJ}(m, \lambda_{m,p}r)r}{\sqrt{\int\limits_0^1 \text{BesselJ}(m, \lambda_{m,n}r)^2 r\, dr}\sqrt{\int\limits_0^1 \text{BesselJ}(m, \lambda_{m,p}r)^2 r\, dr}}\, dr = \delta(n, p) \tag{6.90}$$

> w(theta):=1:Int(Theta[m](theta)*Theta[q](theta)*w(theta),theta=0...b)=delta(m,q);

$$\int\limits_0^\pi \frac{2\sin(m\theta)\sin(q\theta)}{\pi}\, d\theta = \delta(m, q) \tag{6.91}$$

Time-dependent solution

> T[m,n](t):=C(m,n)*exp(−k*lambda[m,n]^2*t);

$$T_{m,n}(t) := C(m, n)\, e^{-\frac{1}{25}\lambda_{m,n}^2 t} \tag{6.92}$$

Generalized series terms

> u[m,n](r,theta,t):=T[m,n](t)*R[m,n](r)*Theta[m](theta);

$$u_{m,n}(r, \theta, t) := \frac{C(m, n)e^{-\frac{1}{25}\lambda_{m,n}^2 t}\,\text{BesselJ}(m, \lambda_{m,n}r)\sqrt{2}\sin(m\theta)}{\sqrt{\int_0^1 \text{BesselJ}(m, \lambda_{m,n}r)^2 r\, dr}\sqrt{\pi}} \tag{6.93}$$

Eigenfunction expansion

> u(r,theta,t):=Sum(Sum(u[m,n](r,theta,t),m=0..infinity),n=0..infinity);

$$u(r, \theta, t) := \sum_{n=0}^{\infty} \left(\sum_{m=0}^{\infty} \frac{C(m, n)e^{-\frac{1}{25}\lambda_{m,n}^2 t}\, \text{BesselJ}(m, \lambda_{m,n}r)\sqrt{2}\sin(m\theta)}{\sqrt{\int_0^1 \text{BesselJ}(m, \lambda_{m,n}r)^2 r\, dr}\sqrt{\pi}} \right) \tag{6.94}$$

The Fourier coefficients $C(m, n)$ are to be determined from the initial condition function $u(r, \theta, 0) = f(r, \theta)$. We consider the special case where

> f(r,theta):=(r−r^3/3)*sin(theta);

$$f(r, \theta) := \left(r - \frac{1}{3}r^3 \right)\sin(\theta) \tag{6.95}$$

At time $t = 0$, we have

> f(r,theta)=eval(subs(t=0,u(r,theta,t)));

$$\left(r - \frac{1}{3}r^3 \right)\sin(\theta) = \sum_{n=0}^{\infty} \left(\sum_{m=0}^{\infty} \frac{C(m, n)\text{BesselJ}(m, \lambda_{m,n}r)\sqrt{2}\sin(m\theta)}{\sqrt{\int_0^1 \text{BesselJ}(m, \lambda_{m,n}r)^2 \, r\, dr}\sqrt{\pi}} \right) \tag{6.96}$$

This is the double Fourier series expansion of $f(r, \theta)$. From Section 6.8, the Fourier coefficients $C(m, n)$ were evaluated by taking the double inner product of the initial condition function $f(r, \theta)$ with respect to the preceding corresponding orthonormalized eigenfunctions.

Because $f(r, \theta)$ depends on $\sin(m\theta)$ for the $m = 1$ case, then, from the given orthogonality statement for θ, only the $m = 1$ solution survives here, and we set

> Theta[1](theta):=subs(m=1,eval(Theta[m](theta)));

$$\Theta_1(\theta) := \frac{\sqrt{2}\,\sin(\theta)}{\sqrt{\pi}} \tag{6.97}$$

Eigenvalues $\lambda_{1,n}$ are the roots of the eigenvalue equation

> subs({m=1,r=a},diff(BesselJ(mu[m],lambda[m,n]*r),r))=0;

$$\left(-\text{BesselJ}(2, \lambda_{1,n}) + \frac{\text{BesselJ}(1, \lambda_{1,n})}{\lambda_{1,n}} \right)\lambda_{1,n} = 0 \tag{6.98}$$

For $m = 1$, we have the Bessel identity

> subs(m=1,BesselJ(m−1,lambda[m,n]*a)=2*m*BesselJ(m,lambda[m,n]*a)/(lambda[m,n]*a)
    −BesselJ(m+1,lambda[m,n]*a));

$$\text{BesselJ}(0, \lambda_{1,n}) = \frac{2\text{BesselJ}(1, \lambda_{1,n})}{\lambda_{1,n}} - \text{BesselJ}(2, \lambda_{1,n}) \tag{6.99}$$

Evaluation of the orthonormalized eigenfunctions for $m = 1$

```
> R[1,n](r):=eval(subs(m=1,BesselJ(mu[m],lambda[m,n]*r)/sqrt(expand(int(subs(m=1,
BesselJ(mu[m],lambda[m,n]*r))^2*r,r=0..a)))));
```

$$R_{1,n}(r) := \frac{2\text{BesselJ}(1, \lambda_{1,n}r)}{\sqrt{2\,\text{BesselJ}(0, \lambda_{1,n})^2 - \dfrac{4\,\text{BesselJ}(0, \lambda_{1,n})\,\text{BesselJ}(1, \lambda_{1,n})}{\lambda_{1,n}} + 2\,\text{BesselJ}(1, \lambda_{1,n})^2}}$$

(6.100)

Substitution of the preceding eigenvalue equation and the Bessel identity simplifies this to

```
> R[1,n](r):=radsimp(subs(BesselJ(0,lambda[1,n]*a)=BesselJ(1,lambda[1,n]*a)/(lambda[1,n]
*a),R[1,n](r)));
```

$$R_{1,n}(r) := \frac{\text{BesselJ}(1, \lambda_{1,n}r)\sqrt{2}\lambda_{1,n}}{\text{BesselJ}(1, \lambda_{1,n})\sqrt{-1 + \lambda_{1,n}^2}}$$

(6.101)

Evaluation of the Fourier coefficients yields

```
> C(1,n):=Int(Int(f(r,theta)*R[1,n](r)*w(r)*Theta[1](theta)*w(theta),r=0..a),theta=0..b);
```

$$C(1, n) := \int_0^\pi \int_0^1 \frac{2\left(r - \frac{1}{3}r^3\right)\sin(\theta)^2\text{BesselJ}(1, \lambda_{1,n}r)\lambda_{1,n}r}{\text{BesselJ}(1, \lambda_{1,n})\sqrt{-1 + \lambda_{1,n}^2}\sqrt{\pi}}\,dr\,d\theta$$

(6.102)

```
> C(1,n):=simplify(subs(BesselJ(0,lambda[1,n]*a)=BesselJ(1,lambda[1,n]*a)/
(lambda[1,n]*a),value(%)));
```

$$C(1, n) := \frac{8}{3}\frac{\sqrt{\pi}}{\lambda_{1,n}^3\sqrt{-1 + \lambda_{1,n}^2}}$$

(6.103)

```
> T[1,n](t):=eval(subs(m=1,T[m,n](t)));
```

$$T_{1,n}(t) := \frac{8}{3}\frac{\sqrt{\pi}\,e^{-\frac{1}{25}\lambda_{1,n}^2 t}}{\lambda_{1,n}^3\sqrt{-1 + \lambda_{1,n}^2}}$$

(6.104)

Generalized series terms

```
> u[1,n](r,theta,t):=T[1,n](t)*R[1,n](r)*Theta[1](theta);
```

$$u_{1,n}(r, \theta, t) := \frac{16}{3}\frac{e^{-\frac{1}{25}\lambda_{1,n}^2 t}\text{BesselJ}(1, \lambda_{1,n}r)\sin(\theta)}{\lambda_{1,n}^2\left(-1 + \lambda_{1,n}^2\right)\text{BesselJ}(1, \lambda_{1,n})}$$

(6.105)

Series solution

> u(r,theta,t):=Sum(u[1,n](r,theta,t),n=1..infinity);

$$u(r, \theta, t) := \sum_{n=1}^{\infty} \frac{16}{3} \frac{e^{-\frac{1}{25}\lambda_{1,n}^2 t} \text{BesselJ}(1, \lambda_{1,n} r) \sin(\theta)}{\lambda_{1,n}^2 \left(-1+\lambda_{1,n}^2\right) \text{BesselJ}(1, \lambda_{1,n})} \tag{6.106}$$

Evaluation of the eigenvalues $\lambda_{1,n}$ from the eigenvalue equation yields

> subs({m=1,r=a},diff(BesselJ(mu[m],lambda[m,n]*r),r))=0;

$$\left(-\text{BesselJ}(2, \lambda_{1,n}) + \frac{\text{BesselJ}(1, \lambda_{1,n})}{\lambda_{1,n}}\right) \lambda_{1,n} = 0 \tag{6.107}$$

> plot(BesselJ(1,v)−v*BesselJ(2,v),v=0..20,thickness=10);

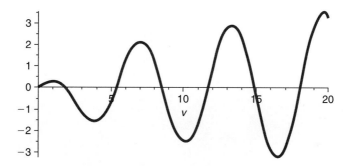

Figure 6.12

If we set $v = \lambda a$, then the eigenvalues are found from the intersection points of the curve $J(1, v) - vJ(2, v)$ with the v-axis shown in Figure 6.12. Using the Maple fsolve command to evaluate the first few eigenvalues yields

> lambda[1,1]:=(1/a)*fsolve(BesselJ(1,v)−v*BesselJ(2,v)=0,v=1..4);

$$\lambda_{1,1} := 1.841183781 \tag{6.108}$$

> lambda[1,2]:=(1/a)*fsolve(BesselJ(1,v)−v*BesselJ(2,v)=0,v=4..7);

$$\lambda_{1,2} := 5.331442774 \tag{6.109}$$

> lambda[1,3]:=(1/a)*fsolve(BesselJ(1,v)−v*BesselJ(2,v)=0,v=7..11);

$$\lambda_{1,3} := 8.536316366 \tag{6.110}$$

Note that we ignore the root $\lambda = 0$, since it leads to a trivial solution.

First few terms in sum

> u(r,theta,t):=sum(u[1,n](r,theta,t),n=1..2):

ANIMATION

> u(x,y,t):=subs({r=sqrt(x^2+y^2),theta=arcsin(y/sqrt(x^2+y^2))},u(r,theta,t))*
  Heaviside(1−sqrt(x^2+y^2)):
> animate3d(u(x,y,t),x=−a..a,y=0..a,t=0..8,axes=framed,thickness=1);

The preceding animation command illustrates the spatial-time-dependent solution for $u(r, \theta, t)$. The animation sequence in Figures 6.13 and 6.14 shows snapshots of the animation at the two different times $t = 0$ and $t = 5$.

ANIMATION SEQUENCE

> u(r,theta,0):=subs(t=0,u(r,theta,t)):cylinderplot([r,theta,u(r,theta,0)],r=0..a,theta=0..b,
  axes=framed,thickness=1);

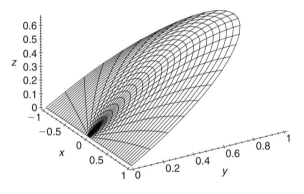

Figure 6.13

> u(r,theta,5):=subs(t=5,u(r,theta,t)):cylinderplot([r,theta,u(r,theta,5)],r=0..a,theta=0..b,
  axes=framed,thickness=1);

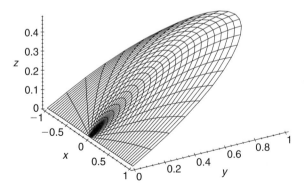

Figure 6.14

EXAMPLE 6.9.3: We consider the temperature distribution in a thin circular plate over the two-dimensional domain $D = \{(r, \theta) \mid 0 < r < 1, -\pi < \theta < \pi\}$. The lateral surfaces of the plate are insulated. The periphery $r = 1$ is held at a fixed temperature of 0, and the θ-dependent solution satisfies periodic boundary conditions. The initial temperature distribution $u(r, \theta, 0) = f(r, \theta)$ is given as follows, and the thermal diffusivity is $k = 1/50$.

SOLUTION: The two-dimensional homogeneous diffusion equation is

$$\frac{\partial}{\partial t} u(r, \theta, t) = k \left(\frac{\frac{\partial}{\partial r} u(r, \theta, t) + r \left(\frac{\partial^2}{\partial r^2} u(r, \theta, t) \right)}{r} + \frac{\frac{\partial^2}{\partial \theta^2} u(r, \theta, t)}{r^2} \right)$$

The boundary conditions are type 1 at $r = 1$, a finite solution at the origin, and periodic boundary conditions on θ:

$$|u(0, \theta, t)| < \infty \quad \text{and} \quad u(1, \theta, t) = 0$$

and

$$u(r, -\pi, t) = u(r, \pi, t) \quad \text{and} \quad u_\theta(r, -\pi, t) = u_\theta(r, \pi, t)$$

The initial condition is

$$u(r, \theta, 0) = \left(r^2 - r^4 \right) \sin(2\theta)$$

The ordinary differential equations from the method of separation of variables are

$$\frac{d}{dt} T(t) + k\lambda^2 T(t) = 0$$

$$r^2 \left(\frac{d^2}{dr^2} R(r) \right) + r \left(\frac{d}{dr} R(r) \right) + \left(\lambda^2 r^2 - \mu^2 \right) R(r) = 0$$

$$\frac{d^2}{d\theta^2} \Theta(\theta) + \mu^2 \Theta(\theta) = 0$$

The boundary conditions on the spatial variables are

$$|R(0)| < \infty \quad \text{and} \quad R(1) = 0$$

and

$$\Theta(-\pi) = \Theta(\pi) \quad \text{and} \quad \Theta_\theta(-\pi) = \Theta_\theta(\pi)$$

Assignment of system parameters

> restart:with(plots):a:=1:b:=Pi:k:=1/50:

Allowed eigenvalues and corresponding orthonormal eigenfunctions are obtained from Examples 2.6.1 and 2.6.2.

> mu[m]:=m*Pi/b;

$$\mu_m := m \tag{6.111}$$

for $m = 1, 2, 3, \ldots$.

Eigenvalues $\lambda_{m,n}$ are the roots of the eigenvalue equation

> BesselJ(mu[m],lambda[m,n]*a)=0;

$$\text{BesselJ}(m, \lambda_{m,n}) = 0 \tag{6.112}$$

for $m = 1, 2, 3, \ldots$, $n = 1, 2, 3, \ldots$.

Orthonormal eigenfunctions

> R[m,n](r):=BesselJ(mu[m],lambda[m,n]*r)/sqrt(Int(BesselJ(mu[m],lambda[m,n]*r)^2*r, r=0..a));R[m,p](r):=subs(n=p,R[m,n](r)):

$$R_{m,n}(r) := \frac{\text{BesselJ}(m, \lambda_{m,n}r)}{\sqrt{\int_0^1 \text{BesselJ}(m, \lambda_{m,n}r)^2 r \, dr}} \tag{6.113}$$

> phi[m](theta):=sqrt(1/Pi)*cos(m*theta);phi[q](theta):=subs(m=q,phi[m](theta)):

$$\phi_m(\theta) := \frac{\cos(m\theta)}{\sqrt{\pi}} \tag{6.114}$$

> psi[m](theta):=sqrt(1/Pi)*sin(m*theta);psi[q](theta):=subs(m=q,psi[m](theta)):

$$\Psi_m(\theta) := \frac{\sin(m\theta)}{\sqrt{\pi}} \tag{6.115}$$

For $m = 0$, the eigenvalues are the roots of the equation

> BesselJ(0,lambda[0,n]*a)=0;

$$\text{BesselJ}(0, \lambda_{0,n}) = 0 \tag{6.116}$$

Orthonormal eigenfunctions

> R[0,n](r):=BesselJ(0,lambda[0,n]*r)/sqrt(Int(BesselJ(0,lambda[0,n]*r)^2*r,r=0..a)); R[0,p](r):=subs(n=p,R[0,n](r)):

$$R_{0,n}(r) := \frac{\text{BesselJ}(0, \lambda_{0,n}r)}{\sqrt{\int_0^1 \text{BesselJ}(0, \lambda_{0,n}r)^2 r \, dr}} \tag{6.117}$$

> phi[0](theta):=1/sqrt(2*Pi);

$$\phi_0(\theta) := \frac{1}{2} \frac{\sqrt{2}}{\sqrt{\pi}} \tag{6.118}$$

Statements of orthonormality with their respective weight functions

> w(r):=r:Int(R[m,n](r)*R[m,p](r)*w(r),r=0..a)=delta(n,p);

$$\int_0^1 \frac{\text{BesselJ}(m, \lambda_{m,n}r)\text{BesselJ}(m, \lambda_{m,p}r)r}{\sqrt{\int_0^1 \text{BesselJ}(m, \lambda_{m,n}r)^2 r\, dr}\sqrt{\int_0^1 \text{BesselJ}(m, \lambda_{m,p}r)^2 r\, dr}}\, dr = \delta(n, p) \qquad (6.119)$$

> w(theta):=1:Int(phi[m](theta)*phi[q](theta)*w(theta),theta=−Pi..Pi)=delta(m,q);

$$\int_{-\pi}^{\pi} \frac{\cos(m\theta)\cos(q\theta)}{\pi}\, d\theta = \delta(m, q) \qquad (6.120)$$

> w(theta):=1:Int(psi[m](theta)*psi[q](theta)*w(theta),theta=−Pi..Pi)=delta(m,q);

$$\int_{-\pi}^{\pi} \frac{\sin(m\theta)\sin(q\theta)}{\pi}\, d\theta = \delta(m, q) \qquad (6.121)$$

Time-dependent solution

> T[m,n](t):=exp(−k*lambda[m,n]^2*t);u[m,n](r,theta,t):=T[m,n](t)*R[m,n](r)*(A(m,n)
  *phi[m](theta)+B(m,n)*psi[m](theta)):

$$T_{m,n}(t) := e^{-\frac{1}{50}\lambda_{m,n}^2 t} \qquad (6.122)$$

> T[0,n](t):=A(0,n)*exp(−k*lambda[0,n]^2*t);u[0,n](r,theta,t):=T[0,n](t)*R[0,n](r)
  *phi[0](theta):

$$T_{0,n}(t) := A(0, n)e^{-\frac{1}{50}\lambda_{0,n}^2 t} \qquad (6.123)$$

Eigenfunction expansion

> u(r,theta,t):=Sum(u[0,n](r,theta,t),n=1..infinity)+Sum(Sum(u[m,n](r,theta,t),m=1..infinity),
  n=1..infinity);

$$u(r, \theta, t) := \sum_{n=1}^{\infty} \frac{1}{2} \frac{A(0, n)e^{-\frac{1}{50}\lambda_{0,n}^2 t}\text{BesselJ}(0, \lambda_{0,n}r)\sqrt{2}}{\sqrt{\int_0^1 \text{BesselJ}(0, \lambda_{0,n}r)^2 r\, dr}\sqrt{\pi}}$$

$$+ \sum_{n=1}^{\infty}\left(\sum_{m=1}^{\infty} \frac{e^{-\frac{1}{50}\lambda_{m,n}^2 t}\text{BesselJ}(m, \lambda_{m,n}r)\left(\frac{A(m,n)\cos(m\theta)}{\sqrt{\pi}} + \frac{B(m,n)\sin(m\theta)}{\sqrt{\pi}}\right)}{\sqrt{\int_0^1 \text{BesselJ}(m, \lambda_{m,n}r)^2 r\, dr}}\right)$$

$$(6.124)$$

The Fourier coefficients $A(m, n)$ and $B(m, n)$ are to be determined from the initial condition function $u(r, \theta, 0) = f(r, \theta)$. We consider the special case where

> f(r,theta):=(r^2−r^4)*sin(2*theta);

$$f(r, \theta) := \left(r^2 - r^4\right)\sin(2\theta) \tag{6.125}$$

At time $t = 0$ we have

> f(r,theta)=eval(subs(t=0,u(r,theta,t)));

$$\left(r^2 - r^4\right)\sin(2\theta) = \sum_{n=1}^{\infty} \frac{1}{2} \frac{A(0, n)\text{BesselJ}(0, \lambda_{0,n}r)\sqrt{2}}{\sqrt{\int_0^1 \text{BesselJ}(0, \lambda_{0,n}r)^2 r\, dr}\sqrt{\pi}}$$

$$+ \sum_{n=1}^{\infty}\left(\sum_{m=1}^{\infty} \frac{\text{BesselJ}(m, \lambda_{m,n}r)\left(\dfrac{A(m, n)\cos(m\theta)}{\sqrt{\pi}} + \dfrac{B(m, n)\sin(m\theta)}{\sqrt{\pi}}\right)}{\sqrt{\int_0^1 \text{BesselJ}(m, \lambda_{m,n}r)^2 r\, dr}}\right) \tag{6.126}$$

This is the double Fourier series expansion of $f(r, \theta)$. From Section 6.8, the Fourier coefficients $A(m, n)$ and $B(m, n)$ are to be evaluated by taking the double inner product of the initial condition function $f(r, \theta)$ with respect to the corresponding orthonormal eigenfunctions given earlier.

Because $f(r, \theta)$ depends on $\sin(m\theta)$ for the case $m = 2$, then, from the given orthogonality statement for θ, only the $m = 2$ solution survives here, and we set

> phi[2](theta):=subs(m=2,eval(phi[m](theta)));

$$\phi_2(\theta) := \frac{\cos(2\theta)}{\sqrt{\pi}} \tag{6.127}$$

> psi[2](theta):=subs(m=2,eval(psi[m](theta)));

$$\psi_2(\theta) := \frac{\sin(2\theta)}{\sqrt{\pi}} \tag{6.128}$$

Eigenvalues $\lambda_{2,n}$ are the roots of the eigenvalue equation

> subs({m=2,r=a},BesselJ(mu[m],lambda[m,n]*r)=0);

$$\text{BesselJ}(2, \lambda_{2,n}) = 0 \tag{6.129}$$

For $m = 2$, we have the Bessel identity

> subs(m=2,BesselJ(0,lambda[m,n]*a)=2*BesselJ(1,lambda[m,n]*a)/(lambda[m,n]*a)
  −BesselJ(2,lambda[m,n]*a));

$$\text{BesselJ}(0, \lambda_{2,n}) = \frac{2\text{BesselJ}(1, \lambda_{2,n})}{\lambda_{2,n}} - \text{BesselJ}(2, \lambda_{2,n}) \tag{6.130}$$

Evaluation of the normalized eigenfunctions for $m = 2$ yields

> R[2,n](r):=eval(subs(m=2,BesselJ(mu[m],lambda[m,n]*r)/sqrt(expand(int(subs(m=2,
  BesselJ(mu[m],lambda[m,n]*r))^2*r,r=0..a)))));

$$R_{2,n}(r) := \frac{2\text{BesselJ}(2, \lambda_{2,n}r)}{\sqrt{2\text{BesselJ}(0, \lambda_{2,n})^2 + 2\text{BesselJ}(1, \lambda_{2,n})^2 - \frac{8\text{BesselJ}(1, \lambda_{2,n})^2}{\lambda_{2,n}^2}}} \tag{6.131}$$

Substitution of the preceding eigenvalue equation and the Bessel identity simplifies this to

> R[2,n](r):=radsimp(subs(BesselJ(0,lambda[2,n]*a)=2*BesselJ(1,lambda[2,n]*a)/
  (lambda[2,n]*a),R[2,n](r)));

$$R_{2,n}(r) := \frac{\text{BesselJ}(2, \lambda_{2,n}r)\sqrt{2}}{\text{BesselJ}(1, \lambda_{2,n})} \tag{6.132}$$

Evaluation of the Fourier coefficients yields

> A(2,n):=Int(Int(f(r,theta)*R[2,n](r)*w(r)*phi[2](theta)*w(theta),r=0..a),theta=−Pi..Pi);

$$A(2, n) := \int_{-\pi}^{\pi} \int_{0}^{1} \frac{\left(r^2 - r^4\right)\sin(2\theta)\,\text{BesselJ}(2, \lambda_{2,n}r)\sqrt{(2)}r\cos(2\theta)}{\text{BesselJ}(1, \lambda_{2,n})\sqrt{\pi}}\,dr\,d\theta \tag{6.133}$$

> A(2,n):=value(%);

$$A(2, n) := 0 \tag{6.134}$$

> A(0,n):=Int(Int(f(r,theta)*R[0,n](r)*w(r)*phi[0](theta)*w(theta),theta=−Pi..Pi),r=0..a);

$$A(0, n) := \int_{0}^{1} \int_{-\pi}^{\pi} \frac{1}{2} \frac{\left(r^2 - r^4\right)\sin(2\theta)\,\text{BesselJ}(0, \lambda_{0,n}r)r\sqrt{2}}{\sqrt{\int_0^1 \text{BesselJ}(0, \lambda_{0,n}r)^2 r\,dr}\sqrt{\pi}}\,d\theta\,dr \tag{6.135}$$

> A(0,n):=value(%);

$$A(0, n) := 0 \tag{6.136}$$

> B(2,n):=Int(Int(f(r,theta)*R[2,n](r)*w(r)*psi[2](theta)*w(theta),r=0..a),theta=−Pi..Pi);

$$B(2, n) := \int\limits_{-\pi}^{\pi} \int\limits_{0}^{1} \frac{\left(r^2 - r^4\right) \sin(2\theta)^2 \, \text{BesselJ}(2, \lambda_{2,n}r)\sqrt{2}r}{\text{BesselJ}(1, \lambda_{2,n})\sqrt{\pi}} \, dr \, d\theta \qquad (6.137)$$

> B(2,n):=subs(BesselJ(0,lambda[2,n]*a)=2*BesselJ(1,lambda[2,n]*a)/(lambda[2,n]*a),
   value(%));

$$B(2, n) := -\frac{12\sqrt{2}\sqrt{\pi}}{\lambda_{2,n}^3} \qquad (6.138)$$

> T[2,n](t):=eval(subs(m=2,T[m,n](t)));

$$T_{2,n}(t) := e^{-\frac{1}{50}\lambda_{2,n}^2 t} \qquad (6.139)$$

Generalized series terms

> u[2,n](r,theta,t):=(T[2,n](t)*R[2,n](r)*(A(2,n)*phi[2](theta)+B(2,n)*psi[2](theta)));

$$u_{2,n}(r, \theta, t) := -\frac{24\,e^{-\frac{1}{50}\lambda_{2,n}^2 t} \, \text{BesselJ}(2, \lambda_{2,n}r) \sin(2\theta)}{\text{BesselJ}(1, \lambda_{2,n})\lambda_{2,n}^3} \qquad (6.140)$$

> T[0,n]:=eval(T[0,n](t));u[0,n](r,theta,t):=T[0,n](t)*R[0,n](r)*phi[0](theta):

$$T_{0,n} := 0 \qquad (6.141)$$

Series solution

> u(r,theta,t):=Sum(u[2,n](r,theta,t),n=1..infinity);

$$u(r, \theta, t) := \sum_{n=1}^{\infty} \left(-\frac{24\,e^{-\frac{1}{50}\lambda_{2,n}^2 t} \, \text{BesselJ}(2, \lambda_{2,n}r) \sin(2\theta)}{\text{BesselJ}(1, \lambda_{2,n})\lambda_{2,n}^3} \right) \qquad (6.142)$$

Evaluation of the eigenvalues $\lambda_{2,n}$ from the roots of the eigenvalue equation yields

> subs({r=a,m=2},BesselJ(mu[m],lambda[m,n]*r))=0;

$$\text{BesselJ}(2, \lambda_{2,n}) = 0 \qquad (6.143)$$

> plot(BesselJ(2,v),v=0..20,thickness=10);

If we set $v = \lambda a$, then the eigenvalues are found from the intersection points of the curve $J(2, v)$ with the v-axis shown in Figure 6.15. A few of these eigenvalues using the Maple fsolve command yields

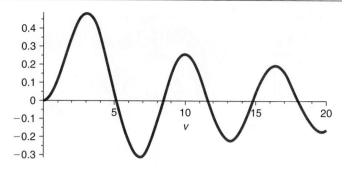

Figure 6.15

> lambda[2,1]:=(1/a)*fsolve(BesselJ(2,v)=0,v=1..6);

$$\lambda_{2,1} := 5.135622302 \qquad\qquad (6.144)$$

> lambda[2,2]:=(1/a)*fsolve(BesselJ(2,v)=0,v=6..10);

$$\lambda_{2,2} := 8.417244140 \qquad\qquad (6.145)$$

> lambda[2,3]:=(1/a)*fsolve(BesselJ(2,v)=0,v=10..14);

$$\lambda_{2,3} := 11.61984117 \qquad\qquad (6.146)$$

First few terms of sum

> u(r,theta,t):=sum(u[2,n](r,theta,t),n=1..1):

ANIMATION

> u(x,y,t):=subs({r=sqrt(x^2+y^2),theta=arctan(y/x)},u(r,theta,t))*Heaviside(1−sqrt(x^2+y^2)):
> animate3d(u(x,y,t),x=−a..a,y=−a..a,t=0..5,axes=framed,thickness=1);

ANIMATION SEQUENCE

The preceding animation illustrates the spatial-time-dependent solution for $u(r, \theta, t)$. The animation sequence in Figures 6.16 and 6.17 shows snapshots of the animation at the two different times $t = 0$ and $t = 4$.

> u(r,theta,0):=subs(t=0,u(r,theta,t)):cylinderplot([r,theta,u(r,theta,0)],r=0..a,theta=−Pi..Pi,
  axes=framed,thickness=1);

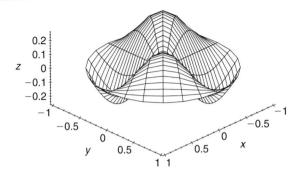

Figure 6.16

> u(r,theta,4):=subs(t=4,u(r,theta,t)):cylinderplot([r,theta,u(r,theta,4)],r=0..a,theta=−Pi..Pi,
axes=framed,thickness=1);

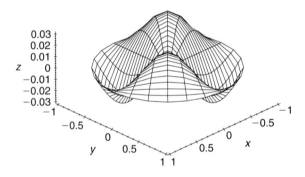

Figure 6.17

Chapter Summary

Nonhomogeneous diffusion equation in two dimensions in rectangular coordinates

$$\frac{\partial}{\partial t}u(x, y, t) = k\left(\frac{\partial^2}{\partial x^2}u(x, y, t) + \frac{\partial^2}{\partial y^2}u(x, y, t)\right) + h(x, y, t)$$

Linear diffusion operator in two dimensions in rectangular coordinates

$$L(u) = \frac{\partial}{\partial t}u(x, y, t) - k\left(\frac{\partial^2}{\partial x^2}u(x, y, t) + \frac{\partial^2}{\partial y^2}u(x, y, t)\right)$$

Method of separation of variables solution

$$u(x, y, t) = X(x)\,Y(y)\,T(t)$$

Eigenfunction expansion solution for rectangular coordinates

$$u(x, y, t) = \sum_{n=0}^{\infty} \left(\sum_{m=0}^{\infty} X_m(x)\, Y_n(y)\, C(m, n) e^{-k\lambda_{m,n}t} \right)$$

Initial condition Fourier coefficients for rectangular coordinates

$$C(m, n) = \int_0^b \int_0^a f(x, y) X_m(x)\, Y_n(y) \mathrm{d}x\, \mathrm{d}y$$

Nonhomogeneous diffusion equation in two dimensions in cylindrical coordinates

$$\frac{\partial}{\partial t} u(r, \theta, t) = k \left(\frac{\frac{\partial}{\partial r} u(r, \theta, t) + r \left(\frac{\partial^2}{\partial r^2} u(r, \theta, t) \right)}{r} + \frac{\frac{\partial^2}{\partial \theta^2} u(r, \theta, t)}{r^2} \right) + h(r, \theta, t)$$

Linear diffusion operator in two dimensions in cylindrical coordinates

$$L(u) = \frac{\partial}{\partial t} u(r, \theta, t) - k \left(\frac{\frac{\partial}{\partial r} u(r, \theta, t) + r \left(\frac{\partial^2}{\partial r^2} u(r, \theta, t) \right)}{r} + \frac{\frac{\partial^2}{\partial \theta^2} u(r, \theta, t)}{r^2} \right)$$

Method of separation of variables solution

$$u(r, \theta, t) = R(r)\Theta(\theta)T(t)$$

Eigenfunction expansion solution in cylindrical coordinates

$$u(r, \theta, t) = \sum_{n=0}^{\infty} \left(\sum_{m=0}^{\infty} R_{m,n}(r)\Theta_m(\theta)C(m, n) e^{-k\lambda_{m,n}^2 t} \right)$$

Initial condition Fourier coefficients for cylindrical coordinates

$$C(m, n) = \int_0^b \int_0^a f(r, \theta) R_{m,n}(r)\Theta_m(\theta) r \mathrm{d}r\, \mathrm{d}\theta$$

We have examined homogeneous partial differential equations describing diffusion phenomena in two spatial dimensions for both the rectangular and the cylindrical coordinate systems.

Exercises

We now consider diffusion or heat equations with homogeneous boundary conditions in two dimensions in both the rectangular coordinate and cylindrical coordinate systems. We use the method of separation of variables and eigenfunction expansions to evaluate the solutions.

Two Dimensions, Rectangular Coordinates

We seek the temperature distribution in thin plates over the two-dimensional domain $D = \{(x, y) \mid 0 < x < 1, 0 < y < 1\}$ whose lateral surfaces are insulated. The two-dimensional homogeneous partial differential equation in rectangular coordinates reads

$$\frac{\partial}{\partial t} u(x, y, t) = k\left(\frac{\partial^2}{\partial x^2} u(x, y, t) + \frac{\partial^2}{\partial y^2} u(x, y, t)\right)$$

For Exercises 6.1 through 6.25, the initial condition $u(x, y, 0) = f(x, y)$ and the boundary conditions are given. You are asked to evaluate the eigenvalues and corresponding two-dimensional orthonormal eigenfunctions and write the general solution. Generate the animated three-dimensional surface solution $u(x, y, t)$, and plot the animated sequence for $0 < t < 5$.

6.1.　Boundaries $x = 0$, $x = 1$, $y = 0$, $y = 1$ are held at fixed temperature 0, $k = 1/10$:

$$u(0, y, t) = 0 \quad \text{and} \quad u(1, y, t) = 0$$
$$u(x, 0, t) = 0 \quad \text{and} \quad u(x, 1, t) = 0$$

Initial condition:

$$f(x, y) = x(1 - x)\, y(1 - y)$$

6.2.　Boundaries $x = 0$, $x = 2$, $y = 0$ are held at fixed temperature 0, and boundary $y = 1$ is insulated, $k = 1/10$:

$$u(0, y, t) = 0 \quad \text{and} \quad u(2, y, t) = 0$$
$$u(x, 0, t) = 0 \quad \text{and} \quad u_y(x, 1, t) = 0$$

Initial condition:

$$f(x, y) = x(2 - x)y\left(1 - \frac{y}{2}\right)$$

6.3.　Boundaries $x = 1$, $y = 0$, $y = 2$ are held at fixed temperature 0, boundary $x = 0$ is insulated, $k = 1/10$:

$$u_x(0, y, t) = 0 \quad \text{and} \quad u(1, y, t) = 0$$
$$u(x, 0, t) = 0 \quad \text{and} \quad u(x, 2, t) = 0$$

Initial condition:

$$f(x, y) = (1 - x^2)y(2 - y)$$

6.4. Boundaries $y = 0$, $y = 2$ are held at fixed temperature 0, boundaries $x = 0$ and $x = 1$ are insulated, $k = 1/10$:

$$u_x(0, y, t) = 0 \quad \text{and} \quad u_x(1, y, t) = 0$$
$$u(x, 0, t) = 0 \quad \text{and} \quad u(x, 2, t) = 0$$

Initial condition:

$$f(x, y) = x^2 \left(1 - \frac{2x}{3}\right) y(2 - y)$$

6.5. Boundaries $x = 1$, $y = 1$ are held at fixed temperature 0, boundaries $x = 0$ and $y = 0$ are insulated, $k = 1/10$:

$$u_x(0, y, t) = 0 \quad \text{and} \quad u(1, y, t) = 0$$
$$u_y(x, 0, t) = 0 \quad \text{and} \quad u(x, 1, t) = 0$$

Initial condition:

$$f(x, y) = (1 - x^2)(1 - y^2)$$

6.6. Boundaries $x = 1$, $y = 0$ are held at fixed temperature 0, boundaries $x = 0$ and $y = 1$ are insulated, $k = 1/10$:

$$u_x(0, y, t) = 0 \quad \text{and} \quad u(1, y, t) = 0$$
$$u(x, 0, t) = 0 \quad \text{and} \quad u_y(x, 1, t) = 0$$

Initial condition:

$$f(x, y) = (1 - x^2)y\left(1 - \frac{y}{2}\right)$$

6.7. Boundary $y = 0$ is held at fixed temperature 0, boundaries $x = 0$, $x = 1$, and $y = 1$ are insulated, $k = 1/10$:

$$u_x(0, y, t) = 0 \quad \text{and} \quad u_x(1, y, t) = 0$$
$$u(x, 0, t) = 0 \quad \text{and} \quad u_y(x, 1, t) = 0$$

Initial condition:

$$f(x, y) = x^2 \left(1 - \frac{2x}{3}\right) y\left(1 - \frac{y}{2}\right)$$

6.8.   Boundaries $x = 0$, $x = 2$, $y = 0$ are held at fixed temperature 0, boundary $y = 1$ is losing heat by convection into a surrounding medium at 0 temperature, $k = 1/10$:

$$u(0, y, t) = 0 \quad \text{and} \quad u(2, y, t) = 0$$
$$u(x, 0, t) = 0 \quad \text{and} \quad u_y(x, 1, t) + u(x, 1, t) = 0$$

Initial condition:

$$f(x, y) = x(2 - x)y\left(1 - \frac{2y}{3}\right)$$

6.9.   Boundaries $x = 1$, $y = 0$ are held at fixed temperature 0, boundary $x = 0$ is insulated, and boundary $y = 1$ is losing heat by convection into a surrounding medium at 0 temperature, $k = 1/10$:

$$u_x(0, y, t) = 0 \quad \text{and} \quad u(1, y, t) = 0$$
$$u(x, 0, t) = 0 \quad \text{and} \quad u_y(x, 1, t) + u(x, 1, t) = 0$$

Initial condition:

$$f(x, y) = (1 - x^2)y\left(1 - \frac{2y}{3}\right)$$

6.10.   Boundaries $x = 1$, $y = 0$ are held at fixed temperature 0, boundary $y = 1$ is insulated, and boundary $x = 0$ is losing heat by convection into a surrounding medium at 0 temperature, $k = 1/10$:

$$u_x(0, y, t) - u(0, y, t) = 0 \quad \text{and} \quad u(1, y, t) = 0$$
$$u(x, 0, t) = 0 \quad \text{and} \quad u_y(x, 1, t) = 0$$

Initial condition:

$$f(x, y) = \left(-\frac{2x^2}{3} + \frac{x}{3} + \frac{1}{3}\right)y(1 - y)$$

6.11.   Boundaries $x = 0$, $y = 1$ are held at fixed temperature 0, boundary $y = 0$ is insulated, and boundary $x = 1$ is losing heat by convection into a surrounding medium at 0 temperature, $k = 1/10$:

$$u(0, y, t) = 0 \quad \text{and} \quad u_x(1, y, t) + u(1, y, t) = 0$$
$$u_y(x, 0, t) = 0 \quad \text{and} \quad u(x, 1, t) = 0$$

Initial condition:

$$f(x, y) = x\left(1 - \frac{2x}{3}\right)(1 - y^2)$$

6.12. Boundaries $y = 0$, $y = 1$ are held at fixed temperature 0, boundary $x = 0$ is insulated, and boundary $x = 1$ is losing heat by convection into a surrounding medium at 0 temperature, $k = 1/10$:

$$u_x(0, y, t) = 0 \quad \text{and} \quad u_x(1, y, t) + u(1, y, t) = 0$$
$$u(x, 0, t) = 0 \quad \text{and} \quad u(x, 1, t) = 0$$

Initial condition:

$$f(x, y) = \left(1 - \frac{x^2}{3}\right)y(1 - y)$$

6.13. Boundaries $x = 0$, $x = 1$ are held at fixed temperature 0, boundary $y = 1$ is insulated, and boundary $y = 0$ is losing heat by convection into a surrounding medium at 0 temperature, $k = 1/10$:

$$u(0, y, t) = 0 \quad \text{and} \quad u(1, y, t) = 0$$
$$u_y(x, 0, t) - u(x, 0, t) = 0 \quad \text{and} \quad u_y(x, 1, t) = 0$$

Initial condition:

$$f(x, y) = x(1 - x)\left(1 - \frac{(y - 1)^2}{3}\right)$$

6.14. Boundaries $x = 0$, $x = 1$, and $y = 0$ are insulated and boundary $y = 1$ is losing heat by convection into a surrounding medium at 0 temperature, $k = 1/10$:

$$u_x(0, y, t) = 0 \quad \text{and} \quad u_x(1, y, t) = 0$$
$$u_y(x, 0, t) = 0 \quad \text{and} \quad u_y(x, 1, t) + u(x, 1, t) = 0$$

Initial condition:

$$f(x, y) = x^2\left(1 - \frac{2x}{3}\right)\left(1 - \frac{y^2}{3}\right)$$

6.15. Boundaries $x = 1$ and $y = 0$ are both insulated and boundaries $x = 0$ and $y = 1$ are losing heat by convection into a surrounding medium at 0 temperature, $k = 1/10$:

$$u_x(0, y, t) - u(0, y, t) = 0 \quad \text{and} \quad u_x(1, y, t) = 0$$
$$u_y(x, 0, t) = 0 \quad \text{and} \quad u_y(x, 1, t) + u(x, 1, t) = 0$$

Initial condition:

$$f(x, y) = \left(1 - \frac{(x-1)^2}{3}\right)\left(1 - \frac{y^2}{3}\right)$$

Example Exercises with Surface Heat Loss

We now consider the temperature distribution in thin plates over the two-dimensional domain $D = \{(x, y) \,|\, 0 < x < 1, 0 < y < 1\}$ whose lateral surfaces are no longer insulated but are now experiencing a heat loss proportional to the plate temperature and the surrounding temperature medium at zero degrees. If we let β account for the heat loss coefficient, then the homogeneous partial differential equation is

$$\frac{\partial}{\partial t}u(x, y, t) = k\left(\frac{\partial^2}{\partial x^2}u(x, y, t) + \frac{\partial^2}{\partial y^2}u(x, y, t)\right) - \beta u(x, y, t)$$

6.16. Boundaries $x = 0$, $x = 1$, $y = 0$, $y = 2$ are held at fixed temperature 0, $k = 1/10$, $\beta = 1/5$:

$$u(0, y, t) = 0 \quad \text{and} \quad u(1, y, t) = 0$$
$$u(x, 0, t) = 0 \quad \text{and} \quad u(x, 2, t) = 0$$

Initial condition:

$$f(x, y) = x(1 - x)y(2 - y)$$

6.17. Boundaries $x = 1$, $y = 0$, $y = 1$ are held at fixed temperature 0, boundary $x = 0$ is insulated, $k = 1/10$, $\beta = 1/5$:

$$u_x(0, y, t) = 0 \quad \text{and} \quad u(1, y, t) = 0$$
$$u(x, 0, t) = 0 \quad \text{and} \quad u(x, 1, t) = 0$$

Initial condition:

$$f(x, y) = (1 - x^2)y(1 - y)$$

6.18. Boundaries $y = 0$, $y = 1$ are held at fixed temperature 0, boundaries $x = 0$ and $x = 1$ are insulated, $k = 1/10$, $\beta = 1/5$:

$$u_x(0, y, t) = 0 \quad \text{and} \quad u_x(1, y, t) = 0$$
$$u(x, 0, t) = 0 \quad \text{and} \quad u(x, 1, t) = 0$$

Initial condition:

$$f(x, y) = x^2\left(1 - \frac{2x}{3}\right)y(1 - y)$$

6.19.   Boundaries $x = 1$, $y = 0$ are held at fixed temperature 0, boundaries $x = 0$ and $y = 1$ are insulated, $k = 1/10$, $\beta = 1/5$:

$$u_x(0, y, t) = 0 \quad \text{and} \quad u(1, y, t) = 0$$
$$u(x, 0, t) = 0 \quad \text{and} \quad u_y(x, 1, t) = 0$$

Initial condition:

$$f(x, y) = (1 - x^2)y\left(1 - \frac{y}{2}\right)$$

6.20   Boundary $y = 0$ is held at fixed temperature 0, boundaries $x = 0$, $x = 1$, and $y = 1$ are insulated, $k = 1/10$, $\beta = 1/5$:

$$u_x(0, y, t) = 0 \quad \text{and} \quad u_x(1, y, t) = 0$$
$$u(x, 0, t) = 0 \quad \text{and} \quad u_y(x, 1, t) = 0$$

Initial condition:

$$f(x, y) = x^2\left(1 - \frac{2x}{3}\right)y\left(1 - \frac{y}{2}\right)$$

6.21.   Boundaries $x = 0$, $x = 1$, $y = 0$ are held at fixed temperature 0, boundary $y = 1$ is losing heat by convection into a surrounding medium at 0 temperature, $k = 1/10$, $\beta = 1/5$:

$$u(0, y, t) = 0 \quad \text{and} \quad u(1, y, t) = 0$$
$$u(x, 0, t) = 0 \quad \text{and} \quad u_y(x, 1, t) + u(x, 1, t) = 0$$

Initial condition:

$$f(x, y) = x(1 - x)y\left(1 - \frac{2y}{3}\right)$$

6.22.   Boundaries $x = 1$, $y = 0$ are held at fixed temperature 0, boundary $x = 0$ is insulated, and boundary $y = 1$ is losing heat by convection into a surrounding medium at 0 temperature, $k = 1/10$, $\beta = 1/5$:

$$u_x(0, y, t) = 0 \quad \text{and} \quad u(1, y, t) = 0$$
$$u(x, 0, t) = 0 \quad \text{and} \quad u_y(x, 1, t) + u(x, 1, t) = 0$$

Initial condition:

$$f(x, y) = (1 - x^2)y\left(1 - \frac{2y}{3}\right)$$

6.23.   Boundaries $x = 1$, $y = 0$ are held at fixed temperature 0, boundary $y = 1$ is insulated, and boundary $x = 0$ is losing heat by convection into a surrounding medium at 0 temperature, $k = 1/10$, $\beta = 1/5$:

$$u_x(0, y, t) - u(0, y, t) = 0 \quad \text{and} \quad u(1, y, t) = 0$$
$$u(x, 0, t) = 0 \quad \text{and} \quad u_y(x, 1, t) = 0$$

Initial condition:

$$f(x, y) = \left(-\frac{2x^2}{3} + \frac{x}{3} + \frac{1}{3} \right) y \left(1 - \frac{y}{2} \right)$$

6.24.   Boundaries $y = 0$, $y = 1$ are held at fixed temperature 0, boundary $x = 0$ is insulated, and boundary $x = 1$ is losing heat by convection into a surrounding medium at 0 temperature, $k = 1/10$, $\beta = 1/5$:

$$u_x(0, y, t) = 0 \quad \text{and} \quad u_x(1, y, t) + u(1, y, t) = 0$$
$$u(x, 0, t) = 0 \quad \text{and} \quad u(x, 1, t) = 0$$

Initial condition:

$$f(x, y) = \left(1 - \frac{x^2}{3} \right) y(1 - y)$$

6.25.   Boundaries $x = 0$, $x = 1$ are held at fixed temperature 0, boundary $y = 1$ is insulated, and boundary $y = 0$ is losing heat by convection into a surrounding medium at 0 temperature, $k = 1/10$, $\beta = 1/5$:

$$u(0, y, t) = 0 \quad \text{and} \quad u(1, y, t) = 0$$
$$u_y(x, 0, t) - u(x, 0, t) = 0 \quad \text{and} \quad u_y(x, 1, t) = 0$$

Initial condition:

$$f(x, y) = x(1 - x) \left(1 - \frac{(y - 1)^2}{3} \right)$$

Two Dimensions, Cylindrical Coordinates

We now seek the temperature distribution in a thin, circular plate over the two-dimensional domain $D = \{(r, \theta) \mid 0 < r < 1, 0 < \theta < b\}$ whose lateral surfaces are insulated. The homogeneous partial differential equation is

$$\frac{\partial}{\partial t} u(r, \theta, t) = k \left(\frac{\frac{\partial}{\partial r} u(r, \theta, t) + r \left(\frac{\partial^2}{\partial r^2} u(r, \theta, t) \right)}{r} + \frac{\frac{\partial^2}{\partial \theta^2} u(r, \theta, t)}{r^2} \right)$$

with initial conditions $u(r, \theta, 0) = f(r, \theta)$. In Exercises 6.26 through 6.37, we seek the temperature solution for various boundary and initial conditions. Evaluate the eigenvalues and corresponding two-dimensional orthonormalized eigenfunctions, and write the general solution. Generate the animated three-dimensional surface solution $u(r, \theta, t)$, and plot the animated sequence for $0 < t < 5$.

6.26. The center of the plate has a finite temperature, and the periphery at $r = 1$ is held at a fixed temperature of 0. The boundaries $\theta = 0$ and $\theta = \pi$ are held at fixed temperature 0, $k = 1/10$:

$$|u(0, \theta, t)| < \infty \quad \text{and} \quad u(1, \theta, t) = 0$$
$$u(r, 0, t) = 0 \quad \text{and} \quad u(r, \pi, t) = 0$$

Initial condition:

$$f(r, \theta) = (r - r^3) \sin(\theta)$$

6.27. The center of the plate has a finite temperature, and the periphery at $r = 1$ is held at a fixed temperature of 0. The boundaries $\theta = 0$ and $\theta = \pi$ are insulated, $k = 1/10$:

$$|u(0, \theta, t)| < \infty \quad \text{and} \quad u(1, \theta, t) = 0$$
$$u_\theta(r, 0, t) = 0 \quad \text{and} \quad u_\theta(r, \pi, t) = 0$$

Initial condition:

$$f(r, \theta) = (r^2 - r^4) \cos(2\theta)$$

6.28. The center of the plate has a finite temperature, and the periphery at $r = 1$ is held at a fixed temperature of 0. The boundaries $\theta = 0$ and $\theta = \pi$ are held at fixed temperature $0, k = 1/10$:

$$|u(0, \theta, t)| < \infty \quad \text{and} \quad u(1, \theta, t) = 0$$
$$u(r, 0, t) = 0 \quad \text{and} \quad u(r, \pi, t) = 0$$

Initial condition:

$$f(r, \theta) = (r^3 - r^5) \sin(3\theta)$$

6.29.   The center of the plate has a finite temperature, and the periphery at $r = 1$ is held at a fixed temperature of 0. The boundaries $\theta = 0$ and $\theta = \pi/2$ are held at fixed temperature $0, k = 1/10$:

$$|u(0, \theta, t)| < \infty \quad \text{and} \quad u(1, \theta, t) = 0$$

$$u(r, 0, t) = 0 \quad \text{and} \quad u\left(r, \frac{\pi}{2}, t\right) = 0$$

Initial condition:

$$f(r, \theta) = (r^2 - r^4)\sin(2\theta)$$

6.30.   The center of the plate has a finite temperature, and the periphery at $r = 1$ is held at a fixed temperature of 0. The boundaries $\theta = 0$ and $\theta = \pi/2$ are insulated, $k = 1/10$:

$$|u(0, \theta, t)| < \infty \quad \text{and} \quad u(1, \theta, t) = 0$$

$$u_\theta(r, 0, t) = 0 \quad \text{and} \quad u_\theta\left(r, \frac{\pi}{2}, t\right) = 0$$

Initial condition:

$$f(r, \theta) = (r^4 - r^6)\cos(4\theta)$$

6.31.   The center of the plate has a finite temperature, and the periphery at $r = 1$ is held at a fixed temperature of 0. The boundary conditions are such that we have periodic boundary conditions on θ, $k = 1/10$:

$$|u(0, \theta, t)| < \infty \quad \text{and} \quad u(1, \theta, t) = 0$$

$$u(r, -\pi, t) = u(r, \pi, t) \quad \text{and} \quad u_\theta(r, -\pi, t) = u_\theta(r, \pi, t)$$

Initial condition:

$$f(r, \theta) = (r - r^3)\cos(\theta)$$

6.32.   The center of the plate has a finite temperature, and the periphery at $r = 1$ is insulated. The boundaries $\theta = 0$ and $\theta = \pi$ are held at fixed temperature $0, k = 1/10$:

$$|u(0, \theta, t)| < \infty \quad \text{and} \quad u_r(1, \theta, t) = 0$$

$$u(r, 0, t) = 0 \quad \text{and} \quad u(r, \pi, t) = 0$$

Initial condition:

$$f(r, \theta) = \left(3r - r^3\right)\sin(\theta)$$

6.33.   The center of the plate has a finite temperature, and the periphery at $r = 1$ is insulated. The boundaries $\theta = 0$ and $\theta = \pi$ are insulated, $k = 1/10$:

$$|u(0, \theta, t)| < \infty \quad \text{and} \quad u_r(1, \theta, t) = 0$$
$$u_\theta(r, 0, t) = 0 \quad \text{and} \quad u_\theta(r, \pi, t) = 0$$

Initial condition:

$$f(r, \theta) = (2r^2 - r^4)\cos(2\theta)$$

6.34.   The center of the plate has a finite temperature, and the periphery at $r = 1$ is insulated. The boundaries $\theta = 0$ and $\theta = \pi$ are held at fixed temperature 0, $k = 1/10$:

$$|u(0, \theta, t)| < \infty \quad \text{and} \quad u_r(1, \theta, t) = 0$$
$$u(r, 0, t) = 0 \quad \text{and} \quad u(r, \pi, t) = 0$$

Initial condition:

$$f(r, \theta) = \left(\frac{5r^3}{3} - r^5\right)\sin(3\theta)$$

6.35.   The center of the plate has a finite temperature, and the periphery at $r = 1$ is insulated. The boundaries $\theta = 0$ and $\theta = \pi/2$ are held at fixed temperature 0, $k = 1/10$:

$$|u(0, \theta, t)| < \infty \quad \text{and} \quad u_r(1, \theta, t) = 0$$
$$u(r, 0, t) = 0 \quad \text{and} \quad u\left(r, \frac{\pi}{2}, t\right) = 0$$

Initial condition:

$$f(r, \theta) = \left(\frac{4r^6}{3} - r^8\right)\sin(6\theta)$$

6.36.   The center of the plate has a finite temperature, and the periphery at $r = 1$ is insulated. The boundaries at $\theta = 0$ and $\theta = \pi/2$ are insulated, $k = 1/10$:

$$|u(0, \theta, t)| < \infty \quad \text{and} \quad u_r(1, \theta, t) = 0$$
$$u_\theta(r, 0, t) = 0 \quad \text{and} \quad u_\theta\left(r, \frac{\pi}{2}, t\right) = 0$$

Initial condition:

$$f(r, \theta) = \left(\frac{3r^4}{2} - r^6\right)\cos(4\theta)$$

6.37. The center of the plate has a finite temperature, and the periphery at $r = 1$ is insulated. The boundary conditions are such that we have periodic boundary conditions on θ, $k = 1/10$:

$$|u(0, \theta, t)| < \infty \quad \text{and} \quad u_r(1, \theta, t) = 0$$

$$u(r, -\pi, t) = u(r, \pi, t) \quad \text{and} \quad u_\theta(r, -\pi, t) = u_\theta(r, \pi, t)$$

Initial condition:

$$f(r, \theta) = (3r - r^3)\sin(\theta)$$

Decay Time for Diffusion Phenomena

The time-dependent portion of the solution to the diffusion equation in two dimensions in the rectangular coordinate system reads

$$T_{m,n}(t) = C(m, n)e^{-k\lambda_{m,n}t}$$

The "decay time" τ at which the (m, n)th mode decays to $(1/e)$ of its initial value is given as

$$\tau_{m,n} = \frac{1}{k\lambda_{m,n}}$$

where

$$\lambda_{m,n} = \alpha_m + \beta_n$$

Here α_m is the eigenvalue corresponding to the x-dependent solution, β_n is the eigenvalue corresponding to the y-dependent solution, and k is the diffusivity of the medium.

For the case of Exercise 6.1, we evaluate $\lambda_{m,n}$ to be

$$\lambda_{m,n} = \frac{m^2\pi^2}{a^2} + \frac{n^2\pi^2}{b^2}$$

for $m = 1, 2, 3, \ldots, n = 1, 2, 3, \ldots$.

From the preceding expressions, we make the following observations: (1) the larger the value of k, the more rapid the decay; (2) the higher-order modes decay most rapidly; (3) the decay slows down as the dimensions of the domain are increased; and (4) as t goes to infinity, all modes decay exponentially to zero. Exercises 6.38 through 6.43 deal with the "decay times" of some of the earlier exercises.

6.38. For Exercise 6.1, evaluate the decay time $\tau(1, 1)$ of the $(1, 1)$ mode. By what factor is $\tau(1, 1)$ larger than $\tau(2, 2)$? If k is doubled, by what factor should the dimensions of the rectangular region be increased to give the same decay time?

6.39. For Exercise 6.3, evaluate the decay time $\tau(1, 1)$ of the $(1, 1)$ mode. By what factor is $\tau(1, 1)$ larger than $\tau(1, 2)$? If k is tripled, by what factor should the dimensions of the rectangular region be increased to give the same decay time?

6.40. For Exercise 6.5, evaluate the decay time $\tau(1, 1)$ of the $(1, 1)$ mode. By what factor is $\tau(1, 1)$ larger than $\tau(3, 2)$? If k is quadrupled, by what factor should the dimensions of the rectangular region be increased to give the same decay time?

6.41. For Exercise 6.7, evaluate the decay time $\tau(1, 1)$ of the $(1, 1)$ mode. By what factor is $\tau(1, 1)$ larger than $\tau(3, 3)$? If k is quadrupled, by what factor should the dimensions of the rectangular region be increased to give the same decay time?

6.42. For Exercise 6.8, evaluate the decay time $\tau(1, 1)$ of the $(1, 1)$ mode. By what factor is $\tau(1, 1)$ larger than $\tau(3, 1)$? If k is tripled, by what factor should the dimensions of the rectangular region be increased to give the same decay time?

6.43. For Exercise 6.9, evaluate the decay time $\tau(1, 1)$ of the $(1, 1)$ mode. By what factor is $\tau(1, 1)$ larger than $\tau(1, 3)$? If k is doubled, by what factor should the dimensions of the rectangular region be increased to give the same decay time?

CHAPTER 7

The Wave Equation in Two Spatial Dimensions

7.1 Introduction

In Chapter 4, we examined the wave partial differential equation in only one spatial dimension. Here, we examine the wave equation in two spatial dimensions. Thus, we will be considering problems with a total of three independent variables: the two spatial variables and time. The method of separation of variables will be used, and we will see the development of two independent Sturm-Liouville problems—one for each of the two spatial variables. The final solutions will have the form of a double Fourier series. We will consider problems in both the rectangular and cylindrical coordinate systems.

7.2 Two-Dimensional Wave Operator in Rectangular Coordinates

In the one-dimensional wave equation, as discussed in Chapter 4, the second-order partial derivative with respect to the spatial variable x can be viewed as the first term of the Laplacian operator. Thus, an extension of this Laplacian operator to higher dimensions is natural, and we show that wave phenomena in two dimensions can be described by the following nonhomogeneous partial differential equation in the rectangular coordinate system:

$$\frac{\partial^2}{\partial t^2} u(x, y, t) = c^2 \left(\frac{\partial^2}{\partial x^2} u(x, y, t) + \frac{\partial^2}{\partial y^2} u(x, y, t) \right) - \gamma \left(\frac{\partial}{\partial t} u(x, y, t) \right) + h(x, y, t)$$

In the preceding equation, for the situation of wave phenomena, $u(x, y, t)$ denotes the spatial-time-dependent wave amplitude, c denotes the wave speed, γ denotes the damping coefficient of the medium, and $h(x, y, t)$ denotes the presence of any external applied forces acting on the system. We have assumed the medium to be isotropic and uniform, so the preceding coefficients are spatially invariant and we write them as constants.

Because we will be considering only example problems that have no extension along the z-axis, we write the Laplacian operator with no z-dependent terms.

The given partial differential equation for wave phenomena is characterized by the fact that it relates the second partial derivative with respect to the time variable with the second partial derivatives with respect to the spatial variables.

We can write this nonhomogeneous equation in terms of the linear wave operator as

$$L(u) = h(x, y, t)$$

where the wave operator in rectangular coordinates in two spatial dimensions reads

$$L(u) = \frac{\partial^2}{\partial t^2} u(x, y, t) - c^2 \left(\frac{\partial^2}{\partial x^2} u(x, y, t) + \frac{\partial^2}{\partial y^2} u(x, y, t) \right) + \gamma \left(\frac{\partial}{\partial t} u(x, y, t) \right)$$

If $h(x, y, t) = 0$, then there are no external applied forces acting on the system and the preceding nonhomogeneous partial differential equation reduces to its corresponding homogeneous equation

$$L(u) = 0$$

A typical illustrative problem for a homogeneous wave equation in two spatial dimensions reads as follows. We consider the transverse vibrations in a thin rectangular membrane over the finite two-dimensional rectangular domain $D = \{(x, y) \mid 0 < x < \pi, 0 < y < \pi\}$. The membrane is in a medium with a small damping coefficient γ. The boundaries are held fixed, and the membrane has an initial displacement function $f(x, y)$ and an initial speed function $g(x, y)$ given as follows. There are no external forces acting on the system, and the wave speed is c. We seek the transverse wave amplitude $u(x, y, t)$.

The homogeneous wave equation for this problem is

$$\frac{\partial^2}{\partial t^2} u(x, y, t) = c^2 \left(\frac{\partial^2}{\partial x^2} u(x, y, t) + \frac{\partial^2}{\partial y^2} u(x, y, t) \right) - \gamma \left(\frac{\partial}{\partial t} u(x, y, t) \right)$$

Since the boundaries of the membrane are held fixed, the boundary conditions on the problem are type 1 at $x = 0$ and $x = \pi$ and type 1 at $y = 0$ and $y = \pi$.

$$u(0, y, t) = 0 \quad \text{and} \quad u(\pi, y, t) = 0$$

and

$$u(x, 0, t) = 0 \quad \text{and} \quad u(x, \pi, t) = 0$$

The initial displacement function is

$$u(x, y, 0) = f(x, y)$$

and the initial speed function is

$$u_t(x, y, 0) = g(x, y)$$

We now examine the solution to this homogeneous partial differential equation over a finite two-dimensional domain with homogeneous boundary conditions using the method of separation of variables.

7.3 Method of Separation of Variables for the Wave Equation

The homogeneous wave equation for a uniform system in two dimensions in rectangular coordinates can be written

$$\frac{\partial^2}{\partial t^2} u(x, y, t) - c^2 \left(\frac{\partial^2}{\partial x^2} u(x, y, t) + \frac{\partial^2}{\partial y^2} u(x, y, t) \right) + \gamma \left(\frac{\partial}{\partial t} u(x, y, t) \right) = 0$$

This can be rewritten in the more familiar form

$$\frac{\partial^2}{\partial t^2} u(x, y, t) + \gamma \left(\frac{\partial}{\partial t} u(x, y, t) \right) = c^2 \left(\frac{\partial^2}{\partial x^2} u(x, y, t) + \frac{\partial^2}{\partial y^2} u(x, y, t) \right)$$

Generally, one seeks a solution to this problem over the finite two-dimensional domain $D = \{(x, y) \mid 0 < x < a, 0 < y < b\}$ subject to the regular homogeneous boundary conditions

$$\kappa_1 u(0, y, t) + \kappa_2 u_x(0, y, t) = 0$$

$$\kappa_3 u(a, y, t) + \kappa_4 u_x(a, y, t) = 0$$

$$\kappa_5 u(x, 0, t) + \kappa_6 u_y(x, 0, t) = 0$$

$$\kappa_7 u(x, b, t) + \kappa_8 u_y(x, b, t) = 0$$

along with the two initial conditions

$$u(x, y, 0) = f(x, y) \quad \text{and} \quad u_t(x, y, 0) = g(x, y)$$

In the method of separation of variables, as in Chapter 4, the character of the equation is such that we can assume a solution in the form of a product

$$u(x, y, t) = X(x)Y(y)T(t)$$

Because the three independent variables in the partial differential equation are the spatial variables x and y, and the time variable t, this assumed solution is written as a product of three functions; one is an exclusive function of x, one is an exclusive function of y, and the last is an exclusive function of t.

Substituting this assumed solution into the partial differential equation, we get

$$X(x)Y(y)\left(\frac{d^2}{dt^2}T(t)\right) + \gamma X(x)Y(y)\left(\frac{d}{dt}T(t)\right) = c^2\left(\left(\frac{d^2}{dx^2}X(x)\right)Y(y)T(t) + X(x)\left(\frac{d^2}{dy^2}Y(y)\right)T(t)\right)$$

Dividing both sides by the product solution yields

$$\frac{\frac{d^2}{dt^2}T(t)+\gamma\left(\frac{d}{dt}T(t)\right)}{c^2 T(t)}=\frac{\left(\frac{d^2}{dx^2}X(x)\right)Y(y)+X(x)\left(\frac{d^2}{dy^2}Y(y)\right)}{X(x)Y(y)}$$

Because the left-hand side of the preceding is an exclusive function of t and the right-hand side an exclusive function of both x and y, and x, y, and t are independent, then the only way this can hold for all values of t, x, and y is to set each side equal to a constant. Doing so, we get the following three ordinary differential equations in terms of the separation constants λ, α, and β:

$$\frac{d^2}{dt^2}T(t)+\gamma\left(\frac{d}{dt}T(t)\right)+c^2\lambda\,T(t)=0$$

$$\frac{d^2}{dx^2}X(x)+\alpha\,X(x)=0$$

$$\frac{d^2}{dy^2}Y(y)+\beta\,Y(y)=0$$

along with the coupling equation

$$\lambda=\alpha+\beta$$

The preceding differential equation in t is an ordinary second-order linear homogeneous differential equation for which we already have the solution from Section 1.4. The second and third differential equations in x and y are ordinary second-order linear, homogeneous differential equations of the Euler type for which we already have the solutions from Section 1.4. We easily recognize each of the two spatial differential equations as being Sturm-Liouville types.

7.4 Sturm-Liouville Problem for the Wave Equation in Two Dimensions

The second and third differential equations in Section 7.3, in the spatial variables x and y, must each be solved subject to their respective homogeneous spatial boundary conditions.

We first consider the x-dependent differential equation over the finite interval $I=\{x\,|\,0<x<a\}$. The Sturm-Liouville problem for the x-dependent portion of the wave equation consists of the ordinary differential equation

$$\frac{d^2}{dx^2}X(x)+\alpha\,X(x)=0$$

along with its respective corresponding regular homogeneous boundary conditions

$$\kappa_1 X(0)+\kappa_2 X_x(0)=0$$

and

$$\kappa_3 X(a) + \kappa_4 X_x(a) = 0$$

We recognize the preceding problem in the spatial variable x to be a regular Sturm-Liouville eigenvalue problem where the allowed values of α are called the system "eigenvalues" and the corresponding solutions are called the "eigenfunctions." Note that the ordinary differential equation given here is of the Euler type, and the weight function is $w(x) = 1$. The indexed eigenvalues and corresponding eigenfunctions are given, respectively, as

$$\alpha_m, X_m(x)$$

for $m = 0, 1, 2, 3, \ldots$.

These eigenfunctions can be normalized, with respect to the weight function $w(x) = 1$, and the corresponding statement of orthonormality reads

$$\int_0^a X_m(x) X_r(x) \mathrm{d}x = \delta(m, r)$$

Similarly, we now focus on the third differential equation in the spatial variable y subject to the homogeneous spatial boundary conditions over the finite interval $I = \{y \mid 0 < y < b\}$. The Sturm-Liouville problem for the y-dependent portion of the wave equation consists of the ordinary differential equation

$$\frac{\mathrm{d}^2}{\mathrm{d}y^2} Y(y) + \beta Y(y) = 0$$

along with its respective corresponding regular homogeneous boundary conditions

$$\kappa_5 Y(0) + \kappa_6 Y_y(0) = 0$$

and

$$\kappa_7 Y(b) + \kappa_8 Y_y(b) = 0$$

We recognize this problem in the spatial variable y as a regular Sturm-Liouville eigenvalue problem where the allowed values of β are called the system "eigenvalues" and the corresponding solutions are called the "eigenfunctions." Note that the ordinary differential equation given here is of the Euler type, and the weight function is $w(y) = 1$. The indexed eigenvalues and corresponding eigenfunctions are given, respectively, as

$$\beta_n, Y_n(y)$$

for $n = 0, 1, 2, 3, \ldots$.

These eigenfunctions can be normalized, with respect to the weight function $w(y) = 1$, and the corresponding statement of orthonormality reads

$$\int_0^b Y_n(y) Y_s(y) dy = \delta(n, s)$$

What was stated in Section 2.2 about regular Sturm-Liouville eigenvalue problems in one dimension can be extended to two dimensions here; that is, the products of the two spatial eigenfunctions form a "complete" set with respect to any piecewise smooth function $f(x, y)$ over the finite two-dimensional domain $D = \{(x, y) \mid 0 < x < a, 0 < y < b\}$.

Finally, the time-dependent differential equation reads

$$\frac{d^2}{dt^2} T(t) + \gamma \left(\frac{d}{dt} T(t) \right) + c^2 \lambda \, T(t) = 0$$

From Section 1.4 on second-order linear differential equations, if we assume the damping coefficient to be small, then the solution to the preceding, in terms of the allowed eigenvalues $\lambda_{m,n}$, is

$$T_{m,n}(t) = e^{-\frac{\gamma t}{2}} \left(A(m,n) \cos\left(\frac{\sqrt{4\lambda_{m,n} c^2 - \gamma^2} t}{2} \right) + B(m,n) \sin\left(\frac{\sqrt{4\lambda_{m,n} c^2 - \gamma^2} t}{2} \right) \right)$$

where the coupling equation now reads

$$\lambda_{m,n} = \alpha_m + \beta_n$$

The coefficients $A(m, n)$ and $B(m, n)$ are unknown arbitrary constants. Throughout most of our example problems, we make the assumption of small damping in the system—that is,

$$\gamma^2 < 4\lambda_{m,n} c^2$$

and the (m, n)th-mode angular frequency is

$$\omega_{m,n} = \frac{\sqrt{4\lambda_{m,n} c^2 - \gamma^2}}{2}$$

Thus, by the method of separation of variables, we arrive at an infinite number of indexed product solutions for the homogeneous wave partial differential equation, over a finite domain, given as

$$u_{m,n}(x, y, t) = X_m(x) Y_n(y) e^{-\frac{\gamma t}{2}} \left(A(m,n) \cos(\omega_{m,n} t) + B(m,n) \sin(\omega_{m,n} t) \right)$$

for $m = 0, 1, 2, 3, \ldots$, and $n = 0, 1, 2, 3, \ldots$.

Since the partial differential operator is linear, any superposition of solutions to the homogeneous equation is also a solution to the problem. Thus, we can write the general solution to the homogeneous partial differential equation as the double-infinite sum

$$u(x, y, t) = \sum_{n=0}^{\infty} \left(\sum_{m=0}^{\infty} X_m(x) Y_n(y) e^{-\frac{\gamma t}{2}} \left(A(m, n) \cos(\omega_{m,n} t) + B(m, n) \sin(\omega_{m,n} t) \right) \right)$$

This equation is the double Fourier series eigenfunction expansion form of the solution to the wave partial differential equation. The terms of this sum form the "basis vectors" of the solution space of the partial differential equation. Similar to a partial differential equation in only one dimension, there are an infinite number of basis vectors in the solution space, and we say the dimension of the solution space is infinite. We demonstrate the preceding concepts with the solution of an illustrative problem.

DEMONSTRATION: We consider the transverse vibrations in a thin rectangular membrane over the finite two-dimensional domain $D = \{(x, y) \mid 0 < x < \pi, 0 < y < \pi\}$. The membrane is in a medium with small damping γ. The boundaries are held fixed, and the membrane has an initial displacement function $f(x, y)$ and an initial speed function $g(x, y)$ given as follows. There are no external forces acting on the system, and the wave speed is c. We seek the transverse wave amplitude $u(x, y, t)$.

SOLUTION: The homogeneous wave equation for this problem is

$$\frac{\partial^2}{\partial t^2} u(x, y, t) = c^2 \left(\frac{\partial^2}{\partial x^2} u(x, y, t) + \frac{\partial^2}{\partial y^2} u(x, y, t) \right) - \gamma \left(\frac{\partial}{\partial t} u(x, y, t) \right)$$

The given boundary conditions on the problem are type 1 at $x = 0$ and $x = \pi$ and type 1 at $y = 0$ and $y = \pi$.

$$u(0, y, t) = 0 \quad \text{and} \quad u(\pi, y, t) = 0$$

and

$$u(x, 0, t) = 0 \quad \text{and} \quad u(x, \pi, t) = 0$$

The initial displacement function is

$$u(x, y, 0) = f(x, y)$$

and the initial speed function is

$$u_t(x, y, 0) = g(x, y)$$

From the method of separation of variables, we obtain the three ordinary differential equations

$$\frac{d^2}{dt^2}T(t) + \gamma\left(\frac{d}{dt}T(t)\right) + c^2\lambda\, T(t) = 0$$

$$\frac{d^2}{dx^2}X(x) + \alpha\, X(x) = 0$$

$$\frac{d^2}{dy^2}Y(y) + \beta\, Y(y) = 0$$

along with the coupling equation

$$\lambda = \alpha + \beta$$

We first consider the solution of the x equation. The differential equation is of the Euler type with weight function $w(x) = 1$, and the corresponding boundary conditions on the x-dependent equation are both type 1:

$$X(0) = 0 \quad \text{and} \quad X(\pi) = 0$$

From the similar problem in Example 2.5.1, the allowed eigenvalues and corresponding orthonormal eigenfunctions are

$$\alpha_m = m^2$$

and

$$X_m(x) = \sqrt{\frac{2}{\pi}}\sin(mx)$$

for $m = 1, 2, 3, \ldots$.

Similarly, for the y equation, the differential equation is of the Euler type with weight function $w(y) = 1$ and the corresponding boundary conditions are, again, both of type 1:

$$Y(0) = 0 \quad \text{and} \quad Y(\pi) = 0$$

The allowed eigenvalues and corresponding orthonormal eigenfunctions from Example 2.5.1 are

$$\beta_n = n^2$$

and

$$Y_n(y) = \sqrt{\frac{2}{\pi}}\sin(ny)$$

for $n = 1, 2, 3, \ldots$.

The corresponding statements of orthonormality for each of the preceding eigenfunctions, with their respective weight functions, read

$$\int_0^\pi \frac{2\sin(mx)\sin(rx)}{\pi} dx = \delta(m, r)$$

and

$$\int_0^\pi \frac{2\sin(ny)\sin(sy)}{\pi} dy = \delta(n, s)$$

for $m, r = 1, 2, 3, \ldots$, and $n, s = 1, 2, 3, \ldots$.

Finally, the time-dependent differential equation has the solution

$$T_{m,n}(t) = e^{-\frac{\gamma t}{2}} \left(A(m, n) \cos(\omega_{m,n} t) + B(m, n) \sin(\omega_{m,n} t) \right)$$

where the damped angular frequency $\omega_{m,n}$ is given as

$$\omega_{m,n} = \frac{\sqrt{4\lambda_{m,n} c^2 - \gamma^2}}{2}$$

and the coupling equation is

$$\lambda_{m,n} = m^2 + n^2$$

for $m = 1, 2, 3, \ldots$, and $n = 1, 2, 3, \ldots$. The terms $A(m, n)$ and $B(m, n)$ are unknown arbitrary constants.

The final eigenfunction expansion form of the solution is constructed from the superposition of the products of the time-dependent solution and the preceding x- and y-dependent eigenfunction. The solution reads

$$u(x, y, t) = \sum_{n=1}^{\infty} \left(\sum_{m=1}^{\infty} 2 e^{-\frac{\gamma t}{2}} \left(A(m, n) \cos(\omega_{m,n} t) + B(m, n) \sin(\omega_{m,n} t) \right) \sqrt{\frac{1}{\pi}} \sin(mx) \sqrt{\frac{1}{\pi}} \sin(ny) \right)$$

The coefficients $A(m, n)$ and $B(m, n)$ are to be determined from the initial conditions.

7.5 Initial Conditions for the Wave Equation in Rectangular Coordinates

We now consider the initial conditions on the problem. At time $t = 0$, we have two conditions. The first condition is the initial displacement function given as

$$u(x, y, 0) = f(x, y)$$

and the second condition is the initial speed function given as

$$u_t(x, y, 0) = g(x, y)$$

Substitution of the first condition into the following general solution

$$u(x, y, t) = \sum_{n=0}^{\infty} \left(\sum_{m=0}^{\infty} X_m(x) Y_n(y) e^{-\frac{\gamma t}{2}} \left(A(m, n) \cos(\omega_{m,n} t) + B(m, n) \sin(\omega_{m,n} t) \right) \right)$$

at time $t = 0$ yields

$$f(x, y) = \sum_{n=0}^{\infty} \left(\sum_{m=0}^{\infty} X_m(x) Y_n(y) A(m, n) \right)$$

Forcing the second condition into the time derivative of the solution at time $t = 0$ yields

$$g(x, y) = \sum_{n=0}^{\infty} \left(\sum_{m=0}^{\infty} \left(B(m, n) \omega_{m,n} - \frac{A(m, n)\gamma}{2} \right) X_m(x) Y_n(y) \right)$$

Because the eigenfunctions of the regular Sturm-Liouville problem form a complete set with respect to piecewise smooth functions over the finite two-dimensional domain, then the preceding sums are the generalized double Fourier series expansions of the functions $f(x, y)$ and $g(x, y)$, respectively, in terms of the allowed eigenfunctions. We can evaluate the unknown Fourier coefficients $A(m, n)$ and $B(m, n)$ by taking advantage of the familiar orthonormality statements.

Using the procedure in Section 2.4, for the generalized Fourier series expansion in one variable, we evaluate the double Fourier coefficients $A(m, n)$ and $B(m, n)$ as follows. We take the double inner products of both sides of the preceding equations with respect to the x- and y-dependent eigenfunctions. Assuming the validity of the interchange between the integration and summation operations, we get for the first condition

$$\int_0^b \int_0^a f(x, y) X_r(x) Y_s(y) dx\, dy = \sum_{n=0}^{\infty} \left(\sum_{m=0}^{\infty} A(m, n) \left(\int_0^a X_m(x) X_r(x) dx \right) \left(\int_0^b Y_n(y) Y_s(y) dy \right) \right)$$

Taking advantage of the statements of orthonormality from Section 7.4 for the two integrals on the right yields

$$\int_0^b \int_0^a f(x, y) X_r(x) Y_s(y) dx\, dy = \sum_{n=0}^{\infty} \left(\sum_{m=0}^{\infty} A(m, n) \delta(m, r) \delta(n, s) \right)$$

Due to the mathematical nature of the Kronecker delta functions in this sum, only the $m = r$, $n = s$ term survives, and the double Fourier coefficient $A(m, n)$ evaluates to

$$A(m, n) = \int_0^b \int_0^a f(x, y) X_m(x) Y_n(y) \, dx \, dy$$

Similarly, for the second condition, we have

$$\int_0^b \int_0^a g(x, y) X_r(x) Y_s(y) \, dx \, dy =$$

$$\sum_{n=0}^{\infty} \left(\sum_{m=0}^{\infty} \left(B(m, n) \omega_{m,n} - \frac{A(m, n)\gamma}{2} \right) \left(\int_0^a X_m(x) X_r(x) \, dx \right) \left(\int_0^b Y_n(y) Y_s(y) \, dy \right) \right)$$

Again, taking advantage of the statements of orthonormality from Section 7.4 for the two integrals on the right yields

$$\int_0^b \int_0^a g(x, y) X_r(x) Y_s(y) \, dx \, dy = \sum_{n=0}^{\infty} \left(\sum_{m=0}^{\infty} \left(B(m, n) \omega_{m,n} - \frac{A(m, n)\gamma}{2} \right) \delta(m, r) \delta(n, s) \right)$$

Again, due to the mathematical nature of the Kronecker delta functions in this sum, only the $m = r, n = s$ term survives the sum, and, combining the previous value for $A(m, n)$, the double Fourier coefficient $B(m, n)$ evaluates to

$$B(m, n) = \int_0^b \int_0^a \frac{\left(\frac{f(x,y)\gamma}{2} + g(x, y) \right) X_m(x) Y_n(y)}{\omega_{m,n}} \, dx \, dy$$

We now demonstrate the evaluation of the Fourier coefficients $A(m, n)$ and $B(m, n)$ for the example problem in Section 7.4.

DEMONSTRATION: We consider the special case where there is no damping in the system ($\gamma = 0$), the wave speed $c = 1$, the initial displacement function is $u(x, y, 0) = f(x, y)$ where

$$f(x, y) = xy(\pi - x)(\pi - y)$$

and the initial speed function is $u_t(x, y, 0) = g(x, y)$ where

$$g(x, y) = 0$$

SOLUTION: Evaluating the wave angular frequency for $c = 1$ and $\gamma = 0$, we get

$$\omega_{m,n} = \sqrt{m^2 + n^2}$$

The Fourier coefficient $A(m, n)$ for the given $f(x, y)$ yields the integral

$$A(m, n) = \int_0^\pi \int_0^\pi x(\pi - x) y(\pi - y) X_m(x) Y_n(y) \, dx \, dy$$

where, from the preceding example, the orthonormal eigenfunctions are

$$X_m(x) = \sqrt{\frac{2}{\pi}} \sin(mx)$$

and

$$Y_n(y) = \sqrt{\frac{2}{\pi}} \sin(ny)$$

for $m = 1, 2, 3, \ldots$, and $n = 1, 2, 3, \ldots$. Insertion of these eigenfunctions into this integral yields

$$A(m, n) = \int_0^\pi \int_0^\pi \frac{2x(\pi - x) y(\pi - y) \sin(mx) \sin(ny)}{\pi} \, dx \, dy$$

which evaluates to

$$A(m, n) = \frac{8 \left((-1)^m - 1 \right) \left((-1)^n - 1 \right)}{n^3 m^3 \pi}$$

Similarly, for $\gamma = 0$ and $g(x, y) = 0$, the integral for $B(m, n)$ evaluates to

$$B(m, n) = 0$$

Thus, with knowledge of the coefficients $A(m, n)$ and $B(m, n)$, the final series form of the eigenfunction expansion solution to the problem, from Section 7.4, reads

$$u(x, y, t) = \sum_{n=1}^\infty \left(\sum_{m=1}^\infty \frac{16 \left((-1)^m - 1 \right) \left((-1)^n - 1 \right) \cos\left(\sqrt{m^2 + n^2} \, t \right) \sin(mx) \sin(ny)}{n^3 m^3 \pi^2} \right)$$

The details for the development of this solution and the graphics are given later in one of the Maple worksheet examples following.

7.6 Example Wave Equation Problems in Rectangular Coordinates

We now consider several examples of partial differential equations for wave phenomena under various homogeneous boundary conditions over finite two-dimensional domains in the rectangular coordinate system. We note that all the spatial ordinary differential equations are of the Euler type with weight functions equal to 1.

EXAMPLE 7.6.1: We seek the wave amplitude $u(x, y, t)$ for the transverse vibrations on a thin rectangular membrane over the finite two-dimensional domain $D = \{(x, y) \mid 0 < x < \pi,$ $0 < y < \pi\}$. The membrane is in a medium with no damping and the wave speed is $c = 1$. The boundaries are held fixed, and the membrane has an initial displacement distribution $f(x, y)$ and an initial speed distribution $g(x, y)$ given as follows.

SOLUTION: The two-dimensional homogeneous wave equation is

$$\frac{\partial^2}{\partial t^2} u(x, y, t) = c^2 \left(\frac{\partial^2}{\partial x^2} u(x, y, t) + \frac{\partial^2}{\partial y^2} u(x, y, t) \right) - \gamma \left(\frac{\partial}{\partial t} u(x, y, t) \right)$$

The boundary conditions are type 1 at $x = 0$, type 1 at $x = \pi$, type 1 at $y = 0$, and type 1 at $y = \pi$:

$$u(0, y, t) = 0 \quad \text{and} \quad u(\pi, y, t) = 0$$

and

$$u(x, 0, t) = 0 \quad \text{and} \quad u(x, \pi, t) = 0$$

The initial conditions are

$$u(x, y, 0) = x(\pi - x)y(\pi - y) \quad \text{and} \quad u_t(x, y, 0) = 0$$

The ordinary differential equations from the method of separation of variables are

$$\frac{d^2}{dt^2} T(t) + \gamma \left(\frac{d}{dt} T(t) \right) + c^2 \lambda\, T(t) = 0$$

$$\frac{d^2}{dx^2} X(x) + \alpha\, X(x) = 0$$

$$\frac{d^2}{dy^2} Y(y) + \beta\, Y(y) = 0$$

The coupling equation reads

$$\lambda = \alpha + \beta$$

The boundary conditions on the spatial equations are

$$X(0) = 0 \quad \text{and} \quad X(\pi) = 0$$

and

$$Y(0) = 0 \quad \text{and} \quad Y(\pi) = 0$$

Assignment of system parameters

> restart:with(plots):a:=Pi:b:=Pi:c:=1:unprotect(gamma):gamma:=0:

Allowed eigenvalues and corresponding orthonormal eigenfunctions are obtained from Example 2.5.1.

> alpha[m]:=(m*Pi/a)^2;beta[n]:=(n*Pi/b)^2;lambda[m,n]:=alpha[m]+beta[n];

$$\alpha_m := m^2$$
$$\beta_n := n^2$$
$$\lambda_{m,n} := m^2 + n^2 \qquad (7.1)$$

for $m = 1, 2, 3, \ldots$, and $n = 1, 2, 3, \ldots$.

> X[m](x):=sqrt(2/a)*sin(m*Pi/a*x);X[r](x):=subs(m=r,X[m](x)):

$$X_m(x) := \frac{\sqrt{2}\,\sin(mx)}{\sqrt{\pi}} \qquad (7.2)$$

> Y[n](y):=sqrt(2/b)*sin(n*Pi/b*y);Y[s](y):=subs(n=s,Y[n](y)):

$$Y_n(y) := \frac{\sqrt{2}\,\sin(ny)}{\sqrt{\pi}} \qquad (7.3)$$

Statements of orthonormality with their respective weight functions

> w(x):=1:Int(X[m](x)*X[r](x)*w(x),x=0..a)=delta(m,r);

$$\int_0^\pi \frac{2\sin(mx)\sin(rx)}{\pi}\,dx = \delta(m, r) \qquad (7.4)$$

> w(y):=1:Int(Y[n](y)*Y[s](y)*w(y),y=0..b)=delta(n,s);

$$\int_0^\pi \frac{2\sin(ny)\sin(sy)}{\pi}\,dy = \delta(n, s) \qquad (7.5)$$

Time-dependent solution

> T[m,n](t):=exp(−gamma/2*t)*(A(m,n)*cos(omega[m,n]*t)+B(m,n)*sin(omega[m,n]*t));

$$T_{m,n}(t) := A(m, n)\cos\big(\omega_{m,n}t\big) + B(m, n)\sin\big(\omega_{m,n}t\big) \qquad (7.6)$$

Generalized series terms

> u[m,n](x,y,t):=T[m,n](t)*X[m](x)*Y[n](y);

$$u_{m,n}(x, y, t) := \frac{2\left(A(m, n)\cos\big(\omega_{m,n}t\big) + B(m, n)\sin\big(\omega_{m,n}t\big)\right)\sin(mx)\sin(ny)}{\pi} \qquad (7.7)$$

Eigenfunction expansion

> u(x,y,t):=Sum(Sum(u[m,n](x,y,t),m=1..infinity),n=1..infinity);

$$u(x, y, t) := \sum_{n=1}^{\infty} \left(\sum_{m=1}^{\infty} \frac{2\left(A(m, n)\cos(\omega_{m,n}t) + B(m, n)\sin(\omega_{m,n}t)\right)\sin(mx)\sin(ny)}{\pi} \right) \quad (7.8)$$

> omega[m,n]:=sqrt(4*lambda[m,n]*c^2−gamma^2)/2;

$$\omega_{m,n} := \sqrt{m^2 + n^2} \quad (7.9)$$

We consider the special case where the initial condition functions $u(x, y, 0) = f(x, y)$ and $u_t(x, y, 0) = g(x, y)$ are given as

> f(x,y):=x*(Pi−x)*y*(Pi−y);

$$f(x, y) := x(\pi - x)y(\pi - y) \quad (7.10)$$

> g(x,y):=0;

$$g(x, y) := 0 \quad (7.11)$$

From Section 7.5, the double Fourier coefficients $A(m, n)$ and $B(m, n)$ are determined from the integrals

> A(m,n):=eval(Int(Int(f(x,y)*X[m](x)*w(x)*Y[n](y)*w(y),x=0..a),y=0..b));A(m,n):
  =expand(value(%)):

$$A(m, n) := \int_{0}^{\pi} \int_{0}^{\pi} \frac{2x(\pi - x)y(\pi - y)\sin(mx)\sin(ny)}{\pi}\,dx\,dy \quad (7.12)$$

> A(m,n):=factor(subs({sin(m*Pi)=0,cos(m*Pi)=(−1)^m,sin(n*Pi)=0,cos(n*Pi)=(−1)^n},
  A(m,n)));

$$A(m, n) := \frac{8\left(-1 + (-1)^n\right)\left(-1 + (-1)^m\right)}{\pi m^3 n^3} \quad (7.13)$$

> B(m,n):=eval(Int(Int((f(x,y)*gamma/2+g(x,y))/(omega[m,n])*X[m](x)*Y[n](y),x=0..a),
  y=0..b));B(m,n):=expand(value(%)):

$$B(m, n) := \int_{0}^{\pi} \int_{0}^{\pi} 0\,dx\,dy \quad (7.14)$$

> B(m,n):=factor(subs({sin(m*Pi)=0,cos(m*Pi)=(−1)^m,sin(n*Pi)=0,cos(n*Pi)=(−1)^n},
  B(m,n)));

$$B(m, n) := 0 \quad (7.15)$$

> T[m,n](t):=eval(T[m,n](t));

$$T_{m,n}(t) := \frac{8\left(-1+(-1)^n\right)\left(-1+(-1)^m\right)\cos\left(\sqrt{m^2+n^2}\,t\right)}{\pi m^3 n^3} \tag{7.16}$$

Generalized series terms

> u[m,n](x,y,t):=eval(T[m,n](t)*X[m](x)*Y[n](y));

$$u_{m,n}(x, y, t) := \frac{16\left(-1+(-1)^n\right)\left(-1+(-1)^m\right)\cos\left(\sqrt{m^2+n^2}\,t\right)\sin(mx)\sin(ny)}{\pi^2 m^3 n^3} \tag{7.17}$$

Series solution

> u(x,y,t):=Sum(Sum(u[m,n](x,y,t),m=1..infinity),n=1..infinity);

$$u(x, y, t) := \sum_{n=1}^{\infty}\left(\sum_{m=1}^{\infty} \frac{16\left(-1+(-1)^n\right)\left(-1+(-1)^m\right)\cos\left(\sqrt{m^2+n^2}\,t\right)\sin(mx)\sin(ny)}{\pi^2 m^3 n^3}\right)$$

$$\tag{7.18}$$

First few terms of sum

> u(x,y,t):=sum(sum(u[m,n](x,y,t),m=1..1),n=1..1):

ANIMATION

> animate3d(u(x,y,t),x=0..a,y=0..b,t=0..3,axes=framed,thickness=1);

The preceding animation command illustrates the spatial-time-dependent solution for $u(x, y, t)$. The animation sequence in Figures 7.1 and 7.2 shows snapshots of the animation at times $t = 0$ and $t = 2$.

ANIMATION SEQUENCE

> u(x,y,0):=subs(t=0,u(x,y,t)):plot3d(u(x,y,0),x=0..a,y=0..b,axes=framed,thickness=1);

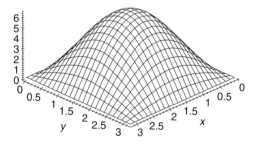

Figure 7.1

> u(x,y,2):=subs(t=2,u(x,y,t)):plot3d(u(x,y,2),x=0..a,y=0..b,axes=framed,thickness=1);

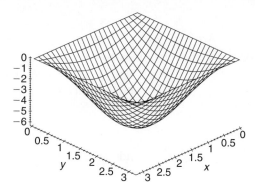

Figure 7.2

EXAMPLE 7.6.2: We seek the wave amplitude $u(x, y, t)$ for the transverse wave propagation on a thin rectangular membrane over the finite two-dimensional domain $D = \{(x, y) \,|\, 0 < x < \pi, 0 < y < \pi\}$. The membrane is in a medium with zero damping, and the wave speed is $c = 1/2$. The boundaries $x = 0$ and $x = \pi$ are unsecured and the boundaries $y = 0$ and $y = \pi$ are held fixed. The initial displacement distribution $f(x, y)$ and the initial speed distribution $g(x, y)$ are given as follows.

SOLUTION: The two-dimensional homogeneous wave equation is

$$\frac{\partial^2}{\partial t^2} u(x, y, t) = c^2 \left(\frac{\partial^2}{\partial x^2} u(x, y, t) + \frac{\partial^2}{\partial y^2} u(x, y, t) \right) - \gamma \left(\frac{\partial}{\partial t} u(x, y, t) \right)$$

The boundary conditions are type 2 at $x = 0$, type 2 at $x = \pi$, type 1 at $y = 0$, and type 1 at $y = \pi$:

$$u_x(0, y, t) = 0 \quad \text{and} \quad u_x(\pi, y, t) = 0$$

and

$$u(x, 0, t) = 0 \quad \text{and} \quad u(x, \pi, t) = 0$$

The initial conditions are

$$u(x, y, 0) = xy(\pi - y) \quad \text{and} \quad u_t(x, y, 0) = 0$$

The ordinary differential equations from the method of separation of variables are

$$\frac{d^2}{dt^2} T(t) + \gamma \left(\frac{d}{dt} T(t) \right) + c^2 \lambda \, T(t) = 0$$

$$\frac{d^2}{dx^2} X(x) + \alpha \, X(x) = 0$$

$$\frac{d^2}{dy^2} Y(y) + \beta \, Y(y) = 0$$

The coupling equation reads

$$\lambda = \alpha + \beta$$

The boundary conditions on the spatial equations are

$$X_x(0) = 0 \quad \text{and} \quad X_x(\pi) = 0$$

and

$$Y(0) = 0 \quad \text{and} \quad Y(\pi) = 0$$

Assignment of system parameters

> restart:with(plots):a:=Pi:b:=Pi:c:=1/2:unprotect(gamma):gamma:=0:

Allowed eigenvalues and orthonormal eigenfunctions are obtained from Examples 2.5.1 and 2.5.3.

> alpha[m]:=(m*Pi/a)^2;beta[n]:=(n*Pi/b)^2;lambda[m,n]:=alpha[m]+beta[n];lambda[0,n]
  :=beta[n];

$$\alpha_m := m^2$$
$$\beta_n := n^2$$
$$\lambda_{m,n} := m^2 + n^2$$
$$\lambda_{0,n} := n^2 \tag{7.19}$$

for $m = 0, 1, 2, 3, \ldots,$ and $n = 1, 2, 3, \ldots.$

Orthonormal eigenfunctions

For $m = 0,$

> X[0](x):=1/sqrt(a);

$$X_0(x) := \frac{1}{\sqrt{\pi}} \tag{7.20}$$

For $m = 1, 2, 3, \ldots,$

> X[m](x):=sqrt(2/a)*cos(m*Pi/a*x);X[r](x):=subs(m=r,X[m](x)):

$$X_m(x) := \frac{\sqrt{2}\cos(mx)}{\sqrt{\pi}} \tag{7.21}$$

For $n = 1, 2, 3, \ldots,$

> Y[n](y):=sqrt(2/b)*sin(n*Pi/b*y);Y[s](y):=subs(n=s,Y[n](y)):

$$Y_n(y) := \frac{\sqrt{2}\sin(ny)}{\sqrt{\pi}} \tag{7.22}$$

Statements of orthonormality with their respective weight functions

> w(x):=1:Int(X[m](x)*X[r](x)*w(x),x=0..a)=delta(m,r);

$$\int_0^\pi \frac{2\cos(mx)\cos(rx)}{\pi}dx = \delta(m,r) \tag{7.23}$$

> w(y):=1:Int(Y[n](y)*Y[s](y)*w(y),y=0..b)=delta(n,s);

$$\int_0^\pi \frac{2\sin(ny)\sin(sy)}{\pi}dy = \delta(n,s) \tag{7.24}$$

Time-dependent solution

> T[m,n](t):=exp(−gamma/2*t)*(A(m,n)*cos(omega[m,n]*t)+B(m,n)*sin(omega[m,n]*t));

$$T_{m,n}(t) := A(m,n)\cos(\omega_{m,n}t) + B(m,n)\sin(\omega_{m,n}t) \tag{7.25}$$

> T[0,n](t):=exp(−gamma/2*t)*(A(0,n)*cos(omega[0,n]*t)+B(0,n)*sin(omega[0,n]*t));

$$T_{0,n}(t) := A(0,n)\cos(\omega_{0,n}t) + B(0,n)\sin(\omega_{0,n}t) \tag{7.26}$$

Generalized series terms

> u[m,n](x,y,t):=T[m,n](t)*X[m](x)*Y[n](y);

$$u_{m,n}(x,y,t) := \frac{2\left(A(m,n)\cos(\omega_{m,n}t) + B(m,n)\sin(\omega_{m,n}t)\right)\cos(mx)\sin(ny)}{\pi} \tag{7.27}$$

> u[0,n](x,y,t):=T[0,n](t)*X[0](x)*Y[n](y);

$$u_{0,n}(x,y,t) := \frac{\left(A(0,n)\cos(\omega_{0,n}t) + B(0,n)\sin(\omega_{0,n}t)\right)\sqrt{2}\sin(ny)}{\pi} \tag{7.28}$$

Eigenfunction expansion

> u(x,y,t):=Sum(u[0,n](x,y,t),n=1..infinity)+Sum(Sum(u[m,n](x,y,t),m=1..infinity),
  n=1..infinity);

$$u(x, y, t) := \sum_{n=1}^{\infty} \frac{\left(A(0, n) \cos(\omega_{0,n} t) + B(0, n) \sin(\omega_{0,n} t)\right) \sqrt{2} \sin(ny)}{\pi}$$

$$+ \sum_{n=1}^{\infty} \left(\sum_{m=1}^{\infty} \frac{2 \left(A(m, n) \cos(\omega_{m,n} t) + B(m, n) \sin(\omega_{m,n} t)\right) \cos(mx) \sin(ny)}{\pi} \right)$$

$$(7.29)$$

> omega[m,n]:=sqrt(4*lambda[m,n]*c^2−gamma^2);

$$\omega_{m,n} := \sqrt{m^2 + n^2} \qquad (7.30)$$

> omega[0,n]:=subs(m=0,omega[m,n]);

$$\omega_{0,n} := \sqrt{n^2} \qquad (7.31)$$

We consider the special case where the initial condition functions $u(x, y, 0) = f(x, y)$ and $u_t(x, y, 0) = g(x, y)$ are given as

> f(x,y):=x*y*(Pi−y);

$$f(x, y) := xy(\pi - y) \qquad (7.32)$$

> g(x,y):=0;

$$g(x, y) := 0 \qquad (7.33)$$

From Section 7.5, the double Fourier coefficients $A(m, n)$ and $B(m, n)$ are determined from the integrals

For $m = 1, 2, 3, \ldots,$ and $n = 1, 2, 3, \ldots,$

> A(m,n):=eval(Int(Int(f(x,y)*X[m](x)*w(x)*Y[n](y)*w(y),x=0..a),y=0..b));A(m,n)
  :=expand(value(%)):

$$A(m, n) := \int_{0}^{\pi} \int_{0}^{\pi} \frac{2xy(\pi - y) \cos(mx) \sin(ny)}{\pi} \, dx \, dy \qquad (7.34)$$

> A(m,n):=factor(subs({sin(m*Pi)=0,cos(m*Pi)=(−1)^m,sin(n*Pi)=0,cos(n*Pi)=(−1)^n},
  A(m,n)));

$$A(m, n) := -\frac{4\left(-1 + (-1)^n\right)\left(-1 + (-1)^m\right)}{\pi m^2 n^3} \qquad (7.35)$$

> B(m,n):=eval(Int(Int((f(x,y)*gamma/2+g(x,y))/(omega[m,n])*X[m](x)*Y[n](y),x=0..a),
  y=0..b));B(m,n):=expand(value(%)):

$$B(m, n) := \int\limits_0^\pi \int\limits_0^\pi 0 \, dx \, dy \qquad (7.36)$$

> B(m,n):=simplify(subs({sin(m*Pi)=0,cos(m*Pi)=(−1)^m,sin(n*Pi)=0,cos(n*Pi)=(−1)^n},
  B(m,n)));

$$B(m, n) := 0 \qquad (7.37)$$

For $m = 0$ and $n = 1, 2, 3, \ldots,$

> A(0,n):=eval(Int(Int(f(x,y)*X[0](x)*w(x)*Y[n](y)*w(y),x=0..a),y=0..b));A(0,n):=
  expand(value(%)):

$$A(0, n) := \int\limits_0^\pi \int\limits_0^\pi \frac{xy(\pi - y)\sqrt{2}\sin(ny)}{\pi} dx \, dy \qquad (7.38)$$

> A(0,n):=simplify(subs({sin(m*Pi)=0,cos(m*Pi)=(−1)^m,sin(n*Pi)=0,cos(n*Pi)=(−1)^n},
  A(0,n)));

$$A(0, n) := -\frac{\pi\sqrt{2}\left(-1+(-1)^n\right)}{n^3} \qquad (7.39)$$

> B(0,n):=eval(Int(Int((f(x,y)*gamma/2+g(x,y))/(omega[0,n])*X[0](x)*Y[n](y),x=0..a),
  y=0..b));B(0,n):=expand(value(%)):

$$B(0, n) := \int\limits_0^\pi \int\limits_0^\pi 0 \, dx \, dy \qquad (7.40)$$

> B(0,n):=simplify(subs({sin(m*Pi)=0,cos(m*Pi)=(−1)^m,sin(n*Pi)=0,cos(n*Pi)=(−1)^n},
  B(0,n)));

$$B(0, n) := 0 \qquad (7.41)$$

> T[m,n](t):=eval(T[m,n](t));

$$T_{m,n}(t) := -\frac{4\left(-1+(-1)^n\right)\left(-1+(-1)^m\right)\cos\left(\sqrt{m^2+n^2}\,t\right)}{\pi m^2 n^3} \qquad (7.42)$$

> T[0,n](t):=eval(T[0,n](t));

$$T_{0,n}(t) := -\frac{\pi\sqrt{2}\left(-1+(-1)^n\right)\cos\left(\sqrt{n^2}\,t\right)}{n^3} \qquad (7.43)$$

Generalized series terms

> u[m,n](x,y,t):=eval(T[m,n](t)*X[m](x)*Y[n](y));

$$u_{m,n}(x,\,y,\,t) := -\frac{8\left(-1+(-1)^n\right)\left(-1+(-1)^m\right)\cos\left(\sqrt{m^2+n^2}\,t\right)\cos(mx)\sin(ny)}{\pi^2 m^2 n^3} \qquad (7.44)$$

> u[0,n](x,y,t):=eval(T[0,n](t)*X[0](x)*Y[n](y));

$$u_{0,n}(x,\,y,\,t) := -\frac{2\left(-1+(-1)^n\right)\cos\left(\sqrt{n^2}\,t\right)\sin(ny)}{n^3} \qquad (7.45)$$

Series solution

> u(x,y,t):=Sum(u[0,n](x,y,t),n=1..infinity)+Sum(Sum(u[m,n](x,y,t),m=1..infinity),
  n=1..infinity);

$$u(x,\,y,\,t) := \sum_{n=1}^{\infty}\left(-\frac{2\left(-1+(-1)^n\right)\cos\left(\sqrt{n^2}\,t\right)\sin(ny)}{n^3}\right) +$$

$$\sum_{n=1}^{\infty}\left(\sum_{m=1}^{\infty}\left(-\frac{8\left(-1+(-1)^n\right)\left(-1+(-1)^m\right)\cos\left(\sqrt{m^2+n^2}\,t\right)\cos(mx)\sin(ny)}{\pi^2 m^2 n^3}\right)\right) \qquad (7.46)$$

First few terms of sum

> u(x,y,t):=sum(u[0,n](x,y,t),n=1..1)+sum(sum(u[m,n](x,y,t),m=1..1),n=1..1):

ANIMATION

> animate3d(u(x,y,t),x=0..a,y=0..b,t=0..3,axes=framed,thickness=1);

The preceding animation command illustrates the spatial-time-dependent solution for $u(x, y, t)$. The animation sequence in Figures 7.3 and 7.4 shows snapshots of the animation at times $t = 0$ and $t = 2$.

ANIMATION SEQUENCE

> u(x,y,0):=subs(t=0,u(x,y,t)):plot3d(u(x,y,0),x=0..a,y=0..b,axes=framed,thickness=1);

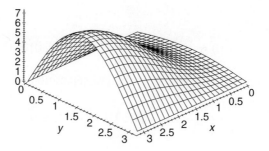

Figure 7.3

> u(x,y,2):=subs(t=2,u(x,y,t)):plot3d(u(x,y,2),x=0..a,y=0..b,axes=framed,thickness=1);

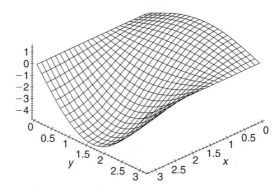

Figure 7.4

EXAMPLE 7.6.3: We seek the wave amplitude $u(x, y, t)$ for the transverse wave propagation on a thin rectangular membrane over the finite two-dimensional domain $D = \{(x, y) \mid 0 < x < 1, 0 < y < 1\}$. The membrane is in a medium with no damping and the wave speed is $c = 1/2$. The boundaries $x = 0$, $x = 1$, and $y = 0$ are held fixed, and the boundary $y = 1$ is constrained to move on an elastic hinge. The initial displacement function $f(x, y)$ and the initial speed function $g(x, y)$ are given as follows.

SOLUTION: The two-dimensional homogeneous wave equation is

$$\frac{\partial^2}{\partial t^2} u(x, y, t) = c^2 \left(\frac{\partial^2}{\partial x^2} u(x, y, t) + \frac{\partial^2}{\partial y^2} u(x, y, t) \right) - \gamma \left(\frac{\partial}{\partial t} u(x, y, t) \right)$$

The boundary conditions are type 1 at $x = 0$, type 1 at $x = 1$, type 1 at $y = 0$, and type 3 at $y = 1$:

$$u(0, y, t) = 0 \quad \text{and} \quad u(1, y, t) = 0$$

and

$$u(x, 0, t) = 0 \quad \text{and} \quad u(x, 1, t) + u_y(x, 1, t) = 0$$

The initial conditions are

$$u(x, y, 0) = \frac{xy(1-x)}{8} \quad \text{and} \quad u_t(x, y, 0) = 0$$

The ordinary differential equations from the method of separation of variables are

$$\frac{d^2}{dt^2} T(t) + \gamma \left(\frac{d}{dt} T(t) \right) + c^2 \lambda \, T(t) = 0$$

$$\frac{d^2}{dx^2} X(x) + \alpha \, X(x) = 0$$

$$\frac{d^2}{dy^2} Y(y) + \beta \, Y(y) = 0$$

The coupling equation reads

$$\lambda = \alpha + \beta$$

The boundary conditions on the spatial equations are

$$X(0) = 0 \quad \text{and} \quad X(1) = 0$$

and

$$Y(0) = 0 \quad \text{and} \quad Y(1) + Y_y(1) = 0$$

Assignment of system parameters

> restart:with(plots):a:=1:b:=1:c:=1/2:unprotect(gamma):gamma:=0:

Allowed eigenvalues and corresponding orthonormal eigenfunctions are obtained from Examples 2.5.1 and 2.5.4.

> alpha[m]:=(m*Pi/a)^2;lambda[m,n]:=alpha[m]+beta[n];

$$\alpha_m := m^2 \pi^2$$

$$\lambda_{m,n} := m^2 \pi^2 + \beta_n \tag{7.47}$$

for $m = 1, 2, 3, \ldots$, and $n = 1, 2, 3, \ldots$, where β_n are the roots of the eigenvalue equation

> tan(sqrt(beta[n])*b)=−sqrt(beta[n]);

$$\tan\left(\sqrt{\beta_n} \right) = -\sqrt{\beta_n} \tag{7.48}$$

Orthonormal eigenfunctions

> X[m](x):=sqrt(2/a)*sin(m*Pi/a*x);X[r](x):=subs(m=r,X[m](x)):

$$X_m(x) := \sqrt{2} \sin(m\pi x) \tag{7.49}$$

> Y[n](y):=sqrt(2)*(1/sqrt((cos(sqrt(beta[n])*b)^2+b)))*sin(sqrt(beta[n])*y);Y[s](y)
  :=subs(n=s,Y[n](y)):

$$Y_n(y) := \frac{\sqrt{2}\sin\left(\sqrt{\beta_n}\, y\right)}{\sqrt{\cos\left(\sqrt{\beta_n}\right)^2 + 1}} \tag{7.50}$$

Statements of orthonormality with their respective weight functions

> w(x):=1:Int(X[m](x)*X[r](x)*w(x),x=0..a)=delta(m,r);

$$\int\limits_0^1 2\sin(m\pi x)\sin(r\pi x)\mathrm{d}x = \delta(m, r) \tag{7.51}$$

> w(y):=1:Int(Y[n](y)*Y[s](y)*w(y),y=0..b)=delta(n,s);

$$\int\limits_0^1 \frac{2\sin\left(\sqrt{\beta_n}\, y\right)\sin\left(\sqrt{\beta_s}\, y\right)}{\sqrt{\cos\left(\sqrt{\beta_n}\right)^2 + 1}\sqrt{\cos\left(\sqrt{\beta_s}\right)^2 + 1}}\mathrm{d}y = \delta(n, s) \tag{7.52}$$

Time-dependent solution

> T[m,n](t):=exp(−gamma/2*t)*(A(m,n)*cos(omega[m,n]*t)+B(m,n)*sin(omega[m,n]*t));

$$T_{m,n}(t) := A(m, n)\cos\left(\omega_{m,n}t\right) + B(m, n)\sin\left(\omega_{m,n}t\right) \tag{7.53}$$

Generalized series terms

> u[m,n](x,y,t):=T[m,n](t)*X[m](x)*Y[n](y);

$$u_{m,n}(x, y, t) := \frac{2\left(A(m, n)\cos\left(\omega_{m,n}t\right) + B(m, n)\sin\left(\omega_{m,n}t\right)\right)\sin(m\pi x)\sin\left(\sqrt{\beta_n}\, y\right)}{\sqrt{\cos\left(\sqrt{\beta_n}\right)^2 + 1}} \tag{7.54}$$

Eigenfunction expansion

> u(x,y,t):=Sum(Sum(u[m,n](x,y,t),m=1..infinity),n=1..infinity);

$$u(x, y, t) := \sum_{n=1}^{\infty}\left(\sum_{m=1}^{\infty} \frac{2\left(A(m, n)\cos\left(\omega_{m,n}t\right) + B(m, n)\sin\left(\omega_{m,n}t\right)\right)\sin(m\pi x)\sin\left(\sqrt{\beta_n}\, y\right)}{\sqrt{\cos\left(\sqrt{\beta_n}\right)^2 + 1}}\right) \tag{7.55}$$

> omega[m,n]:=sqrt(4*lambda[m,n]*c^2−gamma^2);

$$\omega_{m,n} := \sqrt{m^2\pi^2 + \beta_n} \tag{7.56}$$

We consider the special case where the initial condition functions $u(x, y, 0) = f(x, y)$ and $u_t(x, y, 0) = g(x, y)$ are given as

> f(x,y):=x*y*(1−x)/8;

$$f(x, y) := \frac{1}{8}xy(1 - x) \tag{7.57}$$

> g(x,y):=0;

$$g(x, y) := 0 \tag{7.58}$$

From Section 7.5, the double Fourier coefficients $A(m, n)$ and $B(m, n)$ are determined from the integrals:

> A(m,n):=eval(Int(Int(f(x,y)*X[m](x)*w(x)*Y[n](y)*w(y),x=0..a),y=0..b));A(m,n)
  :=expand(value(%)):

$$A(m, n) := \int_0^1 \int_0^1 \frac{1}{4}\frac{xy(1 - x)\sin(m\pi x)\sin(\sqrt{\beta_n}\, y)}{\sqrt{\cos(\sqrt{\beta_n})^2 + 1}}\, dx\, dy \tag{7.59}$$

> A(m,n):=simplify(subs({sin(sqrt(beta[n])*b)=−sqrt(beta[n])*cos(sqrt(beta[n])*b),sin(m*Pi)
  =0,cos(m*Pi)=(−1)^m},A(m,n)));

$$A(m, n) := \frac{\cos(\sqrt{\beta_n})\left(-1 + (-1)^m\right)}{m^3\pi^3\sqrt{\beta_n}\sqrt{\cos(\sqrt{\beta_n})^2 + 1}} \tag{7.60}$$

> B(m,n):=eval(Int(Int((f(x,y)*gamma/2+g(x,y))/(omega[m,n])*X[m](x)*Y[n](y),x=0..a),
  y=0..b));B(m,n):=expand(value(%)):

$$B(m, n) := \int_0^1 \int_0^1 0\, dx\, dy \tag{7.61}$$

> B(m,n):=simplify(subs({sin(sqrt(beta[n])*b)=−sqrt(beta[n])*cos(sqrt(beta[n])*b),sin(m*Pi)
  =0,cos(m*Pi)=(−1)^m},B(m,n)));

$$B(m, n) := 0 \tag{7.62}$$

> T[m,n](t):=eval(T[m,n](t));

$$T_{m,n}(t) := \frac{\cos(\sqrt{\beta_n})\left(-1 + (-1)^m\right)\cos\left(\sqrt{m^2\pi^2 + \beta_n}\, t\right)}{m^3\pi^3\sqrt{\beta_n}\sqrt{\cos(\sqrt{\beta_n})^2 + 1}} \tag{7.63}$$

Generalized series terms

```
> u[m,n](x,y,t):=eval(T[m,n](t)*X[m](x)*Y[n](y));
```

$$u_{m,n}(x, y, t) := \frac{2\cos\left(\sqrt{\beta_n}\right)\left(-1+(-1)^m\right)\cos\left(\sqrt{m^2\pi^2+\beta_n}\,t\right)\sin(m\pi x)\sin\left(\sqrt{\beta_n}\,y\right)}{m^3\pi^3\sqrt{\beta_n}\left(\cos\left(\sqrt{\beta_n}\right)^2+1\right)} \tag{7.64}$$

Series solution

```
> u(x,y,t):=Sum(Sum(u[m,n](x,y,t),m=1..infinity),n=1..infinity);
```

$$u(x, y, t) := \sum_{n=1}^{\infty}\left(\sum_{m=1}^{\infty}\frac{2\cos\left(\sqrt{\beta_n}\right)\left(-1+(-1)^m\right)\cos\left(\sqrt{m^2\pi^2+\beta_n}\,t\right)\sin(m\pi x)\sin\left(\sqrt{\beta_n}\,y\right)}{m^3\pi^3\sqrt{\beta_n}\left(\cos\left(\sqrt{\beta_n}\right)^2+1\right)}\right) \tag{7.65}$$

Evaluation of the eigenvalues β_n from the roots of the eigenvalue equation yields

```
> tan(sqrt(beta[n])*b)+sqrt(beta[n])=0;
```

$$\tan\left(\sqrt{\beta_n}\right)+\sqrt{\beta_n}=0 \tag{7.66}$$

```
> plot({tan(v),-v},v=0..20,y=-20..0,thickness=10);
```

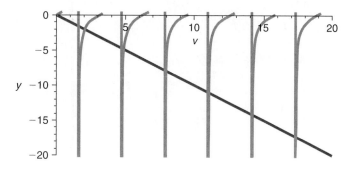

Figure 7.5

If we set $v = \sqrt{\beta}\,b$, then the eigenvalues β_n are found from the intersection points of the curves shown in Figure 7.5. Some of these eigenvalues using the Maple fsolve command are evaluated here:

```
> beta[1]:=(1/b^2)*(fsolve((tan(v)+v),v=1..3))^2;
```

$$\beta_1 := 4.115858365 \tag{7.67}$$

```
> beta[2]:=(1/b^2)*(fsolve((tan(v)+v),v=3..6))^2;
```

$$\beta_2 := 24.13934203 \tag{7.68}$$

> beta[3]:=(1/b^2)*(fsolve((tan(v)+v),v=6..9))^2;

$$\beta_3 := 63.65910654 \qquad\qquad (7.69)$$

First few terms of sum

> u(x,y,t):=eval(sum(sum(u[m,n](x,y,t),m=1..3),n=1..3)):

ANIMATION

> animate3d(u(x,y,t),x=0..a,y=0..b,t=0..3,axes=framed,thickness=1);

The preceding animation command illustrates the spatial-time-dependent solution for $u(x, y, t)$. The animation sequence in Figures 7.6 and 7.7 shows snapshots of the animation at the two times $t = 0$ and $t = 4$.

ANIMATION SEQUENCE

> u(x,y,0):=subs(t=0,u(x,y,t)):plot3d(u(x,y,0),x=0..a,y=0..b,axes=framed,thickness=1);

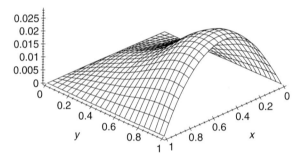

Figure 7.6

> u(x,y,4):=subs(t=4,u(x,y,t)):plot3d(u(x,y,4),x=0..a,y=0..b,axes=framed,thickness=1);

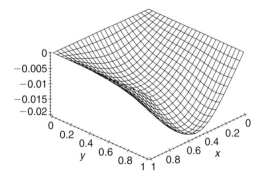

Figure 7.7

7.7 Wave Equation in the Cylindrical Coordinate System

If we replace the Laplacian operator from Section 7.2 in rectangular coordinates in two dimensions to its equivalent in cylindrical coordinates, then the nonhomogeneous partial differential equation for wave phenomena in the cylindrical coordinate system in two dimensions is given as

$$\frac{\partial^2}{\partial t^2}u(r,\theta,t) = c^2 \left(\frac{\frac{\partial}{\partial r}u(r,\theta,t) + r\left(\frac{\partial^2}{\partial r^2}u(r,\theta,t)\right)}{r} + \frac{\frac{\partial^2}{\partial \theta^2}u(r,\theta,t)}{r^2} \right) - \gamma\left(\frac{\partial}{\partial t}u(r,\theta,t)\right) + h(r,\theta,t)$$

where r is the coordinate radius of the system and θ is the polar angle. The γ term takes into account any viscous damping in the medium, and $h(r,\theta,t)$ takes into account any external applied forces acting on the system. There is no z dependence here because we will only examine wave phenomena along very thin membranes or plates that have no extension along the z-axis.

As for the rectangular coordinate system, we can write the preceding equation in terms of the linear operator for the wave equation in cylindrical coordinates as

$$L(u) = h(r,\theta,t)$$

where the wave operator in cylindrical coordinates reads

$$L(u) = \frac{\partial^2}{\partial t^2}u(r,\theta,t) - c^2\left(\frac{\frac{\partial}{\partial r}u(r,\theta,t) + r\left(\frac{\partial^2}{\partial r^2}u(r,\theta,t)\right)}{r} + \frac{\frac{\partial^2}{\partial \theta^2}u(r,\theta,t)}{r^2} \right) + \gamma\left(\frac{\partial}{\partial t}u(r,\theta,t)\right)$$

Generally, one seeks a solution to this equation over the finite two-dimensional domain $D = \{(r,\theta)\,|\,0 < r < a, 0 < \theta < b\}$ subject to the following homogeneous boundary conditions:

$$|u(0,\theta,t)| < \infty$$
$$\kappa_1 u(a,\theta,t) + \kappa_2 u_r(a,\theta,t) = 0$$
$$\kappa_3 u(r,0,t) + \kappa_4 u_\theta(r,0,t) = 0$$
$$\kappa_5 u(r,b,t) + \kappa_6 u_\theta(r,b,t) = 0$$

along with the two initial conditions

$$u(r,\theta,0) = f(r,\theta) \quad \text{and} \quad u_t(r,\theta,0) = g(r,\theta)$$

We now attempt to solve the homogeneous version [no external applied forces, $h(r,\theta,t) = 0$] of this partial differential equation using the method of separation of variables. We set

$$u(r,\theta,t) = R(r)\Theta(\theta)T(t)$$

Substituting this into the preceding homogeneous partial differential equation, we get

$$R(r)\Theta(\theta)\left(\frac{d^2}{dt^2}T(t)\right) + \gamma R(r)\Theta(\theta)\left(\frac{d}{dt}T(t)\right)$$

$$= c^2\left(\frac{\left(\frac{d}{dr}R(r)\right)\Theta(\theta)T(t) + r\left(\frac{d^2}{dr^2}R(r)\right)\Theta(\theta)T(t)}{r} + \frac{R(r)\left(\frac{d^2}{d\theta^2}\Theta(\theta)\right)T(t)}{r^2}\right)$$

Dividing both sides by the product solution yields

$$\frac{\frac{d^2}{dt^2}T(t) + \gamma\left(\frac{d}{dt}T(t)\right)}{c^2 T(t)} = \frac{\Theta(\theta)r\left(\frac{d}{dr}R(r)\right) + \Theta(\theta)r^2\left(\frac{d^2}{dr^2}R(r)\right) + R(r)\left(\frac{d^2}{d\theta^2}\Theta(\theta)\right)}{R(r)\Theta(\theta)r^2}$$

Because the left-hand side is an exclusive function of t and the right-hand side an exclusive function of r and θ, and r, θ, and t are independent, the only way we can ensure equality of the preceding for all values of r, θ, and t is to set each side equal to a constant. Doing so, we arrive at the following three ordinary differential equations in terms of the separation constants (for convenience here, we square the constants to ensure positivity):

$$\frac{d^2}{dt^2}T(t) + \gamma\left(\frac{d}{dt}T(t)\right) + c^2\lambda^2 T(t) = 0$$

$$\frac{d^2}{d\theta^2}\Theta(\theta) + \mu^2\Theta(\theta) = 0$$

$$r^2\left(\frac{d^2}{dr^2}R(r)\right) + r\left(\frac{d}{dr}R(r)\right) + \left(\lambda^2 r^2 - \mu^2\right)R(r) = 0$$

The preceding differential equation in t is an ordinary second-order linear equation for which we already have the solution from Chapter 1. The second differential equation in the variable θ is recognized as being an Euler-type differential equation for which we generated solutions in Section 1.4. The third differential equation is a Bessel-type differential equation whose solutions are Bessel functions of the first kind of order μ as shown in Section 1.11.

We easily recognize each of the two spatial differential equations in θ and r, with their corresponding boundary conditions, as being Sturm-Liouville-type differential equations.

We first consider the differential equation with respect to the variable θ subject to the homogeneous boundary conditions over the finite interval $I = \{\theta \mid 0 < \theta < b\}$. The Sturm-Liouville problem for the θ-dependent portion of the wave equation consists of the ordinary differential equation

$$\frac{d^2}{d\theta^2}\Theta(\theta) + \mu^2\Theta(\theta) = 0$$

along with its respective corresponding regular homogeneous boundary conditions

$$\kappa_3 \Theta(0) + \kappa_4 \Theta_\theta(0) = 0$$

and

$$\kappa_5 \Theta(b) + \kappa_6 \Theta_\theta(b) = 0$$

We recognize this problem in the variable θ to be a regular Sturm-Liouville eigenvalue problem whereby the allowed values of μ are called the system "eigenvalues" and the corresponding solutions are called the "eigenfunctions." Note that the ordinary differential equation given is of the Euler type, and the weight function is $w(\theta) = 1$. The indexed eigenvalues and corresponding eigenfunctions are given, respectively, as

$$\mu_m, \Theta_m(\theta)$$

for $m = 0, 1, 2, 3, \ldots$.

These eigenfunctions can be normalized and the corresponding statement of orthonormality with respect to the weight function $w(\theta) = 1$ reads

$$\int_0^b \Theta_m(\theta) \Theta_q(\theta) d\theta = \delta(m, q)$$

We now consider the r-dependent differential equation over the finite interval $I = \{r \,|\, 0 < r < a\}$. The Sturm-Liouville problem for the r-dependent portion of the wave equation consists of the ordinary differential equation

$$r^2 \left(\frac{d^2}{dr^2} R(r) \right) + r \left(\frac{d}{dr} R(r) \right) + \left(\lambda^2 r^2 - \mu^2 \right) R(r) = 0$$

along with its respective corresponding nonregular homogeneous boundary conditions

$$|R(0)| < \infty$$

and

$$\kappa_1 R(a) + \kappa_2 R_r(a) = 0$$

As discussed in Chapters 1 and 2 on the Bessel differential equation, we recognize this problem in the spatial variable r as a singular Sturm-Liouville eigenvalue problem. This comes about, since the differential equation fails to be "normal" at the origin. Nevertheless, the singular problem can be solved, and the allowed values of λ are called the system "eigenvalues" and the corresponding solutions are called the "eigenfunctions." We note that the allowed values of λ are also dependent on the values of μ. The ordinary differential equation given is of the Bessel type, and, from Section 2.6, the weight function is $w(r) = r$. The indexed eigenvalues and corresponding eigenfunctions are given, respectively, as

$$\lambda_{m,n}, R_{m,n}(r)$$

for $m = 0, 1, 2, 3, \ldots$, and $n = 0, 1, 2, 3, \ldots$.

Again, these eigenfunctions can be normalized and the corresponding statement of orthonormality with respect to the weight function $w(r) = r$ reads

$$\int_0^a R_{m,n}(r) R_{m,p}(r) r \, dr = \delta(n, p)$$

What was stated in Section 2.2 about regular Sturm-Liouville eigenvalue problems in one dimension can be extended to two dimensions here; that is, the product of the above eigenfunctions in r and θ form a "complete" set with respect to any piecewise smooth function $f(r, \theta)$ over the finite two-dimensional domain $D = \{(r, \theta) \,|\, 0 < r < a, 0 < \theta < b\}$.

Finally, we focus on the time-dependent equation, which reads

$$\frac{d^2}{dt^2} T(t) + \gamma \left(\frac{d}{dt} T(t) \right) + c^2 \lambda^2 T(t) = 0$$

For the indexed values of $\lambda_{m,n}$ given, the solution to this second-order differential equation from Section 1.4 is

$$T_{m,n}(t) = e^{-\frac{\gamma t}{2}} \left(A(m, n) \cos \left(\frac{\sqrt{4\lambda_{m,n}^2 c^2 - \gamma^2}\, t}{2} \right) + B(m, n) \sin \left(\frac{\sqrt{4\lambda_{m,n}^2 c^2 - \gamma^2}\, t}{2} \right) \right)$$

where the coefficients $A(m, n)$ and $B(m, n)$ are unknown arbitrary constants.

Throughout most of our example problems, we make the assumption of small damping in the system—that is,

$$\gamma^2 < 4\lambda_{m,n}^2 c^2$$

This assumption yields time-dependent solutions that are oscillatory. The argument of the sine and cosine expressions denotes the angular frequency of the (m, n) term in the system. The damped angular frequency of oscillation of the (m, n)th term or mode is given formally as

$$\omega_{m,n} = \frac{\sqrt{4\lambda_{m,n}^2 c^2 - \gamma^2}}{2}$$

Thus, by the method of separation of variables, we arrive at an infinite number of indexed solutions for the homogeneous wave partial differential equation, over a finite domain. These solutions are

$$(u_{m,n})(r, \theta, t) = R_{m,n}(r) \Theta_m(\theta) e^{-\frac{\gamma t}{2}} \left(A(m, n) \cos(\omega_{m,n} t) + B(m, n) \sin(\omega_{m,n} t) \right)$$

for $m = 0, 1, 2, 3, \ldots$, and $n = 0, 1, 2, 3, \ldots$.

Using arguments similar to those used for the regular Sturm-Liouville problem given earlier, we can write our general solution to the partial differential equation as the superposition of the products of the solutions to each of the preceding ordinary differential equations.

Thus, the generalized solution to the homogeneous wave partial differential equation can be written as the double infinite sum

$$u(r, \theta, t) = \sum_{n=0}^{\infty}\left(\sum_{m=0}^{\infty} R_{m,n}(r)\Theta_m(\theta)e^{-\frac{\gamma t}{2}}\left(A(m, n)\cos(\omega_{m,n}t) + B(m, n)\sin(\omega_{m,n}t)\right)\right) \quad (7.70)$$

The unknown Fourier coefficients $A(m, n)$ and $B(m, n)$ are to be evaluated from the initial conditions on the problem. We now demonstrate an illustrative example problem.

DEMONSTRATION: We seek the wave amplitude $u(r, \theta, t)$ for transverse waves on a thin circular membrane in a viscous medium over the two-dimensional domain $D = \{(r, \theta) \,|\, 0 < r < 1, 0 < \theta < \pi/2\}$. The damping coefficient is zero and the wave speed is $c = 1/2$. The plate is fixed at $\theta = 0$, at $\theta = \pi/2$, and at the outer periphery ($r = 1$). The initial displacement function $f(r, \theta)$ and the initial speed function $g(r, \theta)$ are given as follows.

SOLUTION: The two-dimensional homogeneous wave equation in cylindrical coordinates reads

$$4\left(\frac{\partial^2}{\partial t^2}u(r, \theta, t)\right) = \frac{\frac{\partial}{\partial r}u(r, \theta, t) + r\left(\frac{\partial^2}{\partial r^2}u(r, \theta, t)\right)}{r} + \frac{\frac{\partial^2}{\partial \theta^2}u(r, \theta, t)}{r^2}$$

The boundary conditions are type 1 at $r = 1$, type 1 at $\theta = 0$, type 1 at $\theta = \pi/2$, and we require a finite solution at the origin:

$$u(1, \theta, t) = 0 \quad \text{and} \quad |u(0, \theta, t)| < \infty$$

and

$$u(r, 0, t) = 0 \quad \text{and} \quad u\left(r, \frac{\pi}{2}, t\right) = 0$$

The initial conditions are

$$u(r, \theta, 0) = f(r, \theta) \quad \text{and} \quad u_t(r, \theta, 0) = g(r, \theta)$$

The ordinary differential equations from the method of separation of variables are

$$\frac{d^2}{dt^2}T(t) + \frac{\lambda^2 T(t)}{4} = 0$$

$$\frac{d^2}{d\theta^2}\Theta(\theta) + \mu^2\Theta(\theta) = 0$$

$$r^2\left(\frac{d^2}{dr^2}R(r)\right) + r\left(\frac{d}{dr}R(r)\right) + \left(\lambda^2 r^2 - \mu^2\right)R(r) = 0$$

The boundary conditions on the spatial variables are

$$\Theta(0) = 0 \quad \text{and} \quad \Theta\left(\frac{\pi}{2}\right) = 0$$

and

$$|R(0)| < \infty \quad \text{and} \quad R(1) = 0$$

We consider the θ-dependent equation first. Here, we have a Sturm-Liouville differential equation of the Euler type with weight function $w(\theta) = 1$. The allowed eigenvalues and corresponding orthonormal eigenfunctions, from Example 2.5.1, are

$$\mu_m = 2m$$

and

$$\Theta_m(\theta) = \frac{2\sin(2m\theta)}{\sqrt{\pi}}$$

for $m = 1, 2, 3, \ldots$.

We next consider the r-dependent equation. This is a Sturm-Liouville differential equation of the Bessel type with weight function $w(r) = r$. From Example 2.6.2, the eigenvalues $\lambda_{m,n}$ are the roots of the eigenvalue equation

$$J(2m, \lambda_{m,n}) = 0$$

for $m = 1, 2, 3, \ldots$, and $n = 1, 2, 3, \ldots$. Here, $J(2m, \lambda_{m,n})$ is the Bessel function of the first kind of order $2m$. The corresponding orthonormalized r-dependent eigenfunctions are

$$R_{m,n}(r) = \frac{J(2m, \lambda_{m,n}r)}{\sqrt{\displaystyle\int_0^1 J(2m, \lambda_{m,n}r)^2\, r\, dr}}$$

The statements of orthonormality, with their respective weight functions, are

$$\int_0^1 \frac{J(2m, \lambda_{m,n}r)\, J(2m, \lambda_{m,p}r)\, r}{\sqrt{\displaystyle\int_0^1 J(2m, \lambda_{m,n}r)^2\, r\, dr}\sqrt{\displaystyle\int_0^1 J(2m, \lambda_{m,p}r)^2\, r\, dr}}\, dr = \delta(n, p)$$

and

$$\int_0^{\frac{\pi}{2}} \frac{4\sin(2m\theta)\sin(2q\theta)}{\pi}\, d\theta = \delta(m, q)$$

for $m, q = 1, 2, 3, \ldots$, and $n, p = 1, 2, 3, \ldots$.

Finally, we focus on the time-dependent differential equation, which reads

$$\frac{d^2}{dt^2}T(t) + \frac{\lambda_{m,n}^2 T(t)}{4} = 0$$

The solution to this second-order differential equation from Section 1.4 is

$$T_{m,n} = A(m,n)\cos(\omega_{m,n}t) + B(m,n)\sin(\omega_{m,n}t)$$

where $\omega_{m,n}$ is the undamped angular frequency

$$\omega_{m,n} = \frac{\lambda_{m,n}}{2}$$

Forming the superposition of the product of the solutions to each of the preceding three ordinary differential equations, the general eigenfunction expansion solution to the partial differential equation reads

$$u(r,\theta,t) = \sum_{n=1}^{\infty}\left(\sum_{m=1}^{\infty} \frac{2\left(A(m,n)\cos(\omega_{m,n}t) + B(m,n)\sin(\omega_{m,n}t)\right) J\left(m,\lambda_{m,n}r\right)\sin(2m\theta)}{\sqrt{\pi}\sqrt{\int_0^1 J\left(2m,\lambda_{m,n}r\right)^2 r\,dr}}\right)$$

The Fourier coefficients $A(m,n)$ and $B(m,n)$ are to be determined from the initial condition functions $u(r,\theta,0) = f(r,\theta)$ and $u_t(r,\theta,0) = g(r,\theta)$.

7.8 Initial Conditions for the Wave Equation in Cylindrical Coordinates

We now consider the initial conditions on the problem. At time $t = 0$, let the initial displacement distribution be given as

$$u(r,\theta,0) = f(r,\theta)$$

and the initial speed distribution be

$$u_t(r,\theta,0) = g(r,\theta)$$

Substituting the first condition into the following general solution

$$u(r,\theta,t) = \sum_{n=0}^{\infty}\left(\sum_{m=0}^{\infty} R_{m,n}(r)\Theta_m(\theta)e^{-\frac{\gamma t}{2}}\left(A(m,n)\cos(\omega_{m,n}t) + B(m,n)\sin(\omega_{m,n}t)\right)\right)$$

at time $t = 0$ yields

$$f(r, \theta) = \sum_{n=0}^{\infty} \left(\sum_{m=0}^{\infty} R_{m,n}(r) \Theta_m(\theta) A(m, n) \right)$$

Substituting the second condition into the time derivative of the solution at time $t = 0$ yields

$$g(r, \theta) = \sum_{n=0}^{\infty} \left(\sum_{m=0}^{\infty} \left(B(m, n) \omega_{m,n} - \frac{A(m, n) \gamma}{2} \right) R_{m,n}(r) \Theta_m(\theta) \right)$$

Because the eigenfunctions of the Sturm-Liouville problem form a complete set with respect to piecewise smooth functions over the finite two-dimensional domain, the preceding sums are the generalized double Fourier series expansions of the functions $f(r, \theta)$ and $g(r, \theta)$ in terms of the allowed eigenfunctions. We can evaluate the unknown Fourier coefficients $A(m, n)$ and $B(m, n)$ by taking advantage of the given familiar orthonormality statements.

Similar to what was done in Section 2.4 for the generalized Fourier series expansion in one variable, we evaluate the double Fourier coefficients $A(m, n)$ and $B(m, n)$ as follows. We take the double inner products of both sides of the preceding equations, with respect to the r- and θ-dependent eigenfunctions, and we assume the validity of the interchange between the summation and integration operators. For the first equation we get

$$\int_0^b \int_0^a f(r, \theta) R_{q,p}(r) \Theta_q(\theta) r \, dr \, d\theta =$$

$$\sum_{n=0}^{\infty} \left(\sum_{m=0}^{\infty} A(m, n) \left(\int_0^a R_{q,p}(r) R_{m,n}(r) r \, dr \right) \left(\int_0^b \Theta_m(\theta) \Theta_q(\theta) \, d\theta \right) \right)$$

Taking advantage of the given orthonormality statements, the first condition yields

$$\int_0^b \int_0^a f(r, \theta) R_{q,p}(r) \Theta_q(\theta) r \, dr \, d\theta = \sum_{n=0}^{\infty} \left(\sum_{m=0}^{\infty} A(m, n) \delta(n, p) \delta(m, q) \right)$$

Due to the mathematical nature of the Kronecker delta function, only the $n = p$, $m = q$ term survives the sum and the double Fourier coefficient $A(m, n)$ reads

$$A(m, n) = \int_0^b \int_0^a f(r, \theta) R_{m,n}(r) \Theta_m(\theta) r \, dr \, d\theta$$

For the second condition, we get

$$\int_0^b \int_0^a g(r,\theta) R_{q,p}(r)\Theta_q(\theta) r \, dr \, d\theta =$$

$$\sum_{n=0}^{\infty}\left(\sum_{m=0}^{\infty}\left(B(m,n)\omega_{m,n} - \frac{A(m,n)\gamma}{2}\right)\left(\int_0^a R_{q,p}(r)\,R_{m,n}(r) r \, dr\right)\left(\int_0^b \Theta_m(\theta)\Theta_q(\theta) d\theta\right)\right)$$

Again, taking advantage of the given orthonormality statements, the second condition yields

$$\int_0^b \int_0^a g(r,\theta) R_{q,p}(r)\Theta_q(\theta) r \, dr \, d\theta = \sum_{n=0}^{\infty}\left(\sum_{m=0}^{\infty}\left(B(m,n)\omega_{m,n} - \frac{A(m,n)\gamma}{2}\right)\delta(n,p)\delta(m,q)\right)$$

$$(7.71)$$

Again, due to the mathematical nature of the Kronecker delta function, only the $n = p, m = q$ term survives the sum, and, combining the previous value for $A(m,n)$, the double Fourier coefficient $B(m,n)$ reads

$$B(m,n) = \int_0^b \int_0^a \frac{\left(\frac{f(r,\theta)\gamma}{2} + g(r,\theta)\right) r R_{m,n}(r)\Theta_m(\theta)}{\omega_{m,n}} \, dr \, d\theta$$

We now demonstrate the evaluation of the Fourier coefficients for the example problem from Section 7.7.

DEMONSTRATION: We consider the special case where the initial temperature distribution function $u(r,\theta,0) = f(r,\theta)$ is given as

$$f(r,\theta) = \left(r^2 - r^4\right)\sin(2\theta)$$

and the initial speed distribution function is $u_t(r,\theta,0) = g(r,\theta)$ where

$$g(r,\theta) = 0$$

SOLUTION: From the preceding, the Fourier coefficients $A(m,n)$ and $B(m,n)$ are evaluated by taking the double inner product of the initial condition functions with respect to the corresponding orthonormal eigenfunctions. For the coefficient $A(m,n)$, we have the integral

$$A(m,n) = \int_0^1 \int_0^{\frac{\pi}{2}} \left(r^2 - r^4\right)\sin(2\theta) R_{m,n}(r) r \Theta_m(\theta) d\theta \, dr$$

where, from the example, the orthonormalized eigenfunctions are

$$\Theta_m(\theta) = \frac{2\sin(2m\theta)}{\sqrt{\pi}}$$

and

$$R_{m,n}(r) = \frac{J(2m, \lambda_{m,n}r)}{\sqrt{\displaystyle\int_0^1 J(2m, \lambda_{m,n}r)^2 \, r \, dr}}$$

for $m = 1, 2, 3, \ldots$, and $n = 1, 2, 3, \ldots$. Insertion of these eigenfunctions into the preceding integral yields

$$A(m,n) = \int_0^1 \int_0^{\frac{\pi}{2}} \frac{(r^2 - r^4)\sin(2\theta) R_{m,n}(r) r \cdot 2\sin(2m\theta)}{\sqrt{\pi}} \, d\theta \, dr$$

Integrating first with respect to θ, and noting that $f(r, \theta)$ depends on the term $\sin(2m\theta)$ for the case $m = 1$, then from the statement of orthonormality for θ, we see that only the $m = 1$ term survives here. Thus, we are left with the integral

$$A(1,n) = \int_0^1 \int_0^{\frac{\pi}{2}} \frac{(r^2 - r^4)\sin(2\theta)^2 R_{1,n}(r) r \cdot 2}{\sqrt{\pi}} \, d\theta \, dr$$

where, for $m = 1$,

$$R_{1,n} = \frac{J(2, \lambda_{1,n}r)}{\sqrt{\displaystyle\int_0^1 J(2, \lambda_{1,n}r)^2 r \, dr}}$$

Evaluation of $R_{1,n}$ yields

$$R_{1,n} = \frac{J(2, \lambda_{1,n}r)\sqrt{2}}{J(1, \lambda_{1,n})}$$

Inserting this into $A(1, n)$ and integrating twice yields

$$A(1,n) = -\frac{6\sqrt{2\pi}}{\lambda_{1,n}^3}$$

for $n = 1, 2, 3, \ldots$. Since there is no damping in the system and the initial speed distribution $g(r, \theta) = 0$, the integral for $B(m, n)$ is zero. Thus, the final solution to the problem reads

$$u(r, \theta, t) = \sum_{n=1}^{\infty} \left(-\frac{24\cos\left(\frac{\lambda_{1,n}t}{2}\right) J(2, \lambda_{1,n}r)\sin(2\theta)}{\lambda_{1,n}^3 J(1, \lambda_{1,n})} \right)$$

The preceding solution is completed on evaluation of the eigenvalues $\lambda_{1,n}$ from the eigenvalue equation

$$J(2, \lambda_{1,n}) = 0$$

The details for the development of the solution and the graphics are given later in one of the Maple worksheets.

7.9 Example Wave Equation Problems in Cylindrical Coordinates

We now consider several examples of partial differential equations for wave phenomena under various homogeneous boundary conditions over finite two-dimensional domains in the cylindrical coordinate system. We note that the r-dependent ordinary differential equations in the cylindrical coordinate system are of the Bessel type and the θ-dependent equations are of the Euler type.

EXAMPLE 7.9.1: We seek the wave amplitude $u(r, \theta, t)$ for transverse waves on a thin circular membrane in a viscous medium over the two-dimensional domain $D = \{(r, \theta) | 0 < r < 1, 0 < \theta < \pi/2\}$. The damping coefficient γ is zero and the wave speed is $c = 1/2$. The plate is fixed at $\theta = 0, \theta = \pi/2$, and at the outer periphery $(r = 1)$. The initial displacement function $f(r, \theta)$ and the initial speed distribution $g(r, \theta))$ are given as follows.

SOLUTION: The two-dimensional homogeneous wave equation in cylindrical coordinates reads

$$\frac{\partial^2}{\partial t^2} u(r, \theta, t) + \gamma \left(\frac{\partial}{\partial t} u(r, \theta, t) \right) = c^2 \left(\frac{\frac{\partial}{\partial r} u(r, \theta, t) + r \left(\frac{\partial^2}{\partial r^2} u(r, \theta, t) \right)}{r} + \frac{\frac{\partial^2}{\partial \theta^2} u(r, \theta, t)}{r^2} \right)$$

The boundary conditions are type 1 at $r = 1$, type 1 at $\theta = 0$, type 1 at $\theta = \pi/2$, and we require a finite solution at the origin:

$$u(1, \theta, t) = 0 \quad \text{and} \quad |u(0, \theta, t)| < \infty$$

and

$$u(r, 0, t) = 0 \quad \text{and} \quad u\left(r, \frac{\pi}{2}, t \right) = 0$$

The initial conditions are

$$u(r, \theta, 0) = \left(r^2 - r^4 \right) \sin(2\theta) \quad \text{and} \quad u_t(r, \theta, 0) = 0$$

The ordinary differential equations from the method of separation of variables are

$$\frac{d^2}{dt^2}T(t) + \gamma\left(\frac{d}{dt}T(t)\right) + c^2\lambda^2 T(t) = 0$$

$$\frac{d^2}{d\theta^2}\Theta(\theta) + \mu^2\Theta(\theta) = 0$$

$$r^2\left(\frac{d^2}{dr^2}R(r)\right) + r\left(\frac{d}{dr}R(r)\right) + \left(\lambda^2 r^2 - \mu^2\right)R(r) = 0$$

The boundary conditions on the spatial variables are

$$\Theta(0) = 0 \quad \text{and} \quad \Theta\left(\frac{\pi}{2}\right) = 0$$

and

$$|R(0)| < \infty \quad \text{and} \quad R(1) = 0$$

Assignment of system parameters

> restart:with(plots):a:=1:b:=Pi/2:c:=1/2:unprotect(gamma):gamma:=0:

Allowed eigenvalues and corresponding orthonormal eigenfunctions are obtained from Examples 2.5.1 and 2.6.2.

> mu[m]:=m*Pi/b;

$$\mu_m := 2m \tag{7.72}$$

Eigenvalues $\lambda_{m,n}$ are the roots of the eigenvalue equation

> BesselJ(mu[m],lambda[m,n]*a)=0;

$$\text{BesselJ}(2m, \lambda_{m,n}) = 0 \tag{7.73}$$

for $m = 1, 2, 3, \ldots$, and $n = 1, 2, 3, \ldots$.

Orthonormal eigenfunctions

> Theta[m](theta):=sqrt(2/b)*sin(m*Pi/b*theta);Theta[q](theta):=subs(m=q,Theta[m](theta)):

$$\Theta_m(\theta) := \frac{2\sin(2m\theta)}{\sqrt{\pi}} \tag{7.74}$$

> R[m,n](r):=BesselJ(mu[m],lambda[m,n]*r)/sqrt(Int(BesselJ(mu[m],lambda[m,n]*r)^2*r,
  r=0..a));R[m,p](r):=subs(n=p,R[m,n](r)):

$$R_{m,n}(r) := \frac{\text{BesselJ}\left(2m, \lambda_{m,n}r\right)}{\sqrt{\displaystyle\int_0^1 \text{BesselJ}\left(2m, \lambda_{m,n}r\right)^2 r\, dr}} \tag{7.75}$$

Statements of orthonormality with their respective weight functions

> w(r):=r:Int(R[m,n](r)*R[m,p](r)*w(r),r=0..a)=delta(n,p);

$$\int_0^1 \frac{\text{BesselJ}(2m, \lambda_{m,n}r)\text{BesselJ}(2m, \lambda_{m,p}r)r}{\sqrt{\int_0^1 \text{BesselJ}(2m, \lambda_{m,n}r)^2 r\,dr}\sqrt{\int_0^1 \text{BesselJ}(2m, \lambda_{m,p}r)^2 r\,dr}}\,dr = \delta(n, p) \qquad (7.76)$$

> w(theta):=1:Int(Theta[m](theta)*Theta[q](theta)*w(theta),theta=0..b)=delta(m,q);

$$\int_0^{\frac{1}{2}\pi} \frac{4\sin(2m\theta)\sin(2q\theta)}{\pi}\,d\theta = \delta(m, q) \qquad (7.77)$$

Time-dependent solution

> T[m,n](t):=exp(−gamma/2*t)*(A(m,n)*cos(omega[m,n]*t)+B(m,n)*sin(omega[m,n]*t));

$$T_{m,n}(t) := A(m, n)\cos(\omega_{m,n}t) + B(m, n)\sin(\omega_{m,n}t) \qquad (7.78)$$

Generalized series terms

> u[m,n](r,theta,t):=T[m,n](t)*R[m,n](r)*Theta[m](theta);

$$u_{m,n}(r, \theta, t) := \frac{2\left(A(m, n)\cos(\omega_{m,n}t) + B(m, n)\sin(\omega_{m,n}t)\right)\text{BesselJ}(2m, \lambda_{m,n}r)\sin(2m\theta)}{\sqrt{\int_0^1 \text{BesselJ}(2m, \lambda_{m,n}r)^2 r\,dr}\sqrt{\pi}}$$

$$(7.79)$$

Eigenfunction expansion

> u(r,theta,t):=Sum(Sum(u[m,n](r,theta,t),m=1..infinity),n=1..infinity);

$$u(r, \theta, t) := \sum_{n=1}^{\infty}\left(\sum_{m=1}^{\infty} \frac{2\left(A(m, n)\cos(\omega_{m,n}t) + B(m, n)\sin(\omega_{m,n}t)\right)\text{BesselJ}(2m, \lambda_{m,n}r)\sin(2m\theta)}{\sqrt{\int_0^1 \text{BesselJ}(2m, \lambda_{m,n}r)^2 r\,dr}\sqrt{\pi}}\right)$$

$$(7.80)$$

> omega[m,n]:=sqrt(4*lambda[m,n]^2*c^2−gamma^2)/2;

$$\omega_{m,n} := \frac{1}{2}\sqrt{\lambda_{m,n}^2} \qquad (7.81)$$

We consider the special case where the initial condition functions $u(r, \theta, 0) = f(r, \theta)$ and $u_t(r, \theta, 0) = g(r, \theta)$ are given as

> f(r,theta):=(r^2−r^4)*sin(2*theta);

$$f(r, \theta) := \left(r^2 - r^4\right) \sin(2\theta) \tag{7.82}$$

> g(r,theta):=0;

$$g(r, \theta) := 0 \tag{7.83}$$

Because $f(r, \theta)$ depends on the term $\sin(2m\theta)$ for the case where $m = 1$, then from the given orthonormality statement for θ, only the $m = 1$ term survives here and we set

> Theta[1](theta):=subs(m=1,eval(Theta[m](theta)));

$$\Theta_1(\theta) := \frac{2 \sin(2\theta)}{\sqrt{\pi}} \tag{7.84}$$

> omega[1,n]:=subs(m=1,omega[m,n]);

$$\omega_{1,n} := \frac{1}{2}\sqrt{\lambda_{1,n}^2} \tag{7.85}$$

Eigenvalues $\lambda_{1,n}$ are the roots of the eigenvalue equation

> subs({m=1,r=a},BesselJ(mu[m],lambda[1,n]*r))=0;

$$\text{BesselJ}\left(2, \lambda_{1,n}\right) = 0 \tag{7.86}$$

For $m = 1$, we have the Bessel identity

> subs(m=1,BesselJ(0,lambda[m,n]*a)=2*BesselJ(1,lambda[m,n]*a)/(lambda[m,n]*a)
−BesselJ(2,lambda[m,n]*a));

$$\text{BesselJ}\left(0, \lambda_{1,n}\right) = \frac{2\text{BesselJ}\left(1, \lambda_{1,n}\right)}{\lambda_{1,n}} - \text{BesselJ}\left(2, \lambda_{1,n}\right) \tag{7.87}$$

Evaluation of the eigenfunctions and the Fourier coefficients with the preceding substitutions yields

> R[1,n](r):=eval(subs(m=1,BesselJ(mu[m],lambda[m,n]*r)/sqrt(expand(int(subs(m=1,BesselJ
(mu[m],lambda[m,n]*r))^2*r,r=0..a)))));

$$R_{1,n}(r) := \frac{2\text{BesselJ}\left(2, \lambda_{1,n}r\right)}{\sqrt{2\text{BesselJ}\left(0, \lambda_{1,n}\right)^2 + 2\text{BesselJ}\left(1, \lambda_{1,n}\right)^2 - \frac{8\text{BesselJ}(1,\lambda_{1,n})^2}{\lambda_{1,n}^2}}} \tag{7.88}$$

Substitution of the eigenvalue equation and the preceding Bessel identity simplifies this

> R[1,n](r):=radsimp(subs(BesselJ(0,lambda[1,n]*a)=2*BesselJ(1,lambda[1,n]*a)/
(lambda[1,n]*a),R[1,n](r)));

$$R_{1,n}(r) := \frac{\text{BesselJ}\left(2, \lambda_{1,n}r\right)\sqrt{2}}{\text{BesselJ}\left(1, \lambda_{1,n}\right)} \tag{7.89}$$

From Section 7.8, the double Fourier coefficients $A(m, n)$ and $B(m, n)$ are determined from the integrals

> A(1,n):=Int(Int(f(r,theta)*R[1,n](r)*Theta[1](theta)*w(r),r=0..a),theta=0..b);

$$A(1, n) := \int_{0}^{\frac{1}{2}\pi} \int_{0}^{1} \frac{2\left(r^2 - r^4\right)\sin(2\theta)^2 \text{BesselJ}\left(2, \lambda_{1,n}r\right)\sqrt{2}r}{\text{BesselJ}\left(1, \lambda_{1,n}\right)\sqrt{\pi}} \, dr \, d\theta \tag{7.90}$$

> A(1,n):=radsimp(subs(BesselJ(0,lambda[1,n]*a)=2*BesselJ(1,lambda[1,n]*a)/
(lambda[1,n]*a),value(%)));

$$A(1, n) := -\frac{6\sqrt{\pi}\sqrt{2}}{\lambda_{1,n}^3} \tag{7.91}$$

> B(1,n):=Int(Int((f(r,theta)*gamma/2+g(r,theta))/omega[1,n]*R[1,n](r)*Theta[1](theta)*w(r),
r=0..a),theta=0..b);

$$B(1, n) := \int_{0}^{\frac{1}{2}\pi} \int_{0}^{1} 0 \, dr \, d\theta \tag{7.92}$$

> B(1,n):=radsimp(subs(BesselJ(0,lambda[1,n]*a)=2*BesselJ(1,lambda[1,n]*a)/
(lambda[1,n]*a),value(%)));

$$B(1, n) := 0 \tag{7.93}$$

> T[1,n](t):=eval(subs(m=1,T[m,n](t)));

$$T_{1,n}(t) := -\frac{6\sqrt{\pi}\sqrt{2}\cos\left(\frac{1}{2}\sqrt{\lambda_{1,n}^2}t\right)}{\lambda_{1,n}^3} \tag{7.94}$$

Generalized series terms

> u[1,n](r,theta,t):=eval(T[1,n](t)*R[1,n](r)*Theta[1](theta));

$$u_{1,n}(r, \theta, t) := -\frac{24\cos\left(\frac{1}{2}\sqrt{\lambda_{1,n}^2}t\right)\text{BesselJ}\left(2, \lambda_{1,n}r\right)\sin(2\theta)}{\lambda_{1,n}^3 \text{BesselJ}\left(1, \lambda_{1,n}\right)} \tag{7.95}$$

Series solution

```
> u(r,theta,t):=Sum(u[1,n](r,theta,t),n=1..infinity);
```

$$u(r, \theta, t) := \sum_{n=1}^{\infty} \left(-\frac{24 \cos\left(\frac{1}{2}\sqrt{\lambda_{1,n}^2}\, t\right) \text{BesselJ}(2, \lambda_{1,n} r) \sin(2\theta)}{\lambda_{1,n}^3 \text{BesselJ}(1, \lambda_{1,n})} \right) \tag{7.96}$$

Evaluation of the eigenvalues from the roots of the eigenvalue equation yields

```
> subs(m=1,BesselJ(mu[m],lambda[m,n]*a)=0);
```

$$\text{BesselJ}(2, \lambda_{1,n}) = 0 \tag{7.97}$$

```
> plot(BesselJ(2,v),v=0..20,thickness=10);
```

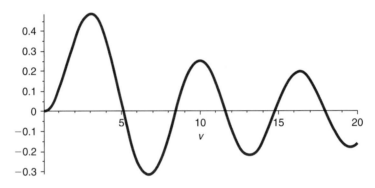

Figure 7.8

If we set $v = \lambda a$, then the eigenvalues $\lambda_{1,n}$ are found from the intersection points of the curves of $J(2, v)$ with the v-axis shown in Figure 7.8. Some of these eigenvalues are calculated here using the Maple fsolve command

```
> lambda[1,1]:=(1/a)*fsolve(BesselJ(2,v)=0,v=1..6);
```

$$\lambda_{1,1} := 5.135622302 \tag{7.98}$$

```
> lambda[1,2]:=(1/a)*fsolve(BesselJ(2,v)=0,v=6..10);
```

$$\lambda_{1,2} := 8.417244140 \tag{7.99}$$

```
> lambda[1,3]:=(1/a)*fsolve(BesselJ(2,v)=0,v=10..14);
```

$$\lambda_{1,3} := 13.32369194 \tag{7.100}$$

First few terms in sum

```
> u(r,theta,t):=sum(u[1,n](r,theta,t),n=1..1):
```

ANIMATION

> u(x,y,t):=subs({r=sqrt(x^2+y^2),theta=arctan(y/x)},u(r,theta,t))*Heaviside(1−sqrt(x^2+y^2)):

> animate3d(u(x,y,t),x=0..a,y=0..a,t=0..5,axes=framed,thickness=1);

The preceding animation command illustrates the spatial-time-dependent solution for $u(r, \theta, t)$. The animation sequence in Figures 7.9 and 7.10 shows snapshots of the animation at the times $t = 0$ and $t = 1$.

ANIMATION SEQUENCE

> u(r,theta,0):=subs(t=0,u(r,theta,t)):cylinderplot([r,theta,u(r,theta,0)],r=0..a,theta=0..b,
axes=framed,thickness=1);

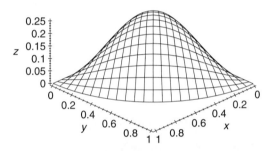

Figure 7.9

> u(r,theta,1):=subs(t=1,u(r,theta,t)):cylinderplot([r,theta,u(r,theta,1)],r=0..a,theta=0..b,
axes=framed,thickness=1);

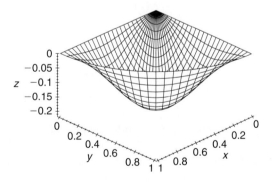

Figure 7.10

EXAMPLE 7.9.2: We seek the transverse wave amplitude $u(r, \theta, t)$ for waves on a thin circular membrane over the two-dimensional domain $D = \{(r, \theta) \mid 0 < r < 1, 0 < \theta < \pi\}$. There is no damping in the system and the wave speed is $c = 1/2$. The sides $\theta = 0$ and $\theta = \pi$ are free, and the edge $r = 1$ is held fixed. The initial displacement function $f(r, \theta)$ and the initial speed function $g(r, \theta)$ are given as follows.

SOLUTION: The two-dimensional homogeneous wave equation in cylindrical coordinates reads

$$\frac{\partial^2}{\partial t^2} u(r, \theta, t) + \gamma \left(\frac{\partial}{\partial t} u(r, \theta, t) \right) = c^2 \left(\frac{\frac{\partial}{\partial r} u(r, \theta, t) + r \left(\frac{\partial^2}{\partial r^2} u(r, \theta, t) \right)}{r} + \frac{\frac{\partial^2}{\partial \theta^2} u(r, \theta, t)}{r^2} \right)$$

The boundary conditions are type 1 at $r = 1$, type 2 at $\theta = 0$, type 2 at $\theta = \pi$, and we require a finite solution at the origin:

$$u(1, \theta, t) = 0 \quad \text{and} \quad |u(0, \theta, t)| < \infty$$

and

$$u_\theta(r, 0, t) = 0 \quad \text{and} \quad u_\theta(r, \pi, t) = 0$$

The initial conditions are

$$u(r, \theta, 0) = r^2 \cos(2\theta) \quad \text{and} \quad u_t(r, \theta, 0) = 0$$

The ordinary differential equations from the method of separation of variables are

$$\frac{d^2}{dt^2} T(t) + \gamma \left(\frac{d}{dt} T(t) \right) + c^2 \lambda^2 T(t) = 0$$

$$\frac{d^2}{d\theta^2} \Theta(\theta) + \mu^2 \Theta(\theta) = 0$$

$$r^2 \left(\frac{d^2}{dr^2} R(r) \right) + r \left(\frac{d}{dr} R(r) \right) + \left(\lambda^2 r^2 - \mu^2 \right) R(r) = 0$$

The boundary conditions on the spatial variables are

$$\Theta_\theta(0) = 0 \quad \text{and} \quad \Theta_\theta(\pi) = 0$$

and

$$|R(0)| < \infty \quad \text{and} \quad R(1) = 0$$

Assignment of system parameters

> restart:with(plots):a:=1:b:=Pi:c:=1/2:unprotect(gamma):gamma:=0:

Allowed eigenvalues and corresponding orthonormal eigenfunctions are obtained from Examples 2.5.3 and 2.6.2. For $m = 1, 2, 3, \ldots, n = 1, 2, 3, \ldots,$

> mu[m]:=m*Pi/b;

$$\mu_m := m \tag{7.101}$$

Eigenvalues $\lambda_{m,n}$ are the roots of the eigenvalue equation

> BesselJ(mu[m],lambda[m,n]*a)=0;

$$\text{BesselJ}\left(m, \lambda_{m,n}\right) = 0 \tag{7.102}$$

Orthonormal eigenfunctions

> R[m,n](r):=BesselJ(mu[m],lambda[m,n]*r)/sqrt(Int(BesselJ(mu[m],lambda[m,n]*r)^2*r,
  r=0..a));R[m,p](r):=subs(n=p,R[m,n](r)):

$$R_{m,n}(r) := \frac{\text{BesselJ}\left(m, \lambda_{m,n}r\right)}{\sqrt{\displaystyle\int_0^1 \text{BesselJ}\left(m, \lambda_{m,n}r\right)^2 r \, dr}} \tag{7.103}$$

> Theta[m](theta):=sqrt(2/b)*cos(m*Pi/b*theta);Theta[q](theta):=subs(m=q,Theta[m](theta)):

$$\Theta_m(\theta) := \frac{\sqrt{2}\cos(m\theta)}{\sqrt{\pi}} \tag{7.104}$$

For $m = 0, n = 1, 2, 3, \ldots, \lambda_{0,n}$ are the roots of the equation

> BesselJ(0,lambda[0,n]*a)=0;

$$\text{BesselJ}\left(0, \lambda_{0,n}\right) = 0 \tag{7.105}$$

Orthonormal eigenfunctions

> R[0,n](r):=BesselJ(0,lambda[0,n]*r)/sqrt(Int(BesselJ(0,lambda[0,n]*r)^2*r,r=0..a));
  R[0,p](r):=subs(n=p,R[0,n](r)):

$$R_{0,n}(r) := \frac{\text{BesselJ}\left(0, \lambda_{0,n}r\right)}{\sqrt{\displaystyle\int_0^1 \text{BesselJ}\left(0, \lambda_{0,n}r\right)^2 r \, dr}} \tag{7.106}$$

> Theta[0](theta):=sqrt(1/b);

$$\Theta_0(\theta) := \frac{1}{\sqrt{\pi}} \tag{7.107}$$

Statements of orthonormality with their respective weight functions

> w(r):=r:Int(R[m,n](r)*R[m,p](r)*w(r),r=0..a)=delta(n,p);

$$\int_0^1 \frac{\text{BesselJ}\left(m, \lambda_{m,n}r\right)\text{BesselJ}\left(m, \lambda_{m,p}r\right) r}{\sqrt{\displaystyle\int_0^1 \text{BesselJ}\left(m, \lambda_{m,n}r\right)^2 r \, dr}\sqrt{\displaystyle\int_0^1 \text{BesselJ}\left(m, \lambda_{m,p}r\right)^2 r \, dr}} \, dr = \delta(n, p) \tag{7.108}$$

> w(theta):=1:Int(Theta[m](theta)*Theta[q](theta)*w(theta),theta=0..b)=delta(m,q);

$$\int_0^\pi \frac{2\cos(m\theta)\cos(q\theta)}{\pi}\,d\theta = \delta(m,q) \tag{7.109}$$

Time-dependent solution

> T[m,n](t):=exp(−gamma/2*t)*(A(m,n)*cos(omega[m,n]*t)+B(m,n)*sin(omega[m,n]*t));
  u[m,n](r,theta,t):=eval(T[m,n](t)*R[m,n](r)*Theta[m](theta)):

$$T_{m,n}(t) := A(m,n)\cos(\omega_{m,n}t) + B(m,n)\sin(\omega_{m,n}t) \tag{7.110}$$

> T[0,n](t):=exp(−gamma/2*t)*(A(0,n)*cos(omega[0,n]*t)+B(0,n)*sin(omega[0,n]*t));
  u[0,n](r,theta,t):=T[0,n](t)*R[0,n](r)*Theta[0](theta):

$$T_{0,n}(t) := A(0,n)\cos(\omega_{0,n}t) + B(0,n)\sin(\omega_{0,n}t) \tag{7.111}$$

Eigenfunction expansion

> u(r,theta,t):=Sum(u[0,n](r,theta,t),n=1..infinity)+Sum(Sum(u[m,n](r,theta,t),m=1..infinity),
  n=1..infinity);

$$u(r,\theta,t) := \sum_{n=1}^\infty \frac{\left(A(0,n)\cos(\omega_{0,n}t) + B(0,n)\sin(\omega_{0,n}t)\right)\mathrm{BesselJ}(0,\lambda_{0,n}r)}{\sqrt{\int_0^1 \mathrm{BesselJ}(0,\lambda_{0,n}r)^2\,r\,dr}\,\sqrt{\pi}}$$

$$+\sum_{n=1}^\infty\left(\sum_{m=1}^\infty \frac{\left(A(m,n)\cos(\omega_{m,n}t) + B(m,n)\sin(\omega_{m,n}t)\right)\mathrm{BesselJ}(m,\lambda_{m,n}r)\,\sqrt{2}\cos(m\theta)}{\sqrt{\int_0^1 \mathrm{BesselJ}(m,\lambda_{m,n}r)^2\,r\,dr}\,\sqrt{\pi}}\right)$$

$$\tag{7.112}$$

> omega[m,n]:=sqrt(4*lambda[m,n]^2*c^2−gamma^2)/2;

$$\omega_{m,n} := \frac{1}{2}\sqrt{\lambda_{m,n}^2} \tag{7.113}$$

We consider the special case where the initial condition functions $u(r,\theta,0) = f(r,\theta)$ and $u_t(r,\theta,0) = g(r,\theta)$ are given as

> f(r,theta):=r^2*cos(2*theta);

$$f(r,\theta) := r^2\cos(2\theta) \tag{7.114}$$

> g(r,theta):=0;

$$g(r, \theta) := 0 \qquad (7.115)$$

Since $f(r, \theta)$ depends on the term $\cos(m\theta)$ for the case where $m = 2$, then from the given orthonormality statement for θ, only the $m = 2$ term survives here and we set

> Theta[2](theta):=subs(m=2,eval(Theta[m](theta)));

$$\Theta_2(\theta) := \frac{\sqrt{2}\cos(2\theta)}{\sqrt{\pi}} \qquad (7.116)$$

> omega[2,n]:=subs(m=2,omega[m,n]);

$$\omega_{2,n} := \frac{1}{2}\sqrt{\lambda_{2,n}^2} \qquad (7.117)$$

Eigenvalues $\lambda_{2,n}$ are the roots of the eigenvalue equation

> subs(m=2,BesselJ(mu[m],lambda[m,n]*a)=0);

$$\text{BesselJ}\left(2, \lambda_{2,n}\right) = 0 \qquad (7.118)$$

To simplify our answer, we take advantage of the Bessel identity

> BesselJ(0,lambda[2,n]*a)=2*BesselJ(1,lambda[2,n]*a)/(lambda[2,n]*a)−BesselJ
  (2,lambda[2,n]*a);

$$\text{BesselJ}\left(0, \lambda_{2,n}\right) = \frac{2\text{BesselJ}\left(1, \lambda_{2,n}\right)}{\lambda_{2,n}} - \text{BesselJ}\left(2, \lambda_{2,n}\right) \qquad (7.119)$$

Evaluation of the eigenfunctions and the Fourier coefficients with the preceding substitutions yields

> R[2,n](r):=eval(subs(m=2,BesselJ(mu[m],lambda[m,n]*r)/sqrt(expand(int(subs(m=2,
  BesselJ(mu[m],lambda[m,n]*r))^2*r,r=0..a)))));

$$R_{2,n}(r) := \frac{2\text{BesselJ}\left(2, \lambda_{2,n}r\right)}{\sqrt{2\text{BesselJ}\left(0, \lambda_{2,n}\right)^2 + 2\text{BesselJ}\left(1, \lambda_{2,n}\right)^2 - \frac{8\text{BesselJ}\left(1,\lambda_{2,n}\right)^2}{\lambda_{2,n}^2}}} \qquad (7.120)$$

Substitution of the eigenvalue equation and the preceding Bessel identity simplifies this

> R[2,n](r):=radsimp(subs(BesselJ(0,lambda[2,n]*a)=2*BesselJ(1,lambda[2,n]*a)/
  (lambda[2,n]*a),R[2,n](r)));

$$R_{2,n}(r) := \frac{\text{BesselJ}\left(2, \lambda_{2,n}r\right)\sqrt{2}}{\text{BesselJ}\left(1, \lambda_{2,n}\right)} \qquad (7.121)$$

From Section 7.8, the double Fourier coefficients $A(m, n)$ and $B(m, n)$ are determined from the integrals

> A(2,n):=Int(Int(f(r,theta)*R[2,n](r)*Theta[2](theta)*w(r),r=0..a),theta=0..b);

$$A(2, n) := \int_0^{\pi} \int_0^1 \frac{2r^3 \cos(2\theta)^2 \, \text{BesselJ}\left(2, \lambda_{2,n}r\right)}{\text{BesselJ}\left(1, \lambda_{2,n}\right) \sqrt{\pi}} \, dr \, d\theta \tag{7.122}$$

> A(2,n):=subs(BesselJ(0,lambda[2,n]*a)=2*BesselJ(1,lambda[2,n]*a)/(lambda[2,n]*a),
  value(%));

$$A(2, n) := -\frac{\sqrt{\pi}}{\lambda_{2,n}} \tag{7.123}$$

> B(2,n):=Int(Int((f(r,theta)*gamma/2+g(r,theta))/(omega[2,n])*R[2,n](r)*Theta[2](theta)*
  w(r),r=0..a),theta=0..b);

$$B(2, n) := \int_0^{\pi} \int_0^1 0 \, dr \, d\theta \tag{7.124}$$

> B(2,n):=value(%);

$$B(2, n) := 0 \tag{7.125}$$

> T[2,n](t):=eval(subs(m=2,T[m,n](t)));

$$T_{2,n}(t) := -\frac{\sqrt{\pi} \cos\left(\frac{1}{2}\sqrt{\lambda_{2,n}^2}\, t\right)}{\lambda_{2,n}} \tag{7.126}$$

Generalized series terms

> u[2,n](r,theta,t):=eval(T[2,n](t)*R[2,n](r)*Theta[2](theta));

$$u_{2,n}(r, \theta, t) := -\frac{2\cos\left(\frac{1}{2}\sqrt{\lambda_{2,n}^2}\, t\right) \text{BesselJ}\left(2, \lambda_{2,n}r\right)\cos(2\theta)}{\lambda_{2,n} \text{BesselJ}\left(1, \lambda_{2,n}\right)} \tag{7.127}$$

Series solution

> u(r,theta,t):=Sum(u[2,n](r,theta,t),n=1..infinity);

$$u(r, \theta, t) := \sum_{n=1}^{\infty} \left(-\frac{2\cos\left(\frac{1}{2}\sqrt{\lambda_{2,n}^2}\, t\right) \text{BesselJ}\left(2, \lambda_{2,n}r\right)\cos(2\theta)}{\lambda_{2,n} \text{BesselJ}\left(1, \lambda_{2,n}\right)} \right) \tag{7.128}$$

Evaluation of the eigenvalues $\lambda_{2,n}$ from the roots of the eigenvalue equation yields

> subs({r=a,m=2},BesselJ(mu[m],lambda[m,n]*r))=0;

$$\text{BesselJ}\left(2, \lambda_{2,n}\right) = 0 \tag{7.129}$$

> plot(BesselJ(2,v),v=0..20,thickness=10);

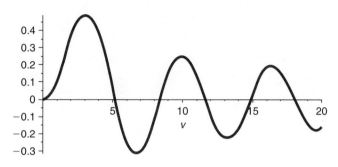

Figure 7.11

If we set $v = \lambda a$, then the eigenvalues $\lambda_{2,n}$ are the roots of the equation $J(2, v)$ shown in Figure 7.11. A few of these eigenvalues are evaluated here using the Maple fsolve command

> lambda[2,1]:=(1/a)*fsolve(BesselJ(2,v)=0,v=1..6);

$$\lambda_{2,1} := 5.135622302 \tag{7.130}$$

> lambda[2,2]:=(1/a)*fsolve(BesselJ(2,v)=0,v=6..10);

$$\lambda_{2,2} := 8.417244140 \tag{7.131}$$

> lambda[2,3]:=(1/a)*fsolve(BesselJ(2,v)=0,v=10..14);

$$\lambda_{2,3} := 11.61984117 \tag{7.132}$$

First few terms in sum

> u(r,theta,t):=sum(u[2,n](r,theta,t),n=1..1):

ANIMATION

> u(x,y,t):=subs({r=sqrt(x^2+y^2),theta=arccos(x/sqrt(x^2+y^2))},u(r,theta,t))*
    Heaviside(1-sqrt(x^2+y^2)):

> animate3d(u(x,y,t),x=-a..a,y=0..a,t=0..5,axes=framed,thickness=1);

The preceding animation command illustrates the spatial-time-dependent solution for $u(r, \theta, t)$. The animation sequence in Figures 7.12 and 7.13 shows snapshots of the animation at the times $t = 0$ and $t = 4$.

ANIMATION SEQUENCE

> u(r,theta,0):=subs(t=0,u(r,theta,t)):cylinderplot([r,theta,u(r,theta,0)],r=0..a,theta=0..b, axes=framed,thickness=1);

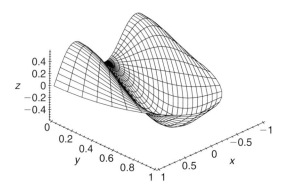

Figure 7.12

> u(r,theta,4):=subs(t=4,u(r,theta,t)):cylinderplot([r,theta,u(r,theta,4)],r=0..a,theta=0..b, axes=framed,thickness=1);

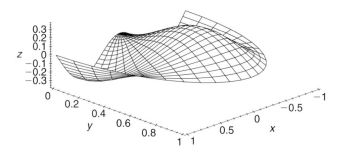

Figure 7.13

EXAMPLE 7.9.3: We seek the wave amplitude $u(r, \theta, t)$ for transverse waves on a thin circularly symmetric membrane in a viscous medium over the two-dimensional domain $D = \{(r, \theta) \mid 0 < r < 1, 0 < \theta < 2\pi\}$. The damping coefficient γ is very small and the wave speed is $c = 1$. The membrane is clamped at its circular edge. The initial displacement function $f(r)$ and the initial speed function $g(r)$ are given as follows. Because the membrane is circularly symmetric and the initial distributions are θ-independent, then the solution is also θ-independent.

SOLUTION: The two-dimensional homogeneous wave equation in cylindrical coordinates for a θ-independent system reads

$$\frac{\partial^2}{\partial t^2} u(r, t) + \gamma \left(\frac{\partial}{\partial t} u(r, t) \right) = \frac{c^2 \left(\frac{\partial}{\partial r} u(r, t) + r \left(\frac{\partial^2}{\partial r^2} u(r, t) \right) \right)}{r}$$

The boundary conditions are type 1 at $r = 1$, and we require a finite solution at the origin:

$$u(1, t) = 0 \quad \text{and} \quad |u(0, t)| < \infty$$

The initial conditions are

$$u(r, 0) = 1 - r^2 \quad \text{and} \quad u_t(r, 0) = 0$$

The ordinary differential equations from the method of separation of variables are

$$\frac{d^2}{dt^2} T(t) + \gamma \left(\frac{d}{dt} T(t) \right) + c^2 \lambda^2 T(t) = 0$$

$$r^2 \left(\frac{d^2}{dr^2} R(r) \right) + r \left(\frac{d}{dr} R(r) \right) + \lambda^2 r^2 R(r) = 0$$

The boundary conditions on the r equation are

$$|R(0)| < \infty \quad \text{and} \quad R(1) = 0$$

Assignment of system parameters

> restart:with(plots):a:=1:c:=1:unprotect(gamma):gamma:=1/2:

The r-dependent equation is a Bessel differential equation of the first kind of order $m = 0$. The allowed eigenvalues and corresponding orthonormal eigenfunctions are obtained from Example 2.6.2. The eigenvalues $\lambda_{0,n}$ are the roots of the eigenvalue equation

> subs(r=a,BesselJ(0,lambda[0,n]*r))=0;

$$\text{BesselJ}\left(0, \lambda_{0,n}\right) = 0 \tag{7.133}$$

for $n = 1, 2, 3, \ldots$.

Orthonormal eigenfunctions

> R[0,n](r):=BesselJ(0,lambda[0,n]*r)/sqrt(Int(BesselJ(0,lambda[0,n]*r)^2*r,r=0..a));
  R[0,p](r):=subs(n=p,R[0,n](r)):

$$R_{0,n}(r) := \frac{\text{BesselJ}\left(0, \lambda_{0,n} r\right)}{\sqrt{\displaystyle\int_0^1 \text{BesselJ}\left(0, \lambda_{0,n} r\right)^2 r\, dr}} \tag{7.134}$$

Statement of orthonormality with weight function $w(r) = r$

> w(r):=r:Int(R[0,n](r)*R[0,p](r)*w(r),r=0..a)=delta(n,p);

$$\int_0^1 \frac{\text{BesselJ}\left(0, \lambda_{0,n} r\right) \text{BesselJ}\left(0, \lambda_{0,p} r\right) r}{\sqrt{\displaystyle\int_0^1 \text{BesselJ}\left(0, \lambda_{0,n} r\right)^2 r\, dr}\sqrt{\displaystyle\int_0^1 \text{BesselJ}\left(0, \lambda_{0,p} r\right)^2 r\, dr}}\, dr = \delta(n, p) \tag{7.135}$$

Time-dependent solution

> T[0,n](t):=exp(−gamma/2*t)*(A(0,n)*cos(omega[0,n]*t)+B(0,n)*sin(omega[0,n]*t));

$$T_{0,n}(t) := e^{-\frac{1}{4}t}\left(A(0,n)\cos\left(\omega_{0,n}t\right) + B(0,n)\sin\left(\omega_{0,n}t\right)\right) \tag{7.136}$$

Generalized series terms

> u[0,n](r,t):=T[0,n](t)*R[0,n](r);

$$u_{0,n}(r,t) := \frac{e^{-\frac{1}{4}t}\left(A(0,n)\cos\left(\omega_{0,n}t\right) + B(0,n)\sin\left(\omega_{0,n}t\right)\right)\text{BesselJ}\left(0,\lambda_{0,n}r\right)}{\sqrt{\int_0^1 \text{BesselJ}\left(0,\lambda_{0,n}r\right)^2 r\,dr}} \tag{7.137}$$

Eigenfunction expansion

> u(r,t):=Sum(u[0,n](r,t),n=1..infinity);

$$u(r,t) := \sum_{n=1}^{\infty} \frac{e^{-\frac{1}{4}t}\left(A(0,n)\cos\left(\omega_{0,n}t\right) + B(0,n)\sin\left(\omega_{0,n}t\right)\right)\text{BesselJ}\left(0,\lambda_{0,n}r\right)}{\sqrt{\int_0^1 \text{BesselJ}\left(0,\lambda_{0,n}r\right)^2 r\,dr}} \tag{7.138}$$

> omega[0,n]:=sqrt(4*lambda[0,n]^2*c^2−gamma^2)/2;

$$\omega_{0,n} := \frac{1}{4}\sqrt{16\lambda_{0,n}^2 - 1} \tag{7.139}$$

We consider the special case where the initial condition functions $u(r,\theta,0) = f(r,\theta)$ and $u_t(r,\theta,0) = g(r,\theta)$ are given as

> f(r):=1−r^2;

$$f(r) := 1 - r^2 \tag{7.140}$$

> g(r):=0;

$$g(r) := 0 \tag{7.141}$$

Eigenvalues $\lambda_{0,n}$ are the roots of the eigenvalue equation

> subs(r=a,BesselJ(0,lambda[0,n]*r))=0;

$$\text{BesselJ}\left(0,\lambda_{0,n}\right) = 0 \tag{7.142}$$

Evaluation of the eigenfunctions yields

> R[0,n](r):=eval(BesselJ(0,lambda[0,n]*r)/sqrt(int(BesselJ(0,lambda[0,n]*r)^2*r,r=0..a)));

$$R_{0,n}(r) := \frac{\text{BesselJ}\left(0,\lambda_{0,n}r\right)\sqrt{2}}{\sqrt{\dfrac{\sqrt{\pi}\,\lambda_{0,n}\text{BesselJ}(0,\lambda_{0,n})^2 + \sqrt{\pi}\,\lambda_{0,n}\text{BesselJ}(1,\lambda_{0,n})^2}{\sqrt{\pi}\,\lambda_{0,n}}}} \tag{7.143}$$

Substitution of the eigenvalue equation in the preceding simplifies this

```
> R[0,n](r):=radsimp(subs(BesselJ(0,lambda[0,n]*a)=0,R[0,n](r)));
```

$$R_{0,n}(r) := \frac{\text{BesselJ}\left(0, \lambda_{0,n} r\right) \sqrt{2}}{\text{BesselJ}\left(1, \lambda_{0,n}\right)} \tag{7.144}$$

From Section 7.8, the double Fourier coefficients $A(m, n)$ and $B(m, n)$ are determined from the integrals:

```
> A(0,n):=Int(f(r)*R[0,n](r)*w(r),r=0..a);
```

$$A(0, n) := \int_0^1 \frac{\left(1 - r^2\right) \text{BesselJ}\left(0, \lambda_{0,n} r\right) \sqrt{2} r}{\text{BesselJ}\left(1, \lambda_{0,n}\right)} \, dr \tag{7.145}$$

```
> A(0,n):=simplify(subs(BesselJ(0,lambda[0,n]*a)=0,value(%)));
```

$$A(0, n) := \frac{4\sqrt{2}}{\lambda_{0,n}^3} \tag{7.146}$$

```
> B(0,n):=Int((f(r)*gamma/2+g(r))/omega[0,n]*R[0,n](r)*w(r),r=0..a);
```

$$B(0, n) := \int_0^1 \frac{4\left(\frac{1}{4} - \frac{1}{4}r^2\right) \text{BesselJ}\left(0, \lambda_{0,n} r\right) \sqrt{2} r}{\sqrt{16\lambda_{0,n}^2 - 1} \, \text{BesselJ}\left(1, \lambda_{0,n}\right)} \, dr \tag{7.147}$$

```
> B(0,n):=simplify(subs(BesselJ(0,lambda[0,n]*a)=0,value(%)));
```

$$B(0, n) := \frac{4\sqrt{2}}{\lambda_{0,n}^3 \sqrt{16\lambda_{0,n}^2 - 1}} \tag{7.148}$$

```
> T[0,n](t):=(eval(T[0,n](t)));
```

$$T_{0,n}(t) := e^{-\frac{1}{4}t} \left(\frac{4\sqrt{2} \cos\left(\frac{1}{4}\sqrt{16\lambda_{0,n}^2 - 1}\, t\right)}{\lambda_{0,n}^3} + \frac{4\sqrt{2} \sin\left(\frac{1}{4}\sqrt{16\lambda_{0,n}^2 - 1}\, t\right)}{\lambda_{0,n}^3 \sqrt{16\lambda_{0,n}^2 - 1}} \right) \tag{7.149}$$

Generalized series terms

> u[0,n](r,t):=eval(T[0,n](t)*R[0,n](r));

$$u_{0,n}(r,t) := \frac{1}{\mathrm{BesselJ}(1,\lambda_{0},n)}\left(e^{-\frac{1}{4}t}\left(\frac{4\sqrt{2}\cos\left(\frac{1}{4}\sqrt{16\lambda_{0,n}^{2}-1}\,t\right)}{\lambda_{0,n}^{3}}\right.\right.$$

$$\left.\left.+\frac{4\sqrt{2}\sin\left(\frac{1}{4}\sqrt{16\lambda_{0,n}^{2}-1}\,t\right)}{\lambda_{0,n}^{3}\sqrt{16\lambda_{0,n}^{2}-1}}\right)\mathrm{BesselJ}(0,\lambda_{0,n}r)\sqrt{2}\right) \tag{7.150}$$

Series solution

> u(r,t):=Sum(u[0,n](r,t),n=1..infinity);

$$u(r,t) := \sum_{n=1}^{\infty}\frac{1}{\mathrm{BesselJ}(1,\lambda_{0,n})}\left(e^{-\frac{1}{4}t}\left(\frac{4\sqrt{2}\cos\left(\frac{1}{4}\sqrt{16\lambda_{0,n}^{2}-1}\,t\right)}{\lambda_{0,n}^{3}}\right.\right.$$

$$\left.\left.+\frac{4\sqrt{2}\sin\left(\frac{1}{4}\sqrt{16\lambda_{0,n}^{2}-1}\,t\right)}{\lambda_{0,n}^{3}\sqrt{16\lambda_{0,n}^{2}-1}}\right)\mathrm{BesselJ}(0,\lambda_{0,n}r)\sqrt{2}\right) \tag{7.151}$$

Evaluation of eigenvalues $\lambda_{0,n}$ from the roots of the eigenvalue equation yields

> subs(r=a,BesselJ(0,lambda[0,n]*r)=0);

$$\mathrm{BesselJ}(0,\lambda_{0,n}) = 0 \tag{7.152}$$

> plot(BesselJ(0,v),v=0..20,thickness=10);

If we set $v = \lambda\,a$, then the eigenvalues $\lambda_{0,n}$ are the roots of the equation $J(0, v)$ shown in Figure 7.14. Evaluation of some of these eigenvalues using the Maple fsolve command yields

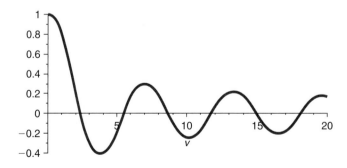

Figure 7.14

> lambda[0,1]:=(1/a)*fsolve(BesselJ(0,v)=0,v=1..4);

$$\lambda_{0,1} := 2.404825558 \tag{7.153}$$

> lambda[0,2]:=(1/a)*fsolve(BesselJ(0,v),v=4..7);

$$\lambda_{0,2} := 5.520078110 \tag{7.154}$$

> lambda[0,3]:=(1/a)*fsolve(BesselJ(0,v)=0,v=7..11);

$$\lambda_{0,3} := 8.653727913 \tag{7.155}$$

First few terms in sum

> u(r,t):=sum(u[0,n](r,t),n=1..3):

ANIMATION

> u(x,y,t):=subs(r=sqrt(x^2+y^2),u(r,t))*Heaviside(1−sqrt(x^2+y^2)):

> animate3d(u(x,y,t),x=−a..a,y=−a..a,t=0..3,axes=framed,thickness=1);

The preceding animation command illustrates the spatial-time-dependent solution for $u(r, t)$. The animation sequence in Figures 7.15 and 7.16 shows snapshots of the animation at the times $t = 0$ and $t = 2$.

ANIMATION SEQUENCE

> u(r,0):=subs(t=0,u(r,t)):cylinderplot([r,theta,u(r,0)],r=0..a,theta=0..2*Pi,axes=framed,
   thickness=1);

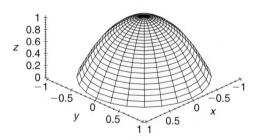

Figure 7.15

> u(r,2):=subs(t=2,u(r,t)):cylinderplot([r,theta,u(r,2)],r=0..a,theta=0..2*Pi,axes=framed,
   thickness=1);

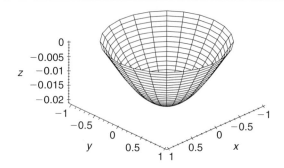

Figure 7.16

Chapter Summary

Nonhomogeneous wave equation in two dimensions in rectangular coordinates

$$\frac{\partial^2}{\partial t^2} u(x, y, t) = c^2 \left(\frac{\partial^2}{\partial x^2} u(x, y, t) + \frac{\partial^2}{\partial y^2} u(x, y, t) \right) - \gamma \left(\frac{\partial}{\partial t} u(x, y, t) \right) + h(x, y, t)$$

Linear wave operator in two dimensions in rectangular coordinates

$$L(u) = \frac{\partial^2}{\partial t^2} u(x, y, t) + \gamma \left(\frac{\partial}{\partial t} u(x, y, t) \right) - c^2 \left(\frac{\partial^2}{\partial x^2} u(x, y, t) + \frac{\partial^2}{\partial y^2} u(x, y, t) \right)$$

The method of separation of variables solution

$$u(x, y, t) = X(x) Y(y) T(t)$$

Eigenfunction expansion solution for rectangular coordinates

$$u(r, y, t) = \sum_{n=0}^{\infty} \left(\sum_{m=0}^{\infty} X_m(x) Y_n(y) e^{-\frac{\gamma t}{2}} \left(A(m, n) \cos(\omega_{m,n} t) + B(m, n) \sin(\omega_{m,n} t) \right) \right)$$

Fourier coefficients from initial conditions for rectangular coordinates

$$A(m, n) = \int_0^b \int_0^a f(x, y) X_m(x) Y_n(y) dx\, dy$$

and

$$B(m, n) = \int_0^b \int_0^a \frac{\left(\frac{f(x,y)\gamma}{2} + g(x, y) \right) X_m(x) Y_n(y)}{\omega_{m,n}} dx\, dy$$

Nonhomogeneous wave equation in two dimensions in cylindrical coordinates

$$\frac{\partial^2}{\partial t^2}u(r, \theta, t) = c^2\left(\frac{\frac{\partial}{\partial r}u(r, \theta, t) + r\left(\frac{\partial^2}{\partial r^2}u(r, \theta, t)\right)}{r} + \frac{\frac{\partial^2}{\partial \theta^2}u(r, \theta, t)}{r^2}\right) - \gamma\left(\frac{\partial}{\partial t}u(r, \theta, t)\right) + h(r, \theta, t)$$

Linear wave operator in two dimensions in cylindrical coordinates

$$L(u) = \frac{\partial^2}{\partial t^2}u(r, \theta, t) + \gamma\left(\frac{\partial}{\partial t}u(r, \theta, t)\right) - c^2\left(\frac{\frac{\partial}{\partial r}u(r, \theta, t) + r\left(\frac{\partial^2}{\partial r^2}u(r, \theta, t)\right)}{r} + \frac{\frac{\partial^2}{\partial \theta^2}u(r, \theta, t)}{r^2}\right)$$

The method of separation of variables solution

$$u(r, \theta, t) = R(r)\,\Theta(\theta)\,T(t)$$

Eigenfunction expansion solution in cylindrical coordinates

$$u(r, \theta, t) = \sum_{n=0}^{\infty}\left(\sum_{m=0}^{\infty} R_{m,n}(r)\Theta_m(\theta)e^{-\frac{\gamma t}{2}}\left(A(m, n)\cos(\omega_{m,n}t) + B(m, n)\sin(\omega_{m,n}t)\right)\right)$$

Initial condition Fourier coefficients for cylindrical coordinates

$$A(m, n) = \int_0^b\int_0^a f(r, \theta)\,R_{m,n}(r)\Theta_m(\theta)\,r\,dr\,d\theta$$

and

$$B(m, n) = \int_0^b\int_0^a \frac{\left(\frac{f(r,\theta)\gamma}{2} + g(r, \theta)\right)R_{m,n}(r)\,\Theta_m(\theta)r}{\omega_{m,n}}\,dr\,d\theta$$

We have examined homogeneous partial differential equations describing wave phenomena in two spatial dimensions for both the rectangular and the cylindrical coordinate systems.

Exercises

We now consider problems for the wave equation with homogeneous boundary conditions in two dimensions in both the rectangular and cylindrical coordinate systems. We use the method of separation of variables and eigenfunction expansions to evaluate the solutions.

Two Dimensions, Rectangular Coordinates

We consider the transverse wave distribution in thin membranes over the two-dimensional domain $D = \{(x, y)\,|\,0 < x < a, 0 < y < b\}$ with a small amount of damping. The two-dimensional homogeneous wave partial differential equation with damping in rectangular coordinates reads

$$\frac{\partial^2}{\partial t^2} u(x, y, t) = c^2 \left(\frac{\partial^2}{\partial x^2} u(x, y, t) + \frac{\partial^2}{\partial y^2} u(x, y, t) \right) - \gamma \left(\frac{\partial}{\partial t} u(x, y, t) \right)$$

For the wave equation, we have to consider the two initial conditions

$$u(x, y, 0) = f(x, y) \quad \text{and} \quad u_t(x, y, 0) = g(x, y)$$

In Exercises 7.1 to 7.15, we seek solutions to the preceding problem for various given boundary and initial conditions. Evaluate the eigenvalues and corresponding two-dimensional orthonormalized eigenfunctions, and write the general solution. Generate the animated three-dimensional surface solution $u(x, y, t)$, and plot the animated sequence for $0 < t < 5$.

7.1. Boundaries $x = 0$, $x = 1$, $y = 0$, $y = 1$ are held secure, $c = 1/2$, $\gamma = 1/10$:

$$u(0, y, t) = 0 \quad \text{and} \quad u(1, y, t) = 0$$

$$u(x, 0, t) = 0 \quad \text{and} \quad u(x, 1, t) = 0$$

Initial conditions:

$$f(x, y) = x(1 - x)y(1 - y) \quad \text{and} \quad g(x, y) = 0$$

7.2. Boundaries $x = 0$, $y = 0$ are held secure, boundaries $x = 1$, $y = 1$ are unsecure, $c = 1/2$, $\gamma = 1/10$:

$$u(0, y, t) = 0 \quad \text{and} \quad u_x(1, y, t) = 0$$

$$u(x, 0, t) = 0 \quad \text{and} \quad u_y(x, 1, t) = 0$$

Initial conditions:

$$f(x, y) = 0 \quad \text{and} \quad g(x, y) = x\left(1 - \frac{x}{2}\right)y\left(1 - \frac{y}{2}\right)$$

7.3. Boundaries $x = 0$, $y = 1$ are held secure, boundaries $x = 1$, $y = 0$ are unsecure, $c = 1/2$, $\gamma = 1/10$:

$$u(0, y, t) = 0 \quad \text{and} \quad u_x(1, y, t) = 0$$

$$u_y(x, 0, t) = 0 \quad \text{and} \quad u(x, 1, t) = 0$$

Initial conditions:

$$f(x, y) = x\left(1 - \frac{x}{2}\right)(1 - y^2) \quad \text{and} \quad g(x, y) = 0$$

7.4. Boundaries $x = 0$, $x = 1$ are held secure, boundaries $y = 0$, $y = 1$ are unsecure, $c = 1/2$, $\gamma = 1/10$:

$$u(0, y, t) = 0 \quad \text{and} \quad u(1, y, t) = 0$$
$$u_y(x, 0, t) = 0 \quad \text{and} \quad u_y(x, 1, t) = 0$$

Initial conditions:

$$f(x, y) = 0 \quad \text{and} \quad g(x, y) = x(1 - x)y^2 \left(1 - \frac{2y}{3}\right)$$

7.5. Boundaries $x = 0$, $x = 1$, $y = 1$ are held secure, boundary $y = 0$ is unsecure, $c = 1/2$, $\gamma = 1/10$:

$$u(0, y, t) = 0 \quad \text{and} \quad u(1, y, t) = 0$$
$$u_y(x, 0, t) = 0 \quad \text{and} \quad u(x, 1, t) = 0$$

Initial conditions:

$$f(x, y) = x(1 - x)(1 - y^2) \quad \text{and} \quad g(x, y) = 0$$

7.6. Boundaries $x = 0$, $x = 1$, $y = 0$ are unsecure, boundary $y = 1$ is held secure, $c = 1/2$, $\gamma = 1/10$:

$$u_x(0, y, t) = 0 \quad \text{and} \quad u_x(1, y, t) = 0$$
$$u_y(x, 0, t) = 0 \quad \text{and} \quad u(x, 1, t) = 0$$

Initial conditions:

$$f(x, y) = 0 \quad \text{and} \quad g(x, y) = x^2 \left(1 - \frac{2x}{3}\right)(1 - y^2)$$

7.7. Boundaries $x = 0$, $x = 1$, $y = 1$ are held secure, boundary $y = 0$ is attached to an elastic hinge, $c = 1/2$, $\gamma = 1/10$:

$$u(0, y, t) = 0 \quad \text{and} \quad u(1, y, t) = 0$$
$$u_y(x, 0, t) - u(x, 0, t) = 0 \quad \text{and} \quad u(x, 1, t) = 0$$

Initial conditions:

$$f(x, y) = x(1 - x)\left(-\frac{2y^2}{3} + \frac{y}{3} + \frac{1}{3}\right) \quad \text{and} \quad g(x, y) = 0$$

7.8.   Boundaries $x = 0$, $y = 0$, $y = 1$ are held secure, boundary $x = 1$ is attached to an elastic hinge, $c = 1/2$, $\gamma = 1/10$:

$$u(0, y, t) = 0 \quad \text{and} \quad u_x(1, y, t) + u(1, y, 0) = 0$$
$$u(x, 0, t) = 0 \quad \text{and} \quad u(x, 1, t) = 0$$

Initial conditions:

$$f(x, y) = 0 \quad \text{and} \quad g(x, y) = x\left(1 - \frac{2x}{3}\right)y(1 - y)$$

7.9.   Boundaries $x = 1$, $y = 0$, $y = 1$ are held secure, boundary $x = 0$ is attached to an elastic hinge, $c = 1/2$, $\gamma = 1/10$:

$$u_x(0, y, t) - u(0, y, t) = 0 \quad \text{and} \quad u(1, y, t) = 0$$
$$u(x, 0, t) = 0 \quad \text{and} \quad u(x, 1, t) = 0$$

Initial conditions:

$$f(x, y) = \left(-\frac{2x^2}{3} + \frac{x}{3} + \frac{1}{3}\right)y(1 - y) \quad \text{and} \quad g(x, y) = 0$$

7.10.   Boundaries $x = 1$, $y = 1$ are held secure, boundary $y = 0$ is unsecure, and boundary $x = 0$ is attached to an elastic hinge, $c = 1/2$, $\gamma = 1/10$:

$$u_x(0, y, t) - u(0, y, t) = 0 \quad \text{and} \quad u(1, y, t) = 0$$
$$u_y(x, 0, t) = 0 \quad \text{and} \quad u(x, 1, t) = 0$$

Initial conditions:

$$f(x, y) = 0 \quad \text{and} \quad g(x, y) = \left(-\frac{2x^2}{3} + \frac{x}{3} + \frac{1}{3}\right)(1 - y^2)$$

7.11.   Boundaries $x = 0$, $x = 1$ are held secure, boundary $y = 0$ is unsecure, and boundary $y = 1$ is attached to an elastic hinge, $c = 1/2$, $\gamma = 1/10$:

$$u(0, y, t) = 0 \quad \text{and} \quad u(1, y, t) = 0$$
$$u_y(x, 0, t) = 0 \quad \text{and} \quad u_y(x, 1, t) + u(x, 1, t) = 0$$

Initial conditions:

$$f(x, y) = x(1 - x)\left(1 - \frac{y^2}{3}\right) \quad \text{and} \quad g(x, y) = 0$$

7.12. Boundaries $x = 0$, $y = 0$ are held secure, boundary $y = 1$ is unsecure, and boundary $x = 1$ is attached to an elastic hinge, $c = 1/2$, $\gamma = 1/10$:

$$u(0, y, t) = 0 \quad \text{and} \quad u_x(1, y, t) + u(1, y, t) = 0$$
$$u(x, 0, t) = 0 \quad \text{and} \quad u_y(x, 1, t) = 0$$

Initial conditions:

$$f(x, y) = 0 \quad \text{and} \quad g(x, y) = x\left(1 - \frac{2x}{3}\right) y \left(1 - \frac{y}{2}\right)$$

7.13. Boundaries $x = 1$, $y = 1$ are held secure, boundary $x = 0$ is unsecure, and boundary $y = 0$ is attached to an elastic hinge, $c = 1/2$, $\gamma = 1/10$:

$$u_x(0, y, t) = 0 \quad \text{and} \quad u(1, y, t) = 0$$
$$u_y(x, 0, t) - u(x, 0, t) = 0 \quad \text{and} \quad u(x, 1, t) = 0$$

Initial conditions:

$$f(x, y) = \left(1 - x^2\right)\left(-\frac{2y^2}{3} + \frac{y}{3} + \frac{1}{3}\right) \quad \text{and} \quad g(x, y) = 0$$

7.14. Boundaries $y = 0$, $y = 1$ are both unsecure, boundary $x = 0$ is secure, and boundary $x = 1$ is attached to an elastic hinge, $c = 1/2$, $\gamma = 1/10$:

$$u(0, y, t) = 0 \quad \text{and} \quad u_x(1, y, t) + u(1, y, t) = 0$$
$$u_y(x, 0, t) = 0 \quad \text{and} \quad u_y(x, 1, t) = 0$$

Initial conditions:

$$f(x, y) = x\left(1 - \frac{2x}{3}\right) y^2 \left(1 - \frac{2y}{3}\right) \quad \text{and} \quad g(x, y) = 1$$

7.15. Boundaries $x = 0$, $y = 1$ are both unsecure, boundaries $x = 1$, $y = 0$ are both attached to an elastic hinge, $c = 1/2$, $\gamma = 1/10$:

$$u_x(0, y, t) = 0 \quad \text{and} \quad u_x(1, y, t) + u(1, y, t) = 0$$
$$u_y(x, 0, t) - u(x, 0, t) = 0 \quad \text{and} \quad u_y(x, 1, t) = 0$$

Initial conditions:

$$f(x, y) = \left(1 - \frac{x^2}{3}\right)\left(1 - \frac{(y-1)^2}{3}\right) \quad \text{and} \quad g(x, y) = 0$$

Angular Frequency and Multiplicity

From Section 7.4, for small damping in the system, the damped angular frequency of vibration for the (m, n) mode was given as

$$\omega_{m,n} = \frac{\sqrt{4\lambda_{m,n}c^2 - \gamma^2}}{2}$$

where c is the propagation speed of the wave, γ is the damping coefficient, and $\lambda_{m,n}$ is given as

$$\lambda_{m,n} = \alpha_m + \beta_n$$

Here, α_m is the eigenvalue corresponding to the x-dependent eigenvalue problem, and β_n is the eigenvalue corresponding to the y-dependent eigenvalue problem. For a two-dimensional membrane, the propagation speed of the wave is given as

$$c = \sqrt{\frac{T}{\rho}}$$

where T is the tension per unit length in the membrane and ρ is the area density (mass/area) of the membrane.

For the special case of no damping in the system, the undamped angular frequency is given by

$$\omega_{m,n} = c\sqrt{\alpha_m + \beta_n}$$

From Example 7.6.1 in Section 7.6, for generalized coordinates $x = a$, $y = b$ of the rectangular membrane, the values for α_m, β_n, and $\omega_{m,n}$ are given, respectively, as

$$\alpha_m = \frac{m^2\pi^2}{a^2}$$

$$\beta_n = \frac{n^2\pi^2}{b^2}$$

and

$$\omega_{m,n} = c\sqrt{\frac{m^2\pi^2}{a^2} + \frac{n^2\pi^2}{b^2}}$$

for $m = 1, 2, 3, \ldots$, and $n = 1, 2, 3, \ldots$.

The corresponding allowed eigenfunctions are

$$X_m(x)Y_n(y) = \sin\left(\frac{m\pi x}{a}\right)\sin\left(\frac{n\pi y}{b}\right)$$

It is obvious from the preceding equations that two different values of m and n could give rise to two different eigenfunctions corresponding to the same angular vibration frequency.

For example, from Exercise 7.1, $a = 1, b = 1$, we see that the $\omega(1, 2)$ mode angular frequency is identical to that of the $\omega(2, 1)$; yet the eigenfunctions are different. Thus, we get a "multiplicity" of the same eigenvalues corresponding to different eigenfunctions.

In Exercises 7.16 through 7.20, you are asked to evaluate the generalized expression for the undamped ($\gamma = 0$) angular frequency $\omega_{m,n}$ and to evaluate the "fundamental" angular frequency corresponding to the "smallest" allowed values of m and n; generally, the "fundamental" angular frequency is the $\omega(1, 1)$ mode.

7.16. Reconsider Exercise 7.1 for the case of no damping ($\gamma = 0$). (a) Evaluate α_m, β_n, and $\omega_{m,n}$ for the given constants. (b) Evaluate the "fundamental" frequency. (c) Find an example of multiplicity in the system by seeking two different sets of values (m, n) that give rise to the same angular frequency.

7.17. Reconsider Exercise 7.2 for the case of no damping ($\gamma = 0$). (a) Evaluate α_m, β_n, and $\omega_{m,n}$ for the given constants. (b) Evaluate the "fundamental" frequency. (c) Find an example of multiplicity in the system by seeking two different sets of values (m, n) that give rise to the same angular frequency.

7.18. Reconsider Exercise 7.4 for the case of no damping ($\gamma = 0$). (a) Evaluate α_m, β_n, and $\omega_{m,n}$ for the given constants. (b) Evaluate the "fundamental" frequency. (c) Find an example of multiplicity in the system by seeking two different sets of values (m, n) that give rise to the same angular frequency.

7.19. Reconsider Exercise 7.5 for the case of no damping ($\gamma = 0$). (a) Evaluate α_m, β_n, and $\omega_{m,n}$ for the given constants. (b) Evaluate the "fundamental" frequency. (c) Find an example of multiplicity in the system by seeking two different sets of values (m, n) that give rise to the same angular frequency.

7.20. Reconsider Exercise 7.6 for the case of no damping ($\gamma = 0$). (a) Evaluate α_m, β_n, and $\omega_{m,n}$ for the given constants. (b) Evaluate the "fundamental" frequency. (c) Find an example of multiplicity in the system by seeking two different sets of values (m, n) that give rise to the same angular frequency.

Two Dimensions, Cylindrical Coordinates

We now consider the propagation of transverse waves in circular membranes over the two-dimensional domain $D = \{(r, \theta) \mid 0 < r < a, 0 < \theta < b\}$ with a small amount of damping. The homogeneous partial differential equation reads as

$$\frac{\partial^2}{\partial t^2} u(r, \theta, t) = c^2 \left(\frac{\frac{\partial}{\partial r} u(r, \theta, t) + r \left(\frac{\partial^2}{\partial r^2} u(r, \theta, t) \right)}{r} + \frac{\frac{\partial^2}{\partial \theta^2} u(r, \theta, t)}{r^2} \right) - \gamma \left(\frac{\partial}{\partial t} u(r, \theta, t) \right)$$

with the two initial conditions

$$u(r, \theta, 0) = f(r, \theta) \quad \text{and} \quad u_t(r, \theta, 0) = g(r, \theta)$$

In Exercises 7.21 to 7.32, we seek solutions to the preceding problem for various given boundary and initial conditions. Evaluate the eigenvalues and corresponding two-dimensional orthonormalized eigenfunctions, and write out the general solution. Generate the animated three-dimensional surface solution $u(r, \theta, t)$, and plot the animated sequence for $0 < t < 5$.

7.21.   The center of the membrane has a finite displacement, and the periphery at $r = 1$ is held secure. The boundaries $\theta = 0$ and $\theta = \pi$ are held secure, $c = 1/2, \gamma = 1/10$:

$$|u(0, \theta, t)| < \infty \quad \text{and} \quad u(1, \theta, t) = 0$$

$$u(r, 0, t) = 0 \quad \text{and} \quad u(r, \pi, t) = 0$$

Initial conditions:

$$f(r, \theta) = \left(r - r^3\right) \sin(\theta) \quad \text{and} \quad g(r, \theta) = 0$$

7.22.   The center of the membrane has a finite displacement, and the periphery at $r = 1$ is held secure. The boundaries $\theta = 0$ and $\theta = \pi$ are unsecure, $c = 1/2, \gamma = 1/10$:

$$|u(0, \theta, t)| < \infty \quad \text{and} \quad u(1, \theta, t) = 0$$

$$u_\theta(r, 0, t) = 0 \quad \text{and} \quad u_\theta(r, \pi, t) = 0$$

Initial conditions:

$$f(r, \theta) = 0 \quad \text{and} \quad g(r, \theta) = \left(r^2 - r^4\right) \cos(2\theta)$$

7.23.   The center of the membrane has a finite displacement, and the periphery at $r = 1$ is held secure. The boundaries $\theta = 0$ and $\theta = \pi$ are held secure, $c = 1/2, \gamma = 1/10$:

$$|u(0, \theta, t)| < \infty \quad \text{and} \quad u(1, \theta, t) = 0$$

$$u(r, 0, t) = 0 \quad \text{and} \quad u(r, \pi, t) = 0$$

Initial conditions:

$$f(r, \theta) = \left(r^3 - r^5\right) \sin(3\theta) \quad \text{and} \quad g(r, \theta) = 0$$

7.24.   The center of the membrane has a finite displacement, and the periphery at $r = 1$ is held secure. The boundaries $\theta = 0$ and $\theta = \pi/2$ are held secure, $c = 1/2, \gamma = 1/10$:

$$|u(0, \theta, t)| < \infty \quad \text{and} \quad u(1, \theta, t) = 0$$

$$u(r, 0, t) = 0 \quad \text{and} \quad u\left(r, \frac{\pi}{2}, t\right) = 0$$

Initial conditions:

$$f(r, \theta) = 0 \quad \text{and} \quad g(r, \theta) = \left(r^2 - r^4\right) \sin(2\theta)$$

7.25.   The center of the membrane has a finite displacement, and the periphery at $r = 1$ is held secure. The boundaries $\theta = 0$ and $\theta = \pi/2$ are unsecure, $c = 1/2, \gamma = 1/10$:

$$|u(0, \theta, t)| < \infty \quad \text{and} \quad u(1, \theta, t) = 0$$

$$u_\theta(r, 0, t) = 0 \quad \text{and} \quad u_\theta\left(r, \frac{\pi}{2}, t\right) = 0$$

Initial conditions:

$$f(r, \theta) = \left(r^4 - r^6\right) \cos(4\theta) \quad \text{and} \quad g(r, \theta) = 0$$

7.26.   The center of the membrane has a finite displacement, and the periphery at $r = 1$ is held secure. The boundary conditions are such that we have periodic boundary conditions on $\theta, c = 1/2, \gamma = 1/10$:

$$|u(0, \theta, t)| < \infty \quad \text{and} \quad u(1, \theta, t) = 0$$

$$u(r, -\pi, t) = u(r, \pi, t) \quad \text{and} \quad u_\theta(r, -\pi, t) = u_\theta(r, \pi, t)$$

Initial conditions:

$$f(r, \theta) = \left(r - r^3\right) \cos(\theta) \quad \text{and} \quad g(r, \theta) = 0$$

7.27.   The center of the membrane has a finite displacement, and the periphery at $r = 1$ is unsecure. The boundaries $\theta = 0$ and $\theta = \pi$ are held secure, $c = 1/2, \gamma = 1/10$:

$$|u(0, \theta, t)| < \infty \quad \text{and} \quad u_r(1, \theta, t) = 0$$

$$u(r, 0, t) = 0 \quad \text{and} \quad u(r, \pi, t) = 0$$

Initial conditions:

$$f(r, \theta) = \left(3r - r^3\right) \sin(\theta) \quad \text{and} \quad g(r, \theta) = 0$$

7.28.   The center of the membrane has a finite displacement, and the periphery at $r = 1$ is unsecure. The boundaries $\theta = 0$ and $\theta = \pi$ are unsecure, $c = 1/2, \gamma = 1/10$:

$$|u(0, \theta, t)| < \infty \quad \text{and} \quad u_r(1, \theta, t) = 0$$

$$u_\theta(r, 0, t) = 0 \quad \text{and} \quad u_\theta(r, \pi, t) = 0$$

Initial conditions:

$$f(r, \theta) = \left(2r^2 - r^4\right) \cos(2\theta) \quad \text{and} \quad g(r, \theta) = 0$$

7.29. The center of the membrane has a finite displacement, and the periphery at $r = 1$ is unsecure. The boundaries $\theta = 0$ and $\theta = \pi$ are held secure, $c = 1/2, \gamma = 1/10$:

$$|u(0, \theta, t)| < \infty \quad \text{and} \quad u_r(1, \theta, t) = 0$$

$$u(r, 0, t) = 0 \quad \text{and} \quad u(r, \pi, t) = 0$$

Initial conditions:

$$f(r, \theta) = 0 \quad \text{and} \quad g(r, \theta) = \left(\frac{5r^3}{3} - r^5 \right) \sin(3\theta)$$

7.30. The center of the membrane has a finite displacement, and the periphery at $r = 1$ is unsecure. The boundaries $\theta = 0$ and $\theta = \pi/2$ are held secure, $c = 1/2, \gamma = 1/10$:

$$|u(0, \theta, t)| < \infty \quad \text{and} \quad u_r(1, \theta, t) = 0$$

$$u(r, 0, t) = 0 \quad \text{and} \quad u\left(r, \frac{\pi}{2}, t \right) = 0$$

Initial conditions:

$$f(r, \theta) = \left(\frac{4r^6}{3} - r^8 \right) \sin(6\theta) \quad \text{and} \quad g(r, \theta) = 0$$

7.31. The center of the membrane has a finite displacement, and the periphery at $r = 1$ is unsecure. The boundaries $\theta = 0$ and $\theta = \pi/2$ are unsecure, $c = 1/2, \gamma = 1/10$:

$$|u(0, \theta, t)| < \infty \quad \text{and} \quad u_r(1, \theta, t) = 0$$

$$u_\theta(r, 0, t) = 0 \quad \text{and} \quad u_\theta\left(r, \frac{\pi}{2}, t \right) = 0$$

Initial conditions:

$$f(r, \theta) = 0 \quad \text{and} \quad g(r, \theta) = \left(\frac{3r^4}{2} - r^6 \right) \cos(4\theta)$$

7.32. The center of the membrane has a finite displacement, and the periphery at $r = 1$ is unsecure. The boundary conditions are such that we have periodic boundary conditions on $\theta, c = 1/2, \gamma = 1/10$:

$$|u(0, \theta, t)| < \infty \quad \text{and} \quad u_r(1, \theta, t) = 0$$

$$u(r, -\pi, t) = u(r, \pi, t) \quad \text{and} \quad u_\theta(r, -\pi, t) = u_\theta(r, \pi, t)$$

Initial conditions:

$$f(r, \theta) = \left(3r - r^3 \right) \sin(\theta) \quad \text{and} \quad g(r, \theta) = 0$$

CHAPTER 8

Nonhomogeneous Partial Differential Equations

8.1 Introduction

Here we consider the nonhomogeneous counterpart to those partial differential equations discussed in Chapters 3 and 4. In Chapters 3 and 4, all internal heat sources or external driving forces were assumed to have the value zero. In addition, all boundary conditions were assumed to be homogeneous. If either of these two conditions is not present, then the problem becomes nonhomogeneous.

We now consider nonhomogeneous versions of both the diffusion and wave partial differential equations over finite intervals in one spatial dimension. We consider the very generalized nonhomogeneous problems whereby we have both nonzero external driving forces or nonzero internal heat sources in addition to nonhomogeneous boundary conditions.

8.2 Nonhomogeneous Diffusion or Heat Equation

The generalized nonhomogeneous diffusion or heat partial differential equation in one dimension over a finite interval $I = \{x \mid 0 < x < a\}$ has the form

$$\frac{\partial}{\partial t} u(x, t) = k \left(\frac{\partial^2}{\partial x^2} u(x, t) \right) + h(x, t)$$

For heat phenomena, $u(x, t)$ denotes the magnitude of the temperature, k is the thermal diffusivity with dimensions of distance squared per time, and the function $h(x, t)$, with dimensions of temperature per time, accounts for any internal heat sources within the system. We consider this partial differential equation with the time-dependent, nonhomogeneous boundary conditions

$$\kappa_1 u(0, t) + \kappa_2 u_x(0, t) = A(t)$$

and

$$\kappa_3 u(a, t) + \kappa_4 u_x(a, t) = B(t)$$

Here, $A(t)$ and $B(t)$ are the time-dependent, nonhomogeneous source terms at the boundaries of the domain. We seek a solution to the preceding problem subject to the initial condition

$$u(x, 0) = f(x)$$

Our solution procedure is driven by our desire to find a complete set of orthonormalized eigenfunctions over the given domain, subject to an equivalent set of homogeneous boundary conditions. To move in this direction, we take advantage of the linearity of the operator, and we use the superposition principle where we assume our solution to be partitioned into two terms

$$u(x, t) = v(x, t) + s(x, t)$$

Depending on the character of the source term $h(x, t)$ and the boundary conditions, the term $v(x, t)$ may exhibit the characteristics of a "variable" or "transient" solution and $s(x, t)$ will exhibit the characteristics of a "steady-state" or "equilibrium" solution.

We now substitute this assumed solution into the partial differential equation and get

$$\frac{\partial}{\partial t} v(x, t) + \frac{\partial}{\partial t} s(x, t) = k \left(\frac{\partial^2}{\partial x^2} v(x, t) + \frac{\partial^2}{\partial x^2} s(x, t) \right) + h(x, t)$$

Doing the same with the boundary conditions, we get

$$A(t) = \kappa_1 v(0, t) + \kappa_2 v_x(0, t) + \kappa_1 s(0, t) + \kappa_2 s_x(0, t)$$

and

$$B(t) = \kappa_3 v(a, t) + \kappa_4 v_x(a, t) + \kappa_3 s(a, t) + \kappa_4 s_x(a, t)$$

Similarly, substituting into the initial condition, we get

$$f(x) = v(x, 0) + s(x, 0)$$

Depending on the character of the boundary conditions, our goal now is to substitute conditions on the partitioned solutions $v(x, t)$ and $s(x, t)$ so

1. $s(x, t)$ (the steady-state portion) will be a linear function of x; thus, all second-order partials with respect to x will be zero

2. $v(x, t)$ (the variable portion) will satisfy the corresponding nonhomogeneous equation with homogeneous boundary conditions

Doing this yields two initial condition-boundary value problems that can be easily resolved.

For $v(x, t)$, the variable portion of the solution, we obtain the nonhomogeneous partial differential equation

$$\frac{\partial}{\partial t} v(x, t) = k \left(\frac{\partial^2}{\partial x^2} v(x, t) \right) - \left(\frac{\partial}{\partial t} s(x, t) \right) + h(x, t)$$

with (the now) homogeneous boundary conditions

$$\kappa_1 v(0, t) + \kappa_2 v_x(0, t) = 0$$

and

$$\kappa_3 v(a, t) + \kappa_4 v_x(a, t) = 0$$

and initial conditions

$$v(x, 0) = f(x) - s(x, 0)$$

Similarly, for $s(x, t)$, the spatially linear portion of the solution, we obtain the ordinary homogeneous differential equation

$$\frac{\partial^2}{\partial x^2} s(x, t) = 0$$

with nonhomogeneous boundary conditions

$$A(t) = \kappa_1 s(0, t) + \kappa_2 s_x(0, t)$$

and

$$B(t) = \kappa_3 s(a, t) + \kappa_4 s_x(a, t)$$

We now focus on finding solutions for the variable and linear portions of the problem.

Solution for the Linear Portion

We first solve for $s(x, t)$. From the earlier ordinary differential equation for $s(x, t)$, the two basis vectors with respect to the variable x are

$$s1(x) = 1 \quad \text{and} \quad s2(x) = x$$

Thus, the general solution to $s(x, t)$ is the linear (with respect to x) equation

$$s(x, t) = m(t)x + b(t)$$

Note that the coefficients $m(t)$ and $b(t)$ in the preceding linear equation are exclusively time dependent. Substituting the preceding nonhomogeneous boundary conditions yields

$$\kappa_1 b(t) + \kappa_2 m(t) = A(t)$$

and

$$\kappa_3(m(t)a + b(t)) + \kappa_4 m(t) = B(t)$$

Solving the preceding for $m(t)$ and $b(t)$, we get

$$b(t) = \frac{\kappa_3 a A(t) + A(t)\kappa_4 - B(t)\kappa_2}{\kappa_1 \kappa_3 a + \kappa_1 \kappa_4 - \kappa_2 \kappa_3}$$

and

$$m(t) = \frac{\kappa_1 B(t) - A(t)\kappa_3}{\kappa_1 \kappa_3 a + \kappa_1 \kappa_4 - \kappa_2 \kappa_3}$$

Thus, the final form for the linear (steady-state) portion of the solution $s(x, t)$ reads

$$s(x, t) = \frac{(\kappa_1 B(t) - A(t)\kappa_3)\, x}{\kappa_1 \kappa_3 a + \kappa_1 \kappa_4 - \kappa_2 \kappa_3} + \frac{\kappa_3 a A(t) + A(t)\kappa_4 - B(t)\kappa_2}{\kappa_1 \kappa_3 a + \kappa_1 \kappa_4 - \kappa_2 \kappa_3}$$

for

$$\kappa_1 \kappa_3 a + \kappa_1 \kappa_4 - \kappa_2 \kappa_3 \neq 0$$

This is the linear or steady-state portion of the solution to the problem.

Solution for the Variable Portion

We now focus on finding the solution to the remaining partial differential equation for $v(x, t)$. If we write

$$q(x, t) = h(x, t) - \left(\frac{\partial}{\partial t} s(x, t) \right)$$

then the earlier partial differential equation in $v(x, t)$ takes on the nonhomogeneous form

$$\frac{\partial}{\partial t} v(x, t) = k \left(\frac{\partial^2}{\partial x^2} v(x, t) \right) + q(x, t)$$

with the homogeneous boundary conditions

$$\kappa_1 v(0, t) + \kappa_2 v_x(0, t) = 0$$

and

$$\kappa_3 v(a, t) + \kappa_4 v_x(a, t) = 0$$

and initial condition

$$v(x, 0) = f(x) - s(x, 0)$$

Note that the boundary conditions on $v(x, t)$ are now homogeneous. The partial differential equation continues to be nonhomogeneous because of the presence of the source term $q(x, t)$.

We solve the preceding nonhomogeneous partial differential equation for $v(x, t)$ by the method of "eigenfunction expansion" where the eigenfunctions are those associated with the corresponding homogeneous equation. The procedure goes as follows.

To solve for $v(x, t)$, we temporarily set $q(x, t) = 0$. This yields the corresponding homogeneous partial differential equation

$$\frac{\partial}{\partial t} v(x, t) = k \left(\frac{\partial^2}{\partial x^2} v(x, t) \right)$$

with homogeneous boundary conditions

$$\kappa_1 v(0, t) + \kappa_2 v_x(0, t) = 0$$

and

$$\kappa_3 v(a, t) + \kappa_4 v_x(a, t) = 0$$

By the method of separation of variables, as done in Chapter 3, we set $v(x, t)$ to be a product of two functions: one that is an exclusive function of x and one that is an exclusive function of t:

$$v(x, t) = X(x)T(t) \tag{8.1}$$

The resulting eigenvalue problem in the variable x has the form

$$\frac{d^2}{dx^2} X(x) + \lambda\, X(x) = 0$$

with regular homogeneous boundary conditions

$$\kappa_1 X(0) + \kappa_2 X_x(0) = 0$$

and

$$\kappa_3 X(a) + \kappa_4 X_x(a) = 0$$

This is a familiar regular Sturm-Liouville eigenvalue problem where the differential equation is of the Euler type, and the boundary conditions are homogeneous. We solve this equation for the eigenvalues and corresponding eigenfunctions, and we denote the indexed eigenvalues and corresponding eigenfunctions, respectively, as

$$\lambda_n, X_n(x)$$

for $n = 0, 1, 2, 3, \dots$.

These eigenfunctions can be orthonormalized over the interval I with respect to the weight function $w(x) = 1$. The resulting statement of orthonormality reads

$$\int_0^a X_n(x)X_m(x)\mathrm{d}x = \delta(n,m)$$

for $n, m = 0, 1, 2, 3, \ldots$. Recall, $\delta(n,m)$ is the familiar Kronecker delta function.

Since the preceding boundary value problem is a regular Sturm-Liouville problem over the finite interval $I = \{x \mid 0 < x < a\}$, then the eigenfunctions form a "complete" set with respect to any piecewise smooth function over this interval. Using similar arguments from Section 2.4, if we assume the solution $v(x, t)$ to be a piecewise smooth function with respect to the variable x over this interval, then the eigenfunction expansion solution of $v(x, t)$, in terms of the "complete" set of orthonormalized eigenfunctions $X_n(x)$, reads

$$v(x, t) = \sum_{n=0}^{\infty} T_n(t)X_n(x)$$

Here, the terms $T_n(t)$ are the time-dependent Fourier coefficients in the expansion of $v(x, t)$.

To determine the Fourier coefficients, we face having to substitute the series solution for $v(x, t)$ into the partial differential equation. Thus, we will be differentiating the infinite series, and this can be a tricky process. Note, however, that the preceding series is an expansion in terms of eigenfunctions $X_n(x)$, which satisfy the same boundary conditions as $v(x, t)$. This condition is in our favor for making the term-by-term differentiation of the series legitimate. Without proof, it can be simply stated that if $v(x, t)$ and its derivatives with respect to both x and t are continuous, and if $v(x, t)$ satisfies the same spatial boundary conditions as $X_n(x)$, then term by term differentiation of the infinite series is justified.

We now deal with the nonhomogeneous term $q(x, t)$. In a similar manner, if we assume $q(x, t)$ to be a piecewise smooth function with respect to the variable x over this interval, then the eigenfunction expansion of $q(x, t)$, in terms of the "complete" set of orthonormalized eigenfunctions $X_n(x)$, reads

$$q(x, t) = \sum_{n=0}^{\infty} Q_n(t)X_n(x)$$

Here, the terms $Q_n(t)$ are the time-dependent Fourier coefficients of $q(x, t)$. Since $q(x, t)$ is already determined from knowledge of the two functions $h(x, t)$ and $s(x, t)$ as shown earlier, then we can evaluate the Fourier coefficents $Q_n(t)$ by taking the inner product of the preceding with respect to the orthonormal eigenfunctions and the weight function $w(x) = 1$. Doing so yields

$$\int_0^a q(x, t)X_m(x)\mathrm{d}x = \sum_{n=0}^{\infty} Q_n(t)\left(\int_0^a X_m(x)X_n(x)\mathrm{d}x \right)$$

Taking advantage of the orthonormality of the eigenfunctions, this reduces to

$$\int_0^a q(x, t) X_m(x) dx = \sum_{n=0}^{\infty} Q_n(t) \delta(m, n)$$

Due to the mathematical nature of the Kronecker delta function, only one term survives the sum, and we evaluate the coefficients $Q_n(t)$ to be

$$Q_n(t) = \int_0^a q(x, t) X_n(x) dx$$

We now focus on determining the time-dependent portion of the problem. The original nonhomogeneous partial differential equation for the variable portion of the solution $v(x, t)$ reads

$$\frac{\partial}{\partial t} v(x, t) = k \left(\frac{\partial^2}{\partial x^2} v(x, t) \right) + q(x, t)$$

Substituting the eigenfunction expansions for $v(x, t)$

$$v(x, t) = \sum_{n=0}^{\infty} T_n(t) X_n(x)$$

and for $q(x, t)$

$$q(x, t) = \sum_{n=0}^{\infty} Q_n(t) X_n(x)$$

into the preceding nonhomogeneous partial differential equation, and assuming the validity of the formal procedures of interchanging the summation and integration operations, we get the following series equation:

$$\sum_{n=0}^{\infty} \left(\frac{d}{dt} T_n(t) \right) X_n(x) = \sum_{n=0}^{\infty} \left(k T_n(t) \left(\frac{d^2}{dx^2} X_n(x) \right) + Q_n(t) X_n(x) \right)$$

Recall that the spatial eigenfunctions $X_n(x)$ satisfy the eigenvalue equation

$$\frac{d^2}{dx^2} X_n(x) + \lambda_n X_n(x) = 0$$

Making this substitution in the preceding equation yields the series equation

$$\sum_{n=0}^{\infty} \left(\frac{d}{dt} T_n(t) + k \lambda_n T_n(t) \right) X_n(x) = \sum_{n=0}^{\infty} Q_n(t) X_n(x)$$

Since the eigenfunctions are linearly independent (orthogonal sets are linearly independent), the only way this series equation can hold is if the coefficients of the eigenfunctions on the left are equal to the coefficients of the same eigenfunctions on the right. From this, we arrive at the time-dependent differential equation for $T_n(t)$:

$$\frac{d}{dt}T_n(t) + k\lambda_n T_n(t) = Q_n(t)$$

This is a first-order nonhomogeneous differential equation. From Section 1.2, the system basis vector for this equation is

$$T1(t) = e^{-k\lambda_n t}$$

and the corresponding first-order Green's function is

$$G1(t, \tau) = e^{-k\lambda_n(t-\tau)}$$

Thus, from Section 1.3 on first-order initial value problems, the solution to the nonhomogeneous ordinary differential equation in t reads

$$T_n(t) = C(n)e^{-k\lambda_n t} + \int_0^t G1(t, \tau)Q_n(\tau)d\tau$$

where the $C(n)$ are arbitrary constants.

The final solution for $v(x, t)$ can be written as the superposition of the products of the solutions to $X_n(x)$ and $T_n(t)$; that is,

$$v_n(x, t) = X_n(x)T_n(t)$$

for $n = 0, 1, 2, 3, \ldots$. Thus, our solution to the partial differential equation for the variable (transient) portion $v(x, t)$ reads

$$v(x, t) = \sum_{n=0}^{\infty}\left(C(n)e^{-k\lambda_n t} + \int_0^t G1(t, \tau)Q_n(\tau)d\tau\right)X_n(x)$$

We demonstrate these concepts with the solution of an illustrative problem.

DEMONSTRATION: We seek the temperature distribution $u(x, t)$ in a thin rod whose lateral surface is insulated over the interval $I = \{x \mid 0 < x < 1\}$. The left end of the rod is up against a temperature bath with an oscillatory temperature variation, and the right end is held at the fixed temperature of zero. There is no internal heat source ($h(x, t) = 0$), the initial temperature distribution $u(x, 0) = f(x)$ is given as follows, and the diffusivity is $k = 1/4$.

SOLUTION: The system partial differential equation is

$$\frac{\partial}{\partial t}u(x,t) = \frac{\frac{\partial^2}{\partial x^2}u(x,t)}{4}$$

The initial condition temperature distribution is

$$u(x,0) = f(x)$$

where

$$f(x) = x(1-x)$$

Since the left end of the rod is in contact with an oscillatory heat bath and the right end is held at the fixed temperature zero, the boundary conditions on the problem are

$$u(0,t) = \sin(t) \quad \text{and} \quad u(1,t) = 0$$

In seeking the linear portion of the solution, we first compare the boundary conditions with the generalized nonhomogeneous boundary conditions that read

$$\kappa_1 u(0,t) + \kappa_2 u_x(0,t) = A(t)$$

and

$$\kappa_3 u(a,t) + \kappa_4 u_x(a,t) = B(t)$$

Comparing the boundary conditions with these generalized conditions, we identify the following:

$$A(t) = \sin(t), \ B(t) = 0$$
$$\kappa_1 = 1, \ \kappa_2 = 0, \ \kappa_3 = 1, \ \kappa_4 = 0$$

With the given values for κ_n, $A(t)$, and $B(t)$, we use the previous generalized equations for $b(t)$ and $m(t)$ to solve for the components of the linear portion, and we get

$$b(t) = \sin(t)$$
$$m(t) = -\sin(t)$$

Thus, the linear portion of the solution reads

$$s(x,t) = -\sin(t)x + \sin(t)$$

For the variable portion of the problem, the partial differential equation for $v(x,t)$ reads

$$\frac{\partial}{\partial t}v(x,t) = \frac{\frac{\partial^2}{\partial x^2}v(x,t)}{4} + q(x,t)$$

where, from the preceding,

$$q(x, t) = -\left(\frac{\partial}{\partial t} s(x, t)\right)$$

Because $h(x, t) = 0$, this evaluates to

$$q(x, t) = x \cos(t) - \cos(t)$$

The corresponding homogeneous boundary conditions on $v(x, t)$ are

$$v(0, t) = 0 \quad \text{and} \quad v(1, t) = 0$$

and the initial conditions are

$$v(x, 0) = f(x) - s(x, 0)$$

which evaluates to

$$v(x, 0) = x(1 - x)$$

To solve the $v(x, t)$ equation, we temporarily set $q(x, t) = 0$, and we assume a solution of the form

$$v(x, t) = X(x)T(t)$$

From the method of separation of variables, the corresponding spatial-dependent equation is the Euler differential equation

$$\frac{d^2}{dx^2} X(x) + \lambda X(x) = 0$$

with the corresponding homogeneous boundary conditions (type 1 at $x = 0$ and type 1 at $x = 1$)

$$X(0) = 0 \quad \text{and} \quad X(1) = 0$$

This eigenvalue problem was examined previously in Example 2.5.1, and the allowed eigenvalues and corresponding orthonormal eigenfunctions are

$$\lambda_n = n^2 \pi^2$$

and

$$X_n(x) = \sqrt{2} \sin(n \pi x)$$

for $n = 1, 2, 3, \dots$.

The corresponding statement of orthonormality, with the weight function $w(x) = 1$, reads

$$\int_0^1 2 \sin(n\pi x) \sin(m\pi x) \mathrm{d}x = \delta(n, m)$$

We now focus on the nonhomogeneous time-dependent differential equation, which reads

$$\frac{\mathrm{d}}{\mathrm{d}t} T_n(t) + k\lambda_n T_n(t) = Q_n(t)$$

where $Q_n(t)$ is the inner product of $q(x, t)$ with the spatial eigenfunctions; that is,

$$Q_n(t) = \int_0^1 (\cos(t)x - \cos(t)) \sqrt{2} \sin(n\pi x) \mathrm{d}x$$

Evaluation of this inner product yields

$$Q_n(t) = -\frac{\cos(t)\sqrt{2}}{n\pi}$$

Thus, the time-dependent differential equation reads

$$\frac{\mathrm{d}}{\mathrm{d}t} T_n(t) + \frac{n^2\pi^2 T_n(t)}{4} = -\frac{\sqrt{2}\cos(t)}{n\pi}$$

A basis vector for this differential equation, from Section 1.3, reads

$$T1(t) = \mathrm{e}^{-\frac{n^2\pi^2 t}{4}}$$

Evaluation of the corresponding first-order Green's function yields

$$G1(t, \tau) = \mathrm{e}^{\frac{n^2\pi^2(-t+\tau)}{4}}$$

Thus, the time-dependent solution for the variable portion $v(x, t)$ is

$$T_n(t) = C(n)\mathrm{e}^{-\frac{n^2\pi^2 t}{4}} + \int_0^t \left(-\frac{\mathrm{e}^{\frac{n^2\pi^2(-t+\tau)}{4}} \cos(\tau)\sqrt{2}}{n\pi} \right) \mathrm{d}\tau$$

for $n = 1, 2, 3, \ldots$. Evaluating the preceding integral yields

$$T_n(t) = C(n)\mathrm{e}^{-\frac{n^2\pi^2 t}{4}} - \frac{4\sqrt{2}\left(n^2\pi^2 \cos(t) + 4 \sin(t)\right)}{(n^4\pi^4 + 16) n\pi} + \frac{4\sqrt{2}n^2\pi^2 \mathrm{e}^{-\frac{n^2\pi^2 t}{4}}}{n\pi \left(n^4\pi^4 + 16\right)}$$

Finally, the solution for the variable portion of the problem is constructed from the super-position of the products of the time-dependent solutions and the given spatial eigenfunctions, and this reads

$$v(x, t) = \sum_{n=1}^{\infty} \left(C(n)e^{-\frac{n^2\pi^2 t}{4}} + \frac{4\sqrt{2}\left(n^2\pi^2 e^{-\frac{n^2\pi^2 t}{4}} - n^2\pi^2 \cos(t) - 4\sin(t)\right)}{n\pi\left(n^4\pi^4 + 16\right)} \right) \sqrt{2}\sin(n\pi x)$$

The arbitrary constants $C(n)$ are determined from the initial conditions for the problem.

8.3 Initial Condition Considerations for the Nonhomogeneous Heat Equation

Substitution of the initial condition into the general solution for the variable portion $v(x, t)$, from Section 8.2, yields

$$v(x, 0) = f(x) - s(x, 0)$$

Substituting the preceding into the following general solution for $v(x, t)$

$$v(x, t) = \sum_{n=0}^{\infty} \left(C(n)e^{-k\lambda_n t} + \int_0^t G1(t, \tau) Q_n(\tau) d\tau \right) X_n(x)$$

at time $t = 0$ yields

$$f(x) - s(x, 0) = \sum_{n=0}^{\infty} C(n) X_n(x)$$

This is the familiar Fourier series expansion of the left-hand side of the equation, and the $C(n)$ are the Fourier coefficients. Taking the inner product of both sides, with respect to the orthonormalized eigenfunctions $X_n(x)$ and the weight function $w(x) = 1$, and assuming validity of the interchange between the summation and integration operations, yields

$$\int_0^a (f(x) - s(x, 0))\, X_m(x)\, dx = \sum_{n=0}^{\infty} C(n) \left(\int_0^a X_m(x) X_n(x) dx \right)$$

From the statement of orthonormality of the spatial eigenfunctions, the preceding equation reduces to

$$\int_0^a (f(x) - s(x, 0))\, X_m(x) dx = \sum_{n=0}^{\infty} C(n)\delta(m, n)$$

Due to the properties of the Kronecker delta function, only one term ($m = n$) survives this sum, and the resulting Fourier coefficient reads

$$C(n) = \int_0^a (f(x) - s(x, 0)) \, X_n(x) \mathrm{d}x \tag{8.2}$$

Now that we have evaluated the coefficients $C(n)$, we can write the final solution to our original nonhomogeneous partial differential equation as the sum of the linear portion plus the transient portion as follows:

$$u(x, t) = \sum_{n=0}^{\infty} \left(C(n) \mathrm{e}^{-k\lambda_n t} + \int_0^t G1(t, \tau) Q_n(\tau) \mathrm{d}\tau \right) X_n(x) + s(x, t)$$

Note that if there is no nonhomogeneous source term $h(x, t)$ in the system, and if the boundary conditions are time invariant, then the transient portion of the solution $v(x, t)$ goes to zero as t goes to infinity, and the only remaining term in the solution is the linear term $s(x, t)$. For this reason, this linear portion of the solution is referred to as the "steady-state" or "equilibrium" solution to the partial differential equation.

We now demonstrate these concepts for the illustrative problem given in Section 8.2.

DEMONSTRATION: From the given initial condition on the problem, we have

$$v(x, 0) = f(x) - s(x, 0)$$

which evaluates to

$$f(x) - s(x, 0) = x(1 - x)$$

SOLUTION: The coefficients $C(n)$ are evaluated from the integral

$$C(n) = \int_0^a (f(x) - s(x, 0)) \, X_n(x) \mathrm{d}x$$

Substituting the preceding into this integral yields

$$C(n) = \int_0^1 x(1 - x)\sqrt{2} \sin(n\pi x) \mathrm{d}x$$

Evaluation of this integral gives

$$C(n) = -\frac{2\sqrt{2}\left((-1)^n - 1\right)}{n^3 \pi^3}$$

for $n = 1, 2, 3, \ldots$. Thus, the solution for the variable portion of the problem reads

$$v(x, t) = \sum_{n=1}^{\infty} \left(\frac{2\sqrt{2}\left(1 - (-1)^n\right)e^{-\frac{n^2\pi^2 t}{4}}}{n^3\pi^3} \right.$$

$$\left. + \frac{4\sqrt{2}\left(n^2\pi^2 e^{-\frac{n^2\pi^2 t}{4}} - n^2\pi^2\cos(t) - 4\sin(t)\right)}{n\pi\left(n^4\pi^4 + 16\right)} \right) \sqrt{2}\sin(n\pi x)$$

Finally, the solution to our problem is the sum of the linear plus the variable portions evaluated earlier—that is, $u(x, t) = v(x, t) + s(x, t)$. This yields

$$u(x, t) = \sin(t)(1 - x) + \sum_{n=1}^{\infty} \left(\frac{2\sqrt{2}\left(1 - (-1)^n\right)e^{-\frac{n^2\pi^2 t}{4}}}{n^3\pi^3} \right.$$

$$\left. + \frac{4\sqrt{2}\left(n^2\pi^2 e^{-\frac{n^2\pi^2 t}{4}} - n^2\pi^2\cos(t) - 4\sin(t)\right)}{n\pi\left(n^4\pi^4 + 16\right)} \right) \sqrt{2}\sin(n\pi x)$$

Note how the solution satisfies the given boundary and initial conditions. The details for the development of this solution along with the graphics are given later in one of the Maple worksheet examples.

8.4 Example Nonhomogeneous Problems for the Diffusion Equation

We now consider solutions to some example nonhomogeneous problems for the heat or diffusion equation using Maple.

EXAMPLE 8.4.1: We seek the temperature distribution $u(x, t)$ in a thin rod whose lateral surface is insulated over the interval $I = \{x \mid 0 < x < 1\}$. The left end of the rod is held at a fixed temperature of 10, and the right end is held at the fixed temperature of 20. There is no internal heat source. The thermal diffusivity of the rod is $k = 1/20$, and the rod has an initial temperature distribution $u(x, 0) = f(x)$ given as follows.

SOLUTION: The nonhomogeneous diffusion equation is

$$\frac{\partial}{\partial t}u(x, t) = k\left(\frac{\partial^2}{\partial x^2}u(x, t)\right) + h(x, t)$$

The boundary conditions are nonhomogeneous type 1 at the left and type 1 at the right:

$$u(0, t) = 10 \quad \text{and} \quad u(1, t) = 20$$

The initial condition is

$$u(x, 0) = 60x - 50x^2 + 10$$

The internal heat source term is

$$h(x, t) = 0$$

The solution is $u(x, t) = v(x, t) + s(x, t)$, where $s(x, t)$ is the linear portion of the solution and $v(x, t)$ is the variable portion of the solution, which satisfies the partial differential equation

$$\frac{\partial}{\partial t} v(x, t) = k \left(\frac{\partial^2}{\partial x^2} v(x, t) \right) + q(x, t)$$

where

$$q(x, t) = h(x, t) - \left(\frac{\partial}{\partial t} s(x, t) \right)$$

Assignment of system parameters

> restart: with(plots):a:=1:k:=1/20:h(x,t):=0:

> A(t):=10:B(t):=20:kappa[1]:=1:kappa[2]:=0:kappa[3]:=1:kappa[4]:=0:

> b(t):=(kappa[3]*a*A(t)+A(t)*kappa[4]−B(t)*kappa[2])/(kappa[1]*kappa[3]*a+
kappa[1]*kappa[4]−kappa[2]*kappa[3]);

$$b(t) := 10 \tag{8.3}$$

> m(t):=(kappa[1]*B(t)−A(t)*kappa[3])/(kappa[1]*kappa[3]*a+kappa[1]*kappa[4]−
kappa[2]*kappa[3]);

$$m(t) := 10 \tag{8.4}$$

Linear portion of solution

> s(x,t):=m(t)*x+b(t);s(x,0):=eval(subs(t=0,s(x,t))):

$$s(x, t) := 10x + 10 \tag{8.5}$$

By the method of separation of variables, the variable solution is

$$v(x, t) = \sum_{n=0}^{\infty} X_n(x) T_n(t)$$

where $T_n(t)$ is the solution to the time-dependent differential equation

$$\frac{d}{dt}T_n(t) + k\lambda_n T_n(t) = Q_n(t)$$

and $X_n(x)$ is the solution to the spatial-dependent eigenvalue equation

$$\frac{d^2}{dx^2}X_n(x) + \lambda_n X_n(x) = 0$$

with boundary conditions

$$X(0) = 0 \quad \text{and} \quad X(1) = 0$$

The corresponding homogeneous eigenfunction problem consists of the given Euler equation, along with the homogeneous boundary conditions that are type 1 at the left and type 1 at the right. The allowed eigenvalues and corresponding orthonormal eigenfunctions are obtained from Example 2.5.1.

> lambda[n]:=(n*Pi/a)^2;

$$\lambda_n := n^2\pi^2 \tag{8.6}$$

> X[n](x):=sqrt(2/a)*sin(n*Pi/a*x);X[m](x):=subs(n=m,X[n](x)):

$$X_n(x) := \sqrt{2}\ \sin(n\pi x) \tag{8.7}$$

for $n = 1, 2, 3, \ldots.$

Statement of orthonormality with the respective weight function $w(x) = 1$

> w(x):=1:Int(X[n](x)*X[m](x)*w(x),x=0..a)=delta(n,m);

$$\int_0^1 2\ \sin(n\pi x)\sin(m\pi x)dx = \delta(n, m) \tag{8.8}$$

Time-dependent equation

> diff(T[n](t),t)+k*lambda[n]*T[n](t)=Q[n](t);

$$\frac{d}{dt}T_n(t) + \frac{1}{20}n^2\pi^2 T_n(t) = Q_n(t) \tag{8.9}$$

> q(x,t):=h(x,t)−diff(s(x,t),t);

$$q(x, t) := 0 \tag{8.10}$$

> Q[n](t):=Int(q(x,t)*X[n](x),x=0..a);Q[n](t):=expand(value(%)):

$$Q_n(t) := \int_0^1 0\,dx \tag{8.11}$$

> Q[n](t):=simplify(factor(subs({sin(n*Pi)=0,cos(n*Pi)=(−1)^n},Q[n](t))));
Q[n](tau):=subs(t=tau,%):

$$Q_n(t) := 0 \tag{8.12}$$

Basis vector

> T1(t):=exp(−k*lambda[n]*t);

$$T1(t) := e^{-\frac{1}{20}n^2\pi^2 t} \tag{8.13}$$

First-order Green's function

> G1(t,tau):=simplify(T1(t)/(subs(t=tau,T1(t))));

$$G1(t, \tau) := e^{-\frac{1}{20}\pi^2 n^2(t-\tau)} \tag{8.14}$$

Time-dependent solution

> T[n](t):=eval(C(n)*T1(t)+Int(G1(t,tau)*Q[n](tau),tau=0..t));T[n](t):=value(%):

$$T_n(t) := C(n)e^{-\frac{1}{20}n^2\pi^2 t} + \int_0^t 0\,d\tau \tag{8.15}$$

Generalized series terms

> v[n](x,t):=simplify(eval(T[n](t)*X[n](x)));

$$v_n(x, t) := C(n)e^{-\frac{1}{20}n^2\pi^2 t}\sqrt{2}\sin(n\pi x) \tag{8.16}$$

Variable portion of solution

> v(x,t):=Sum(v[n](x,t),n=1..infinity);

$$v(x, t) := \sum_{n=1}^{\infty} C(n)e^{-\frac{1}{20}n^2\pi^2 t}\sqrt{2}\sin(n\pi x) \tag{8.17}$$

The Fourier coefficients $C(n)$ are determined from the integral in Section 8.3 for the special case where

> f(x):=60*x−50*x^2+10;

$$f(x) := 60x - 50x^2 + 10 \tag{8.18}$$

Substituting this into the integral for $C(n)$ yields

> C(n):=Int((f(x)−s(x,0))*X[n](x),x=0..a);C(n):=expand(value(%)):

$$C(n) := \int_0^1 (50x - 50x^2)\sqrt{2}\sin(n\pi x)\,dx \tag{8.19}$$

> C(n):=simplify(subs({sin(n*Pi)=0,cos(n*Pi)=(−1)^n},C(n)));

$$C(n) := -\frac{100\sqrt{2}(-1 + (-1)^n)}{n^3\pi^3} \tag{8.20}$$

for $n = 1, 2, 3, \ldots$.

> T[n](t):=(C(n)*T1(t)+int(G1(t,tau)*Q[n](tau),tau=0..t));

$$T_n(t) := -\frac{100\sqrt{2}(-1 + (-1)^n)e^{-\frac{1}{20}n^2\pi^2 t}}{n^3\pi^3} \tag{8.21}$$

Generalized series terms

> v[n](x,t):=simplify(eval(T[n](t)*X[n](x)));

$$v_n(x, t) := -\frac{200(-1 + (-1)^n)e^{-\frac{1}{20}n^2\pi^2 t}\sin(n\pi x)}{n^3\pi^3} \tag{8.22}$$

Final solution (linear plus variable portion)

> u(x,t):=eval(Sum(v[n](x,t),n=1..infinity)+s(x,t));

$$u(x, t) := \sum_{n=1}^{\infty}\left(-\frac{200(-1 + (-1)^n)e^{-\frac{1}{20}n^2\pi^2 t}\sin(n\pi x)}{n^3\pi^3}\right) + 10x + 10 \tag{8.23}$$

First few terms of sum

> u(x,t):=s(x,t)+sum(v[n](x,t),n=1..3):

ANIMATION

> animate(u(x,t),x=0..a,t=0..5,thickness=3);

The preceding animation command displays the spatial-time-dependent solution of $u(x, t)$ for the given boundary conditions and initial conditions. The animation sequence here and in Figure 8.1 shows snapshots of the animation at times $t = 0, 1, 2, 3, 4, 5$. Note how the solution satisfies the given boundary and initial conditions.

ANIMATION SEQUENCE

```
> u(x,0):=subs(t=0,u(x,t)):u(x,1):=subs(t=1,u(x,t)):
> u(x,2):=subs(t=2,u(x,t)):u(x,3):=subs(t=3,u(x,t)):
> u(x,4):=subs(t=4,u(x,t)):u(x,5):=subs(t=5,u(x,t)):
> plot({u(x,0),u(x,1),u(x,2),u(x,3),u(x,4),u(x,5)},x=0..a,thickness=10);
```

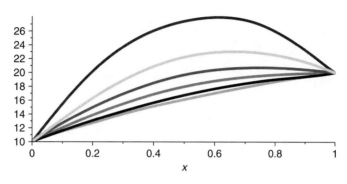

Figure 8.1

EXAMPLE 8.4.2: We seek the temperature distribution $u(x, t)$ in a thin rod whose lateral surface is insulated over the interval $I = \{x \mid 0 < x < 1\}$. The left end of the rod is up against a temperature bath with an oscillatory temperature variation, and the right end is held at the fixed temperature of zero. There is no internal heat source. The thermal diffusivity of the rod is $k = 1/4$, and the rod has an initial temperature distribution $u(x, 0) = f(x)$ given as follows.

SOLUTION: The nonhomogeneous diffusion equation is

$$\frac{\partial}{\partial t} u(x, t) = k \left(\frac{\partial^2}{\partial x^2} u(x, t) \right) + h(x, t)$$

The boundary conditions are nonhomogeneous type 1 at the left and homogeneous type 1 at the right:

$$u(0, t) = \sin(t) \quad \text{and} \quad u(1, t) = 0$$

The initial condition is

$$u(x, 0) = x(1 - x)$$

The internal heat source term is

$$h(x, t) = 0$$

The solution is $u(x, t) = v(x, t) + s(x, t)$, where $s(x, t)$ is the linear portion of the solution and $v(x, t)$ is the variable portion of the solution, which satisfies the partial differential equation

$$\frac{\partial}{\partial t} v(x, t) = k \left(\frac{\partial^2}{\partial x^2} v(x, t) \right) + q(x, t)$$

where

$$q(x, t) = h(x, t) - \left(\frac{\partial}{\partial t} s(x, t) \right)$$

Assignment of system parameters

> restart: with(plots):a:=1:k:=1/4:h(x,t):=0:

> A(t):=sin(t):B(t):=0:kappa[1]:=1:kappa[2]:=0:kappa[3]:=1:kappa[4]:=0:

> b(t):=(kappa[3]*a*A(t)+A(t)*kappa[4]−B(t)*kappa[2])/(kappa[1]*kappa[3]*a+kappa[1]*
kappa[4]−kappa[2]*kappa[3]);

$$b(t) := \sin(t) \tag{8.24}$$

> m(t):=(kappa[1]*B(t)−A(t)*kappa[3])/(kappa[1]*kappa[3]*a+kappa[1]*kappa[4]−
kappa[2]*kappa[3]);

$$m(t) := -\sin(t) \tag{8.25}$$

Linear portion of solution

> s(x,t):=m(t)*x+b(t);s(x,0):=eval(subs(t=0,s(x,t))):

$$s(x, t) := -\sin(t)x + \sin(t) \tag{8.26}$$

By the method of separation of variables, the variable solution is

$$v(x, t) = \sum_{n=0}^{\infty} X_n(x) T_n(t)$$

where $T_n(t)$ is the solution to the time-dependent differential equation

$$\frac{d}{dt} T_n(t) + k\lambda_n T_n(t) = Q_n(t)$$

and $X_n(x)$ is the solution to the spatial-dependent eigenvalue equation

$$\frac{d^2}{dx^2} X_n(x) + \lambda_n X_n(x) = 0$$

with boundary conditions

$$X(0) = 0 \quad \text{and} \quad X(1) = 0$$

The corresponding homogeneous eigenfunction problem consists of the Euler equation, along with the homogeneous boundary conditions, which are type 1 at the left and type 1 at the right. The allowed eigenvalues and corresponding orthonormal eigenfunctions are obtained from Example 2.5.1.

> lambda[n]:=(n*Pi/a)^2;

$$\lambda_n := n^2\pi^2 \tag{8.27}$$

> X[n](x):=sqrt(2/a)*sin(n*Pi/a*x);X[m](x):=subs(n=m,X[n](x)):

$$X_n(x) := \sqrt{2}\sin(n\pi x) \tag{8.28}$$

for $n = 1, 2, 3, \ldots$.

Statement of orthonormality with the respective weight function $w(x) = 1$

> w(x):=1:Int(X[n](x)*X[m](x)*w(x),x=0..a)=delta(n,m);

$$\int_0^1 2\,\sin(n\pi x)\sin(m\pi x)\,\mathrm{d}x = \delta(n, m) \tag{8.29}$$

Time-dependent equation

> diff(T[n](t),t)+k*lambda[n]*T[n](t)=Q[n](t);

$$\frac{\mathrm{d}}{\mathrm{d}t}T_n(t) + \frac{1}{4}n^2\pi^2 T_n(t) = Q_n(t) \tag{8.30}$$

> q(x,t):=h(x,t)-diff(s(x,t),t);

$$q(x, t) := \cos(t)x - \cos(t) \tag{8.31}$$

> Q[n](t):=Int(q(x,t)*X[n](x),x=0..a);Q[n](t):=expand(value(%)):

$$Q_n(t) := \int_0^1 (\cos(t)x - \cos(t))\sqrt{2}\sin(n\pi x)\,\mathrm{d}x \tag{8.32}$$

> Q[n](t):=simplify(factor(subs({sin(n*Pi)=0,cos(n*Pi)=(-1)^n},Q[n](t))));
  Q[n](tau):=subs(t=tau,%):

$$Q_n(t) := -\frac{\sqrt{2}\cos(t)}{\pi n} \tag{8.33}$$

Basis vector

> T1(t):=exp(−k*lambda[n]*t);

$$T1(t) := e^{-\frac{1}{4}n^2\pi^2 t} \tag{8.34}$$

First-order Green's function

> G1(t,tau):=simplify(T1(t)/(subs(t=tau,T1(t))));

$$G1(t, \tau) := e^{-\frac{1}{4}\pi^2 n^2 (t-\tau)} \tag{8.35}$$

Time-dependent solution

> T[n](t):=eval(C(n)*T1(t)+Int(G1(t,tau)*Q[n](tau),tau=0..t)):T[n](t):=value(%);

$$T_n(t) := C(n)e^{-\frac{1}{4}n^2\pi^2 t} - \frac{4\sqrt{2}\left(-n^2\pi^2 e^{-\frac{1}{4}n^2\pi^2 t} + n^2\pi^2\cos(t) + 4\sin(t)\right)}{n\pi\left(n^4\pi^4 + 16\right)} \tag{8.36}$$

Generalized series terms

> v[n](x,t):=(eval(T[n](t)*X[n](x))):

Variable portion of solution

> v(x,t):=Sum(v[n](x,t),n=1..infinity);

$$v(x, t) := \sum_{n=1}^{\infty}\left(C(n)e^{-\frac{1}{4}n^2\pi^2 t} - \frac{4\sqrt{2}\left(-n^2\pi^2 e^{-\frac{1}{4}n^2\pi^2 t} + n^2\pi^2\cos(t) + 4\sin(t)\right)}{n\pi\left(n^4\pi^4 + 16\right)}\right)\sqrt{2}\sin(n\pi x) \tag{8.37}$$

The Fourier coefficients $C(n)$ are determined from the integral in Section 8.3 for the special case where

> f(x):=x*(1−x);

$$f(x) := x(1 - x) \tag{8.38}$$

> C(n):=Int((f(x)−s(x,0))*X[n](x),x=0..a);C(n):=expand(value(%)):

$$C(n) := \int_0^1 x(1 - x)\sqrt{2}\,\sin(n\pi x)\,dx \tag{8.39}$$

> C(n):=simplify(subs({sin(n*Pi)=0,cos(n*Pi)=(−1)^n},C(n)));

$$C(n) := -\frac{2\sqrt{2}\left(-1 + (-1)^n\right)}{\pi^3 n^3} \tag{8.40}$$

for $n = 1, 2, 3, \ldots$.

> T[n](t):=(C(n)*T1(t)+int(G1(t,tau)*Q[n](tau),tau=0..t));

$$T_n(t) := -\frac{2\sqrt{2}\left(-1+(-1)^n\right)e^{-\frac{1}{4}n^2\pi^2 t}}{\pi^3 n^3} - \frac{4\sqrt{2}\left(-n^2\pi^2 e^{-\frac{1}{4}n^2\pi^2 t}+n^2\pi^2\cos(t)+4\sin(t)\right)}{\pi n\left(n^4\pi^4+16\right)}$$

(8.41)

Generalized series terms

> v[n](x,t):=(T[n](t)*X[n](x));

$$v_n(x,t) := \left(-\frac{2\sqrt{2}\left(-1+(-1)^n\right)e^{-\frac{1}{4}n^2\pi^2 t}}{\pi^3 n^3}\right.$$

$$\left.-\frac{4\sqrt{2}\left(-n^2\pi^2 e^{-\frac{1}{4}n^2\pi^2 t}+n^2\pi^2\cos(t)+4\sin(t)\right)}{\pi n\left(n^4\pi^4+16\right)}\right)\sqrt{2}\sin(n\pi x)$$

(8.42)

Final solution (linear plus variable portion)

> u(x,t):=(eval(Sum(v[n](x,t),n=1..infinity)+s(x,t)));

$$u(x,t) := \sum_{n=1}^{\infty}\left(-\frac{2\sqrt{2}\left(-1+(-1)^n\right)e^{-\frac{1}{4}n^2\pi^2 t}}{\pi^3 n^3}\right.$$

(8.43)

$$\left.-\frac{4\sqrt{2}\left(-n^2\pi^2 e^{-\frac{1}{4}n^2\pi^2 t}-n^2\pi^2\cos(t)-4\sin(t)\right)}{\pi n\left(n^4\pi^4+16\right)}\right)\sqrt{2}\,\sin(n\pi x)-\sin(t)x+\sin(t)$$

First few terms of sum

> u(x,t):=s(x,t)+sum(v[n](x,t),n=1..3):

ANIMATION

> animate(u(x,t),x=0..a,t=0..5,thickness=3);

The preceding animation command displays the spatial-time-dependent solution of $u(x,t)$ for the given boundary conditions and initial conditions. The animation sequence here and in Figure 8.2 shows snapshots of the animation at times $t = 0, 1, 2, 3, 4, 5$. Note how the solution satisfies the given boundary and initial conditions.

ANIMATION SEQUENCE

> u(x,0):=subs(t=0,u(x,t)):u(x,1):=subs(t=1,u(x,t)):
> u(x,2):=subs(t=2,u(x,t)):u(x,3):=subs(t=3,u(x,t)):

> u(x,4):=subs(t=4,u(x,t)):u(x,5):=subs(t=5,u(x,t)):
> plot({u(x,0),u(x,1),u(x,2),u(x,3),u(x,4),u(x,5)},x=0..a,thickness=10);

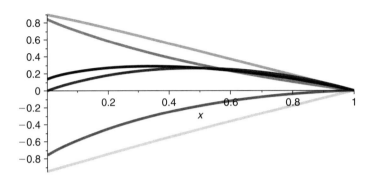

Figure 8.2

EXAMPLE 8.4.3: We seek the temperature distribution $u(x, t)$ in a thin rod whose lateral surface is insulated over the interval $I = \{x \mid 0 < x < 1\}$. The left end of the rod is held at a fixed temperature of 5, and the right end loses heat by convection into a medium whose temperature is 10. In addition, there is an internal heat source term $h(x, t)$ that increases linearly with time. The thermal diffusivity of the rod is $k = 1/20$, and an initial temperature distribution $u(x, 0) = f(x)$ is given as follows.

SOLUTION: The nonhomogeneous diffusion equation is

$$\frac{\partial}{\partial t} u(x, t) = k \left(\frac{\partial^2}{\partial x^2} u(x, t) \right) + h(x, t)$$

The boundary conditions are nonhomogeneous type 1 at the left and nonhomogeneous type 3 at the right:

$$u(0, t) = 5 \quad \text{and} \quad u(1, t) + u_x(1, t) = 10$$

The initial condition is

$$u(x, 0) = -\frac{40x^2}{3} + \frac{45x}{2} + 5$$

The internal heat source term is

$$h(x, t) = t$$

The solution is $u(x, t) = v(x, t) + s(x, t)$, where $s(x, t)$ is the linear portion of the solution and $v(x, t)$ is the variable portion of the solution, which satisfies the partial differential equation

$$\frac{\partial}{\partial t} v(x, t) = k \left(\frac{\partial^2}{\partial x^2} v(x, t) \right) + q(x, t)$$

where

$$q(x, t) = h(x, t) - \left(\frac{\partial}{\partial t} s(x, t) \right)$$

Assignment of system parameters

> restart: with(plots):a:=1:k:=1/20:h(x,t):=t:

> A(t):=5:B(t):=10:kappa[1]:=1:kappa[2]:=0:kappa[3]:=1:kappa[4]:=1:

> b(t):=(kappa[3]*a*A(t)+A(t)*kappa[4]−B(t)*kappa[2])/(kappa[1]*kappa[3]*a+kappa[1]*
kappa[4]−kappa[2]*kappa[3]);

$$b(t) := 5 \tag{8.44}$$

> m(t):=(kappa[1]*B(t)−A(t)*kappa[3])/(kappa[1]*kappa[3]*a+kappa[1]*
kappa[4]−kappa[2]*kappa[3]);

$$m(t) := \frac{5}{2} \tag{8.45}$$

Linear portion of solution

> s(x,t):=m(t)*x+b(t);s(x,0):=eval(subs(t=0,s(x,t))):

$$s(x, t) := \frac{5}{2}x + 5 \tag{8.46}$$

By the method of separation of variables, the variable solution is

$$v(x, t) = \sum_{n=0}^{\infty} X_n(x) T_n(t)$$

where $T_n(t)$ is the solution to the time-dependent differential equation

$$\frac{d}{dt} T_n(t) + k\lambda_n T_n(t) = Q_n(t)$$

and $X_n(x)$ is the solution to the spatial-dependent eigenvalue equation

$$\frac{d^2}{dx^2} X_n(x) + \lambda_n X_n(x) = 0$$

with boundary conditions

$$X(0) = 0 \quad \text{and} \quad X(1) + X_x(1, t) = 0$$

The corresponding homogeneous eigenfunction problem consists of the Euler equation, along with the homogeneous boundary conditions that are type 1 at the left and type 3 at the right.

The allowed eigenvalues and corresponding orthonormal eigenfunctions are obtained from Example 2.5.4. The eigenvalues λ_n are the roots of the eigenvalue equation

> tan(sqrt(lambda[n]*a))=−sqrt(lambda[n]);

$$\tan\left(\sqrt{\lambda_n}\right) = -\sqrt{\lambda_n} \tag{8.47}$$

> X[n](x):=sqrt(2)*sin(sqrt(lambda[n])*x)/sqrt((cos(sqrt(lambda[n])*a)^2+a));X[m](x):=
  subs(n=m,X[n](x)):

$$X_n(x) := \frac{\sqrt{2}\sin\left(\sqrt{\lambda_n}x\right)}{\sqrt{\cos\left(\sqrt{\lambda_n}\right)^2 + 1}} \tag{8.48}$$

for $n = 1, 2, 3, \ldots$.

Statement of orthonormality with the respective weight function $w(x) = 1$

> w(x):=1:Int(X[n](x)*X[m](x),x=0..a)=delta(n,m);

$$\int_0^1 \frac{2\sin\left(\sqrt{\lambda_n}x\right)\sin\left(\sqrt{\lambda_m}x\right)}{\sqrt{\cos\left(\sqrt{\lambda_n}\right)^2 + 1}\sqrt{\cos\left(\sqrt{\lambda_m}\right)^2 + 1}}dx = \delta(n, m) \tag{8.49}$$

Time-dependent equation

> diff(T[n](t),t)+k*lambda[n]*T[n](t)=Q[n](t);

$$\frac{d}{dt}T_n(t) + \frac{1}{20}\lambda_n T_n(t) = Q_n(t) \tag{8.50}$$

> q(x,t):=h(x,t)−diff(s(x,t),t);

$$q(x, t) := t \tag{8.51}$$

> Q[n](t):=Int(q(x,t)*X[n](x),x=0..a);Q[n](t):=expand(value(%)):

$$Q_n(t) := \int_0^1 \frac{t\sqrt{2}\sin\left(\sqrt{\lambda_n}x\right)}{\sqrt{\cos\left(\sqrt{\lambda_n}\right)^2 + 1}}dx \tag{8.52}$$

> Q[n](t):=simplify(factor(subs({sin(n*Pi)=0,cos(n*Pi)=(−1)^n},Q[n](t))));
  Q[n](tau):=subs(t=tau,%):

$$Q_n(t) := -\frac{t\sqrt{2}\left(-1 + \cos\left(\sqrt{\lambda_n}\right)\right)}{\sqrt{\cos\left(\sqrt{\lambda_n}\right)^2 + 1}\sqrt{\lambda_n}} \tag{8.53}$$

Basis vector

> T1(t):=exp(−k*lambda[n]*t);

$$T1(t) := e^{-\frac{1}{20}\lambda_n t} \tag{8.54}$$

First-order Green's function

> G1(t,tau):=simplify(T1(t)/(subs(t=tau,T1(t))));

$$G1(t, \tau) := e^{-\frac{1}{20}\lambda_n (t-\tau)} \tag{8.55}$$

Time-dependent solution

> T[n](t):=eval(C(n)*T1(t)+Int(G1(t,tau)*Q[n](tau),tau=0..t));

$$T_n(t) := C(n)e^{-\frac{1}{20}\lambda_n t} + \int_0^t \left(-\frac{e^{-\frac{1}{20}\lambda_n (t-\tau)}\tau\sqrt{2}\left(-1+\cos\left(\sqrt{\lambda_n}\right)\right)}{\sqrt{\cos\left(\sqrt{\lambda_n}\right)^2+1}\sqrt{\lambda_n}} \right) d\tau \tag{8.56}$$

> T[n](t):=value(%);v[n](x,t):=(eval(T[n](t)*X[n](x))):

$$T_n(t) := C(n)e^{-\frac{1}{20}\lambda_n t} \tag{8.57}$$
$$-\frac{1}{\lambda_n^{5/2}\sqrt{\cos\left(\sqrt{\lambda_n}\right)^2+1}}\left(20\sqrt{2}\left(-20\,e^{-\frac{1}{20}\lambda_n t}+20\,e^{-\frac{1}{20}\lambda_n t}\cos\left(\sqrt{\lambda_n}\right)+20\right.\right.$$
$$\left.\left.-20\cos\left(\sqrt{\lambda_n}\right)-\lambda_n t+\lambda_n t\cos\left(\sqrt{\lambda_n}\right)\right)\right)$$

Variable portion of solution

> v(x,t):=(Sum(v[n](x,t),n=1..infinity));

$$v(x, t) := \sum_{n=1}^{\infty} \frac{1}{\sqrt{\cos\left(\sqrt{\lambda_n}\right)^2+1}}\left(\left(C(n)e^{-\frac{1}{20}\lambda_n t}\right.\right. \tag{8.58}$$
$$-\frac{1}{\lambda_n^{5/2}\sqrt{\cos\left(\sqrt{\lambda_n}\right)^2+1}}\left(20\sqrt{2}\left(-20\,e^{-\frac{1}{20}\lambda_n t}+20\,e^{-\frac{1}{20}\lambda_n t}\cos\left(\sqrt{\lambda_n}\right)+20\right.\right.$$
$$\left.\left.\left.\left.-20\cos\left(\sqrt{\lambda_n}\right)-\lambda_n t+\lambda_n t\cos\left(\sqrt{\lambda_n}\right)\right)\right)\right)\right)\sqrt{2}\sin\left(\sqrt{\lambda_n}x\right)\right)$$

The Fourier coefficients $C(n)$ are determined from the integral in Section 8.3 for the special case where

> f(x):=−40/3*x^2+45/2*x+5;

$$f(x) := -\frac{40}{3}x^2 + \frac{45}{2}x + 5 \tag{8.59}$$

> C(n):=Int((f(x)−s(x,0))*X[n](x),x=0..a);C(n):=expand(value(%)):

$$C(n) := \int_0^1 \frac{\left(-\frac{40}{3}x^2 + 20x\right)\sqrt{2}\sin\left(\sqrt{\lambda_n}x\right)}{\sqrt{\cos\left(\sqrt{\lambda_n}\right)^2 + 1}}\,dx \tag{8.60}$$

> C(n):=simplify(subs({sin(sqrt(lambda[n])*a)=−sqrt(lambda[n])*cos(sqrt(lambda[n])*a),
  sin(n*Pi)=0,cos(n*Pi)=(−1)^n},C(n)));

$$C(n) := -\frac{80}{3}\frac{\sqrt{2}\left(-1 + \cos\left(\sqrt{\lambda_n}\right)\right)}{\lambda_n^{(3/2)}\sqrt{\cos\left(\sqrt{\lambda_n}\right)^2 + 1}} \tag{8.61}$$

for $n = 1, 2, 3, \ldots$.

> T[n](t):=(C(n)*T1(t)+int(G1(t,tau)*Q[n](tau),tau=0..t)):v[n](x,t):=(T[n](t)*X[n](x)):

Final solution (linear plus variable portion)

> u(x,t):=(Sum(v[n](x,t),n=1..infinity)+s(x,t));

$$u(x, t) := \sum_{n=1}^{\infty} \frac{1}{\sqrt{\cos\left(\sqrt{\lambda_n}\right)^2 + 1}}\left(\left(-\frac{80}{3}\frac{\sqrt{2}\left(-1 + \cos\left(\sqrt{\lambda_n}\right)\right)e^{-\frac{1}{20}\lambda_n t}}{\lambda_n^{(3/2)}\sqrt{\cos\left(\sqrt{\lambda_n}\right)^2 + 1}}\right.\right. \tag{8.62}$$

$$-\frac{1}{\lambda_n^{(5/2)}\sqrt{\cos\left(\sqrt{\lambda_n}\right)^2 + 1}}\left(20\sqrt{2}\left(-20\,e^{-\frac{1}{20}\lambda_n t} + 20\,e^{-\frac{1}{20}\lambda_n t}\cos\left(\sqrt{\lambda_n}\right) + 20\right.\right.$$

$$\left.\left.\left.- 20\cos\left(\sqrt{\lambda_n}\right) - \lambda_n t + \lambda_n t\cos\left(\sqrt{\lambda_n}\right)\right)\right)\sqrt{2}\sin\left(\sqrt{\lambda_n}x\right)\right) + \frac{5}{2}x + 5$$

Evaluation of the eigenvalues λ_n from the roots of the eigenvalue equation yields

> tan(sqrt(lambda[n])*a)+sqrt(lambda[n])=0;

$$\tan\left(\sqrt{\lambda_n}\right) + \sqrt{\lambda_n} = 0 \tag{8.63}$$

> plot({tan(z),−z},z=0..20,y=−20..0,thickness=10);

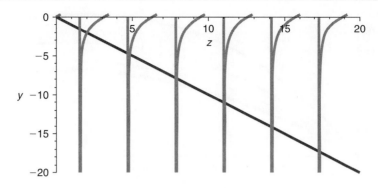

Figure 8.3

If we set $z = \sqrt{\lambda}a$, then the eigenvalues λ_n are the values of z at the intersection points of the curves shown in Figure 8.3. A few of the eigenvalues using the Maple fsolve command are evaluated here:

```
> lambda[1]:=(1/a^2)*(fsolve((tan(z)+z),z=1..3)^2);
```

$$\lambda_1 := 4.115858365 \tag{8.64}$$

```
> lambda[2]:=(1/a^2)*(fsolve((tan(z)+z),z=3..6)^2);
```

$$\lambda_2 := 24.13934203 \tag{8.65}$$

```
> lambda[3]:=(1/a^2)*(fsolve((tan(z)+z),z=6..9)^2);
```

$$\lambda_3 := 63.65910654 \tag{8.66}$$

First few terms of sum

```
> u(x,t):=s(x,t)+sum(v[n](x,t),n=1..3):
```

ANIMATION

```
> animate(u(x,t),x=0..a,t=0..5,thickness=3);
```

The preceding animation command displays the spatial-time-dependent solution of $u(x, t)$ for the given boundary conditions and initial conditions. The animation sequence here and in Figure 8.4 shows snapshots of the animation at times $t = 0, 1, 2, 3, 4, 5$. Note how the solution satisfies the given boundary and initial conditions.

ANIMATION SEQUENCE

```
> u(x,0):=subs(t=0,u(x,t)):u(x,1):=subs(t=1,u(x,t)):
> u(x,2):=subs(t=2,u(x,t)):u(x,3):=subs(t=3,u(x,t)):
> u(x,4):=subs(t=4,u(x,t)):u(x,5):=subs(t=5,u(x,t)):
> plot({u(x,0),u(x,1),u(x,2),u(x,3),u(x,4),u(x,5)},x=0..a,thickness=10);
```

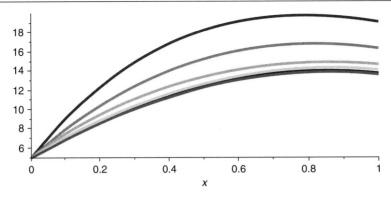

Figure 8.4

EXAMPLE 8.4.4: (Electrical heating of a rod) We seek the temperature distribution in a thin rod whose lateral surface is insulated over the interval $I = \{x \mid 0 < x < 1\}$. The left and right ends of the rod are held at the fixed temperature of zero. The center of the rod is embedded with a wire whose winding pitch decreases linearly with x (the space between windings decreases linearly). The wire is electrically excited by a constant voltage source that is switched into the circuit at time $t = 0$. The resistive heating effect of the wire gives rise to an internal heat source term $h(x, t)$ given as follows. The thermal diffusivity of the rod is $k = 1/40$, and the initial temperature distribution $u(x, 0) = f(x)$ is given here.

SOLUTION: The nonhomogeneous diffusion equation is

$$\frac{\partial}{\partial t} u(x, t) = k \left(\frac{\partial^2}{\partial x^2} u(x, t) \right) + h(x, t)$$

The boundary conditions are homogeneous type 1 at the left and homogeneous type 1 at the right:

$$u(0, t) = 0 \quad \text{and} \quad u(1, t) = 0$$

The initial condition is

$$u(x, 0) = 0$$

The internal heat source term is

$$h(x, t) = xt$$

The solution is $u(x, t) = v(x, t) + s(x, t)$, where $s(x, t)$ is the linear portion of the solution and $v(x, t)$ is the variable portion of the solution, which satisfies the partial differential equation

$$\frac{\partial}{\partial t} v(x, t) = k \left(\frac{\partial^2}{\partial x^2} v(x, t) \right) + q(x, t)$$

where

$$q(x, t) = h(x, t) - \left(\frac{\partial}{\partial t} s(x, t) \right)$$

Assignment of system parameters

> restart: with(plots):a:=1:k:=1/40:h(x,t):=x*t:

> A(t):=0:B(t):=0:kappa[1]:=1:kappa[2]:=0:kappa[3]:=1:kappa[4]:=0:

> b(t):=(kappa[3]*a*A(t)+A(t)*kappa[4]−B(t)*kappa[2])/(kappa[1]*kappa[3]*a+ kappa[1]*kappa[4]−kappa[2]*kappa[3]);

$$b(t) := 0 \tag{8.67}$$

> m(t):=(kappa[1]*B(t)−A(t)*kappa[3])/(kappa[1]*kappa[3]*a+kappa[1]*kappa[4]− kappa[2]*kappa[3]);

$$m(t) := 0 \tag{8.68}$$

Linear portion of solution

> s(x,t):=m(t)*x+b(t);s(x,0):=eval(subs(t=0,s(x,t))):

$$s(x, t) := 0 \tag{8.69}$$

By the method of separation of variables, the variable solution is

$$v(x, t) = \sum_{n=0}^{\infty} X_n(x) T_n(t)$$

where $T_n(t)$ is the solution to the time-dependent differential equation

$$\frac{d}{dt} T_n(t) + k\lambda_n T_n(t) = Q_n(t)$$

and $X_n(x)$ is the solution to the spatial-dependent eigenvalue equation

$$\frac{d^2}{dx^2} X_n(x) + \lambda_n X_n(x) = 0$$

with boundary conditions

$$X(0) = 0 \quad \text{and} \quad X(1) = 0$$

The corresponding homogeneous eigenfunction problem consists of the Euler equation, along with the homogeneous boundary conditions that are type 1 at the left and type 1 at the right. The allowed eigenvalues and corresponding orthonormal eigenfunctions are obtained from Example 2.5.1.

> lambda[n]:=(n*Pi/a)^2;

$$\lambda_n := n^2\pi^2 \tag{8.70}$$

> X[n](x):=sqrt(2/a)*sin(n*Pi/a*x);X[m](x):=subs(n=m,X[n](x)):

$$X_n(x) := \sqrt{2}\,\sin(n\pi x) \tag{8.71}$$

for $n = 1, 2, 3, \ldots$.

Statement of orthonormality with the respective weight function $w(x) = 1$

> w(x):=1:Int(X[n](x)*X[m](x)*w(x),x=0..a)=delta(n,m);

$$\int_0^1 2\,\sin(n\pi x)\sin(m\pi x)\,\mathrm{d}x = \delta(n, m) \tag{8.72}$$

Time-dependent equation

> diff(T[n](t),t)+k*lambda[n]*T[n](t)=Q[n](t);

$$\frac{\mathrm{d}}{\mathrm{d}t}T_n(t) + \frac{1}{40}n^2\pi^2 T_n(t) = Q_n(t) \tag{8.73}$$

> q(x,t):=h(x,t)−diff(s(x,t),t);

$$q(x, t) := xt \tag{8.74}$$

> Q[n](t):=Int(q(x,t)*X[n](x),x=0..a);Q[n](t):=expand(value(%)):

$$Q_n(t) := \int_0^1 xt\sqrt{2}\,\sin(n\pi x)\,\mathrm{d}x \tag{8.75}$$

> Q[n](t):=simplify(factor(subs({sin(n*Pi)=0,cos(n*Pi)=(−1)^n},Q[n](t))));
  Q[n](tau):=subs(t=tau,%):

$$Q_n(t) := -\frac{t\sqrt{2}(-1)^n}{n\pi} \tag{8.76}$$

Basis vector

> T1(t):=exp(−k*lambda[n]*t);

$$T1(t) := e^{-\frac{1}{40}n^2\pi^2 t} \tag{8.77}$$

First-order Green's function

> G1(t,tau):=simplify(T1(t)/(subs(t=tau,T1(t))));

$$G1(t, \tau) := e^{-\frac{1}{40}n^2\pi^2(t-\tau)} \tag{8.78}$$

Time-dependent solution

> T[n](t):=eval(C(n)*T1(t)+Int(G1(t,tau)*Q[n](tau),tau=0..t));

$$T_n(t) := C(n)\, e^{-\frac{1}{40}n^2\pi^2 t} + \int_0^t \left(-\frac{e^{-\frac{1}{40}n^2\pi^2(t-\tau)}\, \tau\sqrt{2}(-1)^n}{n\pi} \right) d\tau \qquad (8.79)$$

> T[n](t):=value(%);v[n](x,t):=simplify(eval(T[n](t)*X[n](x))):

$$T_n(t) := C(n)\, e^{-\frac{1}{40}n^2\pi^2 t} + \frac{40(-1)^{1+n}\sqrt{2}\left(40\, e^{-\frac{1}{40}n^2\pi^2 t} - 40 + n^2\pi^2 t\right)}{n^5\pi^5} \qquad (8.80)$$

Variable portion of solution

> v(x,t):=Sum(v[n](x,t),n=1..infinity);

$$v(x,t) := \sum_{n=1}^{\infty} \frac{1}{n^2\pi^5}\left(\left(C(n)e^{-\frac{1}{40}n^2\pi^2 t}n^5\pi^5 + 1600(-1)^{1+n}e^{-\frac{1}{40}n^2\pi^2 t}\sqrt{2}\right.\right.$$
$$\left.\left. + 1600\sqrt{2}(-1)^n + 40(-1)^{1+n}\sqrt{2}n^2\pi^2 t\right)\sqrt{2}\sin(n\pi x)\right) \qquad (8.81)$$

The Fourier coefficients $C(n)$ are determined from the integral in Section 8.3 for the special case where

> f(x):=0;

$$f(x) := 0 \qquad (8.82)$$

> C(n):=Int((f(x)−s(x,0))*X[n](x),x=0..a);C(n):=expand(value(%)):

$$C(n) := \int_0^1 0\, dx \qquad (8.83)$$

> C(n):=simplify(subs({sin(n*Pi)=0,cos(n*Pi)=(−1)^n},C(n)));

$$C(n) := 0 \qquad (8.84)$$

for $n = 1, 2, 3, \ldots$,

> T[n](t):=(C(n)*T1(t)+int(G1(t,tau)*Q[n](tau),tau=0..t)):v[n](x,t):=(T[n](t)*X[n](x)):

Final solution (linear plus variable portion)

> u(x,t):=eval(Sum(v[n](x,t),n=1..infinity)+s(x,t));

$$u(x,t) := \sum_{n=1}^{\infty} \frac{80(-1)^{1+n}\left(40\, e^{-\frac{1}{40}n^2\pi^2 t} - 40 + n^2\pi^2 t\right)\sin(n\pi x)}{n^5\pi^5} \qquad (8.85)$$

First few terms of sum

```
> u(x,t):=s(x,t)+sum(v[n](x,t),n=1..3):
```

ANIMATION

```
> animate(u(x,t),x=0..a,t=0..5,thickness=3);
```

The preceding animation command displays the spatial-time-dependent solution of $u(x, t)$ for the given boundary conditions and initial conditions. The animation sequence here and in Figure 8.5 shows snapshots of the animation at times $t = 0, 1, 2, 3, 4, 5$. Note how the solution satisfies the given boundary and initial conditions.

ANIMATION SEQUENCE

```
> u(x,0):=subs(t=0,u(x,t)):u(x,1):=subs(t=1,u(x,t)):
> u(x,2):=subs(t=2,u(x,t)):u(x,3):=subs(t=3,u(x,t)):
> u(x,4):=subs(t=4,u(x,t)):u(x,5):=subs(t=5,u(x,t)):
> plot({u(x,0),u(x,1),u(x,2),u(x,3),u(x,4),u(x,5)},x=0..a,thickness=10);
```

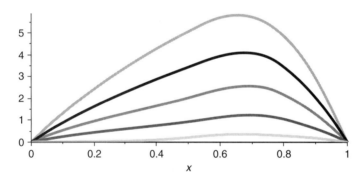

Figure 8.5

8.5 Nonhomogeneous Wave Equation

The generalized nonhomogeneous wave partial differential equation in one dimension over the finite interval $I = \{x \mid 0 < x < a\}$, with no damping in the system, has the form

$$\frac{\partial^2}{\partial t^2} u(x, t) = c^2 \left(\frac{\partial^2}{\partial x^2} u(x, t) \right) + h(x, t)$$

Here, $u(x, t)$ denotes the wave amplitude with dimensions of distance, c is the wave speed with dimensions of distance per time, and $h(x, t)$, with dimensions of force per mass, now accounts for any external forces acting on the system. We consider this partial differential equation along with the time-dependent nonhomogeneous boundary conditions

$$\kappa_1 u(0, t) + \kappa_2 u_x(0, t) = A(t)$$

and

$$\kappa_3 u(a, t) + \kappa_4 u_x(a, t) = B(t)$$

where $A(t)$ and $B(t)$ are the time-dependent nonhomogeneous source terms at the boundaries of the domain. We seek a solution to this problem subject to satisfying the following two initial conditions:

$$u(x, 0) = f(x)$$

and

$$u_t(x, 0) = g(x)$$

Here, $u(x, t) = f(x)$ denotes the initial amplitude distribution of the wave, and $u_t(x, 0) = g(x)$ denotes the initial speed distribution of the wave.

Similar to what we did earlier for the diffusion equation, the solution procedure is driven by our need to find a complete set of orthonormal eigenfunctions corresponding to the equivalent set of homogeneous boundary conditions over the given domain. To move in this direction, we use the superposition principle, and we assume our solution to have the form

$$u(x, t) = v(x, t) + s(x, t)$$

For the types of boundary conditions we will be considering, $s(x, t)$ will continue to have the form of a "linear" solution and $v(x, t)$ will have the form of a "variable" solution. Substituting the preceding assumed solution into the partial differential equation yields

$$\frac{\partial^2}{\partial t^2} v(x, t) + \frac{\partial^2}{\partial t^2} s(x, t) = c^2 \left(\frac{\partial^2}{\partial x^2} v(x, t) + \frac{\partial^2}{\partial x^2} s(x, t) \right) + h(x, t)$$

Doing the same with the boundary conditions, we get

$$A(t) = \kappa_1 v(0, t) + \kappa_2 v_x(0, t) + \kappa_1 s(0, t) + \kappa_2 s_x(0, t)$$

and

$$B(t) = \kappa_3 v(a, t) + \kappa_4 v_x(a, t) + \kappa_3 s(a, t) + \kappa_4 s_x(a, t)$$

Similarly, for the two initial conditions, we have

$$f(x) = v(x, 0) + s(x, 0)$$

and

$$g(x) = v_t(x, 0) + s_t(x, 0)$$

As we did with the diffusion equation, our goal now is to substitute the conditions on the solutions $v(x, t)$ and $s(x, t)$ so

1. $s(x, t)$ (steady-state portion) will be a linear function of x; thus, the second-order partial derivative of $s(x, t)$ with respect to x is zero

2. $v(x, t)$ (the variable portion) will satisfy the corresponding nonhomogeneous equation with homogeneous boundary conditions

Doing so yields two initial condition-boundary value problems that have to be solved. For $v(x, t)$, the variable portion of the solution, we have the nonhomogeneous partial differential equation

$$\frac{\partial^2}{\partial t^2} v(x, t) = c^2 \left(\frac{\partial^2}{\partial x^2} v(x, t) \right) - \left(\frac{\partial^2}{\partial t^2} s(x, t) \right) + h(x, t)$$

with the corresponding homogeneous boundary conditions

$$\kappa_1 v(0, t) + \kappa_2 v_x(0, t) = 0$$

and

$$\kappa_3 v(a, t) + \kappa_4 v_x(a, t) = 0$$

and initial conditions

$$v(x, 0) = f(x) - s(x, 0)$$

and

$$v_t(x, 0) = g(x) - s_t(x, 0)$$

For $s(x, t)$, the spatially linear portion of the solution, we have

$$\frac{\partial^2}{\partial x^2} s(x, t) = 0$$

with nonhomogeneous boundary conditions

$$A(t) = \kappa_1 s(0, t) + \kappa_2 s_x(0, t)$$

and

$$B(t) = \kappa_3 s(a, t) + \kappa_4 s_x(a, t)$$

We solved the exact same problem for the linear portion of the preceding solution $s(x, t)$ for the diffusion equation, and we got

$$s(x, t) = \frac{(\kappa_1 B(t) - A(t)\kappa_3)x}{\kappa_1\kappa_3 a + \kappa_1\kappa_4 - \kappa_2\kappa_3} + \frac{\kappa_3 a A(t) + A(t)\kappa_4 - B(t)\kappa_2}{\kappa_1\kappa_3 a + \kappa_1\kappa_4 - \kappa_2\kappa_3}$$

for

$$\kappa_1\kappa_3 a + \kappa_1\kappa_4 - \kappa_2\kappa_3 \neq 0$$

We now focus on finding the solution to the remaining partial differential equation for $v(x, t)$. Writing

$$q(x, t) = h(x, t) - \left(\frac{\partial^2}{\partial t^2} s(x, t) \right)$$

the preceding partial differential equation for $v(x, t)$ takes on the nonhomogeneous form

$$\frac{\partial^2}{\partial t^2} v(x, t) = c^2 \left(\frac{\partial^2}{\partial x^2} v(x, t) \right) + q(x, t)$$

with the corresponding homogeneous boundary conditions

$$\kappa_1 v(0, t) + \kappa_2 v_x(0, t) = 0$$

and

$$\kappa_3 v(a, t) + \kappa_4 v_x(a, t) = 0$$

and initial conditions

$$v(x, 0) = f(x) - s(x, 0)$$

and

$$v_t(x, 0) = g(x) - s_t(x, 0)$$

Similar to what we did for the diffusion equation, we solve the preceding nonhomogeneous partial differential equation for $v(x, t)$ by the method of "eigenfunction expansion" where the eigenfunctions are those associated with the corresponding homogeneous equation.

To solve the $v(x, t)$ equation, we temporarily set $q(x, t) = 0$ and the partial differential equation becomes

$$\frac{\partial^2}{\partial t^2} v(x, t) = c^2 \left(\frac{\partial^2}{\partial x^2} v(x, t) \right)$$

with homogeneous boundary conditions

$$\kappa_1 v(0, t) + \kappa_2 v_x(0, t) = 0$$

and

$$\kappa_3 v(a, t) + \kappa_4 v_x(a, t) = 0$$

By the method of separation of variables, as done in Chapter 4, we set

$$v(x, t) = X(x)T(t)$$

The corresponding eigenvalue problem in x has the familiar form

$$\frac{d^2}{dx^2}X(x) + \lambda X(x) = 0$$

with regular homogeneous boundary conditions

$$\kappa_1 X(0) + \kappa_2 X_x(0) = 0$$

and

$$\kappa_3 X(a) + \kappa_4 X_x(a) = 0$$

This is a regular Sturm-Liouville eigenvalue problem with an Euler-type operator. The eigenvalues and corresponding eigenfunctions are denoted, respectively, as

$$\lambda_n, X_n(x)$$

for $n = 0, 1, 2, 3, \ldots$.

The corresponding statement of orthonormality, with respect to the weight function $w(x) = 1$ over the interval I, reads—here, $\delta(n, m)$ is the familiar Kronecker delta function—

$$\int_0^a X_n(x)X_m(x)dx = \delta(n, m)$$

The preceding boundary value problem is a regular Sturm-Liouville eigenvalue problem over the finite interval $I = \{x \mid 0 < x < a\}$, and the eigenfunctions form a "complete" set with respect to any piecewise smooth function over this interval. If we assume the solution $v(x, t)$ and the nonhomogeneous function $q(x, t)$ to both be piecewise smooth functions with respect to x over this interval, then the eigenfunction expansions for $v(x, t)$ and $q(x, t)$, in terms of the orthonormal eigenfunctions $X_n(x)$, read, respectively, as

$$v(x, t) = \sum_{n=0}^{\infty} T_n(t)X_n(x)$$

and

$$q(x, t) = \sum_{n=0}^{\infty} Q_n(t)X_n(x)$$

where the terms $T_n(t)$ and $Q_n(t)$ are the time-dependent Fourier coefficients of $v(x, t)$ and $q(x, t)$, respectively.

Because $q(x, t)$ is already determined from knowledge of the functions $h(x, t)$ and $s(x, t)$, we can evaluate the Fourier coefficients $Q_n(t)$ by taking the inner product of the preceding with respect to the orthonormal eigenfunctions and the weight function $w(x) = 1$. Using a procedure similar to that in Section 8.2 yields

$$Q_n(t) = \int_0^a q(x, t) X_n(x) \, dx$$

We now focus on determining the time-dependent portion of the problem. The original nonhomogeneous partial differential equation for the variable portion of the solution $v(x, t)$ reads

$$\frac{\partial^2}{\partial t^2} v(x, t) = c^2 \left(\frac{\partial^2}{\partial x^2} v(x, t) \right) + q(x, t)$$

Substituting the eigenfunction expansions for $v(x, t)$ and $q(x, t)$ into this nonhomogeneous partial differential equation, and assuming the validity of the formal procedures of interchanging the summation and integration operations, yields the following series equation:

$$\sum_{n=0}^{\infty} \left(\frac{d^2}{dt^2} T_n(t) \right) X_n(x) = \sum_{n=0}^{\infty} \left(c^2 T_n(t) \left(\frac{d^2}{dx^2} X_n(x) \right) + Q_n(t) X_n(x) \right)$$

Recall that the spatial eigenfunctions $X_n(x)$ satisfy the eigenvalue equation

$$\frac{d^2}{dx^2} X_n(x) + \lambda_n X_n(x) = 0$$

Making this substitution yields the series equation

$$\sum_{n=0}^{\infty} \left(\frac{d^2}{dt^2} T_n(t) + c^2 \lambda_n T_n(t) \right) X_n(x) = \sum_{n=0}^{\infty} Q_n(t) X_n(x)$$

Since the eigenfunctions are linearly independent, the only way this series equation can hold is if the coefficients of the eigenfunctions on the left are equal to the coefficients of the same eigenfunctions on the right. From this, we arrive at the time-dependent differential equation for $T_n(t)$:

$$\frac{d^2}{dt^2} T_n(t) + c^2 \lambda_n T_n(t) = Q_n(t)$$

From Section 1.4, a set of system basis vectors for the preceding equation (for λ_n positive) is

$$T1(t) = \cos\left(c\sqrt{\lambda_n} t\right) \quad \text{and} \quad T2(t) = \sin\left(c\sqrt{\lambda_n} t\right)$$

From Section 1.8 on second-order initial-value problems, the corresponding second-order Green's function (with no damping in the system) is evaluated to be

$$G2(t, \tau) = \frac{\sin\left(c\sqrt{\lambda_n}(t - \tau)\right)}{c\sqrt{\lambda_n}}$$

The complete solution to this ordinary nonhomogeneous second-order differential equation in t reads

$$T_n(t) = C(n)\cos\left(c\sqrt{\lambda_n}\, t\right) + D(n)\sin\left(c\sqrt{\lambda_n}\, t\right) + \int_0^t G2(t, \tau)Q_n(\tau)\,d\tau$$

Here, $C(n)$ and $D(n)$ are arbitrary unknown coefficients. Similar to our analysis in Chapter 4, examination of the time-dependent behavior of the preceding system, for zero damping, reveals the nth-mode undamped angular frequency of the system to be given as

$$\omega_n = c\sqrt{\lambda_n}$$

Thus, our solution to the partial differential equation for $v(x, t)$ is of the form of a particular solution, and we write it as

$$v(x, t) = \sum_{n=0}^{\infty}\left(C(n)\cos(\omega_n t) + D(n)\sin(\omega_n t) + \int_0^t \frac{\sin(\omega_n(t - \tau))Q_n(\tau)}{\omega_n}d\tau\right)X_n(x)$$

We demonstrate these concepts by solving an illustrative problem.

DEMONSTRATION: We seek the wave amplitude $u(x, t)$ for transverse wave motion on a taut rope over the interval $I = \{x \,|\, 0 < x < 1\}$. The left end of the rope is attached to a vertical oscillatory motor, which provides a sinusoidal motion to the left end of the rope, and the right end of the rope is held fixed. There is no damping in the system, and there is no external applied force acting $[h(x, t) = 0]$. The rope has an initial displacement distribution $f(x)$ and speed distribution $g(x)$ given as follows. The wave speed is $c = 1/2$.

SOLUTION: The system partial differential equation is

$$\frac{\partial^2}{\partial t^2}u(x, t) = \frac{\frac{\partial^2}{\partial x^2}u(x, t)}{4}$$

The initial displacement distribution is

$$u(x, 0) = f(x)$$

and the initial speed distribution is

$$u_t(x, 0) = g(x)$$

Since the left end of the rope is moved sinusoidally and the right end is held secure, the boundary conditions are

$$u(0, t) = \sin(t) \quad \text{and} \quad u(1, t) = 0$$

Comparing the boundary conditions with the generalized boundary conditions shown here,

$$\kappa_1 u(0, t) + \kappa_2 u_x(0, t) = A(t)$$

and

$$\kappa_3 u(a, t) + \kappa_4 u_x(a, t) = B(t)$$

we must set

$$A(t) = \sin(t) \quad \text{and} \quad B(t) = 0$$

and

$$\kappa_1 = 1, \ \kappa_2 = 0, \ \kappa_3 = 1, \ \kappa_4 = 0$$

With the given values for κ_n, $A(t)$, and $B(t)$, the components for the linear portion of the solution are evaluated to be

$$b(t) = \sin(t)$$

and

$$m(t) = -\sin(t)$$

Thus, the linear portion of the solution reads

$$s(x, t) = -x \sin(t) + \sin(t)$$

For the variable (particular) portion of the problem, the partial differential equation for $v(x, t)$ reads

$$\frac{\partial^2}{\partial t^2} v(x, t) = \frac{\frac{\partial^2}{\partial x^2} v(x, t)}{4} + q(x, t)$$

where $q(x, t)$ is

$$q(x, t) = h(x, t) - \left(\frac{\partial^2}{\partial t^2} s(x, t) \right)$$

Because $h(x, t) = 0$, this evaluates to

$$q(x, t) = -x \sin(t) + \sin(t)$$

The corresponding homogeneous boundary conditions on $v(x, t)$ are

$$v(0, t) = 0 \quad \text{and} \quad v(1, t) = 0$$

The initial conditions on $v(x, t)$ are

$$v(x, 0) = f(x) - s(x, 0)$$

which evaluates to

$$v(x, 0) = f(x)$$

and

$$v_t(x, 0) = g(x) - s_t(x, 0)$$

which evaluates to

$$v_t(x, 0) = g(x) + x - 1$$

To solve the $v(x, t)$ equation, we temporarily set $q(x, t) = 0$, and we assume a solution of the form

$$v(x, t) = X(x)T(t)$$

From the method of separation of variables, the corresponding spatial-dependent equation is the Euler-type differential equation

$$\frac{d^2}{dx^2} X(x) + \lambda\, X(x) = 0$$

The corresponding homogeneous boundary conditions are type 1 at $x = 0$ and type 1 at $x = 1$:

$$X(0) = 0 \quad \text{and} \quad X(1) = 0$$

The allowed eigenvalues and corresponding orthonormal eigenfunctions, obtained from Example 2.5.1, are

$$\lambda_n = n^2 \pi^2$$

and

$$X_n(x) = \sqrt{2}\sin(n\pi x)$$

for $n = 1, 2, 3, \ldots$.

The statement of orthonormality with the weight function $w(x) = 1$ reads

$$\int_0^1 2 \sin(n\pi x) \sin(m\pi x)\,dx = \delta(n, m)$$

The time-dependent differential equation reads

$$\frac{d^2}{dt^2}T_n(t) + \frac{\lambda_n T_n(t)}{4} = Q_n(t)$$

where $Q_n(t)$ is the inner product of $q(x, t)$ with the spatial eigenfunction—that is,

$$Q_n(t) = \int_0^1 (-x\sin(t) + \sin(t))\sqrt{2}\sin(n\pi x)dx$$

Evaluation of this integral yields

$$Q_n(t) = \frac{\sqrt{2}\sin(t)}{n\pi}$$

Thus, the nonhomogeneous time-dependent differential equation reads

$$\frac{d^2}{dt^2}T_n(t) + \frac{n^2\pi^2 T_n(t)}{4} = \frac{\sqrt{2}\sin(t)}{n\pi}$$

A set of basis vectors for this differential equation, from Section 1.4, reads

$$T1(t) = \cos\left(\frac{n\pi t}{2}\right) \quad \text{and} \quad T2(t) = \sin\left(\frac{n\pi t}{2}\right)$$

Evaluation of the corresponding second-order Green's function yields

$$G2(t, \tau) = \frac{2\sin\left(\frac{n\pi(t-\tau)}{2}\right)}{n\pi}$$

Thus, the solution to the preceding time-dependent differential equation is

$$T_n(t) = C(n)\cos\left(\frac{n\pi t}{2}\right) + D(n)\sin\left(\frac{n\pi t}{2}\right) + \int_0^t \frac{2\sqrt{2}\sin\left(\frac{n\pi(t-\tau)}{2}\right)\sin(\tau)}{n^2\pi^2}d\tau$$

for $n = 1, 2, 3, \ldots$.

Finally, the solution for the variable or particular portion of the problem is constructed from the superposition of the products of the time-dependent solutions and the spatial eigenfunctions evaluated earlier and this reads

$$v(x, t) = \sum_{n=1}^{\infty}\left(C(n)\cos\left(\frac{n\pi t}{2}\right) + D(n)\sin\left(\frac{n\pi t}{2}\right) + \frac{4\sqrt{2}\sin(t)}{n\pi(n^2\pi^2 - 4)}\right.$$

$$\left. - \frac{8\sqrt{2}\sin\left(\frac{n\pi t}{2}\right)}{n^2\pi^2(n^2\pi^2 - 4)}\right)\sqrt{2}\sin(n\pi x) \tag{8.86}$$

The coefficients $C(n)$ and $D(n)$ are to be determined from the initial conditions on the problem.

8.6 Initial Condition Considerations for the Nonhomogeneous Wave Equation

At time $t = 0$, the two initial conditions on the solution are $u(x, 0) = f(x)$ and $u_t(x, 0) = g(x)$. This converts to the equivalent initial conditions on the variable portion of the solution $v(x, t)$ to read

$$v(x, 0) = f(x) - s(x, 0)$$

and

$$v_t(x, 0) = g(x) - s_t(x, 0)$$

If we set $t = 0$ in the eigenfunction expansion of the variable portion of the solution for $v(x, t)$ shown here

$$v(x, t) = \sum_{n=0}^{\infty} \left(C(n) \cos(\omega_n t) + D(n) \sin(\omega_n t) + \int_0^t \frac{\sin(\omega_n (t - \tau)) Q_n(\tau)}{\omega_n} \, d\tau \right) X_n(x)$$

and its corresponding time derivative, then the initial conditions on $v(x, t)$ lead us to the following two series equations

$$f(x) - s(x, 0) = \sum_{n=0}^{\infty} C(n) X_n(x)$$

and

$$g(x) - s_t(x, 0) = \sum_{n=0}^{\infty} D(n) \omega_n X_n(x)$$

These are the familiar Fourier series expansions of the left-hand functions in terms of the orthonormalized eigenfunctions. To evaluate the Fourier coefficients $C(n)$ and $D(n)$, we take the inner products of both sides with respect to the orthonormal eigenfunctions $X_n(x)$ and the weight function $w(x) = 1$, and we get

$$\int_0^a (f(x) - s(x, 0)) X_m(x) dx = \sum_{n=0}^{\infty} C(n) \left(\int_0^a X_n(x) X_m(x) \, dx \right)$$

and

$$\int_0^a (g(x) - s_t(x, 0)) X_m(x) dx = \sum_{n=0}^{\infty} D(n) \omega_n \left(\int_0^a X_n(x) X_m(x) \, dx \right)$$

Taking advantage of the statement of orthonormality, these two series reduce to

$$\int_0^a (f(x) - s(x, 0)) X_m(x) dx = \sum_{n=0}^{\infty} C(n) \delta(m, n)$$

and

$$\int_0^a (g(x) - s_t(x, 0)) X_m(x) dx = \sum_{n=0}^{\infty} D(n) \omega_n \delta(m, n)$$

Evaluating these sums and taking advantage of the mathematical character of the Kronecker delta functions (only the $m = n$ terms survive), the Fourier coefficients $C(n)$ and $D(n)$ are evaluated from the integrals

$$C(n) = \int_0^a (f(x) - s(x, 0)) X_n(x) \, dx$$

and

$$D(n) = \int_0^a \frac{(g(x) - s_t(x, 0)) X_n(x)}{\omega_n} \, dx$$

Now that we have evaluated the coefficients $C(n)$ and $D(n)$, we can write the final solution to our original nonhomogeneous partial differential equation as the sum of the variable plus the linear solutions

$$u(x, t) = \sum_{n=0}^{\infty} \left(C(n) \cos(\omega_n t) + D(n) \sin(\omega_n t) + \int_0^t \frac{\sin(\omega_n (t - \tau)) Q_n(\tau)}{\omega_n} \, d\tau \right) X_n(x) + s(x, t)$$

$$(8.87)$$

where $C(n)$, $D(n)$, $Q_n(\tau)$, and $s(x, t)$ were all evaluated earlier. We demonstrate the preceding concepts for the illustrative problem from Section 8.5.

DEMONSTRATION: We consider the example problem done in Section 8.5 for the special case where the initial conditions are

$$u(x, 0) = f(x)$$

where

$$f(x) = 0$$

and

$$u_t(x, 0) = g(x)$$

where

$$g(x) = 0$$

SOLUTION: We previously evaluated the linear portion of the solution $s(x, t) = -x\sin(t) + \sin(t)$ and the natural angular frequency $\omega_n = n\pi/2$. The given initial conditions on $u(x, t)$ translate to the corresponding initial conditions on the variable portion of the solution shown here:

$$v(x, 0) = f(x) - s(x, 0)$$
$$v_t(x, 0) = g(x) - s_t(x, 0)$$

Insertion of these values into the integral for $C(n)$

$$C(n) = \int_0^a (f(x) - s(x, 0)) X_n(x) \, dx$$

yields

$$C(n) = \int_0^1 0\sqrt{2}\sin(n\pi x) \, dx$$

which evaluates to

$$C(n) = 0$$

Similarly, the integral for $D(n)$

$$D(n) = \int_0^a \frac{(g(x) - s_t(x, 0)) X_n(x)}{\omega_n} \, dx$$

yields

$$D(n) = 2\left(\int_0^1 \frac{(x - 1)\sqrt{2}\sin(n\pi x)}{n\pi} \, dx \right)$$

which evaluates to

$$D(n) = \frac{-2\sqrt{2}}{n^2\pi^2} \tag{8.88}$$

for $n = 1, 2, 3, \ldots$.

With these values for the Fourier coefficients, the sum of the particular portion plus the linear portion yields the final solution

$$u(x, t) = -4 \left(\sum_{n=1}^{\infty} \frac{\left(n\pi \, \sin\left(\frac{n\pi t}{2}\right) - 2 \, \sin(t)\right) \sin(n\pi x)}{n\pi \left(n^2\pi^2 - 4\right)} \right) + (1 - x)\sin(t)$$

Note how the solution satisfies the given boundary and initial conditions. The details for the development of this solution along with the graphics are given later in one of the Maple worksheet examples.

8.7 Example Nonhomogeneous Problems for the Wave Equation

EXAMPLE 8.7.1: (The whip-rope problem) We seek the amplitude $u(x, t)$ for transverse wave motion along a taut rope over the interval $I = \{x \,|\, 0 < x < 1\}$. The left end of the rope is attached to a vertical oscillatory motor, which provides a sinusoidal motion to the end of the rope, and the right end is held fixed. There is no damping in the system. There is no external force acting on the rope, and the rope has an initial displacement distribution $u(x, 0) = f(x)$ and an initial speed distribution $u_t(x, 0) = g(x)$ given as follows. The wave speed is $c = 1/2$.

SOLUTION: The nonhomogeneous wave equation is

$$\frac{\partial^2}{\partial t^2} u(x, t) = c^2 \left(\frac{\partial^2}{\partial x^2} u(x, t) \right) + h(x, t)$$

The boundary conditions are nonhomogeneous type 1 at the left and homogeneous type 1 at the right:

$$u(0, t) = \sin(t) \quad \text{and} \quad u(1, t) = 0$$

The initial displacement distribution is

$$u(x, 0) = 0$$

and the initial speed distribution is

$$u_t(x, 0) = 0$$

The external applied force is

$$h(x, t) = 0$$

The solution is $u(x, t) = v(x, t) + s(x, t)$, where $s(x, t)$ is the linear portion of the solution and $v(x, t)$ is the variable portion of the solution that satisfies the partial differential equation

$$\frac{\partial^2}{\partial t^2} v(x, t) = c^2 \left(\frac{\partial^2}{\partial x^2} v(x, t) \right) + q(x, t)$$

where

$$q(x, t) = h(x, t) - \left(\frac{\partial^2}{\partial t^2} s(x, t) \right)$$

Assignment of system parameters

> restart: with(plots):a:=1:c:=1/2:h(x,t):=0:

> A(t):=sin(t):B(t):=0:kappa[1]:=1:kappa[2]:=0:kappa[3]:=1:kappa[4]:=0:

> b(t):=(kappa[3]*a*A(t)+A(t)*kappa[4]−B(t)*kappa[2])/(kappa[1]*kappa[3]*a+
kappa[1]*kappa[4]−kappa[2]*kappa[3]);

$$b(t) := \sin(t) \tag{8.89}$$

> m(t):=(kappa[1]*B(t)−A(t)*kappa[3])/(kappa[1]*kappa[3]*a+kappa[1]*kappa[4]−
kappa[2]*kappa[3]);

$$m(t) := -\sin(t) \tag{8.90}$$

Linear portion of solution

> s(x,t):=m(t)*x+b(t);s(x,0):=eval(subs(t=0,s(x,t))):s[t](x,0):=eval(subs(t=0,diff(s(x,t),t))):

$$s(x, t) := -\sin(t)x + \sin(t) \tag{8.91}$$

By the method of separation of variables, the variable solution is

$$v(x, t) = \sum_{n=0}^{\infty} X_n(x) T_n(t)$$

where $T_n(t)$ is the solution to the time-dependent differential equation

$$\frac{d^2}{dt^2} T_n(t) + c^2 \lambda_n T_n(t) = Q_n(t)$$

and $X_n(x)$ is the solution to the spatial-dependent eigenvalue equation

$$\frac{d^2}{dx^2} X_n(x) + \lambda_n X_n(x) = 0$$

with boundary conditions

$$X(0) = 0 \quad \text{and} \quad X(1) = 0$$

The corresponding homogeneous eigenfunction problem consists of the Euler equation, along with the homogeneous boundary conditions that are type 1 at the left and type 1 at the right.

The allowed eigenvalues and corresponding orthonormal eigenfunctions, obtained from Example 2.5.1, are

> lambda[n]:=(n*Pi/a)^2;

$$\lambda_n := n^2\pi^2 \tag{8.92}$$

> X[n](x):=sqrt(2/a)*sin(n*Pi/a*x);X[m](x):=subs(n=m,X[n](x)):

$$X_n(x) := \sqrt{2}\sin(n\pi x) \tag{8.93}$$

for $n = 1, 2, 3, \ldots$.

Statement of orthonormality with the respective weight function $w(x) = 1$

> w(x):=1:Int(X[n](x)*X[m](x)*w(x),x=0..a)=delta(n,m);

$$\int_0^1 2\sin(n\pi x)\sin(m\pi x)\mathrm{d}x = \delta(n,m) \tag{8.94}$$

Time-dependent equation

> diff(T[n](t),t,t)+c^2*lambda[n]*T[n](t)=Q[n](t);

$$\frac{\mathrm{d}^2}{\mathrm{d}t^2}T_n(t) + \frac{1}{4}n^2\pi^2 T_n(t) = Q_n(t) \tag{8.95}$$

> q(x,t):=h(x,t)−diff(s(x,t),t,t);

$$q(x,t) := -\sin(t)x + \sin(t) \tag{8.96}$$

> Q[n](t):=Int(q(x,t)*X[n](x),x=0..a);Q[n](t):=expand(value(%)):

$$Q_n(t) := \int_0^1 (-\sin(t)x + \sin(t))\sqrt{2}\sin(n\pi x)\,\mathrm{d}x \tag{8.97}$$

> Q[n](t):=simplify(factor(subs({sin(n*Pi)=0,cos(n*Pi)=(−1)^n},Q[n](t))));
  Q[n](tau):=subs(t=tau,%):

$$Q_n(t) := \frac{\sqrt{2}\sin(t)}{n\pi} \tag{8.98}$$

Basis vectors

> T1(t):=cos(c*n*Pi*t/a);

$$T1(t) := \cos\left(\frac{1}{2}n\pi t\right) \tag{8.99}$$

> T2(t):=sin(c*n*Pi*t/a);

$$T2(t) := \sin\left(\frac{1}{2}n\pi t\right) \qquad (8.100)$$

Second-order Green's function

> G2(t,tau):=sin(c*n*Pi/a*(t−tau))/(c*n*Pi/a);

$$G2(t, \tau) := \frac{2\sin\left(\frac{1}{2}n\pi(t-\tau)\right)}{n\pi} \qquad (8.101)$$

Time-dependent solution

> T[n](t):=C(n)*T1(t)+D(n)*T2(t)+Int(G2(t,tau)*Q[n](tau),tau=0..t);

$$T_n(t) := C(n)\cos\left(\frac{1}{2}n\pi t\right) + D(n)\sin\left(\frac{1}{2}n\pi t\right) + \int_0^t \frac{2\sin\left(\frac{1}{2}n\pi(t-\tau)\right)\sqrt{2}\sin(\tau)}{n^2\pi^2}\,d\tau \qquad (8.102)$$

> T[n](t):=value(%);v[n](x,t):=(T[n](t)*X[n](x)):

$$T_n(t) := C(n)\cos\left(\frac{1}{2}n\pi t\right) + D(n)\sin\left(\frac{1}{2}n\pi t\right) + \frac{4\sqrt{2}\left(-2\sin\left(\frac{1}{2}n\pi t\right)+\sin(t)n\pi\right)}{n^2\pi^2\left(n^2\pi^2-4\right)} \qquad (8.103)$$

Variable portion of solution

> v(x,t):=Sum(v[n](x,t),n=1..infinity);

$$v(x, t) := \sum_{n=1}^{\infty}\left(C(n)\cos\left(\frac{1}{2}n\pi t\right) + D(n)\sin\left(\frac{1}{2}n\pi t\right)\right.$$
$$\left. + \frac{4\sqrt{2}\left(-2\sin\left(\frac{1}{2}n\pi t\right)+\sin(t)n\pi\right)}{n^2\pi^2\left(n^2\pi^2-4\right)}\right)\sqrt{2}\sin(n\pi x) \qquad (8.104)$$

The Fourier coefficients $C(n)$ and $D(n)$ are to be determined from the initial conditions on $v(x, t)$. We now substitute the initial conditions $v(x, 0) = f(x) - s(x, 0)$ and $v_t(x, 0) = g(x) - s_t(x, 0)$ into the integrals for $C(n)$ and $D(n)$ for the special case where

> f(x):=0;

$$f(x) := 0 \qquad (8.105)$$

> g(x):=0;

$$g(x) := 0 \qquad (8.106)$$

The integrals for $C(n)$ and $D(n)$ are

> C(n):=Int((f(x)−s(x,0))*X[n](x),x=0..a);C(n):=expand(value(%)):

$$C(n) := \int_0^1 0 \, dx \tag{8.107}$$

> C(n):=simplify(subs({sin(n*Pi)=0,cos(n*Pi)=(−1)^n},C(n)));

$$C(n) := 0 \tag{8.108}$$

> D(n):=Int((g(x)−s[t](x,0))/(c*n*Pi/a)*X[n](x),x=0..a);D(n):=expand(value(%)):

$$D(n) := \int_0^1 \frac{2(-1+x)\sqrt{2}\sin(n\pi x)}{n\pi} \, dx \tag{8.109}$$

> D(n):=simplify(subs({sin(n*Pi)=0,cos(n*Pi)=(−1)^n},D(n)));

$$D(n) := -\frac{2\sqrt{2}}{n^2\pi^2} \tag{8.110}$$

for $n = 1, 2, 3, \ldots$.

Final solution (linear plus variable portion)

> u(x,t):=simplify(Sum(eval(v[n](x,t)),n=1..infinity))+s(x,t);

$$u(x, t) := \sum_{n=1}^{\infty} \frac{4\left(-\sin\left(\frac{1}{2}n\pi t\right)n\pi + 2\sin(t)\right)\sin(n\pi x)}{n\pi\left(n^2\pi^2 - 4\right)} - \sin(t)x + \sin(t) \tag{8.111}$$

First few terms of sum

> u(x,t):=simplify(sum(eval(v[n](x,t)),n=1..3)+s(x,t)):

ANIMATION

> animate(u(x,t),x=0..a,t=0..5,thickness=3);

The preceding animation command displays the spatial-time-dependent solution of $u(x, t)$ for the given boundary conditions and initial conditions. The animation sequence here and in Figure 8.6 shows snapshots of the animation at times $t = 0, 1, 2, 3, 4, 5$. Note how the solution satisfies the given boundary and initial conditions.

ANIMATION SEQUENCE

> u(x,0):=subs(t=0,u(x,t)):u(x,1):=subs(t=1,u(x,t)):
> u(x,2):=subs(t=2,u(x,t)):u(x,3):=subs(t=3,u(x,t)):

```
> u(x,4):=subs(t=4,u(x,t)):u(x,5):=subs(t=5,u(x,t)):
> plot({u(x,0),u(x,1),u(x,2),u(x,3),u(x,4),u(x,5)},x=0..a,thickness=10);
```

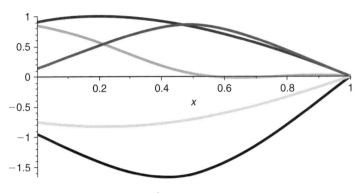

Figure 8.6

EXAMPLE 8.7.2: (Applied force on a magnetic wire) We seek the wave amplitude $u(x, t)$ for transverse wave motion along a taut magnetic wire over the interval $I = \{x \mid 0 < x < 1\}$. Both ends of the wire are held fixed. A nonuniform external applied magnetic force term $h(x, t)$ acts on the wire. There is no damping in the system. The wire has an initial displacement distribution $u(x, 0) = f(x)$ and initial speed distribution $u_t(x, 0) = g(x)$ given as follows. The wave speed is $c = 1/5$.

SOLUTION: The nonhomogeneous wave equation is

$$\frac{\partial^2}{\partial t^2} u(x, t) = c^2 \left(\frac{\partial^2}{\partial x^2} u(x, t) \right) + h(x, t)$$

The boundary conditions are homogeneous type 1 at the left and homogeneous type 1 at the right:

$$u(0, t) = 0 \quad \text{and} \quad u(1, t) = 0$$

The initial displacement distribution is

$$u(x, 0) = x(1 - x)$$

and the initial speed distribution is

$$u_t(x, 0) = x$$

The external applied force is

$$h(x, t) = x \sin(t)$$

The solution is $u(x, t) = v(x, t) + s(x, t)$, where $s(x, t)$ is the linear portion of the solution and $v(x, t)$ is the variable portion of the solution that satisfies the partial differential equation

$$\frac{\partial^2}{\partial t^2} v(x, t) = c^2 \left(\frac{\partial^2}{\partial x^2} v(x, t) \right) + q(x, t)$$

where

$$q(x, t) = h(x, t) - \left(\frac{\partial^2}{\partial t^2} s(x, t) \right)$$

Assignment of system parameters

> restart: with(plots):a:=1:c:=1/5:h(x,t):=x*sin(t):

> A(t):=0:B(t):=0:kappa[1]:=1:kappa[2]:=0:kappa[3]:=1:kappa[4]:=0:

> b(t):=(kappa[3]*a*A(t)+A(t)*kappa[4]−B(t)*kappa[2])/(kappa[1]*kappa[3]*a+kappa[1]*
kappa[4]−kappa[2]*kappa[3]);

$$b(t) := 0 \qquad\qquad (8.112)$$

> m(t):=(kappa[1]*B(t)−A(t)*kappa[3])/(kappa[1]*kappa[3]*a+kappa[1]*kappa[4]
−kappa[2]*kappa[3]);

$$m(t) := 0 \qquad\qquad (8.113)$$

Linear portion of solution

> s(x,t):=m(t)*x+b(t);s(x,0):=eval(subs(t=0,s(x,t))):s[t](x,0):=eval(subs(t=0,diff(s(x,t),t))):

$$s(x, t) := 0 \qquad\qquad (8.114)$$

By the method of separation of variables, the variable solution is

$$v(x, t) = \sum_{n=0}^{\infty} X_n(x) T_n(t)$$

where $T_n(t)$ is the solution to the time-dependent differential equation

$$\frac{d^2}{dt^2} T_n(t) + c^2 \lambda_n T_n(t) = Q_n(t)$$

and $X_n(x)$ is the solution to the spatial-dependent eigenvalue equation

$$\frac{d^2}{dx^2} X_n(x) + \lambda_n X_n(x) = 0$$

with boundary conditions

$$X(0) = 0 \quad \text{and} \quad X(1) = 0$$

The corresponding homogeneous eigenfunction problem consists of the Euler equation, along with the homogeneous boundary conditions that are type 1 at the left and type 1 at the right. The allowed eigenvalues and corresponding orthonormal eigenfunctions, obtained from Example 2.5.1, are

> lambda[n]:=(n*Pi/a)^2;

$$\lambda_n := n^2\pi^2 \tag{8.115}$$

> X[n](x):=sqrt(2/a)*sin(n*Pi/a*x);X[m](x):=subs(n=m,X[n](x)):

$$X_n(x) := \sqrt{2}\sin(n\pi x) \tag{8.116}$$

for $n = 1, 2, 3, \ldots$.

Statement of orthonormality with the respective weight function $w(x) = 1$

> w(x):=1:Int(X[n](x)*X[m](x)*w(x),x=0..a)=delta(n,m);

$$\int_0^1 2\sin(n\pi x)\sin(m\pi x)\,\mathrm{d}x = \delta(n, m) \tag{8.117}$$

Time-dependent equation

> diff(T[n](t),t,t)+c^2*lambda[n]*T[n](t)=Q[n](t);

$$\frac{\mathrm{d}^2}{\mathrm{d}t^2}T_n(t) + \frac{1}{25}n^2\pi^2 T_n(t) = Q_n(t) \tag{8.118}$$

> q(x,t):=h(x,t)−diff(s(x,t),t,t);

$$q(x, t) := x\sin(t) \tag{8.119}$$

> Q[n](t):=Int(q(x,t)*X[n](x),x=0..a);Q[n](t):=expand(value(%)):

$$Q_n(t) := \int_0^1 x\sin(t)\sqrt{2}\sin(n\pi x)\,\mathrm{d}x \tag{8.120}$$

> Q[n](t):=simplify(factor(subs({sin(n*Pi)=0,cos(n*Pi)=(−1)^n},Q[n](t))));Q[n](tau)
  :=subs(t=tau,%):

$$Q_n(t) := -\frac{\sin(t)\sqrt{2}(-1)^n}{n\pi} \tag{8.121}$$

Basis vectors

> T1(t):=cos(c*n*Pi*t/a);

$$T1(t) := \cos\left(\frac{1}{5}n\pi t\right) \tag{8.122}$$

> T2(t):=sin(c*n*Pi*t/a);

$$T2(t) := \sin\left(\frac{1}{5}n\pi t\right) \qquad (8.123)$$

Second-order Green's function

> G2(t,tau):=sin(c*n*Pi/a*(t−tau))/(c*n*Pi/a);

$$G2(t, \tau) := \frac{5 \sin\left(\frac{1}{5}n\pi (t - \tau)\right)}{n\pi} \qquad (8.124)$$

Time-dependent solution

> T[n](t):=C(n)*T1(t)+D(n)*T2(t)+Int(G2(t,tau)*Q[n](tau),tau=0..t);

$$T_n(t) := C(n) \cos\left(\frac{1}{5}n\pi t\right) + D(n) \sin\left(\frac{1}{5}n\pi t\right) + \int_0^t \left(-\frac{5 \sin\left(\frac{1}{5}n\pi(t - \tau)\right) \sin(\tau)\sqrt{2}\,(-1)^n}{n^2\pi^2} \right) d\tau \qquad (8.125)$$

> T[n](t):=value(%);

$$T_n(t) := C(n) \cos\left(\frac{1}{5}n\pi t\right) + D(n) \sin\left(\frac{1}{5}n\pi t\right) + \frac{25(-1)^{1+n}\sqrt{2}\left(-5 \sin\left(\frac{1}{5}n\pi t\right) + \sin(t)n\pi\right)}{n^2\pi^2\left(n^2\pi^2 - 25\right)} \qquad (8.126)$$

Generalized series terms

> v[n](x,t):=(T[n](t)*X[n](x)):

Variable portion of solution

> v(x,t):=Sum(v[n](x,t),n=1..infinity);

$$v(x, t) := \sum_{n=1}^{\infty} \left(C(n) \cos\left(\frac{1}{5}n\pi t\right) + D(n) \sin\left(\frac{1}{5}n\pi t\right) \right. \qquad (8.127)$$

$$\left. + \frac{25(-1)^{1+n}\sqrt{2}\left(-5 \sin\left(\frac{1}{5}n\pi t\right) + \sin(t)n\pi\right)}{n^2\pi^2\left(n^2\pi^2 - 25\right)} \right) \sqrt{2} \sin(n\pi x)$$

The Fourier coefficients $C(n)$ and $D(n)$ are to be determined from the initial conditions on $v(x, t)$. We now substitute the initial conditions $v(x, 0) = f(x) - s(x, 0)$ and $v_t(x, 0) = g(x) - s_t(x, 0)$ into the integrals for $C(n)$ and $D(n)$ for the special case where

```
> f(x):=x*(1−x);
```

$$f(x) := x(1-x) \tag{8.128}$$

```
> g(x):=x;
```

$$g(x) := x \tag{8.129}$$

The integrals for $C(n)$ and $D(n)$ are

```
> C(n):=Int((f(x)−s(x,0))*X[n](x),x=0..a);C(n):=expand(value(%)):
```

$$C(n) := \int_0^1 x(1-x)\sqrt{2}\sin(n\pi x)\,dx \tag{8.130}$$

```
> C(n):=simplify(subs({sin(n*Pi)=0,cos(n*Pi)=(−1)^n},C(n)));
```

$$C(n) := -\frac{2\sqrt{2}\left(-1+(-1)^n\right)}{n^3\pi^3} \tag{8.131}$$

```
> D(n):=Int((g(x)−s[t](x,0))/(c*n*Pi/a)*X[n](x),x=0..a);D(n):=expand(value(%)):
```

$$D(n) := \int_0^1 \frac{5x\sqrt{2}\sin(n\pi x)}{n\pi}\,dx \tag{8.132}$$

```
> D(n):=simplify(subs({sin(n*Pi)=0,cos(n*Pi)=(−1)^n},D(n)));
```

$$D(n) := \frac{5(-1)^{1+n}\sqrt{2}}{n^2\pi^2} \tag{8.133}$$

for $n = 1, 2, 3, \ldots$.

Final solution (linear plus variable portion)

```
> u(x,t):=Sum(eval(v[n](x,t)),n=1..infinity)+s(x,t);
```

$$u(x, t) := \sum_{n=1}^{\infty} \left(-\frac{2\sqrt{2}\left(-1+(-1)^n\right)\cos\left(\frac{1}{5}n\pi t\right)}{n^3\pi^3} + \frac{5(-1)^{1+n}\sqrt{2}\sin\left(\frac{1}{5}n\pi t\right)}{n^2\pi^2} \right.$$
$$\left. + \frac{25(-1)^{1+n}\sqrt{2}\left(-5\sin\left(\frac{1}{5}n\pi t\right)+\sin(t)n\pi\right)}{n^2\pi^2\left(n^2\pi^2-25\right)} \right)\sqrt{2}\sin(n\pi x) \tag{8.134}$$

First few terms of sum

```
> u(x,t):=sum(eval(v[n](x,t)),n=1..3)+s(x,t):
```

ANIMATION

> animate(u(x,t),x=0..a,t=0..5,thickness=3);

The preceding animation command displays the spatial-time-dependent solution of $u(x, t)$ for the given boundary conditions and initial conditions. The animation sequence here and in Figure 8.7 shows snapshots of the animation at times $t = 0, 1, 2, 3, 4, 5$. Note how the solution satisfies the given boundary and initial conditions.

ANIMATION SEQUENCE

> u(x,0):=subs(t=0,u(x,t)):u(x,1):=subs(t=1,u(x,t)):
> u(x,2):=subs(t=2,u(x,t)):u(x,3):=subs(t=3,u(x,t)):
> u(x,4):=subs(t=4,u(x,t)):u(x,5):=subs(t=5,u(x,t)):
> plot({u(x,0),u(x,1),u(x,2),u(x,3),u(x,4),u(x,5)},x=0..a,thickness=10);

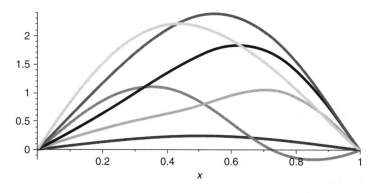

Figure 8.7

EXAMPLE 8.7.3: (Longitudinal wave motion) We seek the wave amplitude $u(x, t)$ for longitudinal wave motion in a rigid bar over the interval $I = \{x \mid 0 < x < 1\}$. The left end of the bar is held fixed, and the right end experiences an oscillatory compression (Young's modulus $E = 1$). There is no external applied force acting on the system. The bar has an initial displacement distribution $f(x)$ and initial speed distribution $g(x)$ given as follows. The wave speed is $c = 1/5$.

SOLUTION: The nonhomogeneous wave equation is

$$\frac{\partial^2}{\partial t^2} u(x, t) = c^2 \left(\frac{\partial^2}{\partial x^2} u(x, t) \right) + h(x, t)$$

The boundary conditions are homogeneous type 1 at the left and nonhomogeneous type 2 at the right:

$$u(0, t) = 0 \quad \text{and} \quad u_x(1, t) = \sin(t)$$

The initial displacement distribution is

$$u(x, 0) = x\left(1 - \frac{x}{2}\right)$$

and the initial speed distribution is

$$u_t(x, 0) = 0$$

The external applied force is

$$h(x, t) = 0$$

The solution is $u(x, t) = v(x, t) + s(x, t)$, where $s(x, t)$ is the linear portion of the solution and $v(x, t)$ is the variable portion of the solution that satisfies the partial differential equation

$$\frac{\partial^2}{\partial t^2} v(x, t) = c^2\left(\frac{\partial^2}{\partial x^2} v(x, t)\right) + q(x, t)$$

where

$$q(x, t) = h(x, t) - \frac{\partial^2}{\partial t^2} s(x, t)$$

Assignment of system parameters

```
> restart: with(plots):a:=1:c:=1/5:h(x,t):=0:
> A(t):=0:B(t):=sin(t):kappa[1]:=1:kappa[2]:=0:kappa[3]:=0:kappa[4]:=1:
> b(t):=(kappa[3]*a*A(t)+A(t)*kappa[4]−B(t)*kappa[2])/(kappa[1]*kappa[3]*a
  +kappa[1]*kappa[4]−kappa[2]*kappa[3]);
```

$$b(t) := 0 \tag{8.135}$$

```
> m(t):=(kappa[1]*B(t)−A(t)*kappa[3])/(kappa[1]*kappa[3]*a+kappa[1]*kappa[4]
  −kappa[2]*kappa[3]);
```

$$m(t) := \sin(t) \tag{8.136}$$

Linear portion of solution

```
> s(x,t):=m(t)*x+b(t);s(x,0):=eval(subs(t=0,s(x,t))):s[t](x,0):=eval(subs(t=0,diff(s(x,t),t))):
```

$$s(x, t) := \sin(t)x \tag{8.137}$$

By the method of separation of variables, the variable solution is

$$v(x, t) = \sum_{n=0}^{\infty} X_n(x) T_n(t)$$

where $T_n(t)$ is the solution to the time-dependent differential equation

$$\frac{d^2}{dt^2} T_n(t) + c^2 \lambda_n T_n(t) = Q_n(t)$$

and $X_n(x)$ is the solution to the spatial-dependent eigenvalue equation

$$\frac{d^2}{dx^2} X_n(x) + \lambda_n X_n(x) = 0$$

with boundary conditions

$$X(0) = 0 \quad \text{and} \quad X_x(1) = 0$$

The corresponding homogeneous eigenfunction problem consists of the Euler equation, along with the homogeneous boundary conditions that are type 1 at the left and type 2 at the right. The allowed eigenvalues and corresponding orthonormal eigenfunctions, obtained from Example 2.5.2, are

> lambda[n]:=((2*n−1)*Pi/(2*a))^2;

$$\lambda_n := \frac{1}{4}(2n-1)^2 \pi^2 \tag{8.138}$$

> X[n](x):=sqrt(2/a)*sin((2*n−1)*Pi/(2*a)*x);X[m](x):=subs(n=m,X[n](x)):

$$X_n(x) := \sqrt{2} \sin\left(\frac{1}{2}(2n-1)\pi x\right) \tag{8.139}$$

for n = 1, 2, 3, ...

Statement of orthonormality with the respective weight function $w(x) = 1$

> w(x):=1:Int(X[n](x)*X[m](x)*w(x),x=0..a)=delta(n,m);

$$\int_0^1 2 \sin\left(\frac{1}{2}(2n-1)\pi x\right) \sin\left(\frac{1}{2}(2m-1)\pi x\right) dx = \delta(n, m) \tag{8.140}$$

Time-dependent equation

> diff(T[n](t),t,t)+c^2*lambda[n]*T[n](t)=Q[n](t);

$$\frac{d^2}{dt^2} T_n(t) + \frac{1}{100}(2n-1)^2 \pi^2 T_n(t) = Q_n(t) \tag{8.141}$$

> q(x,t):=h(x,t)−diff(s(x,t),t,t);

$$q(x, t) := \sin(t)x \tag{8.142}$$

> Q[n](t):=Int(q(x,t)*X[n](x),x=0..a);Q[n](t):=expand(value(%)):

$$Q_n(t) := \int_0^1 \sin(t) x \sqrt{2} \sin\left(\frac{1}{2}(2n-1)\pi x\right) dx \qquad (8.143)$$

> Q[n](t):=simplify(factor(subs({sin(n*Pi)=0,cos(n*Pi)=(−1)^n},Q[n](t))));Q[n](tau):=
  subs(t=tau,%):

$$Q_n(t) := \frac{4(-1)^{1+n}\sin(t)\sqrt{2}}{\pi^2(2n-1)^2} \qquad (8.144)$$

Basis vectors

> T1(t):=cos(c*(2*n−1)*Pi*t/(2*a));

$$T1(t) := \cos\left(\frac{1}{10}(2n-1)\pi t\right) \qquad (8.145)$$

> T2(t):=sin(c*(2*n−1)*Pi*t/(2*a));

$$T2(t) := \sin\left(\frac{1}{10}(2n-1)\pi t\right) \qquad (8.146)$$

Second-order Green's function

> G2(t,tau):=sin(c*(2*n−1)*Pi/(2*a)*(t−tau))/(c*(2*n−1)*Pi/(2*a));

$$G2(t,\tau) := \frac{10\sin\left(\frac{1}{10}(2n-1)\pi(t-\tau)\right)}{(2n-1)\pi} \qquad (8.147)$$

Time-dependent solution

> T[n](t):=C(n)*T1(t)+D(n)*T2(t)+Int(G2(t,tau)*Q[n](tau),tau=0..t);T[n](t):=value(%):
  v[n](x,t):=(T[n](t)*X[n](x)):

$$T_n(t) := C(n)\cos\left(\frac{1}{10}(2n-1)\pi t\right) + D(n)\sin\left(\frac{1}{10}(2n-1)\pi t\right) \qquad (8.148)$$

$$+ \int_0^t \frac{40\sin\left(\frac{1}{10}(2n-1)\pi(t-\tau)\right)(-1)^{1+n}\sin(\tau)\sqrt{2}}{(2n-1)^3\pi^3} d\tau$$

Variable portion of solution

> v(x,t):=Sum(v[n](x,t),n=1..infinity);

$$v(x, t) := \sum_{n=1}^{\infty} \left(C(n) \cos\left(\frac{1}{10}(2n - 1)\pi t \right) + D(n) \sin\left(\frac{1}{10}(2n - 1)\pi t \right) \right.$$
$$+ \left(400(-1)^{1+n} \left(2\sin(t)\pi n - \sin(t) - 10\sin\left(\frac{1}{5}\pi n t \right) \cos\left(\frac{1}{10}\pi t \right) \right) \sqrt{2} \right) /$$
$$\left(\left(16n^4\pi^2 - 32n^3\pi^2 + 24\pi^2 n^2 - 8\pi^2 n + \pi^2 - 400n^2 + 400\,n - 100 \right) \right.$$
$$\left. \left. (2n - 1)\pi^3 \right) \right) \sqrt{2}\sin\left(\frac{1}{2}(2n - 1)\pi x \right)$$

$$(8.149)$$

The Fourier coefficients $C(n)$ and $D(n)$ are to be determined from the initial conditions on $v(x, t)$. We now substitute the initial conditions $v(x, 0) = f(x) - s(x, 0)$ and $v_t(x, 0) = g(x) - s_t(x, 0)$ into the integrals for $C(n)$ and $D(n)$ for the special case where

> f(x):=x*(1−x/2);

$$f(x) := x \left(1 - \frac{1}{2}x \right) \qquad (8.150)$$

> g(x):=0;

$$g(x) := 0 \qquad (8.151)$$

The integrals for $C(n)$ and $D(n)$ are

> C(n):=Int((f(x)−s(x,0))*X[n](x),x=0..a);C(n):=expand(value(%)):

$$C(n) := \int_0^1 x \left(1 - \frac{1}{2}x \right) \sqrt{2}\sin\left(\frac{1}{2}(2n - 1)\pi x \right) dx \qquad (8.152)$$

> C(n):=simplify(subs({sin(n*Pi)=0,cos(n*Pi)=(−1)^n},C(n)));

$$C(n) := \frac{8\sqrt{2}}{\pi^3 \left(8n^3 - 12n^2 + 6n - 1 \right)} \qquad (8.153)$$

> D(n):=Int((g(x)−s[t](x,0))/(c*(2*n−1)*Pi/(2*a))*X[n](x),x=0..a);D(n):=expand(value(%)):

$$D(n) := \int_0^1 \left(-\frac{10x\sqrt{2}\sin\left(\frac{1}{2}(2n - 1)\pi x \right)}{(2n - 1)\pi} \right) dx \qquad (8.154)$$

> D(n):=simplify(subs({sin(n*Pi)=0,cos(n*Pi)=(−1)^n},D(n)));

$$D(n) := \frac{40\sqrt{2}(-1)^n}{\pi^3 \left(8n^3 - 12n^2 + 6n - 1\right)} \tag{8.155}$$

for $n = 1, 2, 3, \ldots$.

Final solution (linear plus variable portion)

> u(x,t):=Sum(eval(v[n](x,t)),n=1..infinity)+s(x,t);

$$u(x, t) := \sum_{n=1}^{\infty} \left(\frac{8\sqrt{2}\cos\left(\frac{1}{10}(2n-1)\pi t\right)}{\pi^3 \left(8n^3 - 12n^2 + 6n - 1\right)} + \frac{40\sqrt{2}(-1)^n \sin\left(\frac{1}{10}(2n-1)\pi t\right)}{\pi^3 \left(8n^3 - 12n^2 + 6n - 1\right)} \right.$$

$$+ \left(400(-1)^{1+n} \left(2\sin(t) - 10\sin\left(\frac{1}{5}\pi n t\right)\cos\left(\frac{1}{10}\pi t\right)\right)\sqrt{2}\right) \Big/$$

$$\left(\left(16n^4\pi^2 - 32n^3\pi^2 + 24\pi^2 n^2 - 8\pi^2 n + \pi^2 - 400n^2 + 400n - 100\right)(2n-1)\pi^3 \right)$$

$$\times \sqrt{2}\sin\left(\frac{1}{2}(2n-1)\pi x\right) + \sin(t)x \tag{8.156}$$

First few terms of sum

> u(x,t):=sum(eval(v[n](x,t)),n=1..3)+s(x,t):

ANIMATION

> animate(u(x,t),x=0..a,t=0..5,thickness=3);

The preceding animation command displays the spatial-time-dependent solution of $u(x, t)$ for the given boundary conditions and initial conditions. The animation sequence here and in Figure 8.8 shows snapshots of the animation at times $t = 0, 1, 2, 3, 4, 5$. Note how the solution satisfies the given boundary and initial conditions.

ANIMATION SEQUENCE

> u(x,0):=subs(t=0,u(x,t)):u(x,1):=subs(t=1,u(x,t)):
> u(x,2):=subs(t=2,u(x,t)):u(x,3):=subs(t=3,u(x,t)):
> u(x,4):=subs(t=4,u(x,t)):u(x,5):=subs(t=5,u(x,t)):
> plot({u(x,0),u(x,1),u(x,2),u(x,3),u(x,4),u(x,5)},x=0..a,thickness=10);

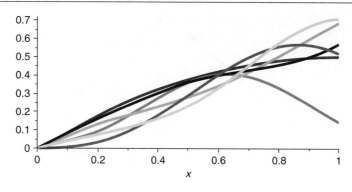

Figure 8.8

EXAMPLE 8.7.4: (Resonance on a plucked magnetic wire) We again consider transverse wave motion along the same taut magnetic wire as discussed in Example 8.7.2. Both ends of the wire are held fixed. An external applied magnetic force acts on the wire. The wire has an initial displacement distribution $f(x)$, corresponding to that of a wire plucked at the center, and an initial speed distribution $g(x)$. The external force term $h(x, t)$ acting upon the wire has a center-symmetric x dependence and a sinusoidal time dependence (angular frequency equals ω), which matches a natural vibration frequency within the system, thus producing resonance. There is no damping in the system and the wave speed is $c = 1$.

SOLUTION: The nonhomogeneous wave equation is

$$\frac{\partial^2}{\partial t^2} u(x, t) = c^2 \left(\frac{\partial^2}{\partial x^2} u(x, t) \right) + h(x, t)$$

The boundary conditions are homogeneous type 1 at the left and homogeneous type 1 at the right:

$$u(0, t) = 0 \quad \text{and} \quad u(1, t) = 0$$

The initial displacement distribution is that of a center-plucked wire

$$u(x, 0) = \frac{x H \left(\frac{1}{2} - x \right) + (1 - x) H \left(x - \frac{1}{2} \right)}{5}$$

where $H(x)$ denotes the Heaviside function. Recall the Heaviside function $H(x)$ has the value 1 for $0 < x$ and the value 0 for $x < 0$. A plot of this initial displacement is shown in Figure 8.9.

The initial speed distribution is

$$u_t(x, 0) = 0$$

and the external applied force is

$$h(x, t) = 10 x(1 - x) \cos(\omega t)$$

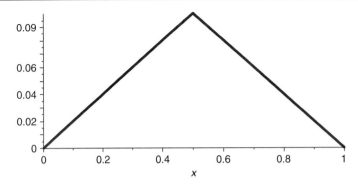

Figure 8.9

The solution is $u(x, t) = v(x, t) + s(x, t)$, where $s(x, t)$ is the linear portion of the solution and $v(x, t)$ is the variable portion of the solution that satisfies the partial differential equation

$$\frac{\partial^2}{\partial t^2} v(x, t) = c^2 \left(\frac{\partial^2}{\partial x^2} v(x, t) \right) + q(x, t)$$

where

$$q(x, t) = h(x, t) - \left(\frac{\partial^2}{\partial t^2} s(x, t) \right)$$

Assignment of system parameters

```
> restart:with(plots):a:=1:c:=1:h(x,t):=10*x*(1−x)*cos(omega*t):
> A(t):=0:B(t):=0:kappa[1]:=1:kappa[2]:=0:kappa[3]:=1:kappa[4]:=0:
> b(t):=(kappa[3]*a*A(t)+A(t)*kappa[4]−B(t)*kappa[2])/(kappa[1]*kappa[3]*a+kappa[1]*
  kappa[4]−kappa[2]*kappa[3]);
```

$$b(t) := 0 \qquad (8.157)$$

```
> m(t):=(kappa[1]*B(t)−A(t)*kappa[3])/(kappa[1]*kappa[3]*a+kappa[1]*kappa[4]
  −kappa[2]*kappa[3]);
```

$$m(t) := 0 \qquad (8.158)$$

Linear portion of solution

```
> s(x,t):=m(t)*x+b(t);s(x,0):=eval(subs(t=0,s(x,t)));s[t](x,0):=eval(subs(t=0,diff(s(x,t),t))):
```

$$s(x, t) := 0 \qquad (8.159)$$

By the method of separation of variables, the variable solution is

$$v(x, t) = \sum_{n=0}^{\infty} X_n(x) T_n(t)$$

where $T_n(t)$ is the solution to the time-dependent differential equation

$$\frac{d^2}{dt^2}T_n(t) + c^2\lambda_n T_n(t) = Q_n(t)$$

and $X_n(x)$ is the solution to the spatial-dependent eigenvalue equation

$$\frac{d^2}{dx^2}X_n(x) + \lambda_n X_n(x) = 0$$

with boundary conditions

$$X(0) = 0 \quad \text{and} \quad X(1) = 0$$

The corresponding homogeneous eigenfunction problem consists of the Euler equation, along with the homogeneous boundary conditions that are type 1 at the left and type 1 at the right. The allowed eigenvalues and corresponding orthonormal eigenfunctions, obtained from Example 2.5.1, are

> lambda[n]:=(n*Pi/a)^2;

$$\lambda_n := n^2\pi^2 \tag{8.160}$$

> X[n](x):=sqrt(2/a)*sin(n*Pi/a*x);X[m](x):=subs(n=m,X[n](x)):

$$X_n(x) := \sqrt{2}\sin(n\pi x) \tag{8.161}$$

for $n = 1, 2, 3, \ldots$.

Statement of orthonormality with the respective weight function $w(x) = 1$

> w(x):=1:Int(X[n](x)*X[m](x)*w(x),x=0..a)=delta(n,m);

$$\int_0^1 2\sin(n\pi x)\sin(m\pi x)dx = \delta(n, m) \tag{8.162}$$

Time-dependent equation

> diff(T[n](t),t,t)+c^2*lambda[n]*T[n](t)=Q[n](t);

$$\frac{d^2}{dt^2}T_n(t) + n^2\pi^2 T_n(t) = Q_n(t) \tag{8.163}$$

> q(x,t):=h(x,t)−diff(s(x,t),t,t);

$$q(x, t) := 10x(1 - x)\cos(\omega t) \tag{8.164}$$

> Q[n](t):=Int(q(x,t)*X[n](x),x=0..a);Q[n](t):=expand(value(%)):

$$Q_n(t) := \int_0^1 10x(1 - x)\cos(\omega t)\sqrt{2}\sin(n\pi x)\,dx \tag{8.165}$$

> Q[n](t):=simplify(factor(subs({sin(n*Pi)=0,cos(n*Pi)=(−1)^n},Q[n](t))));Q[n](tau):=
  subs(t=tau,%):

$$Q_n(t) := -\frac{20\cos(\omega t)\sqrt{2}\left(-1+(-1)^n\right)}{n^3\pi^3} \tag{8.166}$$

Basis vectors

> T1(t):=cos(c*n*Pi*t/a);

$$T1(t) := \cos(n\pi t) \tag{8.167}$$

> T2(t):=sin(c*n*Pi*t/a);

$$T2(t) := \sin(n\pi t) \tag{8.168}$$

Second-order Green's function

> G2(t,tau):=sin(c*n*Pi/a*(t−tau))/(c*n*Pi/a);

$$G2(t, \tau) := \frac{\sin(n\pi(t-\tau))}{n\pi} \tag{8.169}$$

Time-dependent solution

> T[n](t):=C(n)*T1(t)+D(n)*T2(t)+Int(G2(t,tau)*Q[n](tau),tau=0..t);

$$T_n(t) := C(n)\cos(n\pi t) + D(n)\sin(n\pi t) + \int_0^t \left(-\frac{20\sin(n\pi(t-\tau))\cos(\omega\tau)\sqrt{2}\left(-1+(-1)^n\right)}{n^4\pi^4}\right)d\tau \tag{8.170}$$

> T[n](t):=value(%);v[n](x,t):=(T[n](t)*X[n](x)):

$$T_n(t) := C(n)\cos(n\pi t) + D(n)\sin(n\pi t) - \frac{20\sqrt{2}\left(-1+(-1)^n\right)\left(-\cos(n\pi t)+\cos(\omega t)\right)}{n^3\pi^3\left(n^2\pi^2-\omega^2\right)} \tag{8.171}$$

Variable portion of solution

> v(x,t):=Sum(v[n](x,t),n=1..infinity);

$$v(x, t) := \sum_{n=1}^{\infty}\left(C(n)\cos(n\pi t) + D(n)\sin(n\pi t)\right.$$

$$\left. -\frac{20\sqrt{2}\left(-1+(-1)^n\right)\left(-\cos(n\pi t)+\cos(\omega t)\right)}{n^3\pi^3\left(n^2\pi^2-\omega^2\right)}\right)\sqrt{2}\sin(n\pi x) \tag{8.172}$$

The Fourier coefficients $C(n)$ and $D(n)$ are to be determined from the initial conditions on $v(x, t)$. We now substitute the initial conditions $v(x, 0) = f(x) - s(x, 0)$ and $v_t(x, 0) = g(x) - s_t(x, 0)$ into the integrals for $C(n)$ and $D(n)$ for the special case where

```
> f(x):=1/5*(x*Heaviside(1/2−x)+(1−x)*Heaviside(x−1/2));
```

$$f(x) := \frac{1}{5}x \, \text{Heaviside}\left(\frac{1}{2} - x\right) + \frac{1}{5}(1 - x) \, \text{Heaviside}\left(x - \frac{1}{2}\right) \tag{8.173}$$

```
> g(x):=0;
```

$$g(x) := 0 \tag{8.174}$$

The integrals for $C(n)$ and $D(n)$ are

```
> C(n):=Int((f(x)−s(x,0))*X[n](x),x=0..a);C(n):=expand(value(%)):
```

$$C(n) := \int_0^1 \left(\frac{1}{5}x \, \text{Heaviside}\left(\frac{1}{2} - x\right) + \frac{1}{5}(1 - x) \, \text{Heaviside}\left(x - \frac{1}{2}\right)\right) \sqrt{2}\sin(n\pi x)\, dx$$

$$\tag{8.175}$$

```
> C(n):=simplify(subs({sin(n*Pi)=0,cos(n*Pi)=(−1)^n},C(n)));
```

$$C(n) := \frac{2}{5}\frac{\sqrt{2}\sin\left(\frac{1}{2}n\pi\right)}{n^2\pi^2} \tag{8.176}$$

```
> D(n):=Int((g(x)−s[t](x,0))/(c*n*Pi/a)*X[n](x),x=0..a);D(n):=expand(value(%)):
```

$$D(n) := \int_0^1 0\, dx \tag{8.177}$$

```
> D(n):=simplify(subs({sin(n*Pi)=0,cos(n*Pi)=(−1)^n},D(n)));
```

$$D(n) := 0 \tag{8.178}$$

for $n = 1, 2, 3, \ldots$.

Final solution (linear plus variable portion)

```
> u(x,t):=Sum(eval(v[n](x,t)),n=1..infinity)+s(x,t);
```

$$u(x, t) := \sum_{n=1}^{\infty} \left(\frac{2}{5}\frac{\sqrt{2}\sin\left(\frac{1}{2}n\pi\right)\cos(n\pi t)}{n^2\pi^2}\right.$$

$$\left. - \frac{20\sqrt{2}\left(-1 + (-1)^n\right)\left(-\cos(n\pi t) + \cos(\omega t)\right)}{n^3\pi^3\left(n^2\pi^2 - \omega^2\right)}\right) \sqrt{2}\sin(n\pi x) \tag{8.179}$$

The Special Case for Resonance in the System

To demonstrate resonance, we consider the special case where the source angular frequency (ω) is identical to one of the natural vibration frequencies of the system—that is, the case where $\omega = 3\pi$. We set

> omega:=3*Pi;

$$\omega := 3\pi \tag{8.180}$$

From the earlier solution, note that several terms in the denominator equal 0 for $n = 3$ when $\omega = 3\pi$; this is a condition that gives rise to resonance in the system at that particular frequency. In the evaluation of the preceding sum in the solution for $u(x, t)$, we encounter a numerical problem because of the vanishing of these denominator terms. To evaluate the special term when $n = 3$, we must use l'Hopital's rule, which takes a limit on the term. We make a separate evaluation of the limit for the $n = 3$ term, $T_3(t)$:

> T[3](t):=limit(eval(T[n](t)),n=3);v[3](x,t):=T[3](t)*subs(n=3,X[n](x)):

$$T_3(t) := \frac{1}{405}\,\frac{400\sqrt{2}t\sin(\pi t)\cos(\pi t)^2 - 100\sqrt{2}t\sin(\pi t) - 72\sqrt{2}\cos(\pi t)^3\pi^2 + 54\sqrt{2}\cos(\pi t)\pi^2}{\pi^4}$$

$$\tag{8.181}$$

Note that the $T_3(t)$ term has components that increase linearly with time. This is due to the fact that we are driving the wire with a force whose frequency dependence is identical to the natural vibration frequency of the $n = 3$ mode. We plot the $T_3(t)$ term in Figure 8.10, and we see the linear increase in time of the amplitude of this particular component of the Fourier series expansion. This increase in amplitude is the manifestation of what is called "resonance."

> plot(T[3](t),t=0..10,thickness=3);

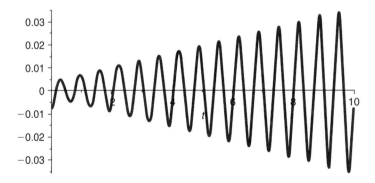

Figure 8.10

Final solution (particular plus linear portion) demonstrating resonance

> u(x,t):=eval(v(x,t))+s(x,t);

$$u(x,t) := \sum_{n=1}^{\infty} \left(\frac{2}{5} \frac{\sqrt{2}\sin\left(\frac{1}{2}n\pi\right)\cos(n\pi t)}{n^2\pi^2} \right.$$

$$\left. - \frac{20\sqrt{2}(-1+(-1)^n)(-\cos(n\pi t)+\cos(3\pi t))}{n^3\pi^3(n^2\pi^2-9\pi^2)} \right)\sqrt{2}\sin(n\pi x) \qquad (8.182)$$

First few terms of sum

> u(x,t):=s(x,t)+sum(eval(v[n](x,t)),n=1..2)+v[3](x,t):

ANIMATION

> animate(u(x,t),x=0..a,t=0..50,frames=50,thickness=3);

The preceding animation command displays the spatial-time-dependent solution of $u(x,t)$ for the given boundary conditions and initial conditions. The animation sequence here and in Figure 8.11 shows snapshots of the animation at times $t = 0, 10/4, 20/4, 30/4, 40/4, 50/4$. As time increases, the $n = 3$ term dominates the solution to the problem, and this is why we see three antinodes in the display of the wave amplitude.

ANIMATION SEQUENCE

> u(x,0):=subs(t=0,u(x,t)):u(x,10/4):=subs(t=10/4,u(x,t)):
> u(x,20/4):=subs(t=20/4,u(x,t)):u(x,30/4):=subs(t=30/4,u(x,t)):
> u(x,40/4):=subs(t=40/4,u(x,t)):u(x,50/4):=subs(t=50/4,u(x,t)):
> plot({u(x,0),u(x,10/4),u(x,20/4),u(x,30/4),u(x,40/4),u(x,50/4)},x=0..a,thickness=10);

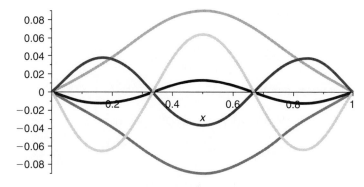

Figure 8.11

Chapter Summary

The nonhomogeneous diffusion or heat equation

$$\frac{\partial}{\partial t}u(x, t) = k\left(\frac{\partial^2}{\partial x^2}u(x, t)\right) + h(x, t)$$

The nonhomogeneous boundary conditions

$$\kappa_1 u(0, t) + \kappa_2 u_x(0, t) = A(t)$$

and

$$\kappa_3 u(a, t) + \kappa_4 u_x(a, t) = B(t)$$

Linear (equilibrium) portion of solution

$$s(x, t) = m(t)x + b(t)$$

First-order Green's function

$$G1(t, \tau) = e^{-\lambda_n k(t - \tau)}$$

Transient solution

$$v(x, t) = \sum_{n=0}^{\infty} \left(C(n)e^{-k\lambda_n t} + \int_0^t G1(t, \tau)Q_n(\tau)\,d\tau\right)X_n(x)$$

Final solution for the diffusion equation (variable plus linear portion)

$$u(x, t) = \sum_{n=0}^{\infty} \left(C(n)e^{-k\lambda_n t} + \int_0^t G1(t, \tau)Q_n(\tau)\,d\tau\right)X_n(x) + m(t)x + b(t)$$

The nonhomogeneous wave equation

$$\frac{\partial^2}{\partial t^2}u(x, t) = c^2\frac{\partial^2}{\partial x^2}u(x, t) + h(x, t)$$

Nonhomogeneous boundary conditions

$$\kappa_1 u(0, t) + \kappa_2 u_x(0, t) = A(t)$$

and

$$\kappa_3 u(a, t) + \kappa_4 u_x(a, t) = B(t)$$

General solution

$$u(x, t) = v(x, t) + s(x, t)$$

Linear portion of the solution

$$s(x, t) = m(t)x + b(t)$$

Second-order Green's function

$$G2(t, \tau) = \frac{\sin\left(c\sqrt{\lambda_n}(t - \tau)\right)}{c\sqrt{\lambda_n}}$$

Particular solution

$$v(x, t) = \sum_{n=0}^{\infty} \left(C(n)\cos\left(c\sqrt{\lambda_n}\,t\right) + D(n)\sin\left(c\sqrt{\lambda_n}\,t\right) + \int_0^t G2(t, \tau)Q_n(\tau)\,d\tau \right) X_n(x)$$

Final solution for the nonhomogeneous wave equation (particular plus linear portion)

$$u(x, t) = \sum_{n=0}^{\infty} \left(C(n)\cos\left(c\sqrt{\lambda_n}\,t\right) + D(n)\sin\left(c\sqrt{\lambda_n}\,t\right) \right.$$

$$\left. + \int_0^t G2(t, \tau)Q_n(\tau)\,d\tau \right) X_n(x) + m(t)x + b(t)$$

In this chapter, we examined the nonhomogeneous equivalents to the diffusion and wave equations discussed in Chapters 3 and 4. The partial differential equations were nonhomogeneous either because of the presence of a driving term or because of the presence of a nonhomogeneous boundary condition. The solutions were resolved by breaking up the problem into two portions: a linear (equilibrium) portion and a variable (particular) portion. Using the method of eigenfunction expansions, we were able to obtain final solutions to the nonhomogeneous problem.

Exercises

We consider problems for nonhomogeneous diffusion and wave equations in the rectangular coordinate system in a single dimension. The problems are nonhomogeneous either because there is an external source acting on the system or the boundary conditions are nonhomogeneous, or both of these reasons. Use the method of separation of variables and eigenfunction expansions to evaluate the solutions.

Nonhomogeneous Heat Equations

We consider the temperature distribution along a thin rod whose lateral surface is insulated over the finite interval $I = \{x \mid 0 < x < 1\}$. We allow for an external heat source $h(x, t)$ to act on the system. The nonhomogeneous diffusion partial differential equation in rectangular coordinates reads

$$\frac{\partial}{\partial t} u(x, t) = k \left(\frac{\partial^2}{\partial x^2} u(x, t) \right) + h(x, t)$$

with the initial condition

$$u(x, 0) = f(x)$$

In Exercises 8.1 to 8.16, we seek solutions to the preceding problem for various given initial conditions and boundary conditions. Resolve the solution into two parts: the variable and the linear, as was done in Section 8.4. Determine the eigenfunctions and eigenvalues for the corresponding homogeneous problem, and write the solution as a sum of the partitioned solutions. Generate the animated solution for $u(x, t)$, and plot the animated sequence for $0 < t < 5$. Note whether the solution satisfies the given boundary and initial conditions.

8.1. Boundary $x = 0$ is held at fixed temperature 0, boundary $x = 1$ is up against a sinusoidal temperature reservoir, there is no external heat source, $k = 1/10$:
   Boundary conditions:

$$u(0, t) = 0 \quad \text{and} \quad u(1, t) = \sin(t)$$

   Initial condition:

$$f(x) = x(1 - x)$$

8.2. Boundary $x = 1$ is held at fixed temperature 0, boundary $x = 0$ is up against a sinusoidal temperature reservoir, there is no external heat source, $k = 1/10$:
   Boundary conditions:

$$u(0, t) = \cos(t) \quad \text{and} \quad u(1, t) = 0$$

   Initial condition:

$$f(x) = x(1 - x)$$

8.3. Boundary $x = 0$ is held at fixed temperature 0, boundary $x = 1$ has an external heat flux input, there is no external heat source, $k = 1/10$:
   Boundary conditions:

$$u(0, t) = 0 \quad \text{and} \quad u_x(1, t) = te^{-t}$$

Initial condition:

$$f(x) = x\left(1 - \frac{x}{2}\right)$$

8.4. Boundary $x = 0$ has an external heat flux input, $x = 1$ is held at fixed temperature 2, there is no external heat source, $k = 1/10$:

Boundary conditions:

$$u_x(0, t) = -\cos(t) \quad \text{and} \quad u(1, t) = 2$$

Initial condition:

$$f(x) = 1 - x^2$$

8.5. Boundary $x = 0$ is held at fixed temperature 0, boundary $x = 1$ is losing heat by convection into a time-varying surrounding temperature medium, there is no external heat source, $k = 1/10$:

Boundary conditions:

$$u(0, t) = 0 \quad \text{and} \quad u_x(1, t) + u(1, t) = \sin(t)$$

Initial condition:

$$f(x) = x\left(1 - \frac{2x}{3}\right)$$

8.6. Boundary $x = 0$ is losing heat by convection into a time-varying surrounding temperature medium, boundary $x = 1$ is held at fixed temperature 1, there is no external heat source, $k = 1/10$:

Boundary conditions:

$$u_x(0, t) - u(0, t) = \cos(t) \quad \text{and} \quad u(1, t) = 1$$

Initial condition:

$$f(x) = \frac{1}{3} + \frac{x}{3} - \frac{2x^2}{3}$$

8.7. Boundary $x = 0$ has an external heat flux input, boundary $x = 1$ is losing heat by convection into a time-varying surrounding temperature medium, there is no external heat source, $k = 1/10$:

Boundary conditions:

$$u_x(0, t) = \sin(t) \quad \text{and} \quad u_x(1, t) + u(1, t) = \sin(t)$$

Initial condition:

$$f(x) = 1 - \frac{x^2}{3}$$

8.8. Boundary $x = 0$ is losing heat by convection into a time-varying surrounding temperature medium, boundary $x = 1$ has an input heat flux, there is no external heat source, $k = 1/10$:

Boundary conditions:

$$u_x(0, t) - u(0, t) = te^{-t} \quad \text{and} \quad u_x(1, t) = 1$$

Initial condition:

$$f(x) = 1 - \frac{(x-1)^2}{3}$$

8.9. Boundaries $x = 0$ and $x = 1$ are held at the fixed temperature of 0, there is an external heat source term $h(x, t), k = 1/10$:

Boundary conditions:

$$u(0, t) = 0 \quad \text{and} \quad u(1, t) = 0$$

Initial condition and heat source:

$$f(x) = x(1 - x) \quad \text{and} \quad h(x, t) = \sin(t)$$

8.10. Boundary $x = 0$ is held at fixed temperature 0, boundary $x = 1$ is insulated, there is an external heat source term $h(x, t), k = 1/10$:

Boundary conditions:

$$u(0, t) = 0 \quad \text{and} \quad u_x(1, t) = 0$$

Initial condition and heat source:

$$f(x) = 0 \quad \text{and} \quad h(x, t) = x\left(1 - \frac{x}{2}\right)\sin(t)$$

8.11. Boundary $x = 0$ is insulated, boundary $x = 1$ is held at fixed temperature 0, there is an external heat source term $h(x, t), k = 1/10$:

Boundary conditions:

$$u_x(0, t) = 0 \quad \text{and} \quad u(1, t) = 0$$

Initial condition and heat source:

$$f(x) = 0 \quad \text{and} \quad h(x, t) = \left(1 - x^2\right)\cos(t)$$

8.12. Boundaries $x = 0$ and $x = 1$ are both insulated, there is an external heat source term $h(x, t)$, $k = 1/10$:

Boundary conditions:

$$u_x(0, t) = 0 \quad \text{and} \quad u_x(1, t) = 0$$

Initial condition and heat source:

$$f(x) = 0 \quad \text{and} \quad h(x, t) = te^{-t}x^2\left(1 - \frac{2x}{3}\right)$$

8.13. Boundary $x = 0$ is held at fixed temperature 10, boundary $x = 1$ is held at fixed temperature 5, there is an external heat source term $h(x, t)$, $k = 1/10$:

Boundary conditions:

$$u(0, t) = 10 \quad \text{and} \quad u(1, t) = 5$$

Initial condition and heat source:

$$f(x) = 0 \quad \text{and} \quad h(x, t) = x(1 - x)\sin(t)$$

8.14. Boundary $x = 0$ is held at fixed temperature 0, boundary $x = 1$ has an external heat flux input, there is an external heat source term $h(x, t)$, $k = 1/10$:

Boundary conditions:

$$u(0, t) = 0 \quad \text{and} \quad u_x(1, t) = \sin(t)$$

Initial condition and heat source:

$$f(x) = 0 \quad \text{and} \quad h(x, t) = x\left(1 - \frac{x}{2}\right)\cos(t)$$

8.15. Boundary $x = 0$ is held at fixed temperature 0, boundary $x = 1$ is losing heat by convection into a time-varying surrounding medium, there is an external heat source term $h(x, t)$, $k = 1/10$:

Boundary conditions:

$$u(0, t) = 0 \quad \text{and} \quad u_x(1, t) + u(1, t) = \sin(t)$$

Initial condition and heat source:

$$f(x) = 0 \quad \text{and} \quad h(x, t) = x\left(1 - \frac{2x}{3}\right)\cos(t)$$

8.16. Boundary $x = 0$ is held at fixed temperature 2, boundary $x = 1$ is losing heat by convection into a time-varying surrounding medium, there is an external heat source term $h(x, t)$, $k = 1/10$:

Boundary conditions:

$$u(0, t) = 2 \quad \text{and} \quad u_x(1, t) + u(1, t) = \sin(t)$$

Initial condition and heat source:

$$f(x) = x \quad \text{and} \quad h(x, t) = x\left(1 - \frac{2x}{3}\right)\cos(t)$$

Nonhomogeneous Wave Equations

We consider transverse wave propagation over taut strings and longitudinal wave propagation in rigid bars over the finite, one-dimensional interval $I = \{x \mid 0 < x < 1\}$. We allow for an external force to act on the system. The nonhomogeneous wave partial differential equation in rectangular coordinates reads

$$\frac{\partial^2}{\partial t^2}u(x, t) = c^2\left(\frac{\partial^2}{\partial x^2}u(x, t)\right) + h(x, t)$$

with the two initial conditions

$$u(x, 0) = f(x) \quad \text{and} \quad u_t(x, 0) = g(x)$$

In Exercises 8.17 through 8.23, we seek solutions to the preceding problem for various given initial and boundary conditions. Resolve the solution into two parts: the variable and the linear as was done in Section 8.7. Determine the eigenfunctions and eigenvalues for the corresponding homogeneous problem, and write the solution as a sum of the partitioned solutions. Generate the animated solution for $u(x, t)$, and plot the animated sequence for $0 < t < 5$. Note whether the solution satisfies the given boundary and initial conditions.

8.17.   Consider transverse motion in a taut string, boundary $x = 0$ is held secure, boundary $x = 1$ is attached to a vertical oscillator, there is no external force acting, $c = 1/2$:

Boundary conditions:

$$u(0, t) = 0 \quad \text{and} \quad u(1, t) = \sin(t)$$

Initial conditions:

$$f(x) = x(1 - x) \quad \text{and} \quad g(x) = 0$$

8.18.   Consider transverse motion in a taut string, boundary $x = 0$ is attached to a vertical oscillator, boundary $x = 1$ is held fixed, there is no external force acting, $c = 1/2$:

Boundary conditions:

$$u(0, t) = \cos(t) \quad \text{and} \quad u(1, t) = 0$$

Initial conditions:

$$f(x) = 0 \quad \text{and} \quad g(x) = x(1 - x)$$

8.19. Consider longitudinal waves in a rigid bar whose boundary $x = 0$ is held fixed, boundary $x = 1$ has an applied compression (Young's modulus $E = 1$), there is no external force, $c = 1/2$:

Boundary conditions:

$$u(0, t) = 0 \quad \text{and} \quad u_x(1, t) = 2$$

Initial conditions:

$$f(x) = x\left(1 - \frac{x}{2}\right) \quad \text{and} \quad g(x) = 0$$

8.20. Consider longitudinal waves in a rigid bar whose boundary $x = 0$ has an applied compression, boundary $x = 1$ is held fixed, there is no external force, $c = 1/2$:

Boundary conditions:

$$u_x(0, t) = \cos(t) \quad \text{and} \quad u(1, t) = 0$$

Initial conditions:

$$f(x) = 0 \quad \text{and} \quad g(x) = 1 - x^2$$

8.21. Consider transverse waves in a taut string whose boundary $x = 0$ is held fixed, boundary $x = 1$ is attached to an elastic hinge with a time-varying support, there is no external force, $c = 1/2$:

Boundary conditions:

$$u(0, t) = 0 \quad \text{and} \quad u_x(1, t) + u(1, t) = \sin(t)$$

Initial conditions:

$$f(x) = x\left(1 - \frac{2x}{3}\right) \quad \text{and} \quad g(x) = 0$$

8.22. Consider transverse waves in a taut string whose boundary $x = 0$ is attached to an elastic hinge with a time-varying support, boundary $x = 1$ is held fixed, there is no external force, $c = 1/2$:

Boundary conditions:

$$u_x(0, t) - u(0, t) = \cos(t) \quad \text{and} \quad u(1, t) = 1$$

Initial conditions:

$$f(x) = 0 \quad \text{and} \quad g(x) = \frac{1}{3} + \frac{x}{3} - \frac{2x^2}{3}$$

8.23. Consider longitudinal waves in a rigid bar whose boundary $x = 0$ has a compressive force, boundary $x = 1$ is attached to an elastic hinge with a time-varying support, there is no external force, $c = 1/2$:

Boundary conditions:

$$u_x(0, t) = \cos(t) \quad \text{and} \quad u_x(1, t) + u(1, t) = \sin(t)$$

Initial conditions:

$$f(x) = 1 - \frac{x^2}{3} \quad \text{and} \quad g(x) = 0$$

8.24. Consider longitudinal waves in a rigid bar whose boundary $x = 0$ is attached to an elastic hinge with a time-varying support, boundary $x = 1$ has a compressive force, there is no external force, $c = 1/2$:

Boundary conditions:

$$u_x(0, t) - u(0, t) = te^{-t} \quad \text{and} \quad u_x(1, t) = 1$$

Initial conditions:

$$f(x) = 0 \quad \text{and} \quad g(x) = 1 - \frac{(x - 1)^2}{3}$$

Natural Angular Frequency and Resonance

In Chapter 4, we investigated homogeneous problems for transverse wave propagation on a taut string. Earlier in this chapter, we considered the corresponding nonhomogeneous problem for transverse wave propagation. From our analysis for the case of no damping in the system, we arrived at a natural angular frequency of oscillation of the wave for the nth mode:

$$\omega_n = c\sqrt{\lambda_n}$$

Here, λ_n is the allowed eigenvalue and c is the speed of propagation of the wave with dimensions of distance per time. For a taut string of linear density ρ under tension T, c is given as

$$c = \sqrt{\frac{T}{\rho}}$$

For a string of length l, which is secure at both ends, the allowed eigenvalues for the nth mode are given as

$$\lambda_n = \frac{n^2\pi^2}{l^2}$$

for $n = 1, 2, 3, \ldots$. Thus, the nth-mode natural angular frequency of oscillation for the transverse wave on a taut string is given as

$$\omega_n = \frac{n\pi\sqrt{\frac{T}{\rho}}}{l}$$

In the following exercises, we consider the nonhomogeneous problem for transverse wave propagation on taut strings. We consider source functions whose driving frequency may or may not coincide with the natural angular frequency of the string. In the case where the two are identical, we have what is called "resonance" as demonstrated in Example 8.7.4.

For Exercises 8.25 through 8.30, we consider a taut string of length $l = \pi$, which is secure at both ends. The initial conditions are given as follows along with the boundary conditions and the source function.

Boundary conditions:

$$u(0, t) = 0 \quad \text{and} \quad u(\pi, t) = 0$$

Source function:

$$h(x, t) = x(\pi - x)\sin(\omega t)$$

Initial conditions:

$$f(x) = x(\pi - x) \quad \text{and} \quad g(x) = 0 \tag{8.183}$$

8.25. For the given boundary and initial conditions, let the string tension and linear density be such that $c = 1$. Evaluate the eigenvalues and the corresponding eigenfunctions and the general solution for $u(x, t)$ for a generalized value of ω. What should the value of ω be in order to trigger resonance for the $n = 1$ (fundamental) mode of the system? For this value of ω, generate the animated solution for $u(x, t)$, and develop the animated sequence for $0 < t < 5$.

8.26. For the string in Exercise 8.25, let the tension in the string be increased by a factor of 4, keeping everything else the same. Evaluate the eigenvalues and the corresponding eigenfunctions and the general solution for $u(x, t)$ for a generalized value of ω. What should the value of ω be in order to trigger resonance for the $n = 3$ mode of the system? For this value of ω, generate the animated solution for $u(x, t)$, and develop the animated sequence for $0 < t < 5$.

8.27. For the string in Exercise 8.25, let the linear density of the string be increased by a factor of 4, keeping everything else the same. Evaluate the eigenvalues and the corresponding eigenfunctions and the general solution for $u(x, t)$ for a generalized value of ω. What should the value of ω be in order to trigger resonance for the $n = 1$ mode of the system? For this value of ω, generate the animated solution for $u(x, t)$, and develop the animated sequence for $0 < t < 5$.

8.28. For the string in Exercise 8.25, let the length of the string be decreased by a factor of 3, keeping everything else the same. Evaluate the eigenvalues and the corresponding eigenfunctions and the general solution for $u(x, t)$ for a generalized value of ω. What should the value of ω be in order to trigger resonance for the $n = 5$ mode of the system? For this value of ω, generate the animated solution for $u(x, t)$, and develop the animated sequence for $0 < t < 5$.

8.29. For the string in Exercise 8.25, let the tension in the string be increased by a factor of 4, and let the length be increased by a factor of 2, keeping everything else the same. Evaluate the eigenvalues and the corresponding eigenfunctions and the general solution for $u(x, t)$ for a generalized value of ω. What should the value of ω be in order to trigger resonance for the $n = 3$ mode of the system? For this value of ω, generate the animated solution for $u(x, t)$, and develop the animated sequence for $0 < t < 5$.

8.30. For the string in Exercise 8.25, let the linear density of the string be decreased by a factor of 4, and let the tension be decreased by a factor of 9, keeping everything else the same. Evaluate the eigenvalues and the corresponding eigenfunctions and the general solution for $u(x, t)$ for a generalized value of ω. What should the value of ω be in order to trigger resonance for the $n = 5$ mode of the system? For this value of ω, generate the animated solution for $u(x, t)$, and develop the animated sequence for $0 < t < 5$.

8.31. For the string in Exercise 8.25, let the tension in the string be increased by a factor of 4, the density be increased by a factor of 9, and the length of the string be decreased by a factor of 2, keeping everything else the same. Evaluate the eigenvalues and the corresponding eigenfunctions and the general solution for $u(x, t)$ for a generalized value of ω. What should the value of ω be in order to trigger resonance for the $n = 3$ mode of the system? For this value of ω, generate the animated solution for $u(x, t)$, and develop the animated sequence for $0 < t < 5$.

CHAPTER 9

Infinite and Semi-infinite Spatial Domains

9.1 Introduction

We now look at partial differential equations over spatial domains that are infinite or semi-infinite. These problems definitely do not fall into the category of "regular" Sturm-Liouville problems and are often referred to as "singular" Sturm-Liouville problems. We treat these problems by way of an extension of the concepts of regular problems over finite domains.

We shall see that our eigenvalues and corresponding eigenfunctions are no longer "discrete" and indexed by integers. Instead, for infinite domains, the eigenvalues consist of a "continuum" of real values, and we term the corresponding eigenfunctions as "singular." Further, the extension of the superposition principle for regular problems, where our solutions gave us a "series" that consisted of summations over the discrete, indexed eigenvalues, now gives way to "integrations" over the continuum of eigenvalues.

Finally, we develop an equivalent concept of orthonormality for "singular" eigenfunctions where the Kronecker delta for regular problems gives way to the Dirac delta function for singular problems.

9.2 Fourier Integral

We first consider recollection of the Fourier series for periodic functions over finite domains. From Section 2.6, we saw that such a series was a special case of the generalized Sturm-Liouville series. If the function $f(x)$ was piecewise continuous and it satisfied the periodic boundary conditions over the finite interval $I = \{x \mid -L < x < L\}$, then we could express this function as a series of eigenfunctions that were said to be "complete" with respect to piecewise continuous functions over the interval. In terms of the orthonormalized eigenfunctions, the series was shown to have the "real" form

$$f(x) = \frac{A(0)\sqrt{2}}{2\sqrt{L}} + \sum_{n=1}^{\infty} \left(\frac{A(n)\cos\left(\frac{n\pi x}{L}\right)}{\sqrt{L}} + \frac{B(n)\sin\left(\frac{n\pi x}{L}\right)}{\sqrt{L}} \right)$$

With some manipulation, using the Euler formulas

$$\cos(u) = \frac{e^{iu} + e^{-iu}}{2}$$

and

$$\sin(u) = \frac{e^{iu} - e^{-iu}}{2i}$$

the real form of the preceding series can be modified and rewritten in what is called the "complex" form of the Fourier series and it reads

$$f(x) = \sum_{n=-\infty}^{\infty} \frac{F(n)\,e^{\frac{in\pi x}{L}}}{2L}$$

The terms $F(n)$ are the "complex" Fourier coefficients, and, from the Sturm-Liouville theorem, the preceding series converges to the average of the left- and right-hand limits of the function at any point within the open interval I.

The preceding form of the Fourier series suggests that $f(x)$ is being expanded in terms of a generalized Sturm-Liouville series whose orthonormalized eigenfunctions $\phi_n(x)$ are given as

$$\phi_n(x) = \frac{\sqrt{2}\,e^{\frac{in\pi,x}{L}}}{2\sqrt{L}}$$

for $n = -3, -2, -1, 0, 1, 2, 3 \ldots$.

Because these eigenfunctions arise from the solution of a regular Sturm-Liouville problem with periodic boundary conditions, we could have referred to them as "regular" eigenfunctions. The statement of orthonormality for these complex eigenfunctions is given as the "complex" inner product (the product of the function times its complex conjugate), with respect to the weight function $w(x) = 1$, over the interval I. In terms of the Kronecker delta function (see references for a more complete description of the mathematical chararcteristics of the Kronecker delta function and the Dirac delta function used following), this reads

$$\int_{-L}^{L} \frac{e^{\frac{in\pi x}{L}}\,e^{-\frac{im\pi x}{L}}}{2L}\,dx = \delta(n, m)$$

Thus, to evaluate the Fourier coefficients as we did in Chapter 2, we take the complex inner product of both sides of the preceding series with respect to the orthonormalized eigenfunctions and the weight function $w(x) = 1$, and we get

$$\int_{-L}^{L} f(x) e^{-\frac{im\pi x}{L}} \, dx = \sum_{n=-\infty}^{\infty} F(n) \left(\int_{-L}^{L} \frac{e^{\frac{in\pi x}{L}} e^{-\frac{im\pi x}{L}}}{2L} \, dx \right)$$

Taking advantage of the earlier statement of orthonormality, this series reduces to

$$\int_{-L}^{L} f(x) e^{-\frac{im\pi x}{L}} \, dx = \sum_{n=-\infty}^{\infty} F(n)\delta(n, m)$$

Due to the mathematical nature of the Kronecker delta function, only the $m = n$ term survives the sum on the right, and from this sum, we evaluate the complex Fourier coefficients $F(n)$ to be

$$F(n) = \int_{-L}^{L} f(x) e^{-\frac{in\pi x}{L}} \, dx$$

Many textbooks cover the Fourier integral theorem, which deals with the transition from Fourier series to Fourier integrals. The Fourier integral theorem is a generalization of the Fourier series expansion as we go from finite domains to infinite domains. The theorem states the following (see related references): let $f(x)$ be piecewise smooth and absolutely integrable on the real line. Then $f(x)$ can be represented by the following Fourier integral

$$f(x) = \int_{-\infty}^{\infty} \frac{F(\omega)e^{i\omega x}}{\pi} \, d\omega$$

This integral converges to the average of the left- and right-hand limits of $f(x)$ at any point over the infinite domain.

Comparing the preceding Fourier integral expansion of $f(x)$ over infinite domains with the earlier Fourier series expansion of $f(x)$ over finite domains, we see a remarkable parallel. Similar to Fourier series, the terms $F(\omega)$ are the singular Fourier coefficients of $f(x)$. By convention, these terms $F(\omega)$ are more commonly called the "Fourier transform" of $f(x)$.

We now recognize the transition from Fourier series to Fourier integrals: As the domain goes from finite to infinite, the series summation is replaced by an integration, and the discrete summing index over real integers is replaced by an integration over a continuous variable.

It can be shown (see references) that the corresponding Fourier transform $F(\omega)$ (singular Fourier coefficient) can be found from the following integral:

$$F(\omega) = \int_{-\infty}^{\infty} f(x)\, e^{-i\omega x}\, dx$$

The functions $f(x)$ and $F(\omega)$ constitute what is called a "Fourier transform pair."

In the development of the transition of the concepts of Sturm-Liouville theory for regular eigenvalue problems over finite intervals to that over infinite intervals, we can view the Fourier integral for $f(x)$ as an expansion in terms of the "singular" eigenfunctions given here:

$$\frac{\sqrt{2}\, e^{i\omega x}}{2\sqrt{\pi}}$$

for $-\infty < \omega < \infty$.

The extension of these concepts would be complete except for the lack of an equivalent statement of orthonormality. If we substitute $f(x)$ in the integral for $F(\omega)$ and formally interchange the order of integration, we get

$$F(\omega) = \int_{-\infty}^{\infty} \int_{-\infty}^{\infty} \frac{F(\beta)\, e^{i\beta x}\, e^{-i\omega x}}{2\pi}\, dx\, d\beta$$

We now compare this integral with the formal integral definition of the Dirac delta function $\delta(\beta - \omega)$ shown here:

$$\int_{-\infty}^{\infty} F(\beta)\, \delta(\beta - \omega)\, d\beta = F(\omega)$$

From a comparison of the two preceding integrals, we can identify the previous operation as the integral representation of the Dirac delta function in terms of the singular eigenfunctions—that is,

$$\int_{-\infty}^{\infty} \frac{e^{i\beta x}\, e^{-i\omega x}}{2\pi}\, dx = \delta(\beta - \omega) \tag{9.1}$$

Parallel with the statement of orthonormality for regular eigenfunctions, the preceding equation can be considered to be the equivalent to the statement of orthonormality for the "singular" eigenfunctions given earlier. Similar to the Kronecker delta function, the preceding Dirac delta function has the value zero when $\beta \neq \omega$ and has the value ∞ for $\beta = \omega$. Because of the sharp character of this function, it is often called the "impulse" function. Thus, the parallel between singular eigenfunctions over infinite domains has been shown to be analogous to that of regular eigenfunctions over finite domains.

9.3 Fourier Sine and Cosine Integrals

If $f(x)$ is an even or odd function of x, then the preceding Fourier integral reduces to a slightly different form. We now demonstrate this. Recall the Euler formula:

$$e^{iu} = \cos(u) + i\sin(u)$$

If we use the Euler formula in the expansion of the Fourier integral, we get

$$f(x) = \int_{-\infty}^{\infty} \frac{F(\omega)(\cos(\omega x) + i\sin(\omega x))}{2\pi} \, d\omega$$

Similarly, if we use the Euler formula in the expansion of the Fourier transform, we get

$$F(\omega) = \int_{-\infty}^{\infty} f(x)(\cos(\omega x) - i\sin(\omega x)) \, dx$$

Combining the two expressions and formally interchanging the order of integration yields

$$F(\omega) = \int_{-\infty}^{\infty} \int_{-\infty}^{\infty} \frac{F(\beta)(\cos(\beta x) + i\sin(\beta x))(\cos(\omega x) - i\sin(\omega x))}{2\pi} \, dx \, d\beta$$

To evaluate this integral, we must first recall the properties of integration of even and odd functions of x over symmetric intervals as shown:

$$\int_{-\infty}^{\infty} even(f(x)) \, dx = 2\left(\int_{0}^{\infty} even(f(x)) \, dx \right)$$

and

$$\int_{-\infty}^{\infty} odd(f(x)) \, dx = 0$$

The first states that an integration of an even function of x over a symmetric interval is equal to twice the integral over half that interval. The second states that an integration of an odd function of x over a symmetric interval is equal to zero.

We focus on evaluation of the earlier integral $F(\omega)$ for the two cases where $f(x)$ is an even or odd function of x.

If $f(x)$ is an even function of x, then, from the preceding Fourier transform, $F(\beta)$ is an even function of β, and the term $F(\beta)\cos(\beta x)\cos(\omega x)$ is the only term given that is even with respect to both x and β; all other terms are odd. Thus, the resulting Fourier transform reduces to the following expression:

$$F(\omega) = \int_0^\infty \int_0^\infty \frac{2F(\beta)\cos(\beta x)\cos(\omega x)}{\pi}\, dx\, d\beta$$

If we compare this integral with the formal integral definition of the preceding Dirac delta function, we can write the statement of orthonormality for the "singular" eigenfunctions as

$$\int_0^\infty \frac{2\cos(\beta x)\cos(\omega x)}{\pi}\, dx = \delta(\beta - \omega)$$

If we define the Fourier cosine transform $C(\omega)$ as the integral

$$C(\omega) = \int_0^\infty f(x)\cos(\omega x)\, dx$$

then we can express the even function $f(x)$ as the Fourier cosine integral

$$f(x) = \int_0^\infty \frac{2C(\omega)\cos(\omega x)}{\pi}\, d\omega$$

We view this equation to be the generalized expansion of an even function of x in terms of the singular, orthonormalized eigenfunctions

$$\frac{\sqrt{2}\cos(\omega x)}{\sqrt{\pi}}$$

for $0 < \omega < \infty$. The preceding is the analog of the Fourier cosine series expansion of an even periodic function of x. The evaluation of the Fourier coefficients $C(\omega)$ can be viewed as the result that comes about from taking the inner product of both sides of the preceding with respect to the singular eigenfunctions and taking advantage of the statement of orthonormality.

The integral and its corresponding transform $C(\omega)$ constitute a "Fourier cosine transform" pair. We note here that for even functions of x, the relation between the cosine transform $C(\omega)$ and the regular Fourier transform is

$$C(\omega) = \frac{F(\omega)}{2}$$

where $F(\omega)$ is the Fourier transform of the even extension of $f(x)$.

Similarly, for odd functions, we do the following. If $f(x)$ is an odd function of x, then, from the preceding Fourier transform, the term $F(\beta)$ is an odd function of β, and $F(\beta)\sin(\beta x)\sin(\omega x)$ is the only term given that is even with respect to both β and x; all other terms are odd. Thus, the resulting integral reduces to

$$F(\omega) = \int_0^\infty \int_0^\infty \frac{2F(\beta) \sin(\beta x) \sin(\omega x)}{\pi} \, dx \, d\beta$$

If we compare this integral with the formal integral definition of the Dirac delta function given earlier, we can write the statement of orthonormality for the "singular" eigenfunctions as

$$\int_0^\infty \frac{2 \sin(\beta x) \sin(\omega x)}{\pi} \, dx = \delta(\beta - \omega)$$

If we define the Fourier sine integral $S(\omega)$ as

$$S(\omega) = \int_0^\infty f(x) \sin(\omega x) \, dx$$

then we can express the odd function $f(x)$ as a Fourier sine integral

$$f(x) = \int_0^\infty \frac{2S(\omega) \sin(\omega x)}{\pi} \, d\omega$$

We view this equation to be the generalized expansion of an odd function of x in terms of the singular orthonormalized eigenfunctions

$$\frac{\sqrt{2} \sin(\omega x)}{\sqrt{\pi}}$$

for $0 < \omega < \infty$. The preceding is the analog of the Fourier sine series expansion of an odd periodic function of x. The evaluation of the Fourier coefficients $S(\omega)$ can be viewed as the result that comes about from taking the inner product of both sides of the preceding with respect to the singular eigenfunctions and taking advantage of the statement of orthonormality.

The preceding integral and its corresponding transform $S(\omega)$ constitute a "Fourier sine transform" pair. We note here that for odd functions of x, the relation between the sine transform and the regular transform is

$$S(\omega) = \frac{iF(\omega)}{2}$$

where $F(\omega)$ is the Fourier transform of the odd extension of $f(x)$.

9.4 Nonhomogeneous Diffusion Equation over Infinite Domains

We now use the Fourier integral method (singular eigenfunction expansion) to formulate the solutions to diffusion partial differential equations over infinite or semi-infinite intervals.

We first consider the nonhomogeneous diffusion partial differential equation over the infinite interval $I = \{x \mid -\infty < x < \infty\}$. The equation reads

$$\frac{\partial}{\partial t} u(x, t) = k \left(\frac{\partial^2}{\partial x^2} u(x, t) \right) + h(x, t)$$

with the initial condition

$$u(x, 0) = f(x)$$

Here, $f(x)$ denotes the initial concentration or temperature distribution, and $h(x, t)$ denotes the presence of any internal source terms within the system. The boundary conditions are such that $u(x, t)$ is piecewise smooth and absolutely integrable over the infinite interval $I = \{x \mid -\infty < x < \infty\}$.

As we did in Chapter 8, we first consider the homogeneous version of the given equation. Using the method of separation of variables as in Chapter 3, we arrive at two ordinary differential equations: one in t and one in x. The x-dependent differential equation is of the Euler type, and it reads

$$\frac{d^2}{dx^2} X(x) + \lambda X(x) = 0$$

A set of basis vectors for this differential equation for $\lambda > 0$ is, from Chapter 1,

$$x1(x) = e^{i\sqrt{\lambda} x}$$

and

$$x2(x) = e^{-i\sqrt{\lambda} x}$$

If we set $\lambda = \omega^2$ and we require that our x-dependent portion of the solution be bounded as x goes to plus or minus infinity, then our x-dependent orthonormalized singular eigenfunctions, as shown in Section 9.2, can be written

$$\frac{\sqrt{2}\, e^{i\omega x}}{2\sqrt{\pi}}$$

for $-\infty < \omega < \infty$. From the Fourier integral theorem, these singular eigenfunctions are "complete" with respect to piecewise smooth, absolutely integrable functions over the infinite interval $I = \{x \mid -\infty < x < \infty\}$. Thus, we can write our solution as the superposition of the preceding singular eigenfunctions, or, equivalently, as the Fourier integral

$$u(x, t) = \int_{-\infty}^{\infty} \frac{U(\omega, t) \, e^{i\omega x}}{2\pi} \, d\omega$$

From Section 9.2, $U(\omega, t)$ is the time-dependent Fourier transform of $u(x, t)$. If we substitute the initial condition $u(x, 0) = f(x)$ in this solution, we get

$$f(x) = \int_{-\infty}^{\infty} \frac{U(\omega, 0) \, e^{i\omega x}}{2\pi} \, d\omega$$

Obviously, $U(\omega, 0)$ is the Fourier transform of the initial condition function $f(x)$ where we have assumed that $f(x)$ is piecewise smooth and absolutely integrable over the infinite interval.

We now deal with the nonhomogeneous driving term in the same manner as we did in Chapter 8. If the source term $h(x, t)$ is piecewise smooth and absolutely integrable with respect to x over the infinite domain, then we can express its Fourier integral as

$$h(x, t) = \int_{-\infty}^{\infty} \frac{H(\omega, t) \, e^{i\omega x}}{2\pi} \, d\omega$$

where $H(\omega, t)$ is the Fourier transform of $h(x, t)$ as given here:

$$H(\omega, t) = \int_{-\infty}^{\infty} h(x, t) \, e^{-i\omega x} \, dx$$

We now require that all functions $u(x, t)$, $f(x)$, and $h(x, t)$ be absolutely integrable with respect to x over the infinite interval. We also require that the solution $u(x, t)$ and its first derivatives with respect to x vanish at both positive and negative infinity. With these assumptions, we substitute the assumed solution into the given partial differential equation for $u(x, t)$, and this yields

$$\frac{\partial}{\partial t} \left(\int_{-\infty}^{\infty} \frac{U(\omega, t) \, e^{i\omega x}}{2\pi} \, d\omega \right) = k \frac{\partial^2}{\partial x^2} \left(\int_{-\infty}^{\infty} \frac{U(\omega, t) \, e^{i\omega x}}{2\pi} \, d\omega \right) + \int_{-\infty}^{\infty} \frac{H(\omega, t) \, e^{i\omega x}}{2\pi} \, d\omega$$

Assuming the validity of the formal interchange between the differentiation and the integration operators, the preceding equation reduces to

$$\int_{-\infty}^{\infty} \left(\frac{\partial}{\partial t} U(\omega, t) + k\omega^2 U(\omega, t) \right) e^{i\omega x} \, d\omega = \int_{-\infty}^{\infty} H(\omega, t) \, e^{i\omega x} \, d\omega$$

In order for this equation to hold, due to the linear independence of the singular eigenfunctions, we set the integrands equal to each other, and we arrive at the following first-order nonhomogeneous differential equation in time t:

$$\frac{\partial}{\partial t} U(\omega, t) + k\omega^2 U(\omega, t) = H(\omega, t)$$

with the initial condition

$$U(\omega, 0) = \int\limits_{-\infty}^{\infty} f(x)\, e^{-i\omega x}\, dx$$

This is a first-order initial value problem and, from Chapter 1, Section 1.3, the solution to this differential equation is

$$U(\omega, t) = U(\omega, 0)\, e^{-k\omega^2 t} + \int\limits_{0}^{t} H(\omega, \tau)\, e^{-k\omega^2(t-\tau)}\, d\tau$$

Thus, from knowledge of the Fourier transform, the general form of the solution to our nonhomogeneous diffusion partial differential equation can be written as the Fourier integral

$$u(x, t) = \int\limits_{-\infty}^{\infty} \frac{\left(U(\omega, 0)\, e^{-k\omega^2 t} + \int_{0}^{t} H(\omega, \tau)\, e^{-k\omega^2(t-\tau)}\, d\tau\right) e^{i\omega x}}{2\pi}\, d\omega$$

We demonstrate the preceding concepts with the solution of an illustrative problem.

DEMONSTRATION: Consider a long, thin cylinder over the infinite interval $I = \{x \mid -\infty < x < \infty\}$ containing a fluid that has an initial salt concentration density $f(x)$ given as follows. We seek the spatial and time dependence of the density $u(x, t)$ of the salt as it diffuses into the fluid. There is no source term in the system, and the diffusivity is $k = 1/4$.

SOLUTION: Because there is no source term, the system homogeneous diffusion partial differential equation reads

$$\frac{\partial}{\partial t} u(x, t) = \frac{\frac{\partial^2}{\partial x^2} u(x, t)}{4}$$

The boundary condition requires that the solution be absolutely integrable over the infinite spatial domain—that is,

$$\int\limits_{-\infty}^{\infty} |u(x, t)|\, dx < \infty$$

For this particular problem, the initial condition is $u(x, 0) = f(x)$ where

$$f(x) = e^{-x^2}$$

From the method of separation of variables, the spatial-dependent differential equation yields the singular orthonormal eigenfunctions:

$$\frac{\sqrt{2} \, e^{i\omega x}}{2\sqrt{\pi}}$$

The Fourier integral form of the solution, in terms of these spatial eigenfunctions, is

$$u(x, t) = \int_{-\infty}^{\infty} \frac{U(\omega, t) \, e^{i\omega x}}{2\pi} \, d\omega$$

The Fourier transform of the initial condition term for the given $f(x)$ is

$$U(\omega, 0) = \int_{-\infty}^{\infty} e^{-x^2} e^{-i\omega x} \, dx$$

Evaluation of this integral yields

$$U(\omega, 0) = e^{-\frac{\omega^2}{4}} \sqrt{\pi}$$

From the method of separation of variables, the resultant time-dependent Fourier transform term satisfies the first-order differential equation

$$\frac{\partial}{\partial t} U(\omega, t) + \frac{\omega^2 U(\omega, t)}{4} = 0$$

The solution of this differential equation, which satisfies the preceding initial condition, is

$$U(\omega, t) = e^{-\frac{\omega^2}{4}} \sqrt{\pi} \, e^{-\frac{\omega^2 t}{4}}$$

Thus, we can write the solution to the problem as the Fourier integral

$$u(x, t) = \int_{-\infty}^{\infty} \frac{e^{-\frac{\omega^2}{4}} e^{-\frac{\omega^2 t}{4}} e^{i\omega x}}{2\sqrt{\pi}} \, d\omega$$

Evaluation of this integral yields the final form of the equation for the salt density

$$u(x, t) = \frac{e^{-\frac{x^2}{1+t}}}{\sqrt{1+t}}$$

The details for the development of this solution along with the graphics are given later in one of the Maple worksheet example problems.

9.5 Convolution Integral Solution for the Diffusion Equation

We can rewrite the form of the earlier general solution in a manner that introduces the concept of the convolution integral in addition to the initial and source value Green's functions. We break up the given solution into two parts. The first is due to the initial condition distribution term $f(x)$, and the second is due to the source term $h(x, t)$. For the nonhomogeneous partial differential equation, the general solution was shown earlier to be

$$u(x, t) = \int_{-\infty}^{\infty} \frac{\left(U(\omega, 0)\, e^{-k\omega^2 t} + \int_0^t H(\omega, \tau)\, e^{-k\omega^2(t-\tau)}\, d\tau\right) e^{i\omega x}}{2\pi}\, d\omega$$

This solution can be partitioned into two parts; that is, we can write

$$u(x, t) = u1(x, t) + u2(x, t)$$

where the first or initial condition part is

$$u1(x, t) = \int_{-\infty}^{\infty} \frac{U(\omega, 0)\, e^{-k\omega^2 t}\, e^{i\omega x}}{2\pi}\, d\omega$$

and the second or source term part is

$$u2(x, t) = \int_{-\infty}^{\infty} \frac{\left(\int_0^t H(\omega, \tau)\, e^{-k\omega^2(t-\tau)}\, d\tau\right) e^{i\omega x}}{2\pi}\, d\omega$$

We focus on the initial condition term first. If we make the substitution

$$U(\omega, 0) = \int_{-\infty}^{\infty} f(s)\, e^{-i\omega s}\, ds$$

into the first part of the solution, we get

$$u1(x, t) = \int_{-\infty}^{\infty} \left(\int_{-\infty}^{\infty} \frac{f(s)\, e^{-i\omega s}}{2\pi}\, ds\right) e^{-k\omega^2 t}\, e^{i\omega x}\, d\omega$$

Interchanging the order of integration yields

$$u1(x, t) = \int_{-\infty}^{\infty} f(s) \left(\int_{-\infty}^{\infty} \frac{e^{-k\omega^2 t}\, e^{i\omega(x-s)}}{2\pi}\, d\omega\right) ds$$

We recognize the inner integral as the shifted Fourier integral of the term

$$e^{-k\omega^2 t}$$

Evaluation of this Fourier integral yields what we call the "initial value Green's function" given here:

$$G1(x, t, s) = \frac{e^{-\frac{(x-s)^2}{4kt}}}{2\sqrt{\pi kt}}$$

Thus, we can write our initial value part of the solution as

$$u1(x, t) = \int_{-\infty}^{\infty} \frac{f(s) e^{-\frac{(x-s)^2}{4kt}}}{2\sqrt{\pi kt}} \, ds$$

or, in terms of the initial value Green's function, as

$$u1(x, t) = \int_{-\infty}^{\infty} f(s) \, G1(x, t, s) \, ds$$

We now focus on the second part of the solution due to the source term. Making the substitution

$$H(\omega, \tau) = \int_{-\infty}^{\infty} h(s, \tau) e^{-i\omega s} \, ds$$

into the second part yields

$$u2(x, t) = \int_{-\infty}^{\infty} \left(\int_{0}^{t} \left(\int_{-\infty}^{\infty} \frac{h(s, \tau) e^{-i\omega s}}{2\pi} \, ds \right) e^{-k\omega^2(t-\tau)} \, d\tau \right) e^{i\omega x} \, d\omega$$

Interchanging the order of integration yields

$$u2(x, t) = \int_{0}^{t} \int_{-\infty}^{\infty} h(s, \tau) \left(\int_{-\infty}^{\infty} \frac{e^{-k\omega^2(t-\tau)} e^{i\omega(x-s)}}{\pi} \, d\omega \right) ds \, d\tau$$

Again, we recognize the inner integral as the shifted Fourier integral of the term

$$e^{-k\omega^2(t-\tau)}$$

Evaluating this integral yields what we call the "source value Green's function" given here:

$$G2(x, t, s, \tau) = \frac{e^{-\frac{(x-s)^2}{4k(t-\tau)}}}{2\sqrt{\pi k(t-\tau)}}$$

Thus, we can write the second part of the solution as

$$u2(x, t) = \int\limits_{0}^{t} \int\limits_{-\infty}^{\infty} \frac{e^{-\frac{(x-s)^2}{4k(t-\tau)}} h(s, \tau)}{2\sqrt{\pi k(t-\tau)}} \, ds \, d\tau$$

or, in terms of the source value Green's function, as

$$u2(x, t) = \int\limits_{0}^{t} \int\limits_{-\infty}^{\infty} h(s, \tau) \, G2(x, t, s, \tau) \, ds \, d\tau \tag{9.2}$$

Combining the two parts gives us the final solution to the diffusion equation in the convolution form:

$$u(x, t) = \int\limits_{-\infty}^{\infty} \frac{f(s) e^{-\frac{(x-s)^2}{4kt}}}{2\sqrt{\pi kt}} \, ds + \int\limits_{0}^{t} \int\limits_{-\infty}^{\infty} \frac{e^{-\frac{(x-s)^2}{4k(t-\tau)}} h(s, \tau)}{2\sqrt{\pi k(t-\tau)}} \, ds \, d\tau$$

We will use this convolution form of the solution in some of the example problems using Maple later.

9.6 Nonhomogeneous Diffusion Equation over Semi-infinite Domains

We now consider the same problem as earlier, but over the semi-infinite interval $I = \{x \mid 0 < x < \infty\}$. We continue to assume that all functions are piecewise smooth and absolutely integrable with respect to x over the semi-infinite interval. Depending on the boundary condition at $x = 0$, we utilize those singular eigenfunctions that are consistent with the boundary condition. We consider only homogeneous boundary conditions.

If, at $x = 0$, we have a type 1 boundary condition

$$u(0, t) = 0$$

then the appropriate (since they satisfy the type 1 boundary condition at the origin) singular orthonormalized eigenfunctions are, from Section 9.3,

$$\frac{\sqrt{2} \sin(\omega x)}{\sqrt{\pi}}$$

for $0 < \omega < \infty$. Thus, we can write our solution in terms of these singular eigenfunctions as

$$u(x, t) = \int\limits_{0}^{\infty} \frac{2U(\omega, t) \sin(\omega x)}{\pi} \, d\omega$$

From Section 9.3, we see that $U(\omega, t)$ is the time-dependent Fourier sine transform of $u(x, t)$. If we substitute the initial condition $u(x, 0) = f(x)$ on this solution, we get

$$f(x) = \int_0^\infty \frac{2U(\omega, 0)\sin(\omega x)}{\pi}\, d\omega$$

Here, $U(\omega, 0)$ is the Fourier sine transform of the initial condition function $f(x)$; that is,

$$U(\omega, 0) = \int_0^\infty f(x)\sin(\omega x)\, dx \qquad (9.3)$$

In a similar manner, the source term $h(x, t)$ can be expressed as the Fourier sine integral

$$h(x, t) = \int_0^\infty \frac{2H(\omega, t)\sin(\omega x)}{\pi}\, d\omega$$

where $H(\omega, t)$ is now the Fourier sine transform of $h(x, t)$; that is,

$$H(\omega, t) = \int_0^\infty h(x, t)\sin(\omega x)\, dx$$

Proceeding formally, as we did earlier, $U(\omega, t)$ satisfies the first-order differential equation

$$\frac{\partial}{\partial t}U(\omega, t) + k\omega^2 U(\omega, t) = H(\omega, t)$$

and the solution to this first-order initial value problem, from Section 1.3, is

$$U(\omega, t) = U(\omega, 0)\, e^{-k\omega^2 t} + \int_0^t H(\omega, \tau)\, e^{-k\omega^2 (t-\tau)}\, d\tau$$

Thus, for the case of a semi-infinite interval with a type 1 condition at $x = 0$, the final solution to our nonhomogeneous diffusion partial differential equation can be written as the Fourier sine integral

$$u(x, t) = \int_0^\infty \frac{2\left(U(\omega, 0)\, e^{-k\omega^2 t} + \int_0^t H(\omega, \tau)\, e^{-k\omega^2 (t-\tau)}\, d\tau\right)\sin(\omega x)}{\pi}\, d\omega$$

On the other hand, if, at $x = 0$, we have the type 2 condition

$$u_x(0, t) = 0$$

then the appropriate (since they satisfy the type 2 boundary condition at the origin) singular orthonormalized eigenfunctions are, from Section 9.3,

$$\frac{\sqrt{2}\cos(\omega x)}{\sqrt{\pi}}$$

for $0 < \omega < \infty$. Thus, we can write our solution in terms of the singular eigenfunctions as

$$u(x, t) = \int_0^\infty \frac{2U(\omega, t)\cos(\omega x)}{\pi}\, d\omega$$

From Section 9.3, we see that $U(\omega, t)$ is the time-dependent Fourier cosine transform of $u(x, t)$. If we substitute the initial conditions into the solution, we get

$$f(x) = \int_0^\infty \frac{2U(\omega, 0)\cos(\omega x)}{\pi}\, d\omega$$

Here, $U(\omega, 0)$ is the Fourier cosine transform of the initial condition function $f(x)$; that is,

$$U(\omega, 0) = \int_0^\infty f(x)\cos(\omega x)\, dx$$

In a similar manner, the source term $h(x, t)$ can be expressed as the Fourier cosine integral

$$h(x, t) = \int_0^\infty \frac{2H(\omega, t)\cos(\omega x)}{\pi}\, d\omega$$

where, $H(\omega, t)$ is now the Fourier cosine transform of $h(x, t)$; that is,

$$H(\omega, t) = \int_0^\infty h(x, t)\cos(\omega x)\, dx$$

Again, proceeding formally, as we did earlier, $U(\omega, t)$ satisfies the first-order differential equation

$$\frac{\partial}{\partial t}U(\omega, t) + k\omega^2 U(\omega, t) = H(\omega, t)$$

and the solution to this first-order initial value problem, from Section 1.3, is

$$U(\omega, t) = U(\omega, 0)\,e^{-k\omega^2 t} + \int_0^t H(\omega, \tau)\,e^{-k\omega^2(t-\tau)}\, d\tau$$

Thus, for the case of a semi-infinite interval with a type 2 condition at $x = 0$, the final solution to our nonhomogeneous diffusion partial differential equation can be written as the Fourier cosine integral

$$u(x, t) = \int\limits_0^\infty \frac{2\left(U(\omega, 0)\, e^{-k\omega^2 t} + \int_0^t H(\omega, \tau)\, e^{-k\omega^2(t-\tau)}\, d\tau\right) \cos(\omega x)}{\pi}\, d\omega$$

The Method of Images

In some cases involving semi-infinite domain problems with homogeneous boundary conditions at the origin, it may be advantageous for us to employ what is called the "method of images." Depending on the boundary condition at the origin, we "reflect" the initial condition function $f(x)$ about the u-axis. For a problem with a type 1 condition at the origin, we simply form the "odd" extension of $f(x)$ and develop the rest of the solution using the regular Fourier integral. Similarly, for a problem with a type 2 condition at the origin, we simply form the "even" extension of $f(x)$ and develop the rest of the solution using the regular Fourier integral. We illustrate these techniques in some of the following examples.

9.7 Example Diffusion Problems over Infinite and Semi-infinite Domains

We now use the procedures discussed to investigate solutions to heat and diffusion problems over infinite and semi-infinite domains.

EXAMPLE 9.7.1: (Diffusion of a salt concentration) We consider a long, thin cylinder over the infinite interval $I = \{x \mid -\infty < x < \infty\}$ containing a fluid that initially has a salt concentration density $u(x, 0) = f(x)$ given as follows. We seek the density $u(x, t)$ of the salt as it diffuses into the fluid. The diffusivity of the medium is $k = 1/4$. There is no source term in the system. Solve the problem by the Fourier transform method.

SOLUTION: The diffusion partial differential equation is

$$\frac{\partial}{\partial t} u(x, t) = k\left(\frac{\partial^2}{\partial x^2} u(x, t)\right)$$

The boundary condition is that the solution be absolutely integrable over the interval; that is,

$$\int\limits_{-\infty}^{\infty} |u(x, t)|\, dx < \infty$$

The initial condition is

$$u(x, 0) = e^{-x^2}$$

The Fourier integral form of the solution is

$$u(x, t) = \int_{-\infty}^{\infty} \frac{U(\omega, t) e^{i\omega x}}{2\pi} \, d\omega$$

Here, $U(\omega, t)$ satisfies the differential equation

$$\frac{\partial}{\partial t} U(\omega, t) + k\omega^2 U(\omega, t) = 0$$

Assignment of system parameters

> restart:with(plots):k:=1/4:f(x):=exp(−x^2):

Fourier transform of initial condition term

> U(omega,0):=Int(f(x)*exp(−I*omega*x),x=−infinity..infinity);U(omega,0):=value(%):

$$U(\omega, 0) := \int_{-\infty}^{\infty} e^{-x^2} e^{-I\omega x} \, dx \tag{9.4}$$

> U(omega,0):=simplify(value(%));

$$U(\omega, 0) := e^{-\frac{1}{4}\omega^2} \sqrt{\pi} \tag{9.5}$$

Fourier transform of solution

> U(omega,t):=U(omega,0)*exp(−k*omega^2*t);

$$U(\omega, t) := e^{-\frac{1}{4}\omega^2} \sqrt{\pi} \, e^{-\frac{1}{4}\omega^2 t} \tag{9.6}$$

Fourier integral solution

> u(x,t):=Int(U(omega,t)/(2*Pi)*exp(I*omega*x),omega=−infinity..infinity);assume(t>0):

$$u(x, t) := \int_{-\infty}^{\infty} \frac{1}{2} \frac{e^{-\frac{1}{4}\omega^2} e^{-\frac{1}{4}\omega^2 t} e^{I\omega x}}{\sqrt{\pi}} \, d\omega \tag{9.7}$$

> u(x,t):=simplify(value(%));

$$u(x, t\sim) := \frac{e^{-\frac{x^2}{1+t\sim}}}{\sqrt{1+t\sim}} \tag{9.8}$$

ANIMATION

> animate(u(x,t),x=−5..5,t=0..5,thickness=3);

The preceding animation command shows the spatial-time-dependent solution of the density $u(x, t)$ in the medium. The animation sequence here and in Figure 9.1 shows snapshots of the animation at times $t = 0, 1, 2, 3, 4, 5$.

ANIMATION SEQUENCE

> u(x,0):=subs(t=0,u(x,t)):u(x,1):=subs(t=1,u(x,t)):

> u(x,2):=subs(t=2,u(x,t)):u(x,3):=subs(t=3,u(x,t)):

> u(x,4):=subs(t=4,u(x,t)):u(x,5):=subs(t=5,u(x,t)):

> plot({u(x,0),u(x,1),u(x,2),u(x,3),u(x,4),u(x,5)},x=−5..5,thickness=10);

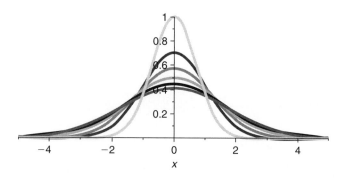

Figure 9.1

EXAMPLE 9.7.2: (Solution by convolution) We consider a long, thin cylinder over the infinite interval $I = \{x \mid -\infty < x < \infty\}$ containing a fluid that initially has a salt concentration density $u(x, 0) = f(x)$ given as follows. We seek the density $u(x, t)$ of the salt as it diffuses into the fluid. The diffusivity of the medium is $k = 1/4$. There is no source term in the system. This is the same problem as the preceding, but we now develop the solution using the method of convolution.

SOLUTION: The diffusion partial differential equation is

$$\frac{\partial}{\partial t}u(x, t) = k\left(\frac{\partial^2}{\partial x^2}u(x, t)\right)$$

The boundary condition is that the solution be absolutely integrable over the interval; that is,

$$\int_{-\infty}^{\infty} |u(x, t)| \, dx < \infty$$

The initial condition is

$$u(x, 0) = e^{-x^2}$$

The convolution form of the solution, from Section 9.5 for no internal source, is

$$u(x, t) = \int_{-\infty}^{\infty} \frac{f(s)\, e^{-\frac{(x-s)^2}{4kt}}}{2\sqrt{\pi kt}}\, ds$$

Assignment of system parameters

> restart:with(plots):k:=1/4:f(x):=exp(−x^2):f(s):=subs(x=s,f(x)):

Convolution of the initial condition contribution

> u(x,t):=Int(f(s)*exp(−(x−s)^2/(4*k*t))/(2*sqrt(Pi*k*t)),s=−infinity..infinity);assume(t>0):

$$u(x, t) := \int_{-\infty}^{\infty} \frac{e^{-s^2}\, e^{-\frac{(x-s)^2}{t}}}{\sqrt{\pi t}}\, ds \tag{9.9}$$

Convolution solution

> u(x,t):=simplify(value(%));

$$u(x, t \sim) := \frac{e^{-\frac{x^2}{t\sim+1}}}{\sqrt{t\sim +1}} \tag{9.10}$$

Comparing our solution with Example 9.7.1, we arrive at the same answer by either the transform method or the convolution method.

EXAMPLE 9.7.3: (The error function solution) We seek the temperature distribution $u(x, t)$ in a thin, long rod over the infinite interval $I = \{x \mid -\infty < x < \infty\}$ whose lateral surface is insulated. The initial temperature distribution $u(x, 0) = f(x)$ is given as follows. There is no source term in the system, and the thermal diffusivity is $k = 1/2$.

SOLUTION: The diffusion partial differential equation is

$$\frac{\partial}{\partial t} u(x, t) = k \left(\frac{\partial^2}{\partial x \partial x} u_{[]}(x, t) \right)$$

The boundary condition is that the solution be absolutely integrable over the interval; that is,

$$\int_{-\infty}^{\infty} |u(x, t)|\, dx < \infty$$

The initial condition is

$$u(x, 0) = H(x)$$

where $H(x)$ is the Heaviside function.

The convolution form of the solution, from Section 9.5 for no internal source, is

$$u(x, t) = \int_{-\infty}^{\infty} \frac{f(s)\, e^{-\frac{(x-s)^2}{4kt}}}{2\sqrt{\pi kt}}\, ds$$

Assignment of system parameters

> restart:with(plots):k:=1/2:f(x):=Heaviside(x):f(s):=subs(x=s,f(x)):assume(t>0):

Convolution of the initial condition contribution

> u(x,t):=Int(f(s)*exp(−(x−s)^2/(4*k*t))/(2*sqrt(Pi*k*t)),s=−infinity..infinity);

$$u(x, t\sim) := \int_{-\infty}^{\infty} \frac{1}{2} \frac{\text{Heaviside}(s)\, e^{-\frac{1}{2}\frac{(x-s)^2}{t\sim}}\, \sqrt{2}}{\sqrt{\pi t\sim}}\, ds \tag{9.11}$$

Convolution solution

> u(x,t):=simplify(value(%));

$$u(x, t\sim) := \frac{1}{2}\, \text{erf}\left(\frac{1}{2}\frac{x\,\sqrt{2}}{\sqrt{t\sim}}\right) + \frac{1}{2} \tag{9.12}$$

where the erf (x) term denotes the error function.

ANIMATION

> animate(u(x,t),x=−5..5,t=1/100..5,thickness=3);

The preceding animation command shows the spatial-time-dependent solution of the temperature $u(x, t)$ in the medium. The animation sequence here and in Figure 9.2 shows snapshots of the animation at times $t = 0, 1, 2, 3, 4, 5$.

ANIMATION SEQUENCE

> u(x,0):=subs(t=1/1000,u(x,t)):u(x,1):=subs(t=1,u(x,t)):

> u(x,2):=subs(t=2,u(x,t)):u(x,3):=subs(t=3,u(x,t)):

> u(x,4):=subs(t=4,u(x,t)):u(x,5):=subs(t=5,u(x,t)):

> plot({u(x,0),u(x,1),u(x,2),u(x,3),u(x,4),u(x,5)}x=−5..5,thickness=10);

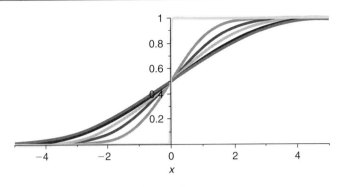

Figure 9.2

We note here that the initial condition function $f(x)$ is not absolutely integrable over the infinite domain, and we would not have been able to solve the preceding by the transform method. The convolution method is more forgiving on the conditions for $f(x)$ in that it just requires that $f(x)$ be finite over the domain. For some problems, the convolution method is the preferred method.

EXAMPLE 9.7.4: (Fourier sine integral solution) We seek the temperature distribution $u(x, t)$ in a thin, long rod over the semi-infinite interval $I = \{x \mid 0 < x < \infty\}$ whose lateral surface is insulated. The end $x = 0$ is held fixed at temperature 0. The initial temperature distribution $u(x, 0) = f(x)$ is given as follows. There is no source term in the system, and the thermal diffusivity is $k = 1/4$.

SOLUTION: The diffusion partial differential equation is

$$\frac{\partial}{\partial t} u(x, t) = k \left(\frac{\partial^2}{\partial x^2} u(x, t) \right)$$

The boundary conditions are that the solution should be absolutely integrable over the interval and we have a type 1 condition at $x = 0$:

$$\int_0^\infty |u(x, t)| \, dx < \infty \quad \text{and} \quad u(0, t) = 0$$

The initial condition is

$$u(x, 0) = 10 x e^{-x^2}$$

Due to the type 1 condition at the origin, we choose the Fourier sine integral form of solution:

$$u(x, t) = \int_0^\infty \frac{2U(\omega, t) \sin(\omega x)}{\pi} \, d\omega$$

Here, $U(\omega, t)$ satisfies the differential equation

$$\frac{\partial}{\partial t} U(\omega, t) + k\omega^2 U(\omega, t) = 0$$

Assignment of system parameters

> restart:with(plots):k:=1/4:f(x):=10*x*exp(−x^2):

Fourier sine transform of initial condition term

> U(omega,0):=Int(f(x)*sin(omega*x),x=0..infinity);

$$U(\omega, 0) := \int_0^\infty 10\,x\,\mathrm{e}^{-x^2}\sin(\omega x)\,\mathrm{d}x \tag{9.13}$$

> U(omega,0):=simplify(value(%));

$$U(\omega, 0) := \frac{5}{2}\sqrt{\pi}\,\omega\,\mathrm{e}^{-\frac{1}{4}\omega^2} \tag{9.14}$$

Fourier sine transform of solution

> U(omega,t):=U(omega,0)*exp(−k*omega^2*t);

$$U(\omega, t) := \frac{5}{2}\sqrt{\pi}\,\omega\,\mathrm{e}^{-\frac{1}{4}\omega^2}\,\mathrm{e}^{-\frac{1}{4}\omega^2 t} \tag{9.15}$$

Fourier sine integral solution

> u(x,t):=Int(U(omega,t)*(2/Pi)*sin(omega*x),omega=0..infinity);assume(t>0):

$$u(x, t) := \int_0^\infty \frac{5\omega\,\mathrm{e}^{-\frac{1}{4}\omega^2}\,\mathrm{e}^{-\frac{1}{4}\omega^2 t}\,\sin(\omega x)}{\sqrt{\pi}}\,\mathrm{d}\omega \tag{9.16}$$

> u(x,t):=simplify(value(%));

$$u(x, t\sim) := \frac{10\,x\,\mathrm{e}^{-\frac{x^2}{1+t\sim}}}{(1+t\sim)^{3/2}} \tag{9.17}$$

ANIMATION

> animate(u(x,t),x=0..5,t=0..5,thickness=3);

The preceding animation command shows the spatial-time-dependent solution of the temperature $u(x, t)$ in the medium. The animation sequence here and in Figure 9.3 shows snapshots of the animation at times $t = 0, 1, 2, 3, 4, 5$.

ANIMATION SEQUENCE

> u(x,0):=subs(t=0,u(x,t)):u(x,1):=subs(t=1,u(x,t)):

> u(x,2):=subs(t=2,u(x,t)):u(x,3):=subs(t=3,u(x,t)):

> u(x,4):=subs(t=4,u(x,t)):u(x,5):=subs(t=5,u(x,t)):

> plot({u(x,0),u(x,1),u(x,2),u(x,3),u(x,4),u(x,5)}x=0..5,thickness=10);

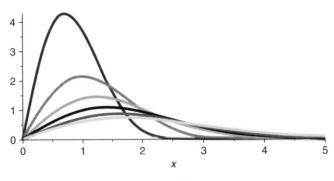

Figure 9.3

EXAMPLE 9.7.5: (Solution by the method of images) Here we consider the same problem as in Example 9.7.4, except now we use the method of images, which allows for use of the regular Fourier integral. Because the boundary conditions are homogeneous and of type 1 at $x = 0$, we reflect the initial condition function $u(x, 0) = f(x)$ about the origin, thus forming its odd extension onto the negative portion of the x-axis. Since we are only interested in the solution for $x > 0$, we will ignore that portion for $x < 0$.

SOLUTION: The diffusion partial differential equation is

$$\frac{\partial}{\partial t} u(x, t) = k \left(\frac{\partial^2}{\partial x^2} u(x, t) \right)$$

The boundary conditions are that the solution be absolutely integrable over the interval, and we have a type 1 condition at $x = 0$:

$$\int_0^\infty |u(x, t)| \, dx < \infty \quad \text{and} \quad u(0, t) = 0$$

The initial condition is

$$u(x, 0) = 10 \, x \, e^{-x^2}$$

By the method of images, we use the Heaviside function $H(x)$ to form the odd extension $fo(x)$ of $f(x)$ with the operation

$$fo(x) = f(x)H(x) - f(-x)H(-x)$$

The Fourier integral form of the solution is

$$u(x, t) = \int_{-\infty}^{\infty} \frac{U(\omega, t)\, e^{i\omega x}}{2\pi}\, d\omega$$

Here, $U(\omega, t)$ satisfies the differential equation

$$\frac{\partial}{\partial t} U(\omega, t) + k\omega^2 U(\omega, t) = 0$$

Assignment of system parameters

> restart:with(plots):k:=1/4:f(x):=10*x*exp(−x^2):f(−x):=subs(x=−x,f(x)):

Formation of the odd extension of $f(x)$

> fo(x):=f(x)*Heaviside(x)−f(−x)*Heaviside(−x);

$$fo(x) := 10x\,e^{-x^2}\,\text{Heaviside}(x) + 10x\,e^{-x^2}\,\text{Heaviside}(-x) \tag{9.18}$$

Fourier transform of odd extension of initial condition term

> U(omega,0):=Int(−f(−x)*exp(−I*omega*x),x=−infinity..0)+Int(f(x)*exp(−I*omega*x), x=0..infinity);

$$U(\omega, 0) := \int_{-\infty}^{0} 10x\,e^{-x^2}\,e^{-I\omega x}\, dx + \int_{0}^{\infty} 10x\,e^{-x^2}\,e^{-I\omega x}\, dx \tag{9.19}$$

> U(omega,0):=simplify(value(%));

$$U(\omega, 0) := -5I\omega\sqrt{\pi}\,e^{-\frac{1}{4}\omega^2} \tag{9.20}$$

Fourier transform of solution

> U(omega,t):=U(omega,0)*exp(−k*omega^2*t);

$$U(\omega, t) := -5I\omega\sqrt{\pi}\,e^{-\frac{1}{4}\omega^2}\,e^{-\frac{1}{4}\omega^2 t} \tag{9.21}$$

Fourier integral solution

> u(x,t):=Int(U(omega,t)*(1/(2*Pi))*exp(I*omega*x),omega=−infinity..infinity);assume(t>0):

$$u(x,t) := \int_{-\infty}^{\infty} \left(-\frac{\frac{5}{2} I \omega\, e^{-\frac{1}{4}\omega^2}\, e^{-\frac{1}{4}\omega^2 t}\, e^{I\omega x}}{\sqrt{\pi}} \right) d\omega \tag{9.22}$$

> u(x,t):=simplify(value(%));

$$u(x,t\sim) := \frac{10\, e^{-\frac{x^2}{1+t\sim}}\, x}{(1+t\sim)^{3/2}} \tag{9.23}$$

Note that we arrive at the exact same solution as we did when using the Fourier sine transform method.

EXAMPLE 9.7.6: (Fourier cosine integral solution) We seek the temperature distribution $u(x, t)$ in a long, thin rod over the semi-infinite interval $I = \{x \mid 0 < x < \infty\}$ whose lateral surface is insulated and whose left end is insulated at $x = 0$. The initial temperature distribution in the rod $u(x, 0) = f(x)$ is given as follows. There is no source term in the system and the thermal diffusivity is $k = 1/4$.

SOLUTION: The diffusion partial differential equation is

$$\frac{\partial}{\partial t} u(x, t) = k \left(\frac{\partial^2}{\partial x^2} u(x, t) \right)$$

The boundary conditions are that the solution be absolutely integrable over the interval, and we have a type 2 condition at $x = 0$:

$$\int_0^\infty |u(x, t)|\, dx < \infty \quad \text{and} \quad u_x(0, t) = 0$$

The initial condition is

$$u(x, 0) = 10\, e^{-x^2}$$

Due to the type 2 condition at the origin, we choose the Fourier cosine integral form of solution:

$$u(x, t) = \int_0^\infty \frac{2U(\omega, t) \cos(\omega x)}{\pi}\, d\omega$$

Here, $U(\omega, t)$ satisfies the differential equation

$$\frac{\partial}{\partial t} U(\omega, t) + k\omega^2 U(\omega, t) = 0$$

Assignment of system parameters

> restart:with(plots):k:=1/4:f(x):=10*exp(−x^2):

Fourier cosine transform of initial condition term

> U(omega,0):=Int(f(x)*cos(omega*x),x=0..infinity);

$$U(\omega, 0) := \int_0^\infty 10 \, e^{-x^2} \cos(\omega x) \, dx \tag{9.24}$$

> U(omega,0):=simplify(value%));

$$U(\omega, 0) := 5 \sqrt{\pi} \, e^{-\frac{1}{4}\omega^2} \tag{9.25}$$

Fourier cosine transform of solution

> U(omega,t):=U(omega,0)*exp(−k*omega^2*t);

$$U(\omega, t) := 5\sqrt{\pi} \, e^{-\frac{1}{4}\omega^2} \, e^{-\frac{1}{4}\omega^2 t} \tag{9.26}$$

Fourier cosine integral solution

> u(x,t):=Int(U(omega,t)*(2/Pi)*cos(omega*x),omega=0..infinity);assume(t>0):

$$u(x, t) := \int_0^\infty \frac{10 \, e^{-\frac{1}{4}\omega^2} \, e^{-\frac{1}{4}\omega^2 t} \cos(\omega x)}{\sqrt{\pi}} \, d\omega \tag{9.27}$$

> u(x,t):=simplify(value(%));

$$u(x, t\sim) := \frac{10 \, e^{-\frac{x^2}{1+t\sim}}}{\sqrt{1+t\sim}} \tag{9.28}$$

ANIMATION

> animate(u(x,t),x=0..5,t=0..5,thickness=3);

The preceding animation command shows the spatial-time-dependent solution of the temperature $u(x, t)$ in the medium. The animation sequence here and in Figure 9.4 shows snapshots of the animation at times $t = 0, 1, 2, 3, 4, 5$.

ANIMATION SEQUENCE

```
> u(x,0):=subs(t=0,u(x,t)):u(x,1):=subs(t=1,u(x,t)):

> u(x,2):=subs(t=2,u(x,t)):u(x,3):=subs(t=3,u(x,t)):

> u(x,4):=subs(t=4,u(x,t)):u(x,5):=subs(t=5,u(x,t)):

> plot({u(x,0),u(x,1),u(x,2),u(x,3),u(x,4),u(x,5)}x=0..5,thickness=10);
```

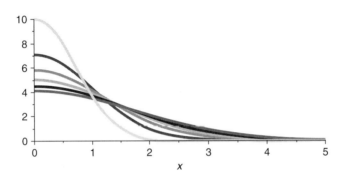

Figure 9.4

EXAMPLE 9.7.7: (Solution in integral form) We seek the temperature distribution $u(x, t)$ in a long, thin rod whose lateral surface is insulated over the infinite interval $I = \{x \mid -\infty < x < \infty\}$. The rod has an internal heat source $h(x, t)$ given as follows. The initial temperature distribution along the rod is 0, and the thermal diffusivity of the rod is $k = 1$.

SOLUTION: The nonhomogeneous diffusion partial differential equation is

$$\frac{\partial}{\partial t} u(x, t) = k \left(\frac{\partial^2}{\partial x^2} u(x, t) \right) + h(x, t)$$

The boundary condition is that the solution be absolutely integrable over the interval; that is,

$$\int_{-\infty}^{\infty} |u(x, t)| \, dx < \infty$$

The initial condition is

$$u(x, 0) = 0$$

The internal source term is

$$h(x, t) = \sin(t) \, e^{-\frac{x^2}{4}}$$

The Fourier integral form of the solution is

$$u(x, t) = \int_{-\infty}^{\infty} \frac{U(\omega, t) \, e^{i\omega x}}{2\pi} \, d\omega$$

Here, $U(\omega, t)$ satisfies the nonhomogeneous differential equation

$$\frac{\partial}{\partial t} U(\omega, t) + k\omega^2 U(\omega, t) = H(\omega, t)$$

Assignment of system parameters

> restart:with(plots):k:=1:h(x,t):=sin(t)*exp(−x^2/4):

Fourier transform of source term

> H(omega,t):=Int(h(x,t)*exp(−I*omega*x),x=−infinity..infinity);U(omega,0):=value(%):

$$H(\omega, t) := \int_{-\infty}^{\infty} \sin(t) \, e^{-\frac{1}{4}x^2} \, e^{-I\omega x} \, dx \tag{9.29}$$

> H(omega,t):=simplify(value(%));H(omega,tau):=subs(t=tau,H(omega,t)):

$$H(\omega, t) := 2 \, e^{-\omega^2} \sin(t) \sqrt{\pi} \tag{9.30}$$

Fourier transform of solution

> U(omega,t):=Int(H(omega,tau)*exp(−k*omega^2*(t−tau)),tau=0..t);

$$U(\omega, t) := \int_{0}^{t} 2 \, e^{-\omega^2} \sin(\tau) \sqrt{\pi} \, e^{-\omega^2(t-\tau)} \, d\tau \tag{9.31}$$

> U(omega,t):=simplify(factor(value(%)));

$$U(\omega, t) := \frac{2\sqrt{\pi}\left(1 - e^{\omega^2 t} \cos(t) + e^{\omega^2 t}\omega^2 \sin(t)\right) e^{-\omega^2(1+t)}}{\omega^4 + 1} \tag{9.32}$$

Fourier integral solution

> u(x,t):=Int(U(omega,t)/(2*Pi)*exp(I*omega*x),omega=−infinity..infinity);

$$u(x, t) := \int_{-\infty}^{\infty} \frac{\left(1 - e^{\omega^2 t} \cos(t) + e^{\omega^2 t}\omega^2 \sin(t)\right) e^{-\omega^2(1+t)} e^{I\omega x}}{\sqrt{\pi}(\omega^4 + 1)} \, d\omega \tag{9.33}$$

We leave this answer in integral form.

9.8 Nonhomogeneous Wave Equation over Infinite Domains

As we did earlier for the diffusion equation, here we use the Fourier integral method (singular eigenfunction expansion) to formulate the solutions to the wave partial differential equations over infinite or semi-infinite intervals.

We first consider the nonhomogeneous wave partial differential equation over the infinite interval $I = \{x \mid -\infty < x < \infty\}$ with no damping in the system

$$\frac{\partial^2}{\partial t^2} u(x, t) = c^2 \left(\frac{\partial^2}{\partial x^2} u(x, t) \right) + h(x, t)$$

with the two initial conditions

$$u(x, 0) = f(x) \quad \text{and} \quad u_t(x, 0) = g(x)$$

where $f(x)$ denotes the initial displacement of the wave, $g(x)$ denotes the initial speed of the wave, and $h(x, t)$ accounts for any external forces acting on the system. The boundary conditions are such that we require $u(x, t)$ to be piecewise smooth and absolutely integrable over the interval I.

As we did in Chapter 8, we first consider the homogeneous version of the preceding equation. Using the method of separation of variables, as in Chapter 3, we arrive at two ordinary differential equations: one in t and one in x. The x-dependent differential equation is, again, of the Euler type and it reads

$$\frac{d^2}{dx^2} X(x) + \lambda X(x) = 0$$

A set of basis vectors of the preceding differential equation, for $\lambda > 0$, is, from Chapter 1,

$$x1(x) = e^{i\sqrt{\lambda} x}$$

and

$$x2(x) = e^{-i\sqrt{\lambda} x}$$

If we set $\lambda = \omega^2$ and we require that our x-dependent portion of the solution be bounded as x goes to plus or minus infinity, then our x-dependent, orthonormalized singular eigenfunctions, as shown in Section 9.2, can be written

$$\frac{\sqrt{2} \, e^{i\omega x}}{2\sqrt{\pi}}$$

for $-\infty < \omega < \infty$.

From the Fourier integral theorem, these singular eigenfunctions are "complete" with respect to piecewise smooth and absolutely integrable functions over the infinite domain. Thus, we can write our solution as the superposition of the given singular eigenfunctions, or, equivalently, as the Fourier integral

$$u(x, t) = \int_{-\infty}^{\infty} \frac{U(\omega, t) \, e^{i\omega x}}{2\pi} \, d\omega$$

From Section 9.2, $U(\omega, t)$ is the time-dependent Fourier transform of $u(x, t)$. If we substitute the two initial conditions into this solution, we get

$$f(x) = \int_{-\infty}^{\infty} \frac{U(\omega, 0) \, e^{i\omega x}}{2\pi} \, d\omega$$

and

$$g(x) = \int_{-\infty}^{\infty} \frac{V(\omega, 0) \, e^{i\omega x}}{2\pi} \, d\omega$$

Here, $U(\omega, 0)$ is the Fourier transform of the initial displacement function $f(x)$ and $V(\omega, 0)$ is the Fourier transform of the initial speed function $g(x)$. We must assume that $f(x)$ and $g(x)$ are both piecewise smooth and absolutely integrable over the infinite interval.

We now deal with the nonhomogeneous driving term in the same manner as we did in Chapter 8. If the source term $h(x, t)$ is piecewise smooth and absolutely integrable with respect to x over the infinite domain, then we can express its Fourier integral as

$$h(x, t) = \int_{-\infty}^{\infty} \frac{H(\omega, t) \, e^{i\omega x}}{2\pi} \, d\omega$$

where $H(\omega, t)$ is the Fourier transform of $h(x, t)$.

We now require that all of the preceding functions behave in the same manner as for the diffusion equation in Section 9.4; that is, if we substitute the assumed solution into the given partial differential equation, and we assume validity of the formal interchange between the differentiation and the integration operators, then we get the following equation:

$$\int_{-\infty}^{\infty} \left(\frac{\partial^2}{\partial t^2} U(\omega, t) + c^2 \omega^2 U(\omega, t) \right) e^{i\omega x} \, d\omega = \int_{-\infty}^{\infty} H(\omega, t) \, e^{i\omega x} \, d\omega$$

In order for this equation to hold, due to the linear independence of the singular eigenfunctions, we set the integrands equal to each other and arrive at the following second-order nonhomogeneous differential equation in the time variable t:

$$\frac{\partial^2}{\partial t^2} U(\omega, t) + c^2 \omega^2 U(\omega, t) = H(\omega, t)$$

with the two initial conditions

$$U(\omega, 0) = \int_{-\infty}^{\infty} f(x) e^{-i\omega x} \, dx$$

and

$$V(\omega, 0) = \int_{-\infty}^{\infty} g(x) e^{-i\omega x} \, dx$$

This is a second-order initial value problem and, from Chapter 1, Section 1.8, the solution to the preceding differential equation is

$$U(\omega, t) = U(\omega, 0) \cos(\omega ct) + \frac{V(\omega, 0) \sin(\omega ct)}{\omega c} + \int_0^t \frac{H(\omega, \tau) \sin(\omega c(t - \tau))}{\omega c} \, d\tau$$

Thus, from the preceding transform, the solution to our nonhomogeneous partial differential equation can be written as the Fourier integral

$$u(x, t) = \int_{-\infty}^{\infty} \frac{U(\omega, t) e^{i\omega x}}{2\pi} \, d\omega$$

where $U(\omega, t)$ is the solution to the preceding time equation.

9.9 Wave Equation over Semi-infinite Domains

We now consider the same problem as in the preceding section, but for a semi-infinite interval $I = \{x \mid 0 < x < \infty\}$. We continue to assume that all functions are piecewise smooth and absolutely integrable with respect to x over the semi-infinite interval. Depending on the boundary condition at $x = 0$, we will utilize those singular eigenfunctions that are consistent with the boundary condition.

If at $x = 0$ we have the type 1 condition

$$u(0, t) = 0$$

then the appropriate singular orthonormalized eigenfunctions, from Section 9.3, are given as

$$\frac{\sqrt{2}\sin(\omega x)}{\sqrt{\pi}}$$

for $0 < \omega < \infty$. Thus, we can write our solution in terms of the singular eigenfunctions as

$$u(x, t) = \int_0^\infty \frac{2U(\omega, t)\sin(\omega x)}{\pi}\, d\omega$$

From Section 9.3, $U(\omega, t)$ is the time-dependent Fourier sine transform of $u(x, t)$. If we substitute the initial conditions into this solution, we get

$$f(x) = \int_0^\infty \frac{2U(\omega, 0)\sin(\omega x)}{\pi}\, d\omega$$

and

$$g(x) = \int_0^\infty \frac{2V(\omega, 0)\sin(\omega x)}{\pi}\, d\omega$$

Here, $U(\omega, 0)$ is the Fourier sine transform of the initial condition function $f(x)$, and $V(\omega, 0)$ is the Fourier sine transform of the initial speed function $g(x)$; that is,

$$U(\omega, 0) = \int_0^\infty f(x)\sin(\omega x)\, dx$$

and

$$V(\omega, 0) = \int_0^\infty g(x)\sin(\omega x)\, dx$$

In a similar manner, the source term $h(x, t)$ can be expressed as the Fourier sine integral

$$h(x, t) = \int_0^\infty \frac{2H(\omega, t)\sin(\omega x)}{\pi}\, d\omega$$

where $H(\omega, t)$ is now the Fourier sine transform of $h(x, t)$; that is,

$$H(\omega, t) = \int_0^\infty h(x, t)\sin(\omega x)\, dx$$

Proceeding formally as we did earlier, the solution for $U(\omega, t)$ is

$$U(\omega, t) = U(\omega, 0)\cos(\omega ct) + \frac{V(\omega, 0)\sin(\omega ct)}{\omega c} + \int_0^t \frac{H(\omega, \tau)\sin(\omega c(t - \tau))}{\omega c}\, d\tau$$

Thus, the final solution to our nonhomogeneous wave partial differential equation over the semi-infinite interval, with type 1 conditions at the origin, can be written as the Fourier sine integral

$$u(x, t) = \frac{\int_0^\infty 2U(\omega, t)\sin(\omega x)\, d\omega}{\pi}$$

where $U(\omega, t)$ was given above.

On the other hand, if, at $x = 0$, we have the type 2 condition

$$u_x(0, t) = 0$$

then the appropriate singular orthonormalized eigenfunctions, from Section 9.3, are given as

$$\frac{\sqrt{2}\cos(\omega x)}{\sqrt{\pi}}$$

for $0 < \omega < \infty$.

Thus, we can write our solution in terms of the singular eigenfunctions as

$$u(x, t) = \int_0^\infty \frac{2U(\omega, t)\cos(\omega x)}{\pi}\, d\omega$$

From Section 9.3, $U(\omega, t)$ is the time-dependent Fourier cosine transform of $u(x, t)$. If we substitute the two initial conditions on this solution, we get

$$f(x) = \int_0^\infty \frac{2U(\omega, 0)\cos(\omega x)}{\pi}\, d\omega$$

and

$$g(x) = \int_0^\infty \frac{2V(\omega, 0)\cos(\omega x)}{\pi}\, d\omega$$

Here, $U(\omega, 0)$ is the Fourier cosine transform of the initial condition function $f(x)$, and $V(\omega, 0)$ is the Fourier cosine transform of the initial speed function $g(x)$; that is,

$$U(\omega, 0) = \int_0^\infty f(x) \cos(\omega x) \, dx$$

and

$$V(\omega, 0) = \int_0^\infty g(x) \cos(\omega x) \, dx$$

In a similar manner, the source term $h(x, t)$ can be expressed as the Fourier cosine integral

$$h(x, t) = \int_0^\infty \frac{2H(\omega, t) \cos(\omega x)}{\pi} \, d\omega$$

where $H(\omega, t)$ is now the Fourier cosine transform of $h(x, t)$; that is,

$$H(\omega, t) = \int_0^\infty h(x, t) \cos(\omega x) \, dx$$

Proceeding formally as we did earlier, the solution for $U(\omega, t)$ is again

$$U(\omega, t) = U(\omega, 0) \cos(\omega c t) + \frac{V(\omega, 0) \sin(\omega c t)}{\omega c} + \int_0^t \frac{H(\omega, \tau) \sin(\omega c (t - \tau))}{\omega c} \, d\tau$$

Thus, the final solution to our nonhomogeneous wave partial differential equation over the semi-infinite interval, with type 2 boundary conditions at the origin, can be written as the Fourier cosine integral

$$u(x, t) = \int_0^\infty \frac{2U(\omega, t) \cos(\omega x)}{\pi} \, d\omega$$

where $U(\omega, t)$ was given above.

Similar to the case for diffusion problems, the method of images can also be used for the wave equation problems over semi-infinite domains. We now demonstrate this concept by solving an illustrative problem.

DEMONSTRATION: We seek the wave amplitude $u(x, t)$ for transverse waves on a long string over the semi-infinite interval $I = \{x \mid 0 < x < \infty\}$, which is unsecured at $x = 0$.

The initial displacement of the string $u(x, 0) = f(x)$ is given as follows, and the initial speed distribution $u_t(x, 0) = g(x)$ has the value 0. There are no external forces acting on the string, and the wave speed is $c = 1/2$.

SOLUTION: For transverse waves on a string, the homogeneous wave partial differential equation is

$$\frac{\partial^2}{\partial t^2} u(x, t) = \frac{\frac{\partial^2}{\partial x^2} u(x, t)}{4}$$

Because the string is unsecured at the end point $x = 0$ and we require the solution to be absolutely integrable over the semi-infinite inteval I, the system boundary conditions are

$$u_x(0, t) = 0 \quad \text{and} \quad \int_0^\infty |u(x, t)|\, dx < \infty$$

The initial displacement function of the string is given as $u(x, 0) = f(x)$ where

$$f(x) = e^{-x^2}$$

and the initial speed distribution is given as $u_t(x, 0) = g(x)$ where

$$g(x) = 0$$

We can solve this problem by using either the method of the Fourier cosine transform, as discussed earlier, or we can use the equivalent method of images. We choose the method of images here.

Using the Heaviside function, we establish the even extension $fe(x)$ of the initial displacement distribution $f(x)$ by reflecting it about the y-axis as follows:

$$fe(x) = e^{-x^2} H(x) + e^{-x^2} H(-x)$$

and the even extension of the initial speed distribution is

$$ge(x) = 0$$

With the even extensions of the initial displacement and speed distributions defined over the infinite interval $I = \{x \mid -\infty < x < \infty\}$, we can now use the method of the Fourier integral and write the solution as

$$u(x, t) = \int_{-\infty}^\infty \frac{U(\omega, t)\, e^{i\omega x}}{2\pi}\, d\omega$$

Substitution of this solution into the partial differential equation, as we did in Section 9.8, yields the time-dependent ordinary differential equation

$$\frac{\partial^2}{\partial t^2}U(\omega, t) + \frac{\omega^2 U(\omega, t)}{4} = 0$$

We must now solve this differential equation subject to the given initial conditions. Evaluation of the transform of the initial displacement function $fe(x)$ (note the function in this case is already even) yields

$$U(\omega, 0) = \int_{-\infty}^{\infty} e^{-x^2} e^{-i\omega x}\, dx$$

which is evaluated to be

$$U(\omega, 0) = \sqrt{\pi} e^{-\frac{\omega^2}{4}}$$

Similarly, evaluation of the transform of the initial speed distribution $ge(x)$ yields

$$V(\omega, 0) = 0$$

Thus, from Section 9.8, the solution to the time-dependent differential equation for the Fourier transform $U(\omega, t)$, subject to the two initial conditions, is

$$U(\omega, t) = \sqrt{\pi} e^{-\frac{\omega^2}{4}} \cos\left(\frac{\omega t}{2}\right)$$

Finally, from this known transform, the Fourier integral form of the solution is given as

$$u(x, t) = \int_{-\infty}^{\infty} \frac{e^{-\frac{\omega^2}{4}} \cos\left(\frac{\omega t}{2}\right)}{2\sqrt{\pi}}\, d\omega$$

Evaluation of this integral yields the solution for the string displacement

$$u(x, t) = \frac{\left(e^{-2xt} + 1\right) e^{-\frac{(2x-t)^2}{4}}}{2}$$

for $x > 0$.

The details for the solution of this problem along with the graphics are given later in one of the Maple worksheet examples.

9.10 Example Wave Equation Problems over Infinite and Semi-infinite Domains

We now consider solutions to some wave equation problems over infinite and semi-infinite domains. We use the procedures from the preceding section to develop our solutions.

EXAMPLE 9.10.1: (Waves on an infinite string) We seek the wave amplitude $u(x, t)$ for transverse wave motion on a long string over the infinite interval $I = \{x \mid -\infty < x < \infty\}$, which has an initial displacement $u(x, 0) = f(x)$ given as follows. The initial speed distribution $u_t(x, 0) = g(x)$ is 0, there are no external forces acting on the system, and the wave speed is $c = 1/2$.

SOLUTION: The wave partial differential equation is

$$\frac{\partial^2}{\partial t^2} u(x, t) = c^2 \left(\frac{\partial^2}{\partial x^2} u(x, t) \right)$$

The boundary condition is that the solution be absolutely integrable over the interval; that is,

$$\int_{-\infty}^{\infty} |u(x, t)| \, dx < \infty$$

The initial conditions are

$$u(x, 0) = e^{-x^2} \text{ and } u_t(x, 0) = 0$$

The Fourier integral form of the solution is

$$u(x, t) = \int_{-\infty}^{\infty} \frac{U(\omega, t) e^{i\omega x}}{2\pi} \, d\omega$$

Here, $U(\omega, t)$ satisfies the differential equation

$$\frac{\partial^2}{\partial t^2} U(\omega, t) + c^2 \omega^2 U(\omega, t) = 0$$

Assignment of system parameters

> restart:with(inttrans):with(plots):c:=1/2:f(x):=exp(−x^2):g(x):=0:

Fourier transform of initial conditions

> U(omega,0):=Int(f(x)*exp(−I*omega*x),x=−infinity..infinity);U(omega,0):=value(%):

$$U(\omega, 0) := \int_{-\infty}^{\infty} e^{-x^2} e^{-I\omega x} \, dx \tag{9.34}$$

> U(omega,0):=simplify(value(%));

$$U(\omega, 0) := e^{-\frac{1}{4}\omega^2}\sqrt{\pi} \tag{9.35}$$

> V(omega,0):=Int(g(x)*exp(−I*omega*x),x=−infinity..infinity);V(omega,0):=value(%):

$$V(\omega, 0) := \int_{-\infty}^{\infty} 0\,dx \tag{9.36}$$

> V(omega,0):=simplify(value(%));

$$V(\omega, 0) := 0 \tag{9.37}$$

Fourier transform of solution

> U(omega,t):=U(omega,0)*cos(omega*c*t)+V(omega,0)*sin(omega*c*t)/(omega*c);

$$U(\omega, t) := e^{-\frac{1}{4}\omega^2}\sqrt{\pi}\cos\left(\frac{1}{2}\omega t\right) \tag{9.38}$$

Fourier integral solution

> u(x,t):=Int(U(omega,t)/(2*Pi)*exp(I*omega*x),omega=−infinity..infinity);

$$u(x, t) := \int_{-\infty}^{\infty} \frac{1}{2} \frac{e^{-\frac{1}{4}\omega^2}\cos\left(\frac{1}{2}\omega t\right)e^{I\omega x}}{\sqrt{\pi}}\,d\omega \tag{9.39}$$

> u(x,t):=invfourier(U(omega,t),omega,x);

$$u(x, t) := \cosh(tx)e^{-\frac{1}{4}t^2 - x^2} \tag{9.40}$$

ANIMATION

> animate(u(x,t),x=−5..5,t=0..5,thickness=3);

The preceding animation command shows the spatial-time-dependent solution of the wave amplitude $u(x, t)$. The animation sequence here and in Figure 9.5 shows snapshots of the animation at times $t = 0, 1, 2, 3, 4, 5$.

ANIMATION SEQUENCE

> u(x,0):=subs(t=0,u(x,t)):u(x,1):=subs(t=1,u(x,t)):

> u(x,2):=subs(t=2,u(x,t)):u(x,3):=subs(t=3,u(x,t)):

> u(x,4):=subs(t=4,u(x,t)):u(x,5):=subs(t=5,u(x,t)):

> plot({u(x,0),u(x,1),u(x,2),u(x,3),u(x,4),u(x,5)},x=−5..5,thickness=10);

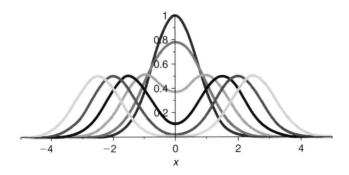

Figure 9.5

EXAMPLE 9.10.2: (Solution by method of images) We seek the wave amplitude $u(x, t)$ for transverse waves over a long string over the semi-infinite interval $I = \{x \mid 0 < x < \infty\}$. The end $x = 0$ is held fixed. The initial displacement distribution $u(x, 0) = f(x)$ is given as follows, and the initial speed distribution $u_t(x, 0) = g(x)$ is 0. There are no external forces acting on the system, and the wave speed is $c = 1/2$. We solve this problem using the method of images by forming the odd extension of $f(x)$ and using the regular Fourier integral.

SOLUTION: The wave partial differential equation is

$$\frac{\partial^2}{\partial t^2} u(x, t) = c^2 \left(\frac{\partial^2}{\partial x^2} u(x, t) \right)$$

The boundary conditions are that the solution be absolutely integrable over the interval and we have a type 1 condition at $x = 0$:

$$\int_0^\infty |u(x, t)| \, dx < \infty \quad \text{and} \quad u(0, t) = 0$$

The initial conditions are

$$u(x, 0) = xe^{-x^2} \quad \text{and} \quad u_t(x, 0) = 0$$

By the method of images, we use the Heaviside function $H(x)$ to form the odd extension $fo(x)$ of $f(x)$ with the operation

$$fo(x) = f(x)H(x) - f(-x)H(-x)$$

The Fourier integral form of the solution is

$$u(x, t) = \int\limits_{-\infty}^{\infty} \frac{U(\omega, t)\, e^{i\omega x}}{2\pi}\, d\omega$$

Here, $U(\omega, t)$ satisfies the differential equation

$$\frac{\partial^2}{\partial t^2} U(\omega, t) + c^2 \omega^2 U(\omega, t) = 0$$

Assignment of system parameters

> restart:with(inttrans):with(plots):c:=1/2:f(x):=x*exp(−x^2):f(−x):=subs(x=−x,f(x)):g(x):=0:
g(−x):=subs(x=−x,g(x)):

Formation of the odd extension of $f(x)$ and $g(x)$

> fo(x):=f(x)*Heaviside(x)−f(−x)*Heaviside(−x);

$$fo(x) := x\,e^{-x^2} \text{Heaviside}(x) + x\,e^{-x^2} \text{Heaviside}(-x) \tag{9.41}$$

> go(x):=g(x)*Heaviside(x)−g(−x)*Heaviside(−x);

$$go(x) := 0 \tag{9.42}$$

Fourier transform of initial conditions

> U(omega,0):=Int(fo(x)*exp(−I*omega*x),x=−infinity..infinity);

$$U(\omega, 0) := \int\limits_{-\infty}^{\infty} \left(x\,e^{-x^2} \text{Heaviside}(x) + x\,e^{-x^2} \text{Heaviside}(-x) \right) e^{-I\omega x}\, dx \tag{9.43}$$

> U(omega,0):=expand(fourier(f(x),x,omega));

$$U(\omega, 0) := -\frac{1}{2} I\omega\, e^{-\frac{1}{4}\omega^2} \sqrt{\pi} \tag{9.44}$$

> V(omega,0):=Int(go(x)*exp(−I*omega*x),x=−infinity..infinity);V(omega,0):=value(%):

$$V(\omega, 0) := \int\limits_{-\infty}^{\infty} 0\, dx \tag{9.45}$$

> V(omega,0):=simplify(value(%));

$$V(\omega, 0) := 0 \tag{9.46}$$

Fourier transform of solution

> U(omega,t):=U(omega,0)*cos(omega*c*t)+V(omega,0)*sin(omega*c*t)/(omega*c);

$$U(\omega, t) := -\frac{1}{2} I \omega \, e^{-\frac{1}{4}\omega^2} \sqrt{\pi} \cos\left(\frac{1}{2}\omega t\right) \tag{9.47}$$

Fourier integral solution

> u(x,t):=Int(U(omega,t)/(2*Pi)*exp(I*omega*x),omega=−infinity..infinity);

$$u(x, t) := \int_{-\infty}^{\infty} \left(-\frac{\frac{1}{4} I \omega \, e^{-\frac{1}{4}\omega^2} \cos\left(\frac{1}{2}\omega t\right) e^{I\omega x}}{\sqrt{\pi}} \right) d\omega \tag{9.48}$$

> u(x,t):=Heaviside(x)*simplify(invfourier(U(omega,t),omega,x));

$$u(x, t) := -\frac{1}{2} \text{Heaviside}(x)(t \sinh(tx) - 2x \cosh(tx)) \, e^{-x^2 - \frac{1}{4}t^2} \tag{9.49}$$

ANIMATION

> animate(u(x,t),x=0..5,t=0..10,thickness=3);

The preceding animation command shows the spatial-time-dependent solution of the wave amplitude $u(x, t)$. The animation sequence here and in Figure 9.6 shows snapshots of the animation at times $t = 0, 1, 2, 3, 4, 5$.

ANIMATION SEQUENCE

> u(x,0):=subs(t=0,u(x,t)):u(x,1):=subs(t=1,u(x,t)):

> u(x,2):=subs(t=2,u(x,t)):u(x,3):=subs(t=3,u(x,t)):

> u(x,4):=subs(t=4,u(x,t)):u(x,5):=subs(t=5,u(x,t)):

> plot({u(x,0),u(x,1),u(x,2),u(x,3),u(x,4),u(x,5)},x=0..5,thickness=10);

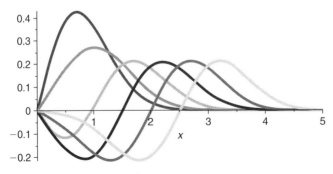

Figure 9.6

EXAMPLE 9.10.3: (Solution by method of images) We seek the wave amplitude $u(x, t)$ for transverse waves on a long string over the semi-infinite interval $I = \{x \,|\, 0 < x < \infty\}$. The end $x = 0$ is unsecured. The initial displacement distribution $u(x, 0) = f(x)$ is given as follows, and the initial speed distibution $u_t(x, 0) = 0$. There are no external forces acting on the system, and the wave speed is $c = 1/2$. We solve this problem using the method of images by forming the even extension of $f(x)$ and using the regular Fourier integral.

SOLUTION: The wave partial differential equation is

$$\frac{\partial^2}{\partial t^2} u(x, t) = c^2 \left(\frac{\partial^2}{\partial x^2} u(x, t) \right)$$

The boundary conditions are that the solution be absolutely integrable over the interval and we have a type 2 condition at $x = 0$:

$$\int_0^\infty |u(x, t)| \, dx < \infty \quad \text{and} \quad u_x(0, t) = 0$$

The initial conditions are

$$u(x, 0) = e^{-x^2} \quad \text{and} \quad u_t(x, 0) = 0$$

By the method of images, we use the Heaviside function $H(x)$ to form the even extension $fe(x)$ of $f(x)$ with the operation

$$fe(x) = f(x)H(x) + f(-x)H(-x)$$

The Fourier integral form of the solution is

$$u(x, t) = \int_{-\infty}^\infty \frac{U(\omega, t) \, e^{i\omega x}}{2\pi} \, d\omega$$

Here, $U(\omega, t)$ satisfies the differential equation

$$\frac{\partial^2}{\partial t^2} U(\omega, t) + c^2 \omega^2 U(\omega, t) = 0$$

Assignment of system parameters

```
>restart:with(inttrans):with(plots):c:=1/2:f(x):=exp(-x^2):f(-x):=subs(x=-x,f(x)):g(x):=0:
g(-x):=subs(x=-x,g(x)):
```

Formation of the even extension of $f(x)$ and $g(x)$

> fe(x):=f(x)*Heaviside(x)+f(−x)*Heaviside(−x);

$$fe(x) := e^{-x^2} \text{Heaviside}(x) + e^{-x^2} \text{Heaviside}(-x) \qquad (9.50)$$

> ge(x):=g(x)*Heaviside(x)+g(−x)*Heaviside(−x);

$$ge(x) := 0 \qquad (9.51)$$

Fourier transform of initial conditions

> U(omega,0):=Int(fe(x)*exp(−I*omega*x),x=−infinity..infinity);U(omega,0):=value(%):

$$U(\omega, 0) := \int_{-\infty}^{\infty} \left(e^{-x^2} \text{Heaviside}(x) + e^{-x^2} \text{Heaviside}(-x) \right) e^{-I\omega x} \, dx \qquad (9.52)$$

> U(omega,0):=simplify(fourier(f(x),x,omega));

$$U(\omega, 0) := \sqrt{\pi} \, e^{-\frac{1}{4}\omega^2} \qquad (9.53)$$

> V(omega,0):=Int(ge(x)*exp(−I*omega*x),x=−infinity..infinity);V(omega,0):=value(%):

$$V(\omega, 0) := \int_{-\infty}^{\infty} 0 \, dx \qquad (9.54)$$

> V(omega,0):=simplify(value(%));

$$V(\omega, 0) := 0 \qquad (9.55)$$

Fourier transform of solution

> U(omega,t):=U(omega,0)*cos(omega*c*t)+V(omega,0)*sin(omega*c*t)/(omega*c);

$$U(\omega, t) := \sqrt{\pi} \, e^{-\frac{1}{4}\omega^2} \cos\left(\frac{1}{2}\omega t\right) \qquad (9.56)$$

Fourier integral solution

> u(x,t):=Int(U(omega,t)/(2*Pi)*exp(I*omega*x),omega=−infinity..infinity);

$$u(x, t) := \int_{-\infty}^{\infty} \frac{1}{2} \frac{e^{-\frac{1}{4}\omega^2} \cos\left(\frac{1}{2}\omega t\right) e^{I\omega x}}{\sqrt{\pi}} \, d\omega \qquad (9.57)$$

> u(x,t):=Heaviside(x)*(invfourier(U(omega,t),omega,x));

$$u(x, t) := \text{Heaviside}(x) \cosh(tx) \, e^{-\frac{1}{4}t^2 - x^2} \tag{9.58}$$

ANIMATION

> animate(u(x,t),x=0..10,t=0..10,thickness=3);

The preceding animation command shows the spatial-time-dependent solution of the wave amplitude $u(x, t)$. The animation sequence here and in Figure 9.7 shows snapshots of the animation at times $t = 0, 1, 2, 3, 4, 5$.

ANIMATION SEQUENCE

> u(x,0):=subs(t=0,u(x,t)):u(x,1):=subs(t=1,u(x,t)):

> u(x,2):=subs(t=2,u(x,t)):u(x,3):=subs(t=3,u(x,t)):

> u(x,4):=subs(t=4,u(x,t)):u(x,5):=subs(t=5,u(x,t)):

> plot({u(x,0),u(x,1),u(x,2),u(x,3),u(x,4),u(x,5)},x=0..5,thickness=10);

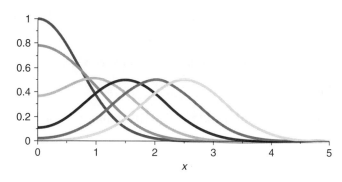

Figure 9.7

EXAMPLE 9.10.4: (Waves on an infinite string with initial speed) We seek the wave amplitude $u(x, t)$ for transverse wave motion on a long string over the infinite interval $I = \{x \mid -\infty < x < \infty\}$. The string has an initial displacement $u(x, 0) = 0$ and an initial speed distribution $u_t(x, 0) = g(x)$ given as follows. There are no external forces acting on the system, and the wave speed is $c = 1/2$.

SOLUTION: The wave partial differential equation is

$$\frac{\partial^2}{\partial t^2} u(x, t) = c^2 \left(\frac{\partial^2}{\partial x^2} u(x, t) \right)$$

The boundary conditions are that the solution be absolutely integrable over the interval; that is,

$$\int_{-\infty}^{\infty} |u(x, t)| \, dx < \infty$$

The initial conditions are

$$u(x, 0) = 0 \quad \text{and} \quad u_t(x, 0) = e^{-x^2}$$

The Fourier integral form of the solution is

$$u(x, t) = \int_{-\infty}^{\infty} \frac{U(\omega, t) \, e^{i\omega x}}{2\pi} \, d\omega$$

Here, $U(\omega, t)$ satisfies the differential equation

$$\frac{\partial^2}{\partial t^2} U(\omega, t) + c^2 \omega^2 U(\omega, t) = 0$$

Assignment of system parameters

> restart:with(inttrans):with(plots):c:=1/2:f(x):=0:g(x):=exp(−x^2):

Fourier transform of initial conditions

> U(omega,0):=Int(f(x)*exp(−I*omega*x),x=−infinity..infinity);U(omega,0):=value(%):

$$U(\omega, 0) := \int_{-\infty}^{\infty} 0 \, dx \tag{9.59}$$

> U(omega,0):=simplify(value(%));

$$U(\omega, 0) := 0 \tag{9.60}$$

> V(omega,0):=Int(g(x)*exp(−I*omega*x),x=−infinity..infinity);V(omega,0):=value(%):

$$V(\omega, 0) := \int_{-\infty}^{\infty} e^{-x^2} e^{-I\omega x} \, dx \tag{9.61}$$

> V(omega,0):=simplify(fourier(g(x),x,omega));

$$V(\omega, 0) := e^{-\frac{1}{4}\omega^2} \sqrt{\pi} \qquad (9.62)$$

Fourier transform of solution

> U(omega,t):=U(omega,0)*cos(omega*c*t)+V(omega,0)*sin(omega*c*t)/(omega*c);

$$U(\omega, t) := \frac{2\, e^{-\frac{1}{4}\omega^2} \sqrt{\pi} \sin\left(\frac{1}{2}\omega t\right)}{\omega} \qquad (9.63)$$

Fourier integral solution

> u(x,t):=Int(U(omega,t)/(2*Pi)*exp(I*omega*x),omega=−infinity..infinity);

$$u(x, t) := \int_{-\infty}^{\infty} \frac{e^{-\frac{1}{4}\omega^2} \sin\left(\frac{1}{2}\omega t\right) e^{I\omega x}}{\sqrt{\pi}\omega} \, d\omega \qquad (9.64)$$

> u(x,t):=invfourier(U(omega,t),omega,x);

$$u(x, t) := \frac{1}{2}\sqrt{\pi}\left(-\mathrm{erf}\left(x - \frac{1}{2}t\right) + \mathrm{erf}\left(x + \frac{1}{2}t\right)\right) \qquad (9.65)$$

ANIMATION

> animate(u(x,t),x=−5..5,t=0..5,thickness=3);

The preceding animation command shows the spatial-time-dependent solution of the wave amplitude $u(x, t)$. The animation sequence here and in Figure 9.8 shows snapshots of the animation at times $t = 0, 1, 2, 3, 4, 5$.

ANIMATION SEQUENCE

> u(x,0):=subs(t=0,u(x,t)):u(x,1):=subs(t=1,u(x,t)):

> u(x,2):=subs(t=2,u(x,t)):u(x,3):=subs(t=3,u(x,t)):

> u(x,4):=subs(t=4,u(x,t)):u(x,5):=subs(t=5,u(x,t)):

> plot({u(x,0),u(x,1),u(x,2),u(x,3),u(x,4),u(x,5)},x=-5..5,thickness=10);

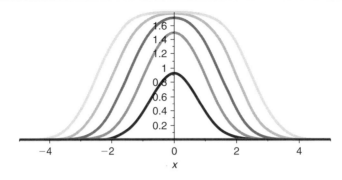

Figure 9.8

EXAMPLE 9.10.5: (Waves on an infinite string with an external applied force) We seek the wave amplitude $u(x, t)$ for transverse wave motion on a long string over the infinite interval $I = \{x \mid -\infty < x < \infty\}$. The string has an initial displacement $u(x, 0) = 0$ and an initial speed distribution $u_t(x, 0) = 0$. There is an external applied force $h(x, t)$ acting on the system, and the wave speed is $c = 1/2$.

SOLUTION: The nonhomogeneous wave partial differential equation is

$$\frac{\partial^2}{\partial t^2} u(x, t) = c^2 \left(\frac{\partial^2}{\partial x^2} u(x, t) \right) + h(x, t)$$

The boundary conditions are that the solution be absolutely integrable over the interval; that is,

$$\int_{-\infty}^{\infty} |u(x, t)| \, dx < \infty$$

The initial conditions are

$$u(x, 0) = 0 \quad \text{and} \quad u_t(x, 0) = 0$$

The external applied force is

$$h(x, t) = e^{-t} e^{-x^2}$$

The Fourier integral form of the solution is

$$u(x, t) = \int_{-\infty}^{\infty} \frac{U(\omega, t) e^{i\omega x}}{2\pi} \, d\omega$$

Here, $U(\omega, t)$ satisfies the nonhomogeneous differential equation

$$\frac{\partial^2}{\partial t^2} U(\omega, t) + c^2 \omega^2 U(\omega, t) = H(\omega, t)$$

Assignment of system parameters

> restart:with(inttrans):with(plots):c:=1/2:f(x):=0:g(x):=0:h(x,t):=exp(−t)*exp(−x^2):

Fourier transform of driving function

> H(omega,t):=Int(h(x,t)*exp(−I*omega*x),x=−infinity..infinity);H(omega,t):=
simplify(fourier(h(x,t),x,omega)):

$$H(\omega, t) := \int\limits_{-\infty}^{\infty} e^{-t} e^{-x^2} e^{-I\omega x} \, dx \tag{9.66}$$

> H(omega,t):=simplify(value(%));H(omega,tau):=subs(t=tau,H(omega,t)):

$$H(\omega, t) := \sqrt{\pi} \, e^{-t - \frac{1}{4}\omega^2} \tag{9.67}$$

Fourier transform of solution

> U(omega,t):=Int(H(omega,tau)*sin(omega*c*(t−tau))/(omega*c),tau=0..t);

$$U(\omega, t) := \int\limits_{0}^{t} \frac{2\sqrt{\pi} \, e^{-\tau - \frac{1}{4}\omega^2} \sin\left(\frac{1}{2}\omega(t - \tau)\right)}{\omega} \, d\tau \tag{9.68}$$

> U(omega,t):=simplify(value(%));

$$U(\omega, t) := -\frac{4\sqrt{\pi} \, e^{-t - \frac{1}{4}\omega^2} \left(e^t \omega \cos\left(\frac{1}{2}\omega t\right) - 2 e^t \sin\left(\frac{1}{2}\omega t\right) - \omega\right)}{\omega(4 + \omega^2)} \tag{9.69}$$

Fourier integral solution

> u(x,t):=Int(U(omega,t)/(2*Pi)*exp(I*omega*x),omega=−infinity..infinity);

$$u(x, t) := \int\limits_{-\infty}^{\infty} \left(-\frac{2 e^{-t - \frac{1}{4}\omega^2} \left(e^t \omega \cos\left(\frac{1}{2}\omega t\right) - 2 e^t \sin\left(\frac{1}{2}\omega t\right) - \omega\right) e^{I\omega x}}{\sqrt{\pi}\omega(4 + \omega^2)} \right) d\omega \tag{9.70}$$

We leave this solution in integral form.

9.11 Laplace Equation over Infinite and Semi-infinite Domains

We now consider the Laplace equation, which is the steady-state version of both the diffusion and the wave partial differential equations. Recall from Chapter 5 that these equations were a result of considering all solutions to be time invariant, thus forcing all time derivative terms equal to 0. We will use the Fourier integral method (singular eigenfunction expansion) to formulate the solutions to these partial differential equations over infinite or semi-infinite intervals.

We consider the homogeneous Laplace partial differential equation in two dimensions in the rectangular coordinate system where one of the coordinates is of infinite or semi-infinite extent.

$$\frac{\partial^2}{\partial x^2} u(x, y) + \frac{\partial^2}{\partial y^2} u(x, y) = 0$$

Using the method of separation of variables as in Chapter 5, we arrive at two ordinary differential equations: one in x and one in y. To obtain solutions that are finite as the variable x or y approaches infinity, we purposely select the format of the differential equations so as to satisfy this condition. Thus, depending on which variable, x or y, is infinite or semi-infinite in extent, we write the separated equations accordingly.

For example, if the domain is infinite or semi-infinite in extent with respect to the x variable, then we write the two separated equations as

$$\frac{d^2}{dx^2} X(x) + \lambda X(x) = 0$$

and

$$\frac{d^2}{dy^2} Y(y) - \lambda Y(y) = 0$$

A set of basis vectors for the x equation for $\lambda > 0$ is, from Chapter 1,

$$x1(x) = e^{i\sqrt{\lambda}x}$$

and

$$x2(x) = e^{-i\sqrt{\lambda}x}$$

If we set $\lambda = \omega^2$, then we satisfy the condition that our x-dependent portion of the solution be bounded as x goes to plus or minus infinity. Therefore, the x-dependent orthonormalized singular eigenfunctions, as shown in Section 9.2, can be written as

$$\frac{\sqrt{2}\,e^{i\omega x}}{2\sqrt{\pi}}$$

for $-\infty < \omega < \infty$.

From the Fourier integral theorem, the singular eigenfunctions are "complete" with respect to piecewise smooth and absolutely integrable functions over the infinite domain. Thus, we can write our solution as the superposition of the singular eigenfunctions just given, or, equivalently, as the Fourier integral

$$u(x, y) = \int_{-\infty}^{\infty} \frac{U(\omega, y)\, e^{i\omega x}}{2\pi}\, d\omega$$

From Section 9.2, $U(\omega, y)$ is the Fourier transform of $u(x, y)$ with respect to the variable x. If we substitute the assumed solution into the partial differential equation, and we assume the validity of the formal interchange between the differentiation and the integration operators, then we get the following differential equation in y:

$$\frac{\partial^2}{\partial y^2} U(\omega, y) - \omega^2 U(\omega, y) = 0$$

From Chapter 1, a set of basis vectors for the preceding corresponding homogeneous differential equation is

$$y1(y) = e^{\omega y}$$

and

$$y2(y) = e^{-\omega y}$$

The solution to the preceding y-dependent differential equation is

$$U(\omega, y) = A(\omega)\, e^{\omega y} + B(\omega)\, e^{-\omega y}$$

Thus, from the given transform, the solution to our homogeneous partial differential equation can be written as the Fourier integral

$$u(x, y) = \int_{-\infty}^{\infty} \frac{\left(A(\omega)\, e^{\omega y} + B(\omega)\, e^{-\omega y}\right) e^{i\omega x}}{2\pi}\, d\omega$$

The Fourier coefficients $A(\omega)$ and $B(\omega)$ are determined from the boundary conditions imposed on the problem.

In a similar manner, if the domain is infinite or semi-infinite in extent with respect to the y variable, then we write the two separated equations as

$$\frac{d^2}{dy^2} Y(y) + \lambda Y(y) = 0$$

and

$$\frac{d^2}{dx^2}X(x) - \lambda X(x) = 0$$

A set of basis vectors for the y equation for $\lambda > 0$ is, from Chapter 1,

$$y1(y) = e^{i\sqrt{\lambda}\,y}$$

and

$$y2(y) = e^{-i\sqrt{\lambda}\,y}$$

If we set $\lambda = \omega^2$, then we satisfy the condition that the y-dependent portion of the solution be bounded as y goes to plus or minus infinity. Thus, the y-dependent orthonormalized singular eigenfunctions, as shown in Section 9.2, can be written as

$$\frac{\sqrt{2}\,e^{i\omega y}}{2\sqrt{\pi}}$$

for $-\infty < \omega < \infty$.

From the Fourier integral theorem, the singular eigenfunctions are "complete" with respect to piecewise smooth and absolutely integrable functions over the infinite domain. Thus, we can write our solution as the superposition of the singular eigenfunctions just given, or, equivalently, as the Fourier integral

$$u(x, y) = \int_{-\infty}^{\infty} \frac{U(\omega, x)\,e^{i\omega y}}{2\pi}\,d\omega$$

From Section 9.2, $U(\omega, x)$ is the Fourier transform of $u(x, y)$ with respect to the variable y. If we substitute the assumed solution into the preceding partial differential equation, and we assume validity of the formal interchange between the differentiation and the integration operators, then we get the following differential equation in x:

$$\frac{\partial^2}{\partial x^2}U(\omega, x) - \omega^2 U(\omega, x) = 0$$

From Chapter 1, a set of basis vectors for the preceding corresponding homogeneous differential equation is

$$x1(x) = e^{\omega x}$$

and

$$x2(x) = e^{-\omega x}$$

The solution to the differential equation in x is

$$U(\omega, x) = A(\omega) e^{\omega x} + B(\omega) e^{-\omega x}$$

Thus, taking the inverse transform, the solution to our homogeneous partial differential equation can be written as the Fourier integral

$$u(x, y) = \int\limits_{-\infty}^{\infty} \frac{\left(A(\omega) e^{\omega x} + B(\omega) e^{-\omega x}\right) e^{i\omega y}}{2\pi} \, d\omega$$

The Fourier coefficients $A(\omega)$ and $B(\omega)$ are determined from the boundary conditions imposed on the problem.

If the domain is semi-infinite in extent with respect to either the x or y variable, then depending on the homogeneous boundary conditions imposed at the origin, we write our solution as either a Fourier sine or Fourier cosine integral, and we proceed as we did in Sections 9.6 and 9.8; the method of images is also available for use for semi-infinite intervals.

In Section 9.5, we discussed the convolution method in solving diffusion equation problems, and we provided several examples using this method. This method can also be used to solve Laplace equation problems. We now demonstrate the convolution method of solution for an illustrative problem.

DEMONSTRATION: We seek the electrostatic potential distribution $u(x, y)$ in a charge-free region over the two-dimensional domain (the right half-plane) $D = \{(x, y) \,|\, 0 < x < \infty, -\infty < y < \infty\}$. The edge $x = 0$ has a potential distribution $f(y)$ given as follows.

SOLUTION: The Laplace partial differential equation is

$$\frac{\partial^2}{\partial x^2} u(x, y) + \frac{\partial^2}{\partial y^2} u(x, y) = 0$$

The domain is infinite in extent with respect to the y variable and semi-infinite in extent with respect to the x variable. We require the solution to be absolutely integrable over this infinite half-plane. In addition, there is a given potential distribution along the line $x = 0$. Thus, the boundary conditions on the problem are

$$u(0, y) = f(y) \quad \text{and} \quad \int\limits_{-\infty}^{\infty} \int\limits_{0}^{\infty} |u(x, y)| \, dx \, dy < \infty$$

Because we have a nonhomogeneous boundary condition with respect to x at the line $x = 0$, and the boundary with respect to y is infinite in extent, then we write the solution $u(x, y)$ as a

Fourier integral with respect to the y variable—that is,

$$u(x, y) = \int\limits_{-\infty}^{\infty} \frac{U(\omega, x)\,e^{i\omega y}}{2\pi}\,d\omega$$

Substitution of this solution into the Laplace partial differential equation yields the ordinary differential equation in x:

$$\frac{\partial^2}{\partial x^2} U(\omega, x) - \omega^2 U(\omega, x) = 0$$

From knowledge of the two basis vectors (see Section 1.4), the general solution of this homogeneous differential equation for the Fourier transform of the solution is

$$U(\omega, x) = A(\omega)\,e^{\omega x} + B(\omega)\,e^{-\omega x}$$

Because the solution must remain bounded as x goes to infinity, and since ω takes on both positive and negative values, we set

$$U(\omega, x) = C(\omega)\,e^{-|\omega|x}$$

where $C(\omega)$ is an unknown arbitrary function of ω.

With knowledge of the given Fourier transform, the Fourier integral of the solution reads

$$u(x, y) = \int\limits_{-\infty}^{\infty} \frac{C(\omega)\,e^{-|\omega|x}\,e^{i\omega y}}{2\pi}\,d\omega$$

Substitution of the boundary condition $u(0, y) = f(y)$ at $x = 0$ yields

$$f(y) = \int\limits_{-\infty}^{\infty} \frac{C(\omega)\,e^{i\omega y}}{2\pi}\,d\omega$$

Thus, $C(\omega)$ is the Fourier transform of $f(y)$—that is,

$$C(\omega) = \int\limits_{-\infty}^{\infty} f(y)\,e^{-i\omega y}\,dy$$

We can combine the two given integrals, as we did in Section 9.5 for the diffusion equation, and our solution becomes

$$u(x, y) = \int\limits_{-\infty}^{\infty} \left(\int\limits_{-\infty}^{\infty} \frac{f(\zeta)\,e^{-i\omega\zeta}}{2\pi}\,d\zeta \right) e^{-|\omega|x}\,e^{i\omega y}\,d\omega$$

If we assume validity of the formal interchange of the order of integration, we get

$$u(x, y) = \int_{-\infty}^{\infty} f(\zeta) \left(\int_{-\infty}^{\infty} \frac{e^{-|\omega|x} e^{i\omega(y-\zeta)}}{2\pi} \, d\omega \right) d\zeta$$

The interior integral is the Fourier integral, with respect to the variable $v = y - \zeta$, of the function

$$g(x, y - \zeta) = \frac{x}{\pi \left(x^2 + (y - \zeta)^2 \right)}$$

Thus, our solution can be written as the convolution integral

$$u(x, y) = \int_{-\infty}^{\infty} f(\zeta) g(x, y - \zeta) \, d\zeta$$

From knowledge of the boundary function $u(0, y) = f(y)$, we can evaluate the final solution.

We now provide an illustrative example where the given potential distribution along the $x = 0$ line is $u(x, 0) = f(y)$, where

$$f(y) = H(y + 1) - H(y - 1)$$

From the definition of the Heaviside functions, this corresponds to a potential on the y-axis, which has a value of 1 for $|y| < 1$ and 0 for $|y| > 1$. For this special case of potential distribution along the y-axis, our integral solution above becomes

$$u(x, y) = \int_{-\infty}^{\infty} \frac{(H(\zeta + 1) - H(\zeta - 1)) x}{\left(x^2 + (y - \zeta)^2 \right) \pi} \, d\zeta$$

This integral is reduced to the equivalent integral

$$u(x, y) = \int_{-1}^{1} \frac{x}{\left(x^2 + (y - \zeta)^2 \right) \pi} \, d\zeta$$

Evaluation of this integral yields the electrostatic potential distribution

$$u(x, y) = \frac{- \arctan\left(\frac{y-1}{x} \right) + \arctan\left(\frac{y+1}{x} \right)}{\pi}$$

The details for the development of this solution along with the graphics are given later in one of the Maple worksheet examples.

9.12 Example Laplace Equation over Infinite and Semi-infinite Domains

We now consider some Laplace equation problems over both infinite and semi-infinite domains.

EXAMPLE 9.12.1: (Temperature distribution along an infinite strip) We seek the steady-state temperature distribution $u(x, y)$ in a thin plate over the domain $D = \{(x, y) \,|\, -\infty < x < \infty,\ 0 < y < 1\}$ whose lateral surfaces are insulated. The edge $y = 0$ is at a fixed temperature of 0, and the edge $y = 1$ has the temperature distribution $f(x)$ given as follows.

SOLUTION: The homogeneous Laplace equation is

$$\frac{\partial^2}{\partial x^2} u(x, y) + \frac{\partial^2}{\partial y^2} u(x, y) = 0$$

The boundary conditions are

$$\int_0^\infty \int_{-\infty}^\infty |u(x, y)| \, dx \, dy < \infty$$

and

$$u(x, 0) = 0 \quad \text{and} \quad u(x, 1) = \frac{1}{1 + x^2}$$

Since the domain is infinite in extent with respect to the x variable, we write the Fourier integral solution with respect to the x variable as

$$u(x, y) = \int_{-\infty}^\infty \frac{U(\omega, y) \, e^{i\omega x}}{2\pi} \, d\omega$$

where the Fourier transform $U(\omega, y)$ satisfies the y-dependent differential equation

$$\frac{\partial^2}{\partial y^2} U(\omega, y) - \omega^2 U(\omega, y) = 0$$

Assignment of system parameters

> restart:with(inttrans):with(plots):b:=1:f(x):=1/(1+x^2):

Solution to the y-dependent equation

> U(omega,y):=A(omega)*cosh(omega*y)+B(omega)*sinh(omega*y);

$$U(\omega, y) := A(\omega) \cosh(\omega y) + B(\omega) \sinh(\omega y) \tag{9.71}$$

Substitution of the boundary condition at $y = 0$ gives $A(\omega) = 0$ and

> U(omega,y):=B(omega)*sinh(omega*y);

$$U(\omega, y) := B(\omega)\sinh(\omega y) \tag{9.72}$$

Fourier integral solution

> u(x,y):=Int(B(omega)*sinh(omega*y)*(1/(2*Pi))*exp(I*omega*x),omega=
  −infinity..infinity);

$$u(x, y) := \int_{-\infty}^{\infty} \frac{1}{2} \frac{B(\omega)\sinh(\omega y)\, e^{I\omega x}}{\pi}\, d\omega \tag{9.73}$$

Substituting the boundary condition $u(x, 1) = f(x)$ at $y = 1$ yields

> f(x)=subs(y=b,u(x,y));

$$\frac{1}{1+x^2} = \int_{-\infty}^{\infty} \frac{B(\omega)\sinh(\omega)\, e^{I\omega x}}{\pi}\, d\omega \tag{9.74}$$

The coefficient $B(\omega)$ is found from the Fourier transform

> B(omega):=Int(f(x)*(1/sinh(omega*b))*exp(−I*omega*x),x=−infinity..infinity);

$$B(\omega) := \int_{-\infty}^{\infty} \frac{e^{-I\omega x}}{\left(1+x^2\right)\sinh(\omega)}\, dx \tag{9.75}$$

Evaluation of this integral yields

> B(omega):=factor(fourier(f(x)/sinh(omega*b),x,omega));

$$B(\omega) := \frac{\pi\left(e^{\omega}\text{Heaviside}(-\omega) + e^{-\omega}\text{Heaviside}(\omega)\right)}{\sinh(\omega)} \tag{9.76}$$

Solution for $U(\omega, y)$

> U(omega,y):=Pi*exp(−abs(omega))*sinh(omega*y)/sinh(omega);

$$U(\omega, y) := \frac{\pi\, e^{-|\omega|}\sinh(\omega y)}{\sinh(\omega)} \tag{9.77}$$

Fourier integral solution

```
> u(x,y):=int(U(omega,y)*(1/(2*Pi))*exp(I*omega*x),omega=−infinity..infinity);
```

$$u(x, y) := \int_{-\infty}^{\infty} \frac{1}{2} \frac{e^{-|\omega|} \sinh(\omega y) e^{I\omega x}}{\sinh(\omega)} d\omega \qquad (9.78)$$

We leave our answer in integral form.

EXAMPLE 9.12.2: (Steady-state temperature in upper half of plane) We seek the steady-state temperature distribution $u(x, y)$ in a thin plate over the domain $D = \{(x, y) \,|-\infty < x < \infty,$ $0 < y < \infty\}$ whose lateral surfaces are insulated. The edge $y = 0$ has a temperature distribution $f(x)$ given as follows.

SOLUTION: The homogeneous Laplace equation is

$$\frac{\partial^2}{\partial x^2} u(x, y) + \frac{\partial^2}{\partial y^2} u(x, y) = 0$$

The boundary conditions are

$$\int_0^{\infty} \int_{-\infty}^{\infty} |u(x, y)| \, dx \, dy < \infty \quad \text{and} \quad u(x, 0) = \frac{16}{16 + x^2}$$

Because the domain is infinite in extent with respect to the x variable, we write the Fourier integral solution with respect to the x variable as

$$u(x, y) = \int_{-\infty}^{\infty} \frac{U(\omega, y) e^{i\omega x}}{2\pi} d\omega$$

where the Fourier transform $U(\omega, y)$ satisfies the y-dependent differential equation

$$\frac{\partial^2}{\partial y^2} U(\omega, y) - \omega^2 U(\omega, y) = 0$$

Assignment of system parameters

```
> restart:with(inttrans):with(plots):f(x):=16/(16+x^2):
```

Solution to the y-dependent equation

```
> U(omega,y):=A(omega)*exp(omega*y)+B(omega)*exp(−omega*y);assume(y>0):
```

$$U(\omega, y) := A(\omega) e^{\omega y} + B(\omega) e^{-\omega y} \qquad (9.79)$$

Because the solution must remain bounded as y goes to plus or minus infinity, and because ω takes on positive and negative values, we set

> U(omega,y):=C(omega)*exp(−abs(omega)*y);

$$U(\omega, y\sim) := C(\omega)\, e^{-|\omega|y\sim} \tag{9.80}$$

Fourier integral solution

> u(x,y):=Int(U(omega,y)/(2*Pi)*exp(I*omega*x),omega=−infinity..infinity);

$$u(x, y\sim) := \int_{-\infty}^{\infty} \frac{1}{2} \frac{C(\omega)\, e^{-|\omega|y\sim}\, e^{I\omega x}}{\pi}\, d\omega \tag{9.81}$$

Substituting the boundary condition $u(x, 0) = f(x)$ at $y = 0$ yields

> f(x)=eval(subs(y=0,u(x,y)));

$$\frac{16}{16 + x^2} = \int_{-\infty}^{\infty} \frac{1}{2} \frac{C(\omega)\, e^{I\omega x}}{\pi}\, d\omega \tag{9.82}$$

Thus, the coefficient $C(\omega)$ is found from the Fourier transform

> C(omega):=Int(f(x)*exp(−I*omega*x),x=−infinity..infinity);

$$C(\omega) := \int_{-\infty}^{\infty} \frac{16\, e^{-I\omega x}}{16 + x^2}\, dx \tag{9.83}$$

Evaluation of this integral using the Maple fourier command yields

> C(omega):=fourier(f(x),x,omega);

$$C(\omega) := 4\pi \left(e^{4\omega}\text{Heaviside}(-\omega) + e^{-4\omega}\text{Heaviside}(\omega) \right) \tag{9.84}$$

Solution for $U(\omega, y)$

> U(omega,y):=C(omega)*exp(−abs(omega)*y);

$$U(\omega, y\sim) := 4\pi \left(e^{4\omega}\text{Heaviside}(-\omega) + e^{-4\omega}\text{Heaviside}(\omega) \right) e^{-|\omega|y\sim} \tag{9.85}$$

Fourier integral solution

> u(x,y):=Int(U(omega,y)*(1/(2*Pi))*exp(I*omega*x),omega=−infinity..infinity);

$$u(x, y\sim) := \int_{-\infty}^{\infty} 2 \left(e^{4\omega}\text{Heaviside}(-\omega) + e^{-4\omega}\text{Heaviside}(\omega) \right) e^{-|\omega|y\sim}\, e^{I\omega x}\, d\omega \tag{9.86}$$

Partitioning the integral over positive and negative values of ω yields

```
> u(x,y):=int(U(omega,y)*(1/(2*Pi))*exp(I*omega*x),omega=−infinity..0)+int(U(omega,y)*
(1/(2*Pi))*exp(I*omega*x),omega=0..infinity):
```

```
> u(x,y):=simplify(evalc(value(%)));
```

$$u(x, y\sim) := \frac{4(4+y\sim)}{16+8y\sim +y\sim^2 +x^2} \tag{9.87}$$

```
> plot3d(u(x,y),x=−5..5,y=0..5,axes=framed,thickness=1);
```

The three-dimensional surface shown in Figure 9.9 illustrates the steady-state temperature distribution $u(x, y)$ in the plate. The temperature isotherms are obtained by clicking on the graph and choosing the "Render the plot using the polygon patch and contour style" option and then clicking the "redraw" button in the graphics bar.

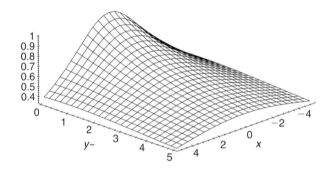

Figure 9.9

EXAMPLE 9.12.3: (Electrostatic potential solved by convolution) We seek the electrostatic potential distribution $u(x, y)$ in a charge-free region over the domain $D = \{(x, y) \mid 0 < x < \infty, -\infty < y < \infty\}$. The edge $x = 0$ has a potential distribution $f(y)$ given as follows. Since the domain is infinite in extent with respect to y, we use a Fourier integral solution in y along with a convolution.

SOLUTION: The homogeneous Laplace equation is

$$\frac{\partial^2}{\partial x^2}u(x, y) + \frac{\partial^2}{\partial y^2}u(x, y) = 0$$

The boundary conditions are

$$\int_{-\infty}^{\infty} \int_{0}^{\infty} |u(x, y)|\,dx\,dy < \infty \quad \text{and} \quad u(0, y) = H(y+1) - H(y-1)$$

From the definition of the Heaviside function, this potential distribution corresponds to a potential of magnitude 1 at $x = 0$ along the y-axis for $|y| < 1$ and 0 for $1 < |y|$.

Because the domain is infinite in extent with respect to the y variable, we write the Fourier integral solution with respect to the y variable as

$$u(x, y) = \int_{-\infty}^{\infty} \frac{U(\omega, x)\, e^{i\omega y}}{2\pi}\, d\omega$$

where the Fourier transform $U(\omega, x)$ satisfies the x-dependent differential equation

$$\frac{\partial^2}{\partial x^2} U(\omega, x) - \omega^2 U(\omega, x) = 0$$

Assignment of system parameters

> restart:with(inttrans):with(plots):

Solution to the x-dependent equation

> U(omega,x):=A(omega)*exp(omega*x)+B(omega)*exp(−omega*x);

$$U(\omega, x) := A(\omega)\, e^{\omega x} + B(\omega)\, e^{-\omega x} \tag{9.88}$$

Because the solution must remain bounded as x goes to infinity, and because ω takes on positive and negative values, we set

> U(omega,x):=C(omega)*exp(−abs(omega)*x);

$$U(\omega, x) := C(\omega)\, e^{-|\omega| x} \tag{9.89}$$

Fourier integral solution

> u(x,y):=Int(U(omega,x)*(1/(2*Pi))*exp(I*omega*y),omega=−infinity..infinity);

$$u(x, y) := \int_{-\infty}^{\infty} \frac{1}{2} \frac{C(\omega)\, e^{-|\omega| x}\, e^{I\omega y}}{\pi}\, d\omega \tag{9.90}$$

Substituting the boundary condition $u(0, y) = f(y)$ at $x = 0$ yields

> f(y)=eval(subs(x=0,u(x,y)));

$$f(y) = \int_{-\infty}^{\infty} \frac{1}{2} \frac{C(\omega)\, e^{I\omega y}}{\pi}\, d\omega \tag{9.91}$$

Thus, the coefficient $C(\omega)$ is found from the following Fourier transform:

> C(omega):=Int(f(y)*exp(−I*omega*y),y=−infinity..infinity);

$$C(\omega) := \int_{-\infty}^{\infty} f(y)\,e^{-I\omega y}\,dy \tag{9.92}$$

To solve by convolution, we can combine the two preceding integrals, as we did in Section 9.5 for the diffusion equation. Doing so, we get

> u(x,y):=Int(Int(f(zeta)/(2*Pi)*exp(−I*omega*zeta),zeta=−infinity..infinity)*
exp(−abs(omega)*x)*exp(I*omega*y),omega=−infinity..infinity);

$$u(x, y) := \int_{-\infty}^{\infty} \left(\int_{-\infty}^{\infty} \frac{1}{2} \frac{f(\zeta)\,e^{-I\omega\zeta}}{\pi}\,d\zeta \right) e^{-|\omega|x}\,e^{I\omega y}\,d\omega \tag{9.93}$$

Assuming the validity of the formal interchange of the order of integration yields

> u(x,y):=Int(f(zeta)*Int(exp(−abs(omega)*x)*(1/(2*Pi))*exp(I*omega*(y−zeta)),omega=
−infinity..infinity),zeta=−infinity..infinity);

$$u(x, y) := \int_{-\infty}^{\infty} f(\zeta) \left(\int_{-\infty}^{\infty} \frac{1}{2} \frac{e^{-|\omega|x}\,e^{I\omega(y-\zeta)}}{\pi}\,d\omega \right) d\zeta \tag{9.94}$$

If we set $v = y - \zeta$, then we recognize the interior integral as the Fourier integral of

> assume(x>0):g(x,y−zeta):=subs(v=y−zeta,invfourier(exp(−abs(omega)*x),omega,v));

$$g(x\sim, y - \zeta) := \frac{x\sim}{\left(x\sim^2 + (y - \zeta)^2\right)\pi} \tag{9.95}$$

Thus, our solution can be written as the convolution

> u(x,y):=Int(f(zeta)*'g(x,y−zeta)',zeta=−infinity..infinity);

$$u(x\sim, y) := \int_{-\infty}^{\infty} f(\zeta)g(x, y - \zeta)\,d\zeta \tag{9.96}$$

We now consider the special case $u(0, y) = f(y)$ where $f(y)$ is given as

> f(y):=Heaviside(y+1)−Heaviside(y−1);f(zeta):=subs(y=zeta,f(y)):

$$f(y) := \text{Heaviside}(y + 1) - \text{Heaviside}(y - 1) \tag{9.97}$$

Substitution of this into the previous integral yields the equivalent integral

> u(x,y):=Int(1*x/(Pi*(x^2+(y−zeta)^2)),zeta=−1..1);

$$u(x{\sim}, y) := \int_{-1}^{1} \frac{x{\sim}}{\left(x{\sim}^2 +(y-\zeta)^2\right)\pi} \, d\zeta \tag{9.98}$$

This integrates to

> u(x,y):=value(%);

$$u(x{\sim}, y) := -\frac{\arctan\left(\frac{y+1}{x{\sim}}\right) - \arctan\left(\frac{y-1}{x{\sim}}\right)}{\pi} \tag{9.99}$$

> plot3d(u(x,y),x=0..10,y=−2..2,axes=framed,thickness=1);

The three-dimensional surface shown in Figure 9.10 illustrates the electrostatic potential distribution $u(x, y)$ over the given region. The electrostatic equipotential lines are obtained by clicking on the graph and choosing the "Render the plot using the polygon patch and contour style" option and then clicking the "redraw" button in the graphics bar.

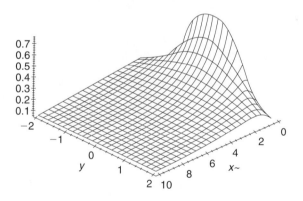

Figure 9.10

Chapter Summary

Fourier series representation of a function $f(x)$ over a finite domain $[-L, L]$

$$f(x) = \sum_{n=-\infty}^{\infty} \frac{F(n)\, e^{\frac{in\pi x}{L}}}{2L}$$

Regular orthonormalized eigenfunctions for Euler operator over a finite domain

$$\phi_n(x) = \frac{\sqrt{2}\,e^{\frac{in\pi x}{L}}}{2\sqrt{L}}$$

for $n = -3, -2, -1, 0, 1, 2, 3 \ldots$.

Statement of orthonormality for the regular eigenfunctions using the Kronecker delta function

$$\int_{-L}^{L} \frac{e^{\frac{in\pi x}{L}}\,e^{-\frac{im\pi x}{L}}}{2L}\,dx = \delta(n,m)$$

Regular Fourier coefficients

$$F(n) = \int_{-L}^{L} f(x)\,e^{-\frac{in\pi}{L}}\,dx$$

Fourier integral representation of $f(x)$ over an infinite domain $(-\infty, \infty)$

$$f(x) = \int_{-\infty}^{\infty} \frac{F(\omega)\,e^{i\omega x}}{2\pi}\,d\omega$$

Singular orthonormalized eigenfunctions for the Euler operator over an infinite domain

$$\frac{\sqrt{2}\,e^{i\omega x}}{2\sqrt{\pi}}$$

for $-\infty < \omega < \infty$.

Statement of orthonormality for singular eigenfunctions using the Dirac delta function

$$\int_{-\infty}^{\infty} \frac{e^{i\beta x}\,e^{-i\omega x}}{2\pi}\,dx = \delta(\beta - \omega)$$

Singular Fourier coefficients (the Fourier integral)

$$F(\omega) = \int_{-\infty}^{\infty} f(x)\,e^{-i\omega x}\,dx$$

The preceding material demonstrated the parallel for the representation of functions over finite and infinite domains. For the case of finite domains, we used the Fourier series representation in terms of the regular orthonormalized eigenfunctions where the regular Fourier coefficients are evaluated using the statement of orthonormality for regular eigenfunctions. For the case of infinite domains, we used the Fourier integral representation in terms of the singular eigenfunctions where the singular Fourier coefficients are evaluated using the equivalent statement of orthonormality for singular eigenfunctions. Comparison of the Fourier representation for finite domains and infinite domains and the corresponding statements of orthonormality for the regular and singular eigenfunctions shown earlier demonstrates dramatic similarities.

Exercises

We first consider nonhomogeneous diffusion and wave equations in a single dimension over infinite and semi-infinite intervals in rectangular coordinates. For the situation of infinite intervals, we use the method of the Fourier integral, which is an extension of the eigenfunction expansion for finite intervals. For semi-infinite intervals, we use the method of Fourier sine or cosine integrals or the method of images.

Fourier Integral Solutions of the Diffusion Equation

We consider the temperature distribution along a thin rod whose lateral surface is insulated over either the infinite interval $I = \{x \mid -\infty < x < \infty\}$ or the semi-infinite interval $I = \{x \mid 0 < x < \infty\}$. We allow for an external heat source to act on the system. The nonhomogeneous diffusion partial differential equation in rectangular coordinates in a single dimension reads

$$\frac{\partial}{\partial t} u(x, t) = k \left(\frac{\partial^2}{\partial x^2} u(x, t) \right) + h(x, t) \tag{9.100}$$

with the initial condition $u(x, 0) = f(x)$. In Exercises 9.1 through 9.23, we seek solutions to the preceding equation in the form of a Fourier integral with respect to x for various initial and boundary conditions over infinite or semi-infinite intervals. In some instances, we use the convolution integral or the method of images to develop the final solution. For the following exercises, whenever possible, generate the animated solution for $u(x, t)$, and plot the animated sequence for $0 < t < 5$.

9.1. Consider a thin cylinder over the infinite interval $I = \{x \mid -\infty < x < \infty\}$ containing a fluid that has an initial salt concentration density given as $f(x)$. There is no source term in the system, and the diffusivity is $k = 1/4$. Evaluate the concentration distribution density $u(x, t)$ in the form of a Fourier integral. Evaluate the inverse by using either the inverse command or a convolution integral.

Boundary condition:

$$\int_{-\infty}^{\infty} |u(x, t)|\, dx < \infty$$

Initial condition:

$$f(x) = e^{-\frac{|x|}{2}}$$

9.2. Consider a thin rod whose lateral surface is insulated over the infinite interval $I = \{x \mid -\infty < x < \infty\}$. The initial temperature distribution in the rod is given as $f(x)$. There is no source term in the system, and the diffusivity is $k = 1/4$. Evaluate the temperature distribution $u(x, t)$ in the form of a Fourier integral. Evaluate the inverse by using either the inverse command or a convolution integral.

Boundary condition:

$$\int_{-\infty}^{\infty} |u(x, t)|\, dx < \infty$$

Initial condition:

$$f(x) = x e^{-\frac{|x|}{2}}$$

9.3. Consider a thin cylinder over the infinite interval $I = \{x \mid -\infty < x < \infty\}$ containing a fluid that has an initial salt concentration density given as $f(x)$. There is no source term in the system, and the diffusivity is $k = 1/4$. Evaluate the concentration distribution density $u(x, t)$ in the form of a Fourier integral. Evaluate the inverse by using either the inverse command or a convolution integral.

Boundary condition:

$$\int_{-\infty}^{\infty} |u(x, t)|\, dx < \infty$$

Initial condition:

$$f(x) = \frac{1}{1 + x^2}$$

9.4. Consider a thin rod whose lateral surface is insulated over the infinite interval $I = \{x \mid -\infty < x < \infty\}$. The initial temperature distribution in the rod is given as $f(x)$. There is no source term in the system, and the diffusivity is $k = 1/4$. Evaluate the temperature distribution density $u(x, t)$ in the form of a Fourier integral. Evaluate the inverse by using either the inverse command or a convolution integral.

Boundary condition:

$$\int_{-\infty}^{\infty} |u(x, t)| \, dx < \infty$$

Initial condition:

$$f(x) = e^{-\frac{x^2}{2}}$$

9.5. Consider a thin cylinder over the infinite interval $I = \{x \,|-\infty < x < \infty\}$ containing a fluid that has an initial salt concentration density given as $f(x)$. There is no source term in the system, and the diffusivity is $k = 1/4$. Evaluate the concentration distribution density $u(x, t)$ in the form of a Fourier integral. Evaluate the inverse by using either the inverse command or a convolution integral.

Boundary condition:

$$\int_{-\infty}^{\infty} |u(x, t)| \, dx < \infty$$

Initial condition:

$$f(x) = x e^{-\frac{x^2}{2}}$$

9.6. Consider a thin rod whose lateral surface is insulated over the infinite interval $I = \{x \,|-\infty < x < \infty\}$. The initial temperature distribution in the rod is given as $f(x)$. There is no source term in the system, and the diffusivity is $k = 1/4$. Evaluate the temperature distribution $u(x, t)$ in the form of a Fourier integral. Evaluate the inverse by using either the inverse command or a convolution integral.

Boundary condition:

$$\int_{-\infty}^{\infty} |u(x, t)| \, dx < \infty$$

Initial condition:

$$f(x) = \frac{x}{(1 + x^2)^2}$$

9.7. Consider a thin cylinder over the semi-infinite interval $I = \{x \,| 0 < x < \infty\}$ containing a fluid that has an initial salt concentration density given as $f(x)$. There is no source term in the system, and the diffusivity is $k = 1/4$. Evaluate the concentration distribution density $u(x, t)$ in the form of a Fourier sine integral. Evaluate the inverse by using either the inverse command or a convolution integral.

Boundary conditions:

$$u(0, t) = 0 \quad \text{and} \quad \int\limits_0^\infty |u(x, t)| \, dx < \infty$$

Initial condition:

$$f(x) = x e^{-\frac{x}{2}}$$

9.8. Consider Exercise 9.7. Develop the odd extension of $f(x)$ and solve the problem by the method of images.

9.9. Consider a thin rod whose lateral surface is insulated over the semi-infinite interval $I = \{x \mid 0 < x < \infty\}$. The initial temperature distribution in the rod is given as $f(x)$. There is no source term in the system, and the diffusivity is $k = 1/4$. Evaluate the temperature distribution $u(x, t)$ in the form of a Fourier sine integral. Evaluate the inverse by using either the inverse command or a convolution integral.

Boundary conditions:

$$u(0, t) = 0 \quad \text{and} \quad \int\limits_0^\infty |u(x, t)| \, dx < \infty$$

Initial condition:

$$f(x) = \frac{x}{1 + x^2}$$

9.10. Consider Exercise 9.9. Develop the odd extension of $f(x)$ and solve the problem by the method of images.

9.11. Consider a thin cylinder over the semi-infinite interval $I = \{x \mid 0 < x < \infty\}$ containing a fluid that has an initial salt concentration density given as $f(x)$. There is no source term in the system, and the diffusivity is $k = 1/4$. Evaluate the concentration distribution density $u(x, t)$ in the form of a Fourier sine integral. Evaluate the inverse by using either the inverse command or a convolution integral.

Boundary conditions:

$$u(0, t) = 0 \quad \text{and} \quad \int\limits_0^\infty |u(x, t)| \, dx < \infty$$

Initial condition:

$$f(x) = x e^{-\frac{x^2}{4}}$$

9.12. Consider Exercise 9.11. Develop the odd extension of $f(x)$ and solve the problem by the method of images.

9.13. Consider a thin rod whose lateral surface is insulated over the semi-infinite interval $I = \{x \mid 0 < x < \infty\}$. The initial temperature distribution in the rod is given as $f(x)$. There is no source term in the system, and the diffusivity is $k = 1/4$. Evaluate the temperature distribution $u(x, t)$ in the form of a Fourier cosine integral. Evaluate the inverse by using either the inverse command or a convolution integral.

Boundary conditions:

$$u_x(0, t) = 0 \quad \text{and} \quad \int_0^\infty |u(x, t)| \, dx < \infty$$

Initial condition:

$$f(x) = e^{-\frac{x}{2}}$$

9.14. Consider Exercise 9.13. Develop the even extension of $f(x)$ and solve the problem by the method of images.

9.15. Consider a thin cylinder over the semi-infinite interval $I = \{x \mid 0 < x < \infty\}$ containing a fluid that has an initial salt concentration density given as $f(x)$. There is no source term in the system, and the diffusivity is $k = 1/4$. Evaluate the concentration distribution density $u(x, t)$ in the form of a Fourier cosine integral. Evaluate the inverse by using either the inverse command or a convolution integral.

Boundary conditions:

$$u_x(0, t) = 0 \quad \text{and} \quad \int_0^\infty |u(x, t)| \, dx < \infty$$

Initial condition:

$$f(x) = \frac{1}{1 + x^2}$$

9.16. Consider Exercise 9.15. Develop the even extension of $f(x)$ and solve the problem by the method of images.

9.17. Consider a thin rod whose lateral surface is insulated over the semi-infinite interval $I = \{x \mid 0 < x < \infty\}$. The initial temperature distribution in the rod is given as $f(x)$. There is no source term in the system, and the diffusivity is $k = 1/4$. Evaluate the temperature distribution $u(x, t)$ in the form of a Fourier cosine integral. Evaluate the inverse by using either the inverse command or a convolution integral.

Boundary conditions:

$$u_x(0, t) = 0 \quad \text{and} \quad \int_0^\infty |u(x, t)| \, dx < \infty$$

Initial condition:

$$f(x) = e^{-\frac{x^2}{4}}$$

9.18. Consider Exercise 9.17. Develop the even extension of $f(x)$ and solve the problem by the method of images.

9.19. Consider a thin rod whose lateral surface is insulated over the infinite interval $I = \{x \mid -\infty < x < \infty\}$. The initial temperature distribution in the rod is given as $f(x)$. There is a heat source term $h(x, t)$ (dimension of degrees per second) in the system, and the diffusivity is $k = 1/4$. Evaluate the temperature distribution $u(x, t)$ in the form of a Fourier integral. If possible, evaluate the inverse by using either the inverse command or a convolution integral.

Boundary condition:

$$\int_{-\infty}^\infty |u(x, t)| \, dx < \infty$$

Initial condition:

$$f(x) = 0$$

Heat source:

$$h(x, t) = e^{-\frac{|x|}{2}} \cos(t)$$

9.20. Consider a thin rod whose lateral surface is insulated over the infinite interval $I = \{x \mid -\infty < x < \infty\}$. The initial temperature distribution in the rod is given as $f(x)$. There is a heat source term $h(x, t)$ (dimension of degrees per second) in the system, and the diffusivity is $k = 1/4$. Evaluate the temperature distribution $u(x, t)$ in the form of a Fourier integral. If possible, evaluate the inverse by using either the inverse command or a convolution integral.

Boundary condition:

$$\int_{-\infty}^\infty |u(x, t)| \, dx < \infty$$

Initial condition:

$$f(x) = 0$$

Heat source:

$$h(x, t) = x e^{-\frac{|x|}{2}} \sin(t)$$

9.21. Consider a thin rod whose lateral surface is insulated over the infinite interval
$I = \{x \mid -\infty < x < \infty\}$. The initial temperature distribution in the rod is given as $f(x)$.
There is a heat source term $h(x, t)$ (dimension of degrees per second) in the system, and
the diffusivity is $k = 1/4$. Evaluate the temperature distribution $u(x, t)$ in the form of a
Fourier integral. If possible, evaluate the inverse by using either the inverse command
or a convolution integral.

Boundary condition:

$$\int_{-\infty}^{\infty} |u(x, t)| \, dx < \infty$$

Initial condition:

$$f(x) = 0$$

Heat source:

$$h(x, t) = \frac{\cos(t)}{1 + x^2}$$

9.22. Consider a thin rod whose lateral surface is insulated over the infinite interval
$I = \{x \mid -\infty < x < \infty\}$. The initial temperature distribution in the rod is given as $f(x)$.
There is a heat source term $h(x, t)$ (dimension of degrees per second) in the system, and
the diffusivity is $k = 1/4$. Evaluate the temperature distribution $u(x, t)$ in the form of a
Fourier integral. If possible, evaluate the inverse by using either the inverse command
or a convolution integral.

Boundary condition:

$$\int_{-\infty}^{\infty} |u(x, t)| \, dx < \infty$$

Initial condition:

$$f(x) = 0$$

Heat source:

$$h(x, t) = e^{-\frac{x^2}{4}} \sin(t)$$

9.23. Consider a thin rod whose lateral surface is insulated over the infinite interval $I = \{x \,|-\infty < x < \infty\}$. The initial temperature distribution in the rod is given as $f(x)$. There is a heat source term $h(x, t)$ (dimension of degrees per second) in the system, and the diffusivity is $k = 1/4$. Evaluate the temperature distribution $u(x, t)$ in the form of a Fourier integral. If possible, evaluate the inverse by using either the inverse command or a convolution integral.

Boundary condition:

$$\int\limits_{-\infty}^{\infty} |u(x, t)| \, dx < \infty$$

Initial condition:

$$f(x) = 0$$

Heat source:

$$h(x, t) = x e^{-\frac{x^2}{4}} \cos(t)$$

Fourier Integral Solutions for the Wave Equation

We now consider transverse wave propagation in taut strings and longitudinal wave propagation in rigid bars over the one-dimensional infinite interval $I = \{x \,|-\infty < x < \infty\}$ or semi-infinite interval $I = \{x \,| 0 < x < \infty\}$. We allow for an external force to act on the system, and there is no damping. The nonhomogeneous partial differential equation in rectangular coordinates is

$$\frac{\partial^2}{\partial t^2} u(x, t) = c^2 \left(\frac{\partial^2}{\partial x^2} u(x, t) \right) + h(x, t)$$

with the two initial conditions

$$u(x, 0) = f(x) \quad \text{and} \quad u_t(x, 0) = g(x)$$

In Exercises 9.24 through 9.42, we seek solutions to the preceding equation in the form of a Fourier integral with respect to x for various initial and boundary conditions. For the following exercises, whenever possible, generate the animated solution for $u(x, t)$ and plot the animated sequence for $0 < t < 5$.

9.24. Consider transverse motion in a taut string over the infinite interval $I = \{x \mid -\infty < x < \infty\}$. The initial displacement distribution $f(x)$ and initial speed distribution $g(x)$ are given following. There is no external force in the system and $c = 1/2$. Evaluate the wave distribution $u(x, t)$ in the form of a Fourier integral. Evaluate the integral either by the inverse command or by direct integration.

Boundary condition:

$$\int_{-\infty}^{\infty} |u(x, t)| \, dx < \infty$$

Initial conditions:

$$f(x) = e^{-\frac{|x|}{2}} \quad \text{and} \quad g(x) = 0$$

9.25. Consider longitudinal motion in a rigid bar over the infinite interval $I = \{x \mid -\infty < x < \infty\}$. The initial displacement distribution $f(x)$ and initial speed distribution $g(x)$ are given following. There is no external force in the system and $c = 1/2$. Evaluate the wave distribution $u(x, t)$ in the form of a Fourier integral. Evaluate the integral either by the inverse command or by direct integration.

Boundary condition:

$$\int_{-\infty}^{\infty} |u(x, t)| \, dx < \infty$$

Initial conditions:

$$f(x) = 0 \quad \text{and} \quad g(x) = x e^{-\frac{|x|}{2}}$$

9.26. Consider transverse motion in a taut string over the infinite interval $I = \{x \mid -\infty < x < \infty\}$. The initial displacement distribution $f(x)$ and initial speed distribution $g(x)$ are given following. There is no external force in the system and $c = 1/2$. Evaluate the wave distribution $u(x, t)$ in the form of a Fourier integral. Evaluate the integral either by the inverse command or by direct integration.

Boundary condition:

$$\int_{-\infty}^{\infty} |u(x, t)| \, dx < \infty$$

Initial conditions:

$$f(x) = \frac{1}{1 + x^2} \quad \text{and} \quad g(x) = 0$$

9.27. Consider longitudinal motion in a rigid bar over the infinite interval $I = \{x \mid -\infty < x < \infty\}$. The initial displacement distribution $f(x)$ and initial speed distribution $g(x)$ are given following. There is no external force in the system and $c = 1/2$. Evaluate the wave distribution $u(x, t)$ in the form of a Fourier integral. Evaluate the integral either by the inverse command or by direct integration.

Boundary condition:

$$\int_{-\infty}^{\infty} |u(x, t)|\, dx < \infty$$

Initial conditions:

$$f(x) = 0 \quad \text{and} \quad g(x) = e^{-\frac{x^2}{4}}$$

9.28. Consider transverse motion in a taut string over the infinite interval $I = \{x \mid -\infty < x < \infty\}$. The initial displacement distribution $f(x)$ and initial speed distribution $g(x)$ are given following. There is no external force in the system and $c = 1/2$. Evaluate the wave distribution $u(x, t)$ in the form of a Fourier integral. Evaluate the integral either by the inverse command or by direct integration.

Boundary condition:

$$\int_{-\infty}^{\infty} |u(x, t)|\, dx < \infty$$

Initial conditions:

$$f(x) = x e^{-\frac{x^2}{4}} \quad \text{and} \quad g(x) = 0$$

9.29. Consider longitudinal motion in a rigid bar over the semi-infinite interval $I = \{x \mid 0 < x < \infty\}$. The initial displacement distribution $f(x)$ and initial speed distribution $g(x)$ are given following. There is no external force in the system and $c = 1/2$. Evaluate the wave distribution $u(x, t)$ in the form of a Fourier sine integral. Evaluate the integral either by the inverse command or by direct integration.

Boundary condition:

$$u(0, t) = 0 \quad \text{and} \quad \int_{0}^{\infty} |u(x, t)|\, dx < \infty$$

Initial conditions:

$$f(x) = x e^{-\frac{x}{2}} \quad \text{and} \quad g(x) = 0$$

9.30. Consider Exercise 9.29. Develop the odd extension of $f(x)$ and solve the problem by the method of images.

9.31. Consider transverse motion in a taut string over the semi-infinite interval $I = \{x \mid 0 < x < \infty\}$. The initial displacement distribution $f(x)$ and initial speed distribution $g(x)$ are given following. There is no external force in the system and $c = 1/2$. Evaluate the wave distribution $u(x, t)$ in the form of a Fourier sine integral. Evaluate the integral either by the inverse command or by direct integration.

Boundary conditions:

$$u(0, t) = 0 \quad \text{and} \quad \int_0^\infty |u(x, t)| \, dx < \infty$$

Initial conditions:

$$f(x) = 0 \quad \text{and} \quad g(x) = \frac{x}{1 + x^2}$$

9.32. Consider Exercise 9.31. Develop the odd extension of $g(x)$ and solve the problem by the method of images.

9.33. Consider longitudinal motion in a rigid bar over the semi-infinite interval $I = \{x \mid 0 < x < \infty\}$. The initial displacement distribution $f(x)$ and initial speed distribution $g(x)$ are given following. There is no external force in the system and $c = 1/2$. Evaluate the wave distribution $u(x, t)$ in the form of a Fourier sine integral. Evaluate the integral either by the inverse command or by direct integration.

Boundary conditions:

$$u(0, t) = 0 \quad \text{and} \quad \int_0^\infty |u(x, t)| \, dx < \infty$$

Initial conditions:

$$f(x) = x e^{-\frac{x^2}{4}} \quad \text{and} \quad g(x) = 0$$

9.34. Consider Exercise 9.33. Develop the odd extension of $f(x)$ and solve the problem by the method of images.

9.35. Consider transverse motion in a taut string over the semi-infinite interval $I = \{x \mid 0 < x < \infty\}$. The initial displacement distribution $f(x)$ and initial speed distribution $g(x)$ are given following. There is no external force acting and $c = 1/2$. Evaluate the wave distribution $u(x, t)$ in the form of a Fourier cosine integral. Evaluate the integral either by the inverse command or by direct integration.

Boundary conditions:

$$u_x(0, t) = 0 \quad \text{and} \quad \int_0^\infty |u(x, t)| \, dx < \infty$$

Initial conditions:

$$f(x) = 0 \quad \text{and} \quad g(x) = \frac{x}{1 + x^2}$$

9.36. Consider Exercise 9.35. Develop the even extension of $g(x)$ and solve the problem by the method of images.

9.37. Consider longitudinal motion in a rigid bar over the semi-infinite interval $I = \{x \mid 0 < x < \infty\}$. The initial displacement distribution $f(x)$ and initial speed distribution $g(x)$ are given following. There is no external force in the system and $c = 1/2$. Evaluate the wave distribution $u(x, t)$ in the form of a Fourier cosine integral. Evaluate the integral either by the inverse command or by direct integration.

Boundary conditions:

$$u_x(0, t) = 0 \quad \text{and} \quad \int_0^\infty |u(x, t)| \, dx < \infty$$

Initial conditions:

$$f(x) = e^{-\frac{x^2}{4}} \quad \text{and} \quad g(x) = 0$$

9.38. Consider Exercise 9.37. Develop the even extension of $f(x)$ and solve the problem by the method of images.

9.39. Consider transverse motion in a taut string over the infinite interval $I = \{x \mid -\infty < x < \infty\}$. The initial displacement distribution $f(x)$ and initial speed distribution $g(x)$ are given following. There is an external force term $h(x, t)$ (dimension of force per unit mass) acting and $c = 1/2$. Evaluate the wave distribution $u(x, t)$ in the form of a Fourier integral. If possible, evaluate the integral either by the inverse command or by direct integration.

Boundary condition:

$$\int_{-\infty}^\infty |u(x, t)| \, dx < \infty$$

Initial conditions:

$$f(x) = 0 \quad \text{and} \quad g(x) = 0$$

External driving force:

$$h(x, t) = e^{-|x|} \sin(t)$$

9.40.   Consider transverse motion in a taut string over the infinite interval $I = \{x \mid -\infty < x < \infty\}$. The initial displacement distribution $f(x)$ and initial speed distribution $g(x)$ are given following. There is an external force term $h(x, t)$ (dimension of force per unit mass) acting and $c = 1/2$. Evaluate the wave distribution $u(x, t)$ in the form of a Fourier integral. If possible, evaluate the integral either by the inverse command or by direct integration.

Boundary condition:

$$\int_{-\infty}^{\infty} |u(x, t)| \, dx < \infty$$

Initial conditions:

$$f(x) = 0 \quad \text{and} \quad g(x) = 0$$

External driving force:

$$h(x, t) = \frac{x \cos(t)}{1 + x^2}$$

9.41.   Consider transverse motion in a taut string over the infinite interval $I = \{x \mid -\infty < x < \infty\}$. The initial displacement distribution $f(x)$ and initial speed distribution $g(x)$ are given following. There is an external force term $h(x, t)$ (dimension of force per unit mass) acting and $c = 1/2$. Evaluate the wave distribution $u(x, t)$ in the form of a Fourier integral. If possible, evaluate the integral either by the inverse command or by direct integration.

Boundary condition:

$$\int_{-\infty}^{\infty} |u(x, t)| \, dx < \infty$$

Initial conditions:

$$f(x) = 0 \quad \text{and} \quad g(x) = 0$$

External driving force:

$$h(x, t) = e^{-\frac{x^2}{4}} \sin(t)$$

9.42.   Consider transverse motion in a taut string over the infinite interval $I = \{x \mid -\infty < x < \infty\}$. The initial displacement distribution $f(x)$ and initial speed distribution $g(x)$ are given following. There is an external force term $h(x, t)$ (dimension of force per unit mass) acting and $c = 1/2$. Evaluate the wave distribution $u(x, t)$ in the form of a Fourier integral. If possible, evaluate the integral either by the inverse command or by direct integration.

Boundary condition:

$$\int_{-\infty}^{\infty} |u(x, t)| \, dx < \infty$$

Initial conditions:

$$f(x) = 0 \quad \text{and} \quad g(x) = 0$$

External driving force:

$$h(x, t) = x \, e^{-\frac{x^2}{4}} \cos(t)$$

Fourier Integral Solutions of the Laplace Equation

We now consider problems involving the Laplace equation over infinite and semi-infinite intervals. Recall that the Laplace equation was the time-invariant or steady-state solution to either the wave or the diffusion equation. The Laplace equation reads

$$\frac{\partial^2}{\partial x^2} u(x, y) + \frac{\partial^2}{\partial y^2} u(x, y) = 0$$

We now consider the solution of this equation, over intervals that may be infinite or semi-infinite, using the method of Fourier transforms. The choice as to which of the two variables we apply to the Fourier transform, with respect to either x or y, depends on the boundary conditions impossed on the problem. Note that the temperature isotherms or the electrostatic equipotential lines can be obtained by clicking on the graph and choosing the "Render the plot using the polygon patch and contour style" option and then clicking the "redraw" button in the graphics bar.

9.43.   Consider the steady-state temperature distribution in a thin plate whose lateral surfaces are insulated over the domain $D = \{(x, y) \mid -\infty < x < \infty, 0 < y < 1\}$. The boundary conditions are

$$u(x, 0) = \frac{1}{1 + x^2} \quad \text{and} \quad u(x, 1) = 0 \quad \text{and} \quad \int_{-\infty}^{\infty} \int_{0}^{1} |u(x, y)| \, dy \, dx < \infty$$

Obtain the solution as a Fourier integral with respect to x. If possible, evaluate the Fourier integral (by the inverse or convolution method) and generate the three-dimensional surface showing the isotherms over a portion of the problem domain.

9.44. Consider the steady-state temperature distribution in a thin plate whose lateral surfaces are insulated over the domain $D = \{(x, y) \mid -\infty < x < \infty, 0 < y < 1\}$. The boundary conditions are

$$u(x, 0) = 0 \quad \text{and} \quad u(x, 1) = e^{-\frac{x^2}{4}} \quad \text{and} \quad \int_{-\infty}^{\infty} \int_{0}^{1} |u(x, y)| \, dy \, dx < \infty$$

Obtain the solution as a Fourier integral with respect to x. If possible, evaluate the Fourier integral (by the inverse or convolution method) and generate the three-dimensional surface showing the isotherms over a portion of the problem domain.

9.45. Consider the steady-state temperature distribution in a thin plate whose lateral surfaces are insulated over the domain $D = \{(x, y) \mid -\infty < x < \infty, 0 < y < 1\}$. The boundary conditions are

$$u(x, 0) = \frac{1}{1 + x^2} \quad \text{and} \quad u(x, 1) = e^{-\frac{x^2}{4}} \quad \text{and} \quad \int_{-\infty}^{\infty} \int_{0}^{1} |u(x, y)| \, dy \, dx < \infty$$

Obtain the solution as a Fourier integral with respect to x. Show that the solution is a superposition of the two solutions obtained in the two previous problems. If possible, evaluate the Fourier integral (by the inverse or convolution method) and generate the three-dimensional surface showing the isotherms over a portion of the problem domain.

9.46. Consider the electrostatic potential distribution in a charge-free region over the domain $D = \{(x, y) \mid -\infty < x < \infty, 0 < y < 1\}$. The boundary conditions are

$$u(x, 0) = x e^{-\frac{x^2}{4}} \quad \text{and} \quad u(x, 1) = 0 \quad \text{and} \quad \int_{-\infty}^{\infty} \int_{\infty}^{\infty} |u(x, y)| \, dy \, dx < \infty$$

Obtain the solution as a Fourier integral with respect to x. If possible, evaluate the Fourier integral (by the inverse or convolution method) and generate the three-dimensional surface showing the equipotential lines over a portion of the problem domain.

9.47. Consider the steady-state temperature distribution in a thin plate whose lateral surfaces are insulated over the domain $D = \{(x, y) \mid 0 < x < \infty, -\infty < y < \infty\}$. The boundary conditions are

$$u(0, y) = e^{-\frac{y^2}{4}} \quad \text{and} \quad \int_{-\infty}^{\infty} \int_{0}^{\infty} |u(x, y)| \, dx \, dy < \infty$$

Obtain the solution as a Fourier integral with respect to y. If possible, evaluate the Fourier integral (by the inverse command or convolution method), and generate the three-dimensional surface showing the isotherms over a portion of the problem domain.

9.48. Consider the electrostatic potential distribution in a charge-free region over the domain $D = \{(x, y) \mid 0 < x < 1, -\infty < y < \infty\}$. The boundary conditions are

$$u(0, y) = 0 \quad \text{and} \quad u(1, y) = \frac{1}{1 + y^2} \quad \text{and} \quad \int_{-\infty}^{\infty} \int_{0}^{1} |u(x, y)| \, dx \, dy < \infty$$

Obtain the solution as a Fourier integral with respect to y. If possible, evaluate the Fourier integral (by the inverse or convolution method), and generate the three-dimensional surface showing the equipotential lines over a portion of the problem domain.

9.49. Consider the steady-state temperature distribution in a thin plate whose lateral surfaces are insulated over the domain $D = \{(x, y) \mid 0 < x < \infty, 0 < y < \infty\}$. The boundary conditions are

$$u(0, y) = 0 \quad \text{and} \quad u(x, 0) = x e^{-x^2} \quad \text{and} \quad \int_{0}^{\infty} \int_{0}^{\infty} |u(x, y)| \, dx \, dy < \infty$$

Obtain the solution as a Fourier sine integral with respect to x. If possible, evaluate the Fourier integral (by the inverse command or convolution method), and generate the three-dimensional surface showing the isotherms over a portion of the problem domain.

9.50. Consider Exercise 9.49. Generate the odd extension of $u(x, 0)$ about the y-axis, and solve by the method of images to obtain the solution as a Fourier integral with respect to x. If possible, evaluate the Fourier integral (by the inverse command or convolution method), and generate the three-dimensional surface showing the isotherms over a portion of the problem domain.

9.51.   Consider the electrostatic potential distribution in a charge-free region over the domain $D = \{(x, y) \mid 0 < x < \infty, 0 < y < \infty\}$. The boundary conditions are

$$u(0, y) = \frac{y}{1+y^2} \quad \text{and} \quad u(x, 0) = 0 \quad \text{and} \quad \int_0^\infty \int_0^\infty |u(x, y)| \, dx \, dy < \infty$$

Obtain the solution as a Fourier sine integral with respect to y. If possible, evaluate the Fourier integral (by the inverse or convolution method), and generate the three-dimensional surface showing the equipotential lines over a portion of the problem domain.

9.52.   Consider Exercise 9.51. Generate the odd extension of $u(0, y)$ about the x-axis, and solve by the method of images to obtain the solution as a Fourier integral with respect to y. If possible, evaluate the Fourier integral (by the inverse command or convolution method) and generate the three-dimensional surface showing the equipotential lines over a portion of the problem domain.

9.53.   Consider the electrostatic potential distribution in a charge-free region over the domain $D = \{(x, y) \mid 0 < x < \infty, 0 < y < \infty\}$. The boundary conditions are

$$u(0, y) = \frac{y}{1+y^2} \quad \text{and} \quad u(x, 0) = x e^{-x^2} \quad \text{and} \quad \int_0^\infty \int_0^\infty |u(x, y)| \, dx \, dy < \infty$$

Note that we have nonhomogeneous boundary conditions with respect to both spatial variables. Use the superposition method as outlined in the exercises of Chapter 5, and show the solution to be a superposition of the two solutions obtained in Exercises 9.49 and 9.51. If possible, evaluate the Fourier integral (by the inverse or convolution method), and generate the three-dimensional surface showing the equipotential lines over a portion of the problem domain.

CHAPTER 10

Laplace Transform Methods for Partial Differential Equations

10.1 Introduction

In many partial differential equation problems that we encounter, such as the diffusion equation and the wave equation, the time variable is generally understood to have the domain $0 < t < \infty$. In the case of partial differential equations, the particular operation called the "Laplace transform" is often used to integrate out the time dependence of the equation, with the result being a partial differential equation with one fewer independent variable that is generally easier to solve.

In the case of ordinary differential equations, the operation reduces the differential equation to an algebraic equation, and, because algebraic equations are generally easier to manipulate and solve, we see an immediate advantage in the use of the Laplace operation.

10.2 Laplace Transform Operator

We begin by defining the Laplace operator as an integration operation with respect to the time variable t. With certain restrictions on the function $f(t)$, we define the Laplace operation or transform of $f(t)$ to be the integral

$$F(s) = \int_0^\infty f(t)\, e^{-st} dt$$

We note that the integration operation is performed on the integrand $f(t)e^{-st}$, and the integration is along the positive real t-axis. The result is a function of the Laplace variable s as indicated by the corresponding capital letter function $F(s)$. The variable s is, in general, a complex variable. The expression e^{-st} is often called the "kernel" of the Laplace transformation.

The preceding integral is an "improper integral," and in order for this integral to converge, the restrictions imposed on $f(t)$ are that it be a "piecewise continuous" function of t and of "exponential order" over the interval $I = \{t \mid 0 < t < \infty\}$.

By "piecewise continuous" we mean $f(t)$ does not have to be continuous; the interval $I = \{t \,|\, 0 < t < \infty\}$ can be divided into a finite number of subintervals such that the function $f(t)$ is continuous in the interior of each subinterval and appoaches finite limits at each end point of the subinterval. By "exponential order" we mean that there is a positive constant σ such that the following product remains bounded by some constant M; that is,

$$e^{-\sigma t} |f(t)| < M$$

as t goes to infinity. The greatest lower bound on the value of σ is called the "abscissa of convergence" of $f(t)$.

Some functions and their evaluated Laplace transforms, that may be of immediate use to us in solving some partial differential equations, are given here:

1.

$$f1(t) = e^{at}\sin(bt), \ F1(s) = \frac{b}{(s-a)^2 + b^2}$$

2.

$$f2(t) = e^{at}\cos(bt), \ F2(s) = \frac{s-a}{(s-a)^2 + b^2}$$

3.

$$f3(t) = e^{at}t^n, \ F3(s) = \frac{\Gamma(n+1)}{(s-a)^{n+1}}$$

In the preceding integrations, we have assumed that a and b are real and the real part of s is larger than a, which is larger than σ—that is, $\mathrm{Re}\{s\} > a > \sigma$. All of these integrations can be confirmed by direct integration of the Laplace transform integral.

Because we will be confronting derivatives in our solutions to partial differential equations, we now consider Laplace transforms of derivatives of functions with respect to the transformed variable t. By integration by parts, it can be shown that the transform of the first derivative reads

$$\int_0^\infty \left(\frac{\mathrm{d}}{\mathrm{d}t}f(t)\right) e^{-st}\,\mathrm{d}t = sF(s) - f(0)$$

and the transform of the second derivative reads

$$\int_0^\infty \left(\frac{\mathrm{d}^2}{\mathrm{d}t^2}f(t)\right) e^{-st}\,\mathrm{d}t = s^2 F(s) - sf(0) - \left(\frac{\mathrm{d}}{\mathrm{d}t}f(0)\right)$$

where $f(0)$ and $\frac{d}{dt}f(0)$ denote the zero and first derivative, respectively, of the function $f(t)$ at time $t = 0$. These results are conditional on $\text{Re}\{s\}$ being greater than the abscissa of convergence of $f(t)$ and its first two derivatives.

10.3 Inverse Transform and Convolution Integral

If the function $f(t)$ satisfies the conditions of exponential order and piecewise continuity, then it can be shown that there exists a uniqueness between the function and its inverse. From complex analysis, it can be shown that we can retrieve the function $f(t)$ from its transform $F(s)$ by performing the following inverse operation:

$$f(t) = \int_{a-i\infty}^{a+i\infty} \frac{F(s)\,e^{st}}{2\pi i}\,ds$$

This inverse operation is an integration over the variable s in the complex s-plane along the line $\text{Re}\{s\} = a$, where a is any point greater than σ, the "abscissa of convergence" of $f(t)$. We do not intend to evaluate inverses from integrations in the complex plane here, since this would take us too far from our immediate intentions.

Generally, inverse transforms are evaluated by doing partial fraction expansions on $F(s)$ and then relating the expansion terms to previously evaluated transforms. An alternative procedure is to use a convolution method similar to that developed in Chapter 9.

For Laplace transforms, the convolution integral reads as follows. Let $F(s)$ be the Laplace transform of $f(t)$ and assume that it can be partitioned into a product of two functions $F1(s)$ and $F2(s)$; that is,

$$F(s) = F1(s)F2(s)$$

where $F1(s)$ is the Laplace transform of $f1(t)$ and $F2(s)$ is the Laplace transform of $f2(t)$:

$$F1(s) = \int_0^\infty f1(t)\,e^{-st}dt$$

and

$$F2(s) = \int_0^\infty f2(t)\,e^{-st}dt$$

Of course, $f1(t)$ and $f2(t)$ must both be piecewise continuous and of exponential order. It can be shown (see related references) that $f(t)$, the inversion of $F(s)$, can be expressed as the convolution integral

$$f(t) = \int_0^t f1(\tau)f2(t-\tau)\,d\tau$$

We take advantage of the convolution integral in evaluating some inverses in some of the following problems.

10.4 Laplace Transform Procedures on the Diffusion Equation

We first consider the nonhomogeneous partial differential equation for diffusion in one spatial dimension in rectangular coordinates as given here:

$$\frac{\partial}{\partial t} u(x, t) = k \left(\frac{\partial^2}{\partial x^2} u(x, t) \right) + h(x, t)$$

with the initial condition

$$u(x, 0) = f(x)$$

where $h(x, t)$ accounts for any internal source terms in the system.

In evaluating the Laplace transform of terms in the given partial differential equation, we have to contend with taking improper integrals, with respect to t, of functions that are dependent on both x and t. We denote the Laplace transform of $u(x, t)$ with respect to the time variable t by the capital letter representation shown here:

$$U(x, s) = \int_0^\infty u(x, t) e^{-st} dt$$

Proceeding formally with the evaluation of transforms of the derivative terms in the given partial differential equation, we get

$$\int_0^\infty \left(\frac{\partial}{\partial t} u(x, t) \right) e^{-st} dt = sU(x, s) - u(x, 0)$$

where $U(x, s)$ denotes the Laplace transform of $u(x, t)$ with respect to t, and $u(x, 0)$ denotes the initial value of $u(x, t)$.

Proceeding formally in evaluating the transforms of the partial derivative terms with respect to the spatial variable x, we get

$$\int_0^\infty \left(\frac{\partial^2}{\partial x^2} u(x, t) \right) e^{-st} dt = \frac{\partial^2}{\partial x^2} U(x, s)$$

This operation suggests that we could interchange the differentiation operation, with respect to x, with the integration operation, with respect to t. The conditions for the validity of doing this are based on the uniform convergence of the improper integrals of the derivative terms. Assuming the validity of these operations, the Laplace transform, with respect to t, of the given partial differential equation yields

$$\frac{\partial^2}{\partial x^2}U(x, s) - \frac{sU(x, s)}{k} = -\frac{f(x)}{k} - \frac{H(x, s)}{k}$$

Here, $H(x, s)$ is the Laplace transform of the source term; that is,

$$H(x, s) = \int_0^\infty h(x, t)\,e^{-st}\,dt$$

The preceding differential equation is an ordinary second-order nonhomogeneous differential equation in the single spatial variable x. We investigated the solutions for this equation in Chapter 1. The solutions are, of course, dependent on the spatial boundary conditions on the problem.

The preceding operations demonstrate the advantage of using the Laplace transform operation on the original partial differential equation. The time dependence of the partial differential equation has been integrated out, and the result is a lower-order differential equation in the single variable x.

We now consider the application of the Laplace transform to evaluate some solutions to the heat equation over semi-infinite domains with respect to the spatial variable x. We focus on those types of problems that have nonhomogeneous boundary conditions. These are problems that cannot ordinarily be solved using the Fourier transform methods as covered in Chapter 9. Further, we do not consider using Laplace transform methods for diffusion problems over finite domains here, since eigenfunction expansion methods, as used in Chapters 3, 6, and 8, are more effective.

In general, we consider time-dependent boundary conditions that are nonhomogeneous at $x = 0$ and that are bounded at $x = \infty$; that is,

$$\kappa_1 u(0, t) + \kappa_2 u_x(0, t) = q(t)$$

and

$$u(\infty, t) < \infty$$

The transform of these boundary conditions gives us

$$\kappa_1 U(0, s) + \kappa_2\left(\frac{\partial}{\partial x}U(0, s)\right) = Q(s)$$

and

$$U(\infty, s) < \infty$$

where the term $\frac{\partial}{\partial x}U(0, s)$ is understood to be the x-dependent derivative of $U(x, s)$ evaluated at $x = 0$ and $Q(s)$ is the Laplace transform of $q(t)$ as shown here:

$$Q(s) = \int_0^\infty q(t)\,e^{-st}\,dt$$

The earlier differential equation is a nonhomogeneous ordinary differential equation with respect to the spatial variable x. We solved similar problems in Chapter 1. To develop a general solution, we must first consider finding a set of basis vectors for the corresponding homogeneous differential equation:

$$\frac{\partial^2}{\partial x^2}U(x, s) - \frac{sU(x, s)}{k} = 0$$

Two basis vectors are

$$\phi_1(x, s) = e^{-\sqrt{\frac{s}{k}}\,x}$$

and

$$\phi_2(x, s) = e^{\sqrt{\frac{s}{k}}\,x}$$

From the two basis vectors, as shown in Section 1.7, we are able to generate the second-order Green's function $G(x, v, s)$:

$$G(x, v, s) = \frac{\phi_1(v, s)\phi_2(x, s) - \phi_1(x, s)\phi_2(v, s)}{W(\phi_1(v, s), \phi_2(v, s))}$$

where the denominator term is the familiar Wronskian, which, for the just given basis vectors, is evaluated to be

$$W(\phi_1(v, s), \phi_2(v, s)) = 2\sqrt{\frac{s}{k}}$$

Thus, from the method of variation of parameters, the particular solution to the nonhomogeneous ordinary differential equation in x is

$$Up(x, s) = -\left(\int \frac{G(x, v, s)(f(v) + H(v, s))}{k}\,dv\right)$$

The general solution to the x-dependent differential equation is the sum of the complementary solution and the preceding particular solution. Combining the two solutions, we get the general solution for the transform $U(x, s)$:

$$U(x, s) = C1(s)\phi_1(x, s) + C2(s)\phi_2(x, s) - \left(\int \frac{G(x, v, s)(f(v) + H(v, s))}{k}\,dv\right)$$

The arbitrary coefficients $C1(s)$ and $C2(s)$ are to be determined from the boundary conditions for the problem. The final solution to our problem is the inverse Laplace transform of $U(x, s)$. We demonstrate these concepts discussed by solving an illustrative problem.

DEMONSTRATION: We seek the temperature distribution $u(x, t)$ in a thin rod over the semi-infinite interval $I = \{x \mid 0 < x < \infty\}$ whose lateral surface is insulated. The left end of the rod is held at the temperature 0, and the rod has the initial temperature distribution $f(x)$ given following. There is no internal heat source in the system, and the diffusivity is $k = 1/5$.

SOLUTION: The diffusion partial differential equation is

$$\frac{\partial}{\partial t} u(x, t) = \frac{\frac{\partial^2}{\partial x^2} u(x, t)}{5}$$

Because the left end of the rod is held at the fixed temperature 0, we have a type 1 condition at $x = 0$. Further, we require the solution to be finite at infinity; thus, the boundary conditions are

$$u(0, t) = 0 \quad \text{and} \quad u(\infty, t) < \infty$$

The rod has an initial temperature distribution $u(x, 0) = f(x)$ given as

$$f(x) = \sin(x)$$

We begin by taking the Laplace transform, with respect to the variable t, of the partial differential equation. This yields the ordinary differential equation in the spatial variable x:

$$\frac{\partial^2}{\partial x^2} U(x, s) - 5sU(x, s) = -5\sin(x)$$

We solve this nonhomogeneous ordinary differential equation, in the variable x, subject to the transformed boundary conditions

$$U(0, s) = 0 \quad \text{and} \quad U(\infty, s) < \infty$$

From Section 1.4, a set of basis vectors is

$$\phi_1(x, s) = e^{-\sqrt{5s}\, x}$$

and

$$\phi_2(x, s) = e^{\sqrt{5s}\, x}$$

The Wronskian of this basis is evaluated to be

$$W\left(e^{-\sqrt{5s}\, v}, e^{\sqrt{5s}\, v}\right) = 2\sqrt{5s}$$

and the corresponding Green's function is

$$G(x, v, s) = \frac{\left(e^{-\sqrt{5s}\,v}\,e^{\sqrt{5s}\,x} - e^{-\sqrt{5s}\,x}\,e^{\sqrt{5s}\,v}\right)\sqrt{5}}{10\sqrt{s}}$$

By the method of variation of parameters, from Section 1.7, the particular solution to the preceding differential equation can be expressed in terms of the Green's function as the integral

$$Up(x, s) = -\left(\int \frac{\left(e^{-\sqrt{5s}\,v}\,e^{\sqrt{5s}\,x} - e^{-\sqrt{5s}\,x}\,e^{\sqrt{5s}\,v}\right)\sqrt{5}\sin(v)}{2\sqrt{s}}\,dv\right)$$

which is evaluated to be

$$Up(x, s) = \frac{5\sin(x)}{5s+1}$$

Thus, the general solution to the transformed equation is

$$U(x, s) = C1(s)\,e^{-\sqrt{5s}\,x} + C2(s)\,e^{\sqrt{5s}\,x} + \frac{5\sin(x)}{5s+1}$$

where $C1(s)$ and $C2(s)$ are arbitrary constants. Substituting the boundary condition at $x = 0$ and requiring the solution to be finite as x approaches infinity, for $\text{Re}\{s\} > 0$, we determine that $C1(s) = 0$ and $C2(s) = 0$, and we arrive at the final transformed solution:

$$U(x, s) = \frac{5\sin(x)}{5s+1}$$

The final solution for the spatial- and time-dependent temperature of the system is evaluated from the inverse Laplace transform of the preceding, and this yields

$$u(x, t) = \sin(x)\,e^{-\frac{t}{5}} \tag{10.1}$$

The details for the development of the solution and the graphics are given later in one of the Maple worksheet examples. Note that this problem could not be solved by using the Fourier integral with respect to the variable x as we did in Chapter 9. The reason for this is that the initial function $f(x)$ is not absolutely integrable over the given interval I.

10.5 Example Laplace Transform Problems for the Diffusion Equation

We now look at some specific applications of the Laplace transform to evaluate solutions to the heat equation over semi-infinite intervals with respect to the spatial variable x. We focus on those types of problems that have nonhomogeneous boundary conditions—problems that cannot ordinarily be solved using Fourier transform methods as covered in Chapter 9.

EXAMPLE 10.5.1: We seek the temperature distribution $u(x, t)$ in a thin rod over the semi-infinite interval $I = \{x \mid 0 < x < \infty\}$ whose lateral surface is insulated. The left end of the rod is held at a fixed temperature given following, and the rod has an initial temperature distribution equal to 0. There is no internal heat source in the system and the thermal diffusivity is $k = 1/10$. We solve by taking Laplace transforms with respect to t.

SOLUTION: The diffusion partial differential equation is

$$\frac{\partial}{\partial t} u(x, t) = k\left(\frac{\partial^2}{\partial x^2} u(x, t)\right)$$

The boundary conditions are

$$u(0, t) = 10 \quad \text{and} \quad u(\infty, t) < \infty$$

The initial condition is

$$u(x, 0) = 0$$

The Laplace transformed equation reads

$$\frac{\partial^2}{\partial x^2} U(x, s) - \frac{sU(x, s)}{k} = -\frac{f(x)}{k}$$

Here, $U(x, s)$ is the Laplace transform of $u(x, t)$ with respect to t. The transformed boundary conditions are

$$U(0, s) = Q(s) \quad \text{and} \quad U(\infty, s) < \infty$$

Assignment of system parameters

> restart:with(plots):with(inttrans):k:=1/10:q(t):=10:f(x):=0:

Transformed equation in x reads

> diff(U(x,s),x,x)−(1/k)*s*U(x,s)=−1/k*f(x);

$$\frac{\partial^2}{\partial x^2} U(x, s) - 10\, sU(x, s) = 0 \tag{10.2}$$

Transformed boundary conditions

> Q(s):=Int(q(t)*exp(−s*t),t=0..infinity);

$$Q(s) := \int_0^\infty 10\, e^{-st}\, dt \tag{10.3}$$

> Q(s):=laplace(q(t),t,s);

$$Q(s) := \frac{10}{s} \tag{10.4}$$

General solution for $U(x, s)$

> assume(x>0):U(x,s):=C1(s)*exp(−sqrt(s/k)*x)+C2(s)*exp(sqrt(s/k)*x);

$$U(x{\sim}, s) := C1(s)\, e^{-\sqrt{10}\sqrt{s}\, x{\sim}} + C2(s)\, e^{\sqrt{10}\sqrt{s}\, x{\sim}} \tag{10.5}$$

Substituting the boundary conditions at $x = 0$ and as x approaches infinity, for $\text{Re}\{s\} > 0$, we get

> C2(s):=0;

$$C2(s) := 0 \tag{10.6}$$

> eq:=Q(s)=eval(subs(x=0,U(x,s)));

$$eq := \frac{10}{s} = C1(s) \tag{10.7}$$

> C1(s):=solve(eq,C1(s));

$$C1(s) := \frac{10}{s} \tag{10.8}$$

Transformed solution

> U(x ,s):=eval(U(x,s));

$$U(x{\sim}, s) := \frac{10\, e^{-\sqrt{10}\sqrt{s}\, x{\sim}}}{s} \tag{10.9}$$

Solution is the inverse Laplace of $U(x, s)$

> u(x,t):=invlaplace(U(x,s),s,t);

$$u(x{\sim}, t) := 10\, \text{erfc}\left(\frac{1}{2} \frac{x{\sim}\sqrt{10}}{\sqrt{t}} \right) \tag{10.10}$$

ANIMATION

> animate(u(x,t),x=0..10,t=0..10,thickness=3);

The preceding animation command shows the spatial-time temperature distribution $u(x, t)$ in the rod. The animation sequence shown here and in Figure 10.1 shows snapshots of the animation at times $t = 1/10, 1, 2, 3, 4, 5.$

ANIMATION SEQUENCE

```
> u(x,0):=subs(t=1/10,u(x,t)):u(x,1):=subs(t=1,u(x,t)):
> u(x,2):=subs(t=2,u(x,t)):u(x,3):=subs(t=3,u(x,t)):
> u(x,4):=subs(t=4,u(x,t)):u(x,5):=subs(t=5,u(x,t)):
> plot({u(x,0),u(x,1),u(x,2),u(x,3),u(x,4),u(x,5)},x=0..2,thickness=10);
```

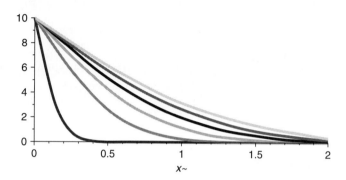

Figure 10.1

EXAMPLE 10.5.2: (Solution by convolution)   We seek the temperature distribution $u(x, t)$ in a thin rod over the semi-infinite interval $I = \{x \mid 0 < x < \infty\}$ whose lateral surface is insulated. The left end of the rod has a temperature constraint $u(0, t) = q(t)$ given following, and the initial temperature distribution is 0. There is no internal heat source in the system and the thermal diffusivity is k. We solve the problem by using the convolution method.

SOLUTION: The diffusion partial differential equation is

$$\frac{\partial}{\partial t} u(x, t) = k\left(\frac{\partial^2}{\partial x^2} u(x, t) \right)$$

The boundary conditions are

$$u(0, t) = q(t) \quad \text{and} \quad u(\infty, t) < \infty$$

The initial condition is

$$u(x, 0) = 0$$

The Laplace transformed equation reads

$$\frac{\partial^2}{\partial x^2} U(x, s) - \frac{sU(x, s)}{k} = -\frac{f(x)}{k}$$

Here, $U(x, s)$ is the Laplace transform of $u(x, t)$ with respect to t. The transformed boundary conditions are

$$U(0, s) = Q(s) \quad \text{and} \quad U(\infty, s) < \infty$$

Assignment of system parameters

> restart:with(plots):with(inttrans):f(x):=0:

Transformed equation in x reads

> diff(U(x,s),x,x)−(1/k)*s*U(x,s)=−1/k*f(x);

$$\frac{\partial^2}{\partial x^2} U(x, s) - \frac{s U(x, s)}{k} = 0 \tag{10.11}$$

Transformed boundary conditions

> Q(s)=Int(q(t)*exp(−s*t),t=0..infinity);

$$Q(s) = \int_0^\infty q(t) e^{-st} dt \tag{10.12}$$

General solution for $U(x, s)$

> assume(x>0):assume(k>0):U(x,s):=C1(s)*exp(−sqrt(s/k)*x)+C2(s)*exp(sqrt(s/k)*x);

$$U(x\sim, s) := C1(s) e^{-\sqrt{\frac{s}{k\sim}} x\sim} + C2(s) e^{\sqrt{\frac{s}{k\sim}} x\sim} \tag{10.13}$$

Substituting the boundary conditions at $x = 0$ and as x approaches infinity, for $\text{Re}\{s\} > 0$, we get

> C2(s):=0;

$$C2(s) := 0 \tag{10.14}$$

> eq:=Q(s)=eval(subs(x=0,U(x,s)));

$$eq := Q(s) = C1(s) \tag{10.15}$$

> C1(s):=solve(eq,C1(s));

$$C1(s) := Q(s) \tag{10.16}$$

Transformed solution

> U(x,s):=eval(U(x,s));

$$U(x\sim, s) := Q(s) e^{-\sqrt{\frac{s}{k\sim}} x\sim} \tag{10.17}$$

Convolution partition

> U1(x,s):=Q(s);

$$U1(x\sim, s) := Q(s) \tag{10.18}$$

> U2(x,s):=U(x,s)/Q(s);

$$U2(x\sim, s) := e^{-\sqrt{\frac{s}{k\sim}}\, x\sim} \tag{10.19}$$

> u1(x,t):=q(t);u1(x,tau):=subs(t=tau,u1(x,t)):

$$u1(x\sim, t) := q(t) \tag{10.20}$$

> u2(x,t):=simplify(invlaplace(U2(x,s),s,t));u2(x,t−tau):=subs(t=t−tau,u2(x,t)):

$$u2(x\sim, t) := \frac{1}{2} \frac{x\sim\, e^{-\frac{1}{4}\frac{x\sim^2}{k\sim t}}}{\sqrt{\pi}\sqrt{k\sim}\, t^{3/2}} \tag{10.21}$$

Convolution integral solution

> u(x,t):=Int(u1(x,tau)*u2(x,t−tau),tau=0..t);u(x,t):=simplify(value(%)):

$$u(x\sim, t) := \int_0^t \frac{1}{2} \frac{q(\tau)x\sim\, e^{-\frac{1}{4}\frac{x\sim^2}{k\sim(t-\tau)}}}{\sqrt{\pi}\sqrt{k\sim}(t-\tau)^{3/2}} \, d\tau \tag{10.22}$$

We leave the solution in the form of a convolution integral. This form of the solution can accommodate any temperature constraint $q(t)$ at $x = 0$. We leave specific cases for the exercises.

EXAMPLE 10.5.3: We seek the temperature distribution $u(x, t)$ in a thin rod over the semi-infinite interval $I = \{x \,|\, 0 < x < \infty\}$ whose lateral surface is insulated. The left end of the rod is held at the temperature 0, and the rod has the initial temperature distribution $f(x)$ given following. There is no internal heat source in the system, and the thermal diffusivity is $k = 1/5$.

SOLUTION: The diffusion partial differential equation is

$$\frac{\partial}{\partial t} u(x, t) = k \left(\frac{\partial^2}{\partial x^2} u(x, t) \right)$$

The boundary conditions are

$$u(0, t) = 0 \quad \text{and} \quad u(\infty, t) < \infty$$

The initial condition is

$$u(x, 0) = \sin(x)$$

The Laplace transformed equation reads

$$\frac{\partial^2}{\partial x^2} U(x, s) - \frac{sU(x, s)}{k} = -\frac{f(x)}{k}$$

Here, $U(x, s)$ is the Laplace transform of $u(x, t)$ with respect to t. The transformed boundary conditions are

$$U(0, s) = Q(s) \quad \text{and} \quad U(\infty, s) < \infty$$

Assignment of system parameters

> restart:with(plots):with(inttrans):k:=1/5:q(t):=0:f(x):=sin(x):f(v):=subs(x=v,f(x)):

Transformed equation in x reads

> diff(U(x,s),x,x)−(1/k)*s*U(x,s)=−1/k*f(x);

$$\frac{\partial^2}{\partial x^2} U(x, s) - 5sU(x, s) = -5\sin(x) \tag{10.23}$$

Transformed boundary conditions

> Q(s):=Int(q(t)*exp(−s*t),t=0..infinity);

$$Q(s) := \int_0^\infty 0 \, dt \tag{10.24}$$

> Q(s):=laplace(q(t),t,s);

$$Q(s) := 0 \tag{10.25}$$

Basis vectors to x equation

> phi[1](x,s):=exp(−sqrt(s/k)*x);phi[1](v,s):=subs(x=v,phi[1](x,s)):phi[2](x,s)
  :=exp(sqrt(s/k)*x);phi[2](v,s):=subs(x=v,phi[2](x,s)):

$$\phi_1(x, s) := e^{-\sqrt{5}\sqrt{s}\,x}$$
$$\phi_2(x, s) := e^{\sqrt{5}\sqrt{s}\,x} \tag{10.26}$$

Evaluation of Wronskian

> W(phi[1](v,s),phi[2](v,s)):=simplify(phi[1](v,s)*diff(phi[2](v,s),v)−phi[2](v,s)*
  diff(phi[1](v,s),v));

$$W\left(e^{-\sqrt{5}\sqrt{s}\,v},\ e^{\sqrt{5}\sqrt{s}\,v}\right):=2\sqrt{5}\sqrt{s} \tag{10.27}$$

Evaluation of Green's function

> G(x,v,s):=(phi[1](v,s)*phi[2](x,s)−phi[1](x,s)*phi[2](v,s))/W(phi[1](v,s),phi[2](v,s));

$$G(x,v,s):=\frac{1}{10}\frac{\left(e^{-\sqrt{5}\sqrt{s}\,v}e^{\sqrt{5}\sqrt{s}\,x}-e^{-\sqrt{5}\sqrt{s}\,x}e^{\sqrt{5}\sqrt{s}\,v}\right)\sqrt{5}}{\sqrt{s}} \tag{10.28}$$

Particular solution to the transformed equation

> Up(x,s):=−Int(G(x,v,s)*f(v)/k,v);Up(x,s):=value(%):

$$Up(x,s):=-\left(\int\frac{1}{2}\frac{\left(e^{-\sqrt{5}\sqrt{s}\,v}e^{\sqrt{5}\sqrt{s}\,x}-e^{-\sqrt{5}\sqrt{s}\,x}e^{\sqrt{5}\sqrt{s}\,v}\right)\sqrt{5}\sin(v)}{\sqrt{s}}\,dv\right) \tag{10.29}$$

> Up(x,s):=simplify(subs(v=x,%));

$$Up(x,s):=\frac{5\sin(x)}{5s+1} \tag{10.30}$$

General solution

> U(x,s):=C1(s)*phi[1](x,s)+C2(s)*phi[2](x,s)+Up(x,s);

$$U(x,s):=C1(s)\,e^{-\sqrt{5}\sqrt{s}\,x}+C2(s)\,e^{\sqrt{5}\sqrt{s}\,x}+\frac{5\sin(x)}{5s+1} \tag{10.31}$$

Substituting the boundary conditions at $x=0$ and as x approaches infinity, for Re{s} > 0, we get

> C2(s):=0;

$$C2(s):=0 \tag{10.32}$$

> eq:=Q(s)=eval(subs(x=0,U(x,s)));

$$eq:=0=C1(s) \tag{10.33}$$

> C1(s):=solve(eq,C1(s));

$$C1(s):=0 \tag{10.34}$$

Transformed solution

> U(x,s):=eval(U(x,s));

$$U(x, s) := \frac{5 \sin(x)}{5s + 1} \tag{10.35}$$

Final solution

> u(x,t):=invlaplace(U(x,s),s,t);

$$u(x, t) := \sin(x)\, e^{-\frac{1}{5} t} \tag{10.36}$$

ANIMATION

> animate(u(x,t),x=0..10,t=0..5,thickness=3);

The preceding animation command shows the spatial-time temperature distribution $u(x, t)$ in the rod. The animation sequence here and in Figure 10.2 shows snapshots of the animation at times $t = 0,\ 1,\ 2,\ 3,\ 4,\ 5$.

ANIMATION SEQUENCE

> u(x,0):=subs(t=0,u(x,t)):u(x,1):=subs(t=1,u(x,t)):
> u(x,2):=subs(t=2,u(x,t)):u(x,3):=subs(t=3,u(x,t)):
> u(x,4):=subs(t=4,u(x,t)):u(x,5):=subs(t=5,u(x,t)):
> plot({u(x,0),u(x,1),u(x,2),u(x,3),u(x,4),u(x,5)},x=0..10,thickness=10);

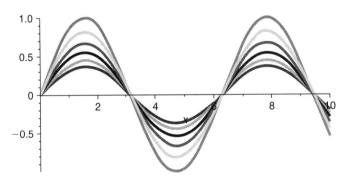

Figure 10.2

EXAMPLE 10.5.4: (Applied Dirac delta function)   We seek the temperature distribution $u(x, t)$ in a thin rod over the interval $I = \{x \mid 0 < x < \infty\}$ whose lateral surface is insulated. The left end of the rod has an applied impulse temperature (Dirac delta function). The rod has an initial temperature distribution $u(x, 0) = 0$, and there are no internal heat sources in the system. The thermal diffusivity is $k = 1$.

SOLUTION: The diffusion partial differential equation is

$$\frac{\partial}{\partial t}u(x, t) = k\left(\frac{\partial^2}{\partial x^2}u(x, t)\right)$$

The boundary conditions are

$$u(0, t) = \delta(t) \quad \text{and} \quad u(\infty, t) < \infty$$

The initial condition is

$$u(x, 0) = 0$$

The Laplace transformed equation reads

$$\frac{\partial^2}{\partial x^2}U(x, s) - \frac{sU(x, s)}{k} = -\frac{f(x)}{k}$$

Here, $U(x, s)$ is the Laplace transform of $u(x, t)$ with respect to t. The transformed boundary conditions are

$$U(0, s) = Q(s) \quad \text{and} \quad U(\infty, s) < \infty$$

Assignment of system parameters

> restart:with(plots):with(inttrans):k:=1:q(t):=Dirac(t):f(x):=0:

Transformed equation in x reads

> diff(U(x,s),x,x)−(1/k)*s*U(x,s)=−1/k*f(x);

$$\frac{\partial^2}{\partial x^2}U(x, s) - sU(x, s) = 0 \tag{10.37}$$

Transformed boundary conditions

> Q(s):=Int(q(t)*exp(−s*t),t=0..infinity);

$$Q(s) := \int_0^\infty \text{Dirac}(t)\,e^{-st}\,dt \tag{10.38}$$

> Q(s):=laplace(q(t),t,s);

$$Q(s) := 1 \tag{10.39}$$

General solution for $U(x, s)$

> assume(x>0):U(x,s):=C1(s)*exp(−sqrt(s/k)*x)+C2(s)*exp(sqrt(s/k)*x);

$$U(x\sim, s) := C1(s)\, e^{-\sqrt{s}\,x\sim} + C2(s)\, e^{\sqrt{s}\,x\sim} \qquad (10.40)$$

Substituting the boundary conditions at $x = 0$ and as x approaches infinity, for $\text{Re}\{s\} > 0$, we get

> C2(s):=0;

$$C2(s) := 0 \qquad (10.41)$$

> eq:=Q(s)=eval(subs(x=0,U(x,s)));

$$eq := 1 = C1(s) \qquad (10.42)$$

> C1(s):=solve(eq,C1(s));

$$C1(s) := 1 \qquad (10.43)$$

Transformed solution

> U(x,s):=eval(U(x,s));

$$U(x\sim, s) := e^{-\sqrt{s}\,x\sim} \qquad (10.44)$$

Solution is the inverse Laplace of $U(x, s)$

> u(x,t):=invlaplace(U(x,s),s,t);

$$u(x\sim, t) := \frac{1}{2}\, \frac{x\sim\, e^{-\frac{1}{4}\frac{x\sim^2}{t}}}{\sqrt{\pi}\, t^{3/2}} \qquad (10.45)$$

ANIMATION

> animate(u(x,t),x=0..10,t=0..5,thickness=3);

The preceding animation command shows the spatial-time temperature distribution $u(x, t)$ in the rod. The animation sequence here and in Figure 10.3 shows snapshots of the animation at times $t = 1/2$, 1, 2, 3, 4, 5.

ANIMATION SEQUENCE

> u(x,0):=subs(t=1/2,u(x,t)):u(x,1):=subs(t=1,u(x,t)):
> u(x,2):=subs(t=2,u(x,t)):u(x,3):=subs(t=3,u(x,t)):
> u(x,4):=subs(t=4,u(x,t)):u(x,5):=subs(t=5,u(x,t)):
> plot({u(x,0),u(x,1),u(x,2),u(x,3),u(x,4),u(x,5)},x=0..10,thickness=10);

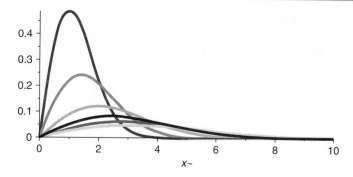

Figure 10.3

EXAMPLE 10.5.5: (End of rod with temperature constraint)   We seek the temperature distribution $u(x, t)$ in a thin rod over the interval $I = \{x \,|\, 0 < x < \infty\}$ whose lateral surface is insulated. The left end of the rod is constrained at a varying temperature $q(t)$ given following, and the rod has an initial temperature distribution $u(x, 0) = 0$. There is no heat source in the system and the thermal diffusivity of the rod is $k = 1/50$.

SOLUTION: The diffusion partial differential equation is

$$\frac{\partial}{\partial t} u(x, t) = k \left(\frac{\partial^2}{\partial x^2} u(x, t) \right)$$

The boundary conditions are

$$u(0, t) = \frac{1}{\sqrt{t}} \quad \text{and} \quad u(\infty, t) < \infty$$

The initial condition is

$$u(x, 0) = 0$$

The Laplace transformed equation reads

$$\frac{\partial^2}{\partial x^2} U(x, s) - \frac{s U(x, s)}{k} = -\frac{f(x)}{k}$$

Here, $U(x, s)$ is the Laplace transform of $u(x, t)$ with respect to t. The transformed boundary conditions are

$$U(0, s) = Q(s) \quad \text{and} \quad U(\infty, s) < \infty$$

Assignment of system parameters

> restart:with(plots):with(inttrans):k:=1/50:q(t):=1/sqrt(t):f(x):=0:

Transformed equation in x reads

> diff(U(x,s),x,x)−(1/k)*s*U(x,s)=−1/k*f(x);

$$\frac{\partial^2}{\partial x^2} U(x, s) - 50sU(x, s) = 0 \tag{10.46}$$

Transformed boundary conditions

> Q(s):=Int(q(t)*exp(−s*t),t=0..infinity);

$$Q(s) := \int_0^\infty \frac{e^{-st}}{\sqrt{t}} \, dt \tag{10.47}$$

> Q(s):=laplace(q(t),t,s);

$$Q(s) := \sqrt{\frac{\pi}{s}} \tag{10.48}$$

General solution for $U(x, s)$

> assume(x>0):U(x,s):=C1(s)*exp(−sqrt(s/k)*x)+C2(s)*exp(sqrt(s/k)*x);

$$U(x\sim, s) := C1(s) \, e^{-5\sqrt{2}\sqrt{s}x\sim} + C2(s) \, e^{5\sqrt{2}\sqrt{s}x\sim} \tag{10.49}$$

Substituting the boundary conditions at $x = 0$ and as x approaches infinity, for $\mathrm{Re}\{s\} > 0$, we get

> C2(s):=0;

$$C2(s) := 0 \tag{10.50}$$

> eq:=Q(s)=eval(subs(x=0,U(x,s)));

$$eq := \sqrt{\frac{\pi}{s}} = C1(s) \tag{10.51}$$

> C1(s):=solve(eq,C1(s));

$$C1(s) := \sqrt{\frac{\pi}{s}} \tag{10.52}$$

Transformed solution

> U(x,s):=eval(U(x,s));

$$U(x\sim, s) := \sqrt{\frac{\pi}{s}} e^{-5\sqrt{2}\sqrt{s}x\sim} \tag{10.53}$$

Solution is the inverse Laplace of $U(x, s)$

> u(x,t):=invlaplace(U(x,s),s,t);

$$u(x{\sim}, t) := \frac{e^{-\frac{25}{2}\frac{x{\sim}^2}{t}}}{\sqrt{t}}$$

(10.54)

ANIMATION

> animate(u(x,t),x=0..1,t=0..5,thickness=3);

The preceding animation command shows the spatial-time temperature distribution $u(x, t)$ in the rod. The animation sequence here and in Figure 10.4 shows snapshots of the animation at times $t = 1/5, 1, 2, 3, 4, 5$.

ANIMATION SEQUENCE

>u(x,0):=subs(t=1/5,u(x,t)):u(x,1):=subs(t=1,u(x,t)):
> u(x,2):=subs(t=2,u(x,t)):u(x,3):=subs(t=3,u(x,t)):
> u(x,4):=subs(t=4,u(x,t)):u(x,5):=subs(t=5,u(x,t)):
> plot({u(x,0),u(x,1),u(x,2),u(x,3),u(x,4),u(x,5)},x=0..10,thickness=10);

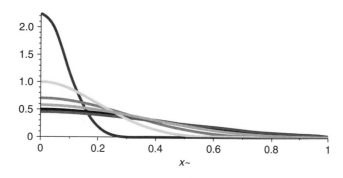

Figure 10.4

EXAMPLE 10.5.6: (Input heat flux)   We seek the temperature distribution $u(x, t)$ in a thin rod over the semi-infinite interval $I = \{x \mid 0 < x < \infty\}$ whose lateral surface is insulated. The left end of the rod has an input heat flux $q(t)$ given following, and the initial temperature distribution is 0. There is no internal heat source in the system, and the thermal diffusivity is $k = 1/20$. We want to solve this problem using convolution methods.

SOLUTION: The diffusion partial differential equation is

$$\frac{\partial}{\partial t}u(x, t) = k\left(\frac{\partial^2}{\partial x^2}u(x, t)\right)$$

The boundary conditions are

$$u_x(0, t) = -10 \quad \text{and} \quad u(\infty, t) < \infty$$

The initial condition is

$$u(x, 0) = 0$$

The Laplace transformed equation reads

$$\frac{\partial^2}{\partial x^2} U(x, s) - \frac{sU(x, s)}{k} = -\frac{f(x)}{k}$$

Here, $U(x, s)$ is the Laplace transform of $u(x, t)$ with respect to t. The transformed boundary conditions are

$$\frac{\partial}{\partial x} U(0, s) = Q(s) \quad \text{and} \quad U(\infty, s) < \infty$$

Assignment of system parameters

> restart:with(plots):with(inttrans):k:=1/20:q(t):=−10:f(x):=0:

Transformed equation in x reads

> diff(U(x,s),x,x)−(1/k)*s*U(x,s)=−1/k*f(x);

$$\frac{\partial^2}{\partial x^2} U(x, s) - 20sU(x, s) = 0 \tag{10.55}$$

Transformed boundary conditions

> Q(s)=Int(q(t)*exp(−s*t),t=0..infinity);

$$Q(s) = \int_0^\infty \left(-10\,e^{-st}\right) dt \tag{10.56}$$

> Q(s):=laplace(q(t),t,s);

$$Q(s) := -\frac{10}{s} \tag{10.57}$$

General solution for $U(x, s)$

> assume(x>0):U(x,s):=C1(s)*exp(−sqrt(s/k)*x)+C2(s)*exp(sqrt(s/k)*x);

$$U(x\sim, s) := C1(s)\,e^{-2\sqrt{5}\sqrt{s}\,x\sim} + C2(s)\,e^{2\sqrt{5}\sqrt{s}\,x\sim} \tag{10.58}$$

Substituting the boundary conditions at $x = 0$ and as x approaches infinity, for $\text{Re}\{s\} > 0$, we get

> C2(s):=0;

$$C2(s) := 0 \tag{10.59}$$

> eq:=Q(s)=eval(subs(x=0,diff(U(x,s),x)));

$$eq := -\frac{10}{s} = -2C1(s)\sqrt{5}\sqrt{s} \tag{10.60}$$

> C1(s):=solve(eq,C1(s));

$$C1(s) := \frac{\sqrt{5}}{s^{3/2}} \tag{10.61}$$

Transformed solution

> U(x,s):=eval(U(x,s));

$$U(x\sim, s) := \frac{\sqrt{5}\,e^{-2\sqrt{5}\sqrt{s}\,x\sim}}{s^{3/2}} \tag{10.62}$$

Convolution partition

> U1(x,s):=U(x,s)*s;

$$U1(x\sim, s) := \frac{\sqrt{5}\,e^{-2\sqrt{5}\sqrt{s}\,x\sim}}{\sqrt{s}} \tag{10.63}$$

> U2(x,s):=1/s;

$$U2(x\sim, s) := \frac{1}{s} \tag{10.64}$$

> u1(x,t):=invlaplace(U1(x,s),s,t);u1(x,tau):=subs(t=tau,u1(x,t)):

$$u1(x\sim, t) := \frac{\sqrt{5}\,e^{-\frac{5x\sim^2}{t}}}{\sqrt{\pi t}} \tag{10.65}$$

> u2(x,t):=simplify(invlaplace(U2(x,s),s,t));u2(x,t−tau):=subs(t=t−tau,u2(x,t)):

$$u2(x \sim, t) := 1 \tag{10.66}$$

Convolution integral solution

> u(x,t):=Int(u1(x,tau)*u2(x,t−tau),tau=0..t);u(x,t):=simplify(value(%)):

$$u(x\sim, t) := \int_0^t \frac{\sqrt{5}\,e^{-\frac{5x\sim^2}{\tau}}}{\sqrt{\pi \tau}}\, d\tau \tag{10.67}$$

Conversion of integral to error function integral

> u(x,t):=simplify(convert(%,erfc));

$$u(x\!\sim, t) := \frac{2\left(-5\sqrt{\pi}\,x\!\sim\sqrt{t} + \sqrt{5}\,e^{-\frac{5x\sim^2}{t}}\,t + 5x\!\sim\sqrt{\pi}\,\mathrm{erf}\left(\frac{\sqrt{5}\,x\sim}{\sqrt{t}}\right)\sqrt{t}\right)}{\sqrt{\pi}\sqrt{t}} \tag{10.68}$$

ANIMATION

> animate(u(x,t),x=0..2,t=1/10..5,thickness=3);

The preceding animation command shows the spatial-time temperature distribution $u(x, t)$ in the rod. The animation sequence here and in Figure 10.5 shows snapshots of the animation at times $t = 1/10,\ 1,\ 2,\ 3,\ 4,\ 5$.

ANIMATION SEQUENCE

> u(x,0):=subs(t=1/10,u(x,t)):u(x,1):=subs(t=1,u(x,t)):
> u(x,2):=subs(t=2,u(x,t)):u(x,3):=subs(t=3,u(x,t)):
> u(x,4):=subs(t=4,u(x,t)):u(x,5):=subs(t=5,u(x,t)):
> plot({u(x,0),u(x,1),u(x,2),u(x,3),u(x,4),u(x,5)},x=0..2,thickness=10);

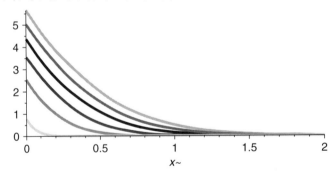

Figure 10.5

EXAMPLE 10.5.7: (Rod with internal heat source)   We seek the temperature distribution $u(x, t)$ in a thin rod over the semi-infinite interval $I = \{x \mid 0 < x < \infty\}$ whose lateral surface is insulated. The left end of the rod is held at the temperature 0, and the rod has an initial temperature distribution of 0. There is an internal heat source $h(x, t)$ in the system, given following, and the thermal diffusivity is $k = 1/5$.

SOLUTION: The nonhomogeneous diffusion partial differential equation is

$$\frac{\partial}{\partial t}u(x, t) = k\left(\frac{\partial^2}{\partial x^2}u(x, t)\right) + h(x, t)$$

The boundary conditions are

$$u(0, t) = 0 \quad \text{and} \quad u(\infty, t) < \infty$$

The initial condition is

$$u(x, 0) = 0$$

and the internal heat source is

$$h(x, t) = \sin(x)\, e^{-t}$$

The Laplace transformed equation reads

$$\frac{\partial^2}{\partial x^2} U(x, s) - \frac{sU(x, s)}{k} = -\frac{f(x)}{k} - \frac{H(x, s)}{k}$$

Here, $U(x, s)$ is the Laplace transform of $u(x, t)$ with respect to t. The transformed boundary conditions are

$$U(0, s) = Q(s) \quad \text{and} \quad U(\infty, s) < \infty$$

Assignment of system parameters

> restart:with(plots):with(inttrans):k:=1/5:q(t):=0:f(x):=0:h(x,t):=sin(x)*exp(−t):

Transformed boundary conditions

> Q(s):=Int(q(t)*exp(−s*t),t=0..infinity);

$$Q(s) := \int_0^\infty 0 \, dt \tag{10.69}$$

> Q(s):=laplace(q(t),t,s);

$$Q(s) := 0 \tag{10.70}$$

Transformed heat source

> H(x,s):=Int(h(x,t)*exp(−s*t),t=0..infinity);

$$H(x, s) := \int_0^\infty \sin(x)\, e^{-t}\, e^{-st}\, dt \tag{10.71}$$

> H(x,s):=laplace(h(x,t),t,s);H(v,s):=subs(x=v,H(x,s)):

$$H(x, s) := \frac{\sin(x)}{1 + s} \tag{10.72}$$

Transformed equation in x reads

> diff(U(x,s),x,x)−(1/k)*s*U(x,s)=−f(x)/k−H(x,s)/k;

$$\frac{\partial^2}{\partial x^2}U(x, s) - 5sU(x, s) = -\frac{5\sin(x)}{1+s} \tag{10.73}$$

Basis vectors to x equation

> phi[1](x,s):=exp(−sqrt(s/k)*x);phi[1](v,s):=subs(x=v,phi[1](x,s)):phi[2](x,s):
=exp(sqrt(s/k)*x);phi[2](v,s):=subs(x=v,phi[2](x,s)):

$$\phi_1(x, s) := e^{-\sqrt{5}\sqrt{s}\,x}$$

$$\phi_2(x, s) := e^{\sqrt{5}\sqrt{s}\,x} \tag{10.74}$$

Evaluation of Wronskian

> W(phi[1](v,s),phi[2](v,s)):=simplify(phi[1](v,s)*diff(phi[2](v,s),v)−phi[2](v,s)*
diff(phi[1](v,s),v));

$$W\left(e^{-\sqrt{5}\sqrt{s}\,v},\ e^{\sqrt{5}\sqrt{s}\,v}\right) := 2\sqrt{5}\sqrt{s} \tag{10.75}$$

Evaluation of Green's function

> G(x,v,s):=(phi[1](v,s)*phi[2](x,s)−phi[1](x,s)*phi[2](v,s))/W(phi[1](v,s),phi[2](v,s));

$$G(x, v, s) := \frac{1}{10}\frac{\left(e^{-\sqrt{5}\sqrt{s}\,v}e^{\sqrt{5}\sqrt{s}\,x} - e^{-\sqrt{5}\sqrt{s}\,x}e^{\sqrt{5}\sqrt{s}\,v}\right)\sqrt{5}}{\sqrt{s}} \tag{10.76}$$

Particular solution to the transformed equation

> Up(x,s):=−Int(G(x,v,s)*H(v,s)/k,v);Up(x,s):=value(%):

$$Up(x, s) := -\left(\int \frac{1}{2}\frac{\left(e^{-\sqrt{5}\sqrt{s}\,v}e^{\sqrt{5}\sqrt{s}\,x} - e^{-\sqrt{5}\sqrt{s}\,x}e^{\sqrt{5}\sqrt{s}\,v}\right)\sqrt{5}\sin(v)}{\sqrt{s}(1+s)}\,dv\right) \tag{10.77}$$

> Up(x,s):=simplify(subs(v=x,%));

$$Up(x, s) := \frac{5\sin(x)}{(1+s)(5s+1)} \tag{10.78}$$

General solution

> U(x,s):=C1(s)*phi[1](x,s)+C2(s)*phi[2](x,s)+Up(x,s);

$$U(x, s) := C1(s)\,e^{-\sqrt{5}\sqrt{s}\,x} + C2(s)\,e^{\sqrt{5}\sqrt{s}\,x} + \frac{5\sin(x)}{(1+s)(5s+1)} \tag{10.79}$$

Substituting the boundary conditions at $x = 0$ and as x approaches infinity, for $\mathrm{Re}\{s\} > 0$, we get

> C2(s):=0;

$$C2(s) := 0 \tag{10.80}$$

> eq:=Q(s)=eval(subs(x=0,U(x,s)));

$$eq := 0 = C1(s) \tag{10.81}$$

> C1(s):=solve(eq,C1(s));

$$C1(s) := 0 \tag{10.82}$$

Transformed solution

> U(x,s):=eval(U(x,s));

$$U(x, s) := \frac{5\sin(x)}{(1+s)(5s+1)} \tag{10.83}$$

Final solution

> u(x,t):=invlaplace(U(x,s),s,t);

$$u(x, t) := \frac{5}{2}\sin(x)\,e^{-\frac{3}{5}t}\sinh\left(\frac{2}{5}t\right) \tag{10.84}$$

ANIMATION

> animate(u(x,t),x=0..10,t=0..5,thickness=3);

The preceding animation command shows the spatial-time temperature distribution $u(x, t)$ in the rod. The animation sequence here and in Figure 10.6 shows snapshots of the animation at times $t = 0, 1, 2, 3, 4, 5$.

ANIMATION SEQUENCE

> u(x,0):=subs(t=0,u(x,t)):u(x,1):=subs(t=1,u(x,t)):
> u(x,2):=subs(t=2,u(x,t)):u(x,3):=subs(t=3,u(x,t)):
> u(x,4):=subs(t=4,u(x,t)):u(x,5):=subs(t=5,u(x,t)):
> plot({u(x,0),u(x,1),u(x,2),u(x,3),u(x,4),u(x,5)},x=0..10,thickness=10);

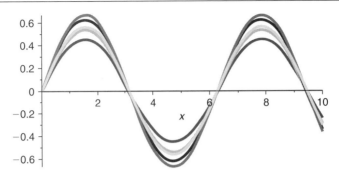

Figure 10.6

10.6 Laplace Transform Procedures on the Wave Equation

We consider the nonhomogeneous partial differential equation for wave phenomena in one spatial dimension in rectangular coordinates as given here:

$$\frac{\partial^2}{\partial t^2}u(x,t) = c^2\left(\frac{\partial^2}{\partial x^2}u(x,t)\right) + h(x,t)$$

with the two initial conditions

$$u(x,0) = f(x) \quad \text{and} \quad u_t(x,0) = g(x)$$

The term $h(x,t)$ accounts for any external sources in the system, $f(x)$ denotes the initial displacment distribution, and $g(x)$ denotes the initial speed distribution. We assume that there is no damping in the system.

In evaluating the Laplace transform of terms in the preceding equation, we have to contend with taking improper integrals with respect to t of functions that are dependent on both x and t. We denote the Laplace transform of $u(x,t)$ with respect to the time variable t by the capital letter representation as shown:

$$\int_0^\infty u(x,t)\,e^{-st}\,dt = U(x,s)$$

Proceeding formally with the evaluation of transforms of the derivative terms in the given partial differential equation, we get

$$\int_0^\infty \left(\frac{\partial^2}{\partial t^2}u(x,t)\right)e^{-st}\,dt = s^2 U(x,s) - su(x,0) - u_t(x,0)$$

In the preceding, $U(x, s)$ denotes the Laplace transform of $u(x, t)$ with respect to t; $u(x, 0)$ denotes the initial value of $u(x, t)$; and $u_t(x, 0)$ denotes the initial speed.

Proceeding formally, in evaluating the transforms of the partial derivative terms with respect to the spatial variable x, we get

$$\int_0^\infty \left(\frac{\partial^2}{\partial x^2} u(x, t) \right) e^{-st} dt = \frac{\partial^2}{\partial x^2} U(x, s)$$

This operation assumes that we could interchange the order of the differentiation operation, with respect to x, with the integration operation, with respect to t. The conditions for the validity of doing this are based on the uniform convergence of the improper integrals of the derivative terms.

Assuming the validity of all these given conditions, then the Laplace transform of the wave partial differential equation gives us

$$\frac{\partial^2}{\partial x^2} U(x, s) - \frac{s^2 U(x, s)}{c^2} = -\frac{sf(x)}{c^2} - \frac{g(x)}{c^2} - \frac{H(x, s)}{c^2}$$

Here, $H(x, s)$ is the Laplace transform of the source term given as

$$H(x, s) = \int_0^\infty h(x, t) e^{-st} dt$$

The preceding differential equation is an ordinary second-order nonhomogeneous differential equation in the single spatial variable x. We investigated the solutions for this equation in Chapter 1. The solutions are, of course, dependent on the spatial boundary conditions on the problem.

We now implement the Laplace transform method to evaluate solutions to the wave equation over semi-infinite domains with respect to the spatial variable x. We focus on those types of problems that may or may not have nonhomogeneous boundary conditions—those problems that cannot ordinarily be solved using Fourier transform methods as covered in Chapter 9. Further, we do not consider using Laplace transform methods for wave problems over finite domains here because eigenfunction expansion methods, as used in Chapters 4, 7, and 8, are more effective.

In general, we consider time-dependent boundary conditions that are nonhomogeneous at $x = 0$ and are bounded as x approaches infinity; that is,

$$\kappa_1 u(0, t) + \kappa_2 u_x(0, t) = q(t)$$

and

$$u(\infty, t) < \infty$$

The transform of these boundary conditions yields

$$\kappa_1 U(0, s) + \kappa_2 \left(\frac{\partial}{\partial x} U(0, s) \right) = Q(s)$$

and

$$U(\infty, s) < \infty$$

where the term $\frac{\partial}{\partial x} U(0, s)$ is understood to be the x-dependent derivative of $U(x, s)$ evaluated at $x = 0$ and $Q(s)$ is the Laplace transform of $q(t)$ as shown here:

$$Q(s) = \int_0^\infty q(t) e^{-st} dt$$

To develop the general solution to the given nonhomogeneous differential equation, we use the method of variation of parameters as outlined in Chapter 1 whereby we must first consider finding a set of basis vectors for the following corresponding homogeneous differential equation:

$$\frac{\partial^2}{\partial x^2} U(x, s) - \frac{s^2 U(x, s)}{c^2} = 0$$

A set of two basis vectors is

$$\phi_1(x, s) = e^{-\frac{sx}{c}}$$

and

$$\phi_2(x, s) = e^{\frac{sx}{c}}$$

From the two basis vectors, we are able to generate the second-order Green's function

$$G(x, v, s) = \frac{\phi_1(v, s)\phi_2(x, s) - \phi_1(x, s)\phi_2(v, s)}{W(\phi_1(v, s), \phi_2(v, s))}$$

where the denominator term is the familiar Wronskian, which, for the basis vectors shown, is evaluated to be

$$W(\phi_1(v, s), \phi_2(v, s)) = \frac{2s}{c}$$

Thus, from the method of variaiation of parameters, the particular solution to the nonhomogeneous ordinary differential equation in x is

$$Up(x, s) = - \left(\int \frac{G(x, v, s)(sf(v) + g(v) + H(v, s))}{c^2} dv \right)$$

The general solution to the x-dependent differential equation is the sum of the complementary solution and the particular solution preceding. Combining the two solutions, we get the general solution for the transform $U(x, s)$:

$$U(x, s) = C1(s) \, e^{-\frac{sx}{c}} + C2(s) \, e^{\frac{sx}{c}} - \left(\int \frac{G(x, v, s)(sf(v) + g(v) + H(v, s))}{c^2} \, dv \right)$$

The arbitrary coefficients $C1(s)$ and $C2(s)$ are determined from the boundary conditions for the problem. The final solution to our problem is the inverse Laplace transform of $U(x, s)$. We now demonstrate these concepts by solving the illustrative problem of a string falling under its own weight.

DEMONSTRATION: We seek the wave amplitude $u(x, t)$ for transverse wave motion along a taut string over the semi-infinite interval $I = \{x \mid 0 < x < \infty\}$. The left end of the string is held secure. The string has zero initial displacement and initial speed. An external force due to gravity (acceleration $= -980 \, \text{cm/sec}^2$) is acting uniformly on the string, and the wave speed constant is $c = 2$. There is no damping in the system.

SOLUTION: The nonhomogeneous wave partial differential equation for the string is

$$\frac{\partial^2}{\partial t^2} u(x, t) = 4 \left(\frac{\partial^2}{\partial x^2} u(x, t) \right) - 980$$

Because the string is secure at the left end and we require the solution to be finite as x approaches infinity, the boundary conditions are

$$u(0, t) = 0 \quad \text{and} \quad u(\infty, t) < \infty$$

The string is initially at rest, and it has no initial displacement; thus, the initial conditions are

$$u(x, 0) = 0 \quad \text{and} \quad u_t(x, 0) = 0$$

The Laplace transform, with respect to t, of the partial differential equation yields the ordinary nonhomogeneous differential equation for the transformed variable $U(x, s)$, which reads

$$\frac{\partial^2}{\partial x^2} U(x, s) - \frac{s^2 U(x, s)}{4} = \frac{245}{s}$$

The corresponding transformed boundary conditions are

$$U(0, s) = 0 \quad \text{and} \quad U(\infty, s) < \infty$$

From Section 1.5, a set of basis vectors for the x-dependent differential equation is

$$\phi_1(x, s) = e^{-\frac{sx}{2}}$$

and

$$\phi_2(x, s) = e^{\frac{sx}{2}}$$

The Wronskian for this basis is

$$W\left(e^{-\frac{sv}{2}}, e^{\frac{sv}{2}}\right) = s$$

From Section 1.7, the corresponding second-order Green's function is

$$G(x, v, s) = \frac{e^{-\frac{sv}{2}} e^{\frac{sx}{2}} - e^{-\frac{sx}{2}} e^{\frac{sv}{2}}}{s}$$

The particular solution to the nonhomogeneous differential equation, for the transformed variable $U(x, s)$, is given as the Green's function integral

$$Up(x, s) = \int \frac{245\left(e^{-\frac{sv}{2}} e^{\frac{sx}{2}} - e^{-\frac{sx}{2}} e^{\frac{sv}{2}}\right)}{s^2} \, dv$$

which is evaluated to be

$$Up(x, s) = -\frac{980}{s^3}$$

Thus, the general solution to the ordinary differential equation for the transformed solution $U(x, s)$ is

$$U(x, s) = C1(s) e^{-\frac{sx}{2}} + C2(s) e^{\frac{sx}{2}} - \frac{980}{s^3} \tag{10.85}$$

Here, $C1(s)$ and $C2(s)$ are arbitrary constants that are determined from the boundary conditions. Substituting the boundary condition at $x = 0$ and as x approaches infinity, for $\text{Re}\{s\} > 0$, we get the transformed solution

$$U(x, s) = \frac{980 e^{-\frac{sx}{2}}}{s^3} - \frac{980}{s^3}$$

Evaluation of the inverse Laplace transform yields the spatial- and time-dependent solution $u(x, t)$ for the falling string:

$$u(x, t) = 490 \, H\left(t - \frac{x}{2}\right)\left(t - \frac{x}{2}\right)^2 - 490 \, t^2$$

where $H\left(t - \frac{x}{2}\right)$ denotes the translated Heaviside function. The details for the development of this solution and the graphics are given later in one of the Maple worksheet examples.

10.7 Example Laplace Transform Problems for the Wave Equation

We now look at some specific applications of the Laplace transform in evaluating solutions to the wave equation over semi-infinite domains with respect to the spatial variable x. We focus on those types of problems that have nonhomogeneous boundary conditions—problems that cannot ordinarily be solved using Fourier transform methods as covered in Chapter 9.

EXAMPLE 10.7.1: (The vibrating string)   We seek the wave amplitude $u(x, t)$ for transverse wave motion along a taut string over the semi-infinite interval $I = \{x \mid 0 < x < \infty\}$. The left end of the string is constrained to move in accordance with the sinusoidal function $q(t)$ given following. The string has an initial displacement distribution $f(x)$ and an initial speed distribution $g(x)$. There is no external source function, and there is no damping in the system. The wave speed is $c = 2$.

SOLUTION: The wave partial differential equation is

$$\frac{\partial^2}{\partial t^2} u(x, t) = c^2 \left(\frac{\partial^2}{\partial x^2} u(x, t) \right)$$

The boundary conditions are

$$u(0, t) = \sin(t) \quad \text{and} \quad u(\infty, t) < \infty$$

The initial conditions are

$$u(x, 0) = 0 \quad \text{and} \quad u_t(x, 0) = 0$$

The Laplace transformed equation reads

$$\frac{\partial^2}{\partial x^2} U(x, s) - \frac{s^2 U(x, s)}{c^2} = -\frac{s f(x)}{c^2} - \frac{g(x)}{c^2}$$

Here, $U(x, s)$ is the Laplace transform of $u(x, t)$ with respect to t. The transformed boundary conditions are

$$U(0, s) = Q(s) \quad \text{and} \quad U(\infty, s) < \infty$$

Assignment of system parameters

> restart:with(plots):with(inttrans):c:=2:q(t):=sin(t):f(x):=0:g(x):=0:

Transformed equation in x reads

> diff(U(x,s),x,x)−s^2/c^2*U(x,s)=−s*f(x)/c^2−g(x)/c^2;

$$\frac{\partial^2}{\partial x^2} U(x, s) - \frac{1}{4} s^2 U(x, s) = 0 \tag{10.86}$$

Transformed boundary conditions

> Q(s):=Int(q(t)*exp(−s*t),t=0..infinity);

$$Q(s) := \int_0^\infty \sin(t)\, e^{-st} dt \tag{10.87}$$

> Q(s):=laplace(q(t),t,s);

$$Q(s) := \frac{1}{s^2 + 1} \tag{10.88}$$

General solution for $U(x, s)$

> assume(x>0):U(x,s):=C1(s)*exp(−s/c*x)+C2(s)*exp(s/c*x);

$$U(x{\sim}, s) := C1(s)\, e^{-\frac{1}{2}sx{\sim}} + C2(s)\, e^{\frac{1}{2}sx{\sim}} \tag{10.89}$$

Substituting the boundary conditions at $x = 0$ and as x approaches infinity, for Re{s} > 0, we get

> C2(s):=0;

$$C2(s) := 0 \tag{10.90}$$

> eq:=Q(s)=eval(subs(x=0,U(x,s)));

$$eq := \frac{1}{s^2 + 1} = C1(s) \tag{10.91}$$

> C1(s):=solve(eq, C1(s));

$$C1(s) := \frac{1}{s^2 + 1} \tag{10.92}$$

Transformed solution

> U(x,s):=eval(U(x,s));

$$U(x{\sim}, s) := \frac{e^{-\frac{1}{2}sx{\sim}}}{s^2 + 1} \tag{10.93}$$

Solution is the inverse Laplace of $U(x, s)$

> u(x,t):=invlaplace(U(x,s),s,t);

$$u(x{\sim}, t) := \text{Heaviside}\left(t - \frac{1}{2}x{\sim}\right) \sin\left(t - \frac{1}{2}x{\sim}\right) \tag{10.94}$$

ANIMATION

> animate(u(x,t),x=0..10,t=0..10,thickness=3);

The preceding animation command shows the spatial-time wave amplitude distribution $u(x, t)$ on the string. The animation sequence here and in Figure 10.7 shows snapshots of the animation at times $t = 0, 1, 2, 3, 4, 5$.

ANIMATION SEQUENCE

> u(x,0):=subs(t=0,u(x,t)):u(x,1):=subs(t=1,u(x,t)):
> u(x,2):=subs(t=2,u(x,t)):u(x,3):=subs(t=3,u(x,t)):
> u(x,4):=subs(t=4,u(x,t)):u(x,5):=subs(t=5,u(x,t)):
> plot({u(x,0),u(x,1),u(x,2),u(x,3),u(x,4),u(x,5)},x=0..10,thickness=10);

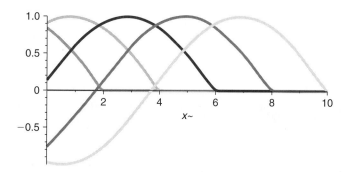

Figure 10.7

EXAMPLE 10.7.2: (String falling under its own weight)   We seek the wave amplitude $u(x, t)$ for transverse wave motion along a taut string over the semi-infinite interval $I = \{x \mid 0 < x < \infty\}$. The left end of the string is held fixed. The string has the initial displacement distribution $f(x) = 0$ and the initial speed distribution $g(x) = 0$. An external force due to gravity (acceleration $= -980 \, \text{cm/sec}^2$) is acting uniformly on the string, and we assume no damping in the system. The wave speed is $c = 2$.

SOLUTION: The nonhomogeneous wave partial differential equation is

$$\frac{\partial^2}{\partial t^2}u(x, t) = c^2 \left(\frac{\partial^2}{\partial x^2}u(x, t) \right) + h(x, t)$$

The boundary conditions are

$$u(0, t) = 0 \quad \text{and} \quad u(\infty, t) < \infty$$

The initial conditions are

$$u(x, 0) = 0 \quad \text{and} \quad u_t(x, 0) = 0$$

The Laplace transformed equation reads

$$\frac{\partial^2}{\partial x^2} U(x, s) - \frac{s^2 U(x, s)}{c^2} = -\frac{sf(x)}{c^2} - \frac{g(x)}{c^2} - \frac{H(x, s)}{c^2}$$

Here, $U(x, s)$ is the Laplace transform of $u(x, t)$ with respect to t. The transformed boundary conditions are

$$U(0, s) = Q(s) \quad \text{and} \quad U(\infty, s) < \infty$$

Assignment of system parameters

> restart:with(plots):with(inttrans):c:=2:q(t):=0:f(x):=0:f(v):=subs(x=v,f(x)):g(x):=0:
  g(v):=subs(x=v,g(x)):h(x,t):=−980:

Transformed equation in x reads

> diff(U(x,s),x,x)−s^2/c^2*U(x,s)=−s*f(x)/c^2−g(x)/c^2−H(x,s)/c^2;

$$\frac{\partial^2}{\partial x^2} U(x, s) - \frac{1}{4} s^2 U(x, s) = -\frac{1}{4} H(x, s) \tag{10.95}$$

Transformed boundary conditions

> Q(s):=Int(q(t)*exp(−s*t),t=0..infinity);

$$Q(s) := \int_0^\infty 0 \, dt \tag{10.96}$$

> Q(s):=value(%);

$$Q(s) := 0 \tag{10.97}$$

Transformed source term

> H(x,s):=Int(h(x,t)*exp(−s*t),t=0..infinity);

$$H(x, s) := \int_0^\infty (-980 \, e^{-st}) \, dt \tag{10.98}$$

> H(x,s):=laplace(h(x,t),t,s);H(v,s):=subs(x=v,H(x,s)):

$$H(x, s) := -\frac{980}{s} \tag{10.99}$$

Basis vectors

> assume(x>0):phi[1](x,s):=exp(−s/c*x);phi[1](v,s):=subs(x=v,phi[1](x,s)):
  phi[2](x,s):=exp(s/c*x);phi[2](v,s):=subs(x=v,phi[2](x,s)):

$$\phi_1(x\sim, s) := e^{-\frac{1}{2}sx\sim}$$

$$\phi_2(x\sim, s) := e^{\frac{1}{2}sx\sim} \tag{10.100}$$

Evaluation of the Wronskian

> W(phi[1](v,s),phi[2](v,s)):=simplify(phi[1](v,s)*diff(phi[2](v,s),v)−phi[2](v,s)*
  diff(phi[1](v,s),v));

$$W\left(e^{-\frac{1}{2}sv}, e^{\frac{1}{2}sv}\right) := s \tag{10.101}$$

Evaluation of the Green's function

> G(x,v,s):=(phi[1](v,s)*phi[2](x,s)−phi[1](x,s)*phi[2](v,s))/W(phi[1](v,s),phi[2](v,s));

$$G(x\sim, v, s) := \frac{e^{-\frac{1}{2}sv} e^{\frac{1}{2}sx\sim} - e^{-\frac{1}{2}sx\sim} e^{\frac{1}{2}sv}}{s} \tag{10.102}$$

Particular solution

> Up(x,s):=−Int(G(x,v,s)*(s*f(v)+g(v)+H(v,s))/c^2,v);

$$Up(x\sim, s) := -\left(\int\left(-\frac{245\left(e^{-\frac{1}{2}sv} e^{\frac{1}{2}sx\sim} - e^{-\frac{1}{2}sx\sim} e^{\frac{1}{2}sv}\right)}{s^2}\right) dv\right) \tag{10.103}$$

> Up(x,s):=simplify(subs(v=x,value(%)));

$$Up(x\sim, s) := -\frac{980}{s^3} \tag{10.104}$$

General solution

> U(x,s):=C1(s)*exp(−s/c*x)+C2(s)*exp(s/c*x)+Up(x,s);

$$U(x\sim, s) := C1(s)e^{-\frac{1}{2}sx\sim} + C2(s)e^{\frac{1}{2}sx\sim} - \frac{980}{s^3} \tag{10.105}$$

Substituting the boundary conditions at $x = 0$ and as x approaches infinity, for $\text{Re}\{s\} > 0$, we get

```
> C2(s):=0;
```

$$C2(s) := 0 \tag{10.106}$$

```
> eq:=Q(s)=eval(subs(x=0,U(x,s)));
```

$$eq := 0 = C1(s) - \frac{980}{s^3} \tag{10.107}$$

```
> C1(s):=solve(eq,C1(s));
```

$$C1(s) := \frac{980}{s^3} \tag{10.108}$$

Transformed solution for $U(x, s)$

```
> U(x,s):=eval(U(x,s));
```

$$U(x\sim, s) := \frac{980\, e^{-\frac{1}{2}sx\sim}}{s^3} - \frac{980}{s^3} \tag{10.109}$$

Solution is the inverse Laplace of $U(x, s)$

```
> u(x,t):=invlaplace(U(x,s),s,t);
```

$$u(x\sim, t) := -490\, t^2 + \frac{245}{2}\text{Heaviside}\left(t - \frac{1}{2}x\sim\right)(2t - x\sim)^2 \tag{10.110}$$

ANIMATION

```
> animate(u(x,t),x=0..12,t=0..5,thickness=3);
```

The preceding animation command shows the spatial-time wave amplitude distribution $u(x, t)$ on the string. The animation sequence here and in Figure 10.8 shows snapshots of the animation at times $t = 0, 1, 2, 3, 4, 5$.

ANIMATION SEQUENCE

```
> u(x,0):=subs(t=0,u(x,t)):u(x,1):=subs(t=1,u(x,t)):
> u(x,2):=subs(t=2,u(x,t)):u(x,3):=subs(t=3,u(x,t)):
> u(x,4):=subs(t=4,u(x,t)):u(x,5):=subs(t=5,u(x,t)):
> plot({u(x,0),u(x,1),u(x,2),u(x,3),u(x,4),u(x,5)},x=0..12,thickness=10);
```

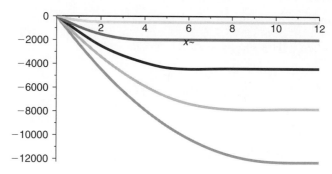

Figure 10.8

EXAMPLE 10.7.3: (Longitudinal vibrations in a bar)   We seek the wave amplitude for longitudinal wave motion in a bar over the semi-infinite interval $I = \{x \mid 0 < x < \infty\}$. The left end of the bar is experiencing an applied stress force $q(t)$ (the Young's modulus $E = 1$). Each cross section of the bar has the initial displacement distribution $f(x) = 0$ and the initial speed distribution $g(x) = 0$. There are no external forces acting upon the bar, and there is no damping. The wave speed is $c = 2$.

SOLUTION: The wave partial differential equation is

$$\frac{\partial^2}{\partial t^2} u(x, t) = c^2 \left(\frac{\partial^2}{\partial x^2} u(x, t) \right)$$

The boundary conditions are

$$u_x(0, t) = -\cos(t) \quad \text{and} \quad u(\infty, t) < \infty$$

The initial conditions are

$$u(x, 0) = 0 \quad \text{and} \quad u_t(x, 0) = 0$$

The Laplace transformed equation reads

$$\frac{\partial^2}{\partial x^2} U(x, s) - \frac{s^2 U(x, s)}{c^2} = -\frac{s f(x)}{c^2} - \frac{g(x)}{c^2}$$

Here, $U(x, s)$ is the Laplace transform of $u(x, t)$ with respect to t. The transformed boundary conditions are

$$\frac{\partial}{\partial x} U(0, s) = -Q(s) \quad \text{and} \quad U(\infty, s) < \infty$$

Assignment of system parameters

> restart:with(plots):with(inttrans):c:=2:q(t):=cos(t):f(x):=0:g(x):=0:

Transformed equation in x reads

> diff(U(x,s),x,x)−s^2/c^2*U(x,s)=−s*f(x)/c^2−g(x)/c^2;

$$\frac{\partial^2}{\partial x^2} U(x, s) - \frac{1}{4} s^2 U(x, s) = 0 \tag{10.111}$$

Transformed boundary conditions

> Q(s):=Int(q(t)*exp(−s*t),t=0..infinity);

$$Q(s) := \int_0^\infty \cos(t)\, e^{-st}\, dt \tag{10.112}$$

> Q(s):=laplace(q(t),t,s);

$$Q(s) := \frac{s}{s^2 + 1} \tag{10.113}$$

General solution for $U(x, s)$

> assume(x>0):U(x,s):=C1(s)*exp(−s/c*x)+C2(s)*exp(s/c*x);

$$U(x\!\sim, s) := C1(s)\, e^{-\frac{1}{2} sx\sim} + C2(s)\, e^{\frac{1}{2} sx\sim} \tag{10.114}$$

Substituting the boundary conditions at $x = 0$ and as x approaches infinity, for $\mathrm{Re}\{s\} > 0$, we get

> C2(s):=0;

$$C2(s) := 0 \tag{10.115}$$

> eq:=Q(s)=−eval(subs(x=0,diff(U(x,s),x)));

$$eq := \frac{s}{s^2 + 1} = \frac{1}{2} C1(s)s \tag{10.116}$$

> C1(s):=solve(eq,C1(s));

$$C1(s) := \frac{2}{s^2 + 1} \tag{10.117}$$

Transformed solution

> U(x,s):=eval(U(x,s));

$$U(x\!\sim, s) := \frac{2\, e^{-\frac{1}{2} sx\sim}}{s^2 + 1} \tag{10.118}$$

Solution is the inverse Laplace of $U(x, s)$

> u(x,t):=simplify(invlaplace(U(x,s),s,t));

$$u(x\sim, t) := -2\sin\left(t - \frac{1}{2}x\sim\right)(-1 + \text{Heaviside}(x\sim - 2t)) \qquad (10.119)$$

ANIMATION

> animate(u(x,t),x=0..10,t=0..10,thickness=3);

The preceding animation command shows the spatial-time wave amplitude distribution $u(x, t)$ in the bar. The animation sequence here and in Figure 10.9 shows snapshots of the animation at times $t = 0, 1, 2, 3, 4, 5$.

ANIMATION SEQUENCE

> u(x,0):=subs(t=0,u(x,t)):u(x,1):=subs(t=1,u(x,t)):
> u(x,2):=subs(t=2,u(x,t)):u(x,3):=subs(t=3,u(x,t)):
> u(x,4):=subs(t=4,u(x,t)):u(x,5):=subs(t=5,u(x,t)):
> plot({u(x,0),u(x,1),u(x,2),u(x,3),u(x,4),u(x,5)}x=0..10,thickness=10);

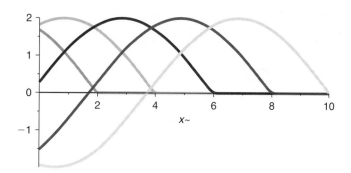

Figure 10.9

EXAMPLE 10.7.4: (Initially displaced string)   We seek the wave amplitude $u(x, t)$ for transverse wave motion along a taut string over the semi-infinite interval $I = \{x \mid 0 < x < \infty\}$. The left end of the string is held fixed. The string has the initial displacement distribution $u(x, 0) = f(x)$ given following, and the initial speed distribution $g(x) = 0$. There are no external forces acting upon the string, and there is no damping in the system. The wave speed is $c = 1$.

SOLUTION: The wave partial differential equation is

$$\frac{\partial^2}{\partial t^2}u(x, t) = c^2\left(\frac{\partial^2}{\partial x^2}u(x, t)\right)$$

The boundary conditions are

$$u(0, t) = 0 \quad \text{and} \quad u(\infty, t) < \infty$$

The initial conditions are

$$u(x, 0) = x e^{-x} \quad \text{and} \quad u_t(x, 0) = 0$$

The Laplace transformed equation reads

$$\frac{\partial^2}{\partial x^2} U(x, s) - \frac{s^2 U(x, s)}{c^2} = -\frac{s f(x)}{c^2} - \frac{g(x)}{c^2}$$

Here, $U(x, s)$ is the Laplace transform of $u(x, t)$ with respect to t. The transformed boundary conditions are

$$U(0, s) = Q(s) \quad \text{and} \quad U(\infty, s) < \infty$$

Assignment of system parameters

> restart:with(plots):with(inttrans):c:=1:q(t):=0:f(x):=x*exp(−x):f(v):=subs(x=v,f(x)):g(x):=0:
  g(v):=subs(x=v,g(x)):

Transformed equation in x reads

> diff(U(x,s),x,x)−s^2/c^2*U(x,s)=−s*f(x)/c^2−g(x)/c^2;

$$\frac{\partial^2}{\partial x^2} U(x, s) - s^2 U(x, s) = -s x e^{-x} \tag{10.120}$$

Transformed boundary conditions

> Q(s):=Int(q(t)*exp(−s*t),t=0..infinity);

$$Q(s) := \int_0^\infty 0 \, dt \tag{10.121}$$

> Q(s):=value(%);

$$Q(s) := 0 \tag{10.122}$$

Basis vectors

> assume(x>0):phi[1](x,s):=exp(−s/c*x):phi[1](v,s):=subs(x=v,phi[1](x,s)):phi[2](x,s)
  :=exp(s/c*x):phi[2](v,s):=subs(x=v,phi[2](x,s)):

$$\phi_1(x\sim, s) := e^{-s x \sim}$$

$$\phi_2(x\sim, s) := e^{s x \sim} \tag{10.123}$$

Evaluation of the Wronskian

> W(phi[1](v,s),phi[2](v,s)):=simplify(phi[1](v,s)*diff(phi[2](v,s),v)−phi[2](v,s)*
diff(phi[1](v,s),v));

$$W(e^{-sv}, e^{sv}) := 2s \qquad (10.124)$$

Evaluation of the Green's function

> G(x,v,s):=(phi[1](v,s)*phi[2](x,s)−phi[1](x,s)*phi[2](v,s))/W(phi[1](v,s),phi[2](v,s));

$$G(x\sim, v, s) := \frac{1}{2} \frac{e^{-sv} e^{sx\sim} - e^{-sx\sim} e^{sv}}{s} \qquad (10.125)$$

Particular solution

> Up(x,s):=−Int(G(x,v,s)*(s*f(v)+g(v))/c^2,v);

$$Up(x\sim, s) := -\left(\int \frac{1}{2} (e^{-sv} e^{sx\sim} - e^{-sx\sim} e^{sv}) v e^{-v} dv \right) \qquad (10.126)$$

> Up(x,s):=simplify(subs(v=x,value(%)));

$$Up(x\sim, s) := \frac{e^{-x\sim} s(-x\sim -2+x\sim s^2)}{(1+s)^2(-1+s)^2} \qquad (10.127)$$

General solution

> U(x,s):=C1(s)*exp(−s/c*x)+C2(s)*exp(s/c*x)+Up(x,s);

$$U(x\sim, s) := C1(s) e^{-sx\sim} + C2(s) e^{sx\sim} + \frac{e^{-x\sim} s(-x\sim -2+x\sim s^2)}{(1+s)^2(-1+s)^2} \qquad (10.128)$$

Substituting the boundary conditions at $x = 0$ and as x approaches infinity, for Re$\{s\} > 0$, we get

> C2(s):=0;

$$C2(s) := 0 \qquad (10.129)$$

> eq:=Q(s)=eval(subs(x=0,U(x,s)));

$$eq := 0 = C1(s) - \frac{2s}{(1+s)^2(-1+s)^2} \qquad (10.130)$$

> C1(s):=solve(eq,C1(s));

$$C1(s) := \frac{2s}{1-2s^2+s^4} \qquad (10.131)$$

Transformed solution for $U(x, s)$

> U(x,s):=eval(U(x,s));

$$U(x\sim, s) := \frac{2se^{-sx\sim}}{1 - 2s^2 + s^4} + \frac{e^{-x\sim}s(-x\sim - 2 + x\sim s^2)}{(1+s)^2(-1+s)^2} \qquad (10.132)$$

Solution is the inverse Laplace of $U(x, s)$

> u(x,t):=invlaplace(U(x,s),s,t);

$$u(x\sim, t) := \text{Heaviside}\ (t - x\sim)(t - x\sim)\sinh(t - x\sim) + e^{-x\sim}\ (x\sim \cosh(t) - t\sinh(t)) \qquad (10.133)$$

ANIMATION

> animate(u(x,t),x=0..12,t=0..5,thickness=3);

The preceding animation command shows the spatial-time amplitude distribution $u(x, t)$ on the string. The animation sequence here and in Figure 10.10 shows snapshots of the animation at times $t = 0, 1, 2, 3, 4, 5$.

ANIMATION SEQUENCE

> u(x,0):=subs(t=0,u(x,t)):u(x,1):=subs(t=1,u(x,t)):
> u(x,2):=subs(t=2,u(x,t)):u(x,3):=subs(t=3,u(x,t)):
> u(x,4):=subs(t=4,u(x,t)):u(x,5):=subs(t=5,u(x,t)):
> plot({u(x,0),u(x,1),u(x,2),u(x,3),u(x,4),u(x,5),},x=0..12,thickness=10);

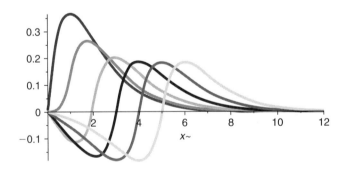

Figure 10.10

EXAMPLE 10.7.5: (String with initial speed)   We seek the wave amplitude $u(x, t)$ for transverse wave motion along a taut string over the semi-infinite interval $I = \{x\,|\,0 < x < \infty\}$. The left end of the string is held fixed. The string has an initial displacement distribution $f(x) = 0$ and an initial speed distribution $u_t(x, 0) = g(x)$ given following. There are no external forces acting upon the string, and there is no damping. The wave speed is $c = 2$.

SOLUTION: The wave partial differential equation is

$$\frac{\partial^2}{\partial t^2}u(x, t) = c^2\left(\frac{\partial^2}{\partial x^2}u(x, t)\right)$$

The boundary conditions are

$$u(0, t) = 0 \quad \text{and} \quad u(\infty, t) < \infty$$

The initial conditions are

$$u(x, 0) = 0 \quad \text{and} \quad u_t(x, 0) = \sin(t)$$

The Laplace transformed equation reads

$$\frac{\partial^2}{\partial x^2}U(x, s) - \frac{s^2 U(x, s)}{c^2} = -\frac{sf(x)}{c^2} - \frac{g(x)}{c^2}$$

Here, $U(x, s)$ is the Laplace transform of $u(x, t)$ with respect to t. The transformed boundary conditions are

$$U(0, s) = Q(s) \quad \text{and} \quad U(\infty, s) < \infty$$

Assignment of system parameters

> restart:with(plots):with(inttrans):c:=2:q(t):=0:f(x):=0:f(v):=subs(x=v,f(x)):g(x):=sin(x):
  g(v):=subs(x=v,g(x)):

Transformed equation in x reads

> diff(U(x,s),x,x)−s^2/c^2*U(x,s)=−s*f(x)/c^2−g(x)/c^2;

$$\frac{\partial^2}{\partial x^2}U(x, s) - \frac{1}{4}s^2 U(x, s) = -\frac{1}{4}\sin(x) \tag{10.134}$$

Transformed boundary conditions

> Q(s):=Int(q(t)*exp(−s*t),t=0..infinity);

$$Q(s) := \int_0^\infty 0\,dt \tag{10.135}$$

> Q(s):=value(%);

$$Q(s) := 0 \tag{10.136}$$

Basis vectors

> assume(x>0):phi[1](x,s):=exp(−s/c*x);phi[1](v,s):=subs(x=v,phi[1](x,s)):
  phi[2](x,s):=exp(s/c*x);phi[2](v,s):=subs(x=v,phi[2](x,s)):

$$\phi_1(x{\sim}, s) := e^{-\frac{1}{2}sx{\sim}}$$
$$\phi_2(x{\sim}, s) := e^{\frac{1}{2}sx{\sim}} \tag{10.137}$$

Evaluation of the Wronskian

> W(phi[1](v,s),phi[2](v,s)):=simplify(phi[1](v,s)*diff(phi[2](v,s),v)
  −phi[2](v,s)*diff(phi[1](v,s),v));

$$W(e^{-\frac{1}{2}sv}, e^{\frac{1}{2}sv}) := s \tag{10.138}$$

Evaluation of the Green's function

> G(x,v,s):=(phi[1](v,s)*phi[2](x,s)−phi[1](x,s)*phi[2](v,s))/W(phi[1](v,s),phi[2](v,s));

$$G(x{\sim}, v, s) := \frac{e^{-\frac{1}{2}sv} e^{\frac{1}{2}sx{\sim}} - e^{-\frac{1}{2}sx{\sim}} e^{\frac{1}{2}sv}}{s} \tag{10.139}$$

Particular solution

> Up(x,s):=−Int(G(x,v,s)*(s*f(v)+g(v))/c^2,v);

$$Up(x{\sim}, s) := -\left(\int \frac{1}{4} \frac{(e^{-\frac{1}{2}sv} e^{\frac{1}{2}sx{\sim}} - e^{-\frac{1}{2}sx{\sim}} e^{\frac{1}{2}sv}) \sin(v)}{s} \, dv \right) \tag{10.140}$$

> Up(x,s):=simplify(subs(v=x,value(%)));

$$Up(x{\sim}, s) := \frac{\sin(x{\sim})}{s^2 + 4} \tag{10.141}$$

General solution

> U(x,s):=C1(s)*exp(−s/c*x)+C2(s)*exp(s/c*x)+Up(x,s);

$$U(x{\sim}, s) := C1(s) e^{-\frac{1}{2}sx{\sim}} + C2(s) e^{\frac{1}{2}sx{\sim}} + \frac{\sin(x{\sim})}{s^2 + 4} \tag{10.142}$$

Substituting the boundary conditions at $x = 0$ and as x approaches infinity, for $\text{Re}\{s\} > 0$, we get

> C2(s):=0;

$$C2(s) := 0 \tag{10.143}$$

> eq:=Q(s)=eval(subs(x=0,U(x,s)));

$$eq := 0 = C1(s) \tag{10.144}$$

> C1(s):=solve(eq,C1(s));

$$C1(s) := 0 \tag{10.145}$$

Transformed solution for $U(x, s)$

> U(x,s):=eval(U(x,s));

$$U(x\sim, s) := \frac{\sin(x\sim)}{s^2 + 4} \tag{10.146}$$

Solution is the inverse Laplace of $U(x, s)$

> u(x,t):=simplify(invlaplace(U(x,s),s,t));

$$u(x\sim, t) := \frac{1}{2} \sin(x\sim) \sin(2t) \tag{10.147}$$

ANIMATION

> animate(u(x,t),x=0..12,t=0..5,thickness=3);

The preceding animation command shows the spatial-time wave amplitude distribution $u(x, t)$ on the string. The animation sequence here and in Figure 10.11 shows snapshots of the animation at times $t = 0, 1, 2, 3, 4, 5$.

ANIMATION SEQUENCE

> u(x,0):=subs(t=0,u(x,t)):u(x,1):=subs(t=1,u(x,t)):
> u(x,2):=subs(t=2,u(x,t)):u(x,3):=subs(t=3,u(x,t)):
> u(x,4):=subs(t=4,u(x,t)):u(x,5):=subs(t=5,u(x,t)):
> plot({u(x,0),u(x,1),u(x,2),u(x,3),u(x,4),u(x,5)},x=0..12,thickness=10);

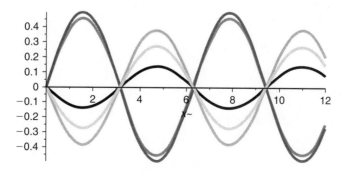

Figure 10.11

EXAMPLE 10.7.6: (Longitudinal vibrations in a bar with impulse force)   We consider longitudinal wave motion in a bar over the semi-infinite interval $I = \{x \mid 0 < x < \infty\}$. The left end of the bar is held fixed. Each cross section of the bar has the initial displacement distribution $f(x) = 0$ and the initial speed distribution $g(x) = 0$. There is an applied impulse force $h(x, t)$, given following, acting on the bar at time $t = 1$. There is no damping in the system, and the wave speed is $c = 1$.

SOLUTION: The nonhomogeneous wave partial differential equation is

$$\frac{\partial^2}{\partial t^2} u(x, t) = c^2 \left(\frac{\partial^2}{\partial x^2} u(x, t) \right) + h(x, t)$$

The boundary conditions are

$$u(0, t) = 0 \quad \text{and} \quad u(\infty, t) < \infty$$

The initial conditions are

$$u(x, 0) = 0 \quad \text{and} \quad u_t(x, 0) = 0$$

The Laplace transformed equation reads

$$\frac{\partial^2}{\partial x^2} U(x, s) - \frac{s^2 U(x, s)}{c^2} = -\frac{sf(x)}{c^2} - \frac{g(x)}{c^2} - \frac{H(x, s)}{c^2}$$

Here, $U(x, s)$ is the Laplace transform of $u(x, t)$ with respect to t. The transformed boundary conditions are

$$U(0, s) = Q(s) \quad \text{and} \quad U(\infty, s) < \infty$$

Assignment of system parameters

> restart:with(plots):with(inttrans):c:=1:q(t):=0:f(x):=0:f(v):=subs(x=v,f(x)):g(x):=0: g(v):=subs(x=v,g(x)):h(x,t):=Dirac(t−1):

Transformed equation in x reads

> diff(U(x,s),x,x)−s^2/c^2*U(x,s)=−s*f(x)/c^2−g(x)/c^2−H(x,s)/c^2;

$$\frac{\partial^2}{\partial x^2} U(x, s) - s^2 U(x, s) = -H(x, s) \tag{10.148}$$

Transformed boundary conditions

> Q(s):=Int(q(t)*exp(−s*t),t=0..infinity);

$$Q(s) := \int_0^\infty 0 \, dt \tag{10.149}$$

> Q(s):=value(%);

$$Q(s) := 0 \tag{10.150}$$

Transformed source term

> H(x,s):=Int(h(x,t)*exp(−s*t),t=0..infinity);

$$H(x, s) := \int_0^\infty \mathrm{Dirac}(t-1)\,\mathrm{e}^{-st}\mathrm{d}t \tag{10.151}$$

> H(x,s):=laplace(h(x,t),t,s);H(v,s):=subs(x=v,H(x,s)):

$$H(x, s) := \mathrm{e}^{-s} \tag{10.152}$$

Basis vectors

> assume(x>0):phi[1](x,s):=exp(−s/c*x);phi[1](v,s):=subs(x=v,phi[1](x,s)):
  phi[2](x,s):=exp(s/c*x);phi[2](v,s):=subs(x=v,phi[2](x,s)):

$$\phi_1(x{\sim}, s) := \mathrm{e}^{-sx{\sim}}$$
$$\phi_2(x{\sim}, s) := \mathrm{e}^{sx{\sim}} \tag{10.153}$$

Evaluation of the Wronskian

> W(phi[1](v,s),phi[2](v,s)):=simplify(phi[1](v,s)*diff(phi[2](v,s),v)−phi[2](v,s)*
  diff(phi[1](v,s),v));

$$W(\mathrm{e}^{-sv}, \mathrm{e}^{sv}) := 2s \tag{10.154}$$

Evaluation of the Green's function

> G(x,v,s):=(phi[1](v,s)*phi[2](x,s)-phi[1](x,s)*phi[2](v,s))/W(phi[1](v,s),phi[2](v,s));

$$G(x{\sim}, v, s) := \frac{1}{2}\frac{\mathrm{e}^{-sv}\,\mathrm{e}^{sx{\sim}} - \mathrm{e}^{-sx{\sim}}\,\mathrm{e}^{sv}}{s} \tag{10.155}$$

Particular solution

> Up(x,s):=−Int(G(x,v,s)*(s*f(v)+g(v)+H(v,s))/c^2,v);

$$Up(x{\sim}, s) := -\left(\int \frac{1}{2}\frac{(\mathrm{e}^{-sv}\,\mathrm{e}^{sx{\sim}} - \mathrm{e}^{-sx{\sim}}\,\mathrm{e}^{sv})\,\mathrm{e}^{-s}}{s}\mathrm{d}v\right) \tag{10.156}$$

> Up(x,s):=simplify(subs(v=x,value(%)));

$$Up(x{\sim}, s) := \frac{\mathrm{e}^{-s}}{s^2} \tag{10.157}$$

General solution

> U(x,s):=C1(s)*exp(−s/c*x)+C2(s)*exp(s/c*x)+Up(x,s);

$$U(x\sim, s) := C1(s)\, e^{-sx\sim} + C2(s)\, e^{sx\sim} + \frac{e^{-s}}{s^2} \tag{10.158}$$

Substituting the boundary conditions at $x = 0$ and as x approaches infinity, for Re$\{s\} > 0$, we get

> C2(s):=0;

$$C2(s) := 0 \tag{10.159}$$

> eq:=Q(s)=eval(subs(x=0,U(x,s)));

$$eq := 0 = C1(s) + \frac{e^{-s}}{s^2} \tag{10.160}$$

> C1(s):=solve(eq,C1(s));

$$C1(s) := -\frac{e^{-s}}{s^2} \tag{10.161}$$

Transformed solution for $U(x, s)$

> U(x,s):=eval(U(x,s));

$$U(x\sim, s) := -\frac{e^{-s}\, e^{-sx\sim}}{s^2} + \frac{e^{-s}}{s^2} \tag{10.162}$$

Solution is the inverse Laplace of $U(x, s)$

> u(x,t):=invlaplace(U(x,s),s,t);

$$u(x\sim, t) := -\text{Heaviside}(t - 1 - x\sim)(t - 1 - x\sim) + \text{Heaviside}(t - 1)(t - 1) \tag{10.163}$$

ANIMATION

> animate(u(x,t),x=0..12,t=0..5,thickness=3);

The preceding animation command shows the spatial-time wave amplitude distribution $u(x, t)$ in the bar. The animation sequence here and in Figure 10.12 shows snapshots of the animation at times $t = 0, 1, 2, 3, 4, 5$.

ANIMATION SEQUENCE

> u(x,0):=subs(t=0,u(x,t)):u(x,1):=subs(t=1,u(x,t)):
> u(x,2):=subs(t=2,u(x,t)):u(x,3):=subs(t=3,u(x,t)):

```
> u(x,4):=subs(t=4,u(x,t)):u(x,5):=subs(t=5,u(x,t)):
> plot({u(x,0),u(x,1),u(x,2),u(x,3),u(x,4),u(x,5)},x=0..10,thickness=10);
```

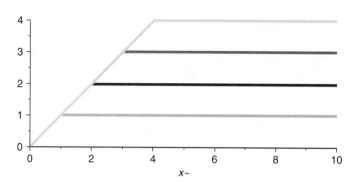

Figure 10.12

Note that for times $t < 1$, the longitudinal displacements of the bar are zero because the impulse does not occur until time $t = 1$.

EXAMPLE 10.7.7: (String with external applied force)   We seek the wave amplitude $u(x, t)$ for transverse wave motion along a taut string over the semi-infinite interval $I = \{x \mid 0 < x < \infty\}$. The left end of the string is held fixed. The string has the initial displacement distribution $f(x) = 0$ and the initial speed distribution $g(x) = 0$. An external applied force $h(x, t)$, given following, acts on the system. There is no damping in the system, and the wave speed is $c = 1$.

SOLUTION: The nonhomogeneous wave partial differential equation is

$$\frac{\partial^2}{\partial t^2} u(x, t) = c^2 \left(\frac{\partial^2}{\partial x^2} u(x, t) \right) + h(x, t)$$

The boundary conditions are

$$u(0, t) = 0 \quad \text{and} \quad u(\infty, t) < \infty$$

The initial conditions are

$$u(x, 0) = 0 \quad \text{and} \quad u_t(x, 0) = 0$$

The Laplace transformed equation reads

$$\frac{\partial^2}{\partial x^2} U(x, s) - \frac{s^2 U(x, s)}{c^2} = -\frac{s f(x)}{c^2} - \frac{g(x)}{c^2} - \frac{H(x, s)}{c^2}$$

Here, $U(x, s)$ is the Laplace transform of $u(x, t)$ with respect to t. The transformed boundary conditions are

$$U(0, s) = Q(s) \quad \text{and} \quad U(\infty, s) < \infty$$

Assignment of system parameters

> restart:with(plots):with(inttrans):c:=1:q(t):=0:f(x):=0:f(v):=subs(x=v,f(x)):g(x):=0:
  g(v):=subs(x=v,g(x)):h(x,t):=exp(−x)*t*exp(−t):

Transformed equation in x reads

> diff(U(x,s),x,x)−s^2/c^2*U(x,s)=−s*f(x)/c^2−g(x)/c^2−H(x,s)/c^2;

$$\frac{\partial^2}{\partial x^2} U(x, s) - s^2 U(x, s) = -H(x, s) \tag{10.164}$$

Transformed boundary conditions

> Q(s):=Int(q(t)*exp(−s*t),t=0..infinity);

$$Q(s) := \int_0^\infty 0 \, dt \tag{10.165}$$

> Q(s):=value(%);

$$Q(s) := 0 \tag{10.166}$$

Transformed source term

> H(x,s):=Int(h(x,t)*exp(−s*t),t=0..infinity);

$$H(x, s) := \int_0^\infty e^{-x} t \, e^{-t} e^{-st} \, dt \tag{10.167}$$

> H(x,s):=laplace(h(x,t),t,s);H(v,s):=subs(x=v,H(x,s)):

$$H(x, s) := \frac{e^{-x}}{(1 + s)^2} \tag{10.168}$$

Basis vectors

> assume(x>0):phi[1](x,s):=exp(−s/c*x);phi[1](v,s):=subs(x=v,phi[1](x,s)):
  phi[2](x,s):=exp(s/c*x);phi[2](v,s):=subs(x=v,phi[2](x,s)):

$$\phi_1(x{\sim}, s) := e^{-sx{\sim}}$$

$$\phi_2(x{\sim}, s) := e^{sx{\sim}} \tag{10.169}$$

Evaluation of the Wronskian

> W(phi[1](v,s),phi[2](v,s)):=simplify(phi[1](v,s)*diff(phi[2](v,s),v)
  −phi[2](v,s)*diff(phi[1](v,s),v));

$$W(e^{-sv}, e^{sv}) := 2s \qquad (10.170)$$

Evaluation of the Green's function

> G(x,v,s):=(phi[1](v,s)*phi[2](x,s)−phi[1](x,s)*phi[2](v,s))/W(phi[1](v,s),phi[2](v,s));

$$G(x\sim, v, s) := \frac{1}{2} \frac{e^{-sv} e^{sx\sim} - e^{-sx\sim} e^{sv}}{s} \qquad (10.171)$$

Particular solution

> Up(x,s):=−Int(G(x,v,s)*(s*f(v)+g(v)+H(v,s))/c^2,v);

$$Up(x\sim, s) := - \left(\int \frac{1}{2} \frac{(e^{-sv} e^{sx\sim} - e^{-sx\sim} e^{sv}) e^{-v}}{s(1+s)^2} dv \right) \qquad (10.172)$$

> Up(x,s):=simplify(subs(v=x,value(%)));

$$Up(x\sim, s) := \frac{e^{-x\sim}}{(1+s)^3(-1+s)} \qquad (10.173)$$

General solution

> U(x,s):=C1(s)*exp(−s/c*x)+C2(s)*exp(s/c*x)+Up(x,s);

$$U(x\sim, s) := C1(s) e^{-sx\sim} + C2(s) e^{sx\sim} + \frac{e^{-x\sim}}{(1+s)^3(-1+s)} \qquad (10.174)$$

Substituting the boundary conditions at $x = 0$ and as x approaches infinity, for Re$\{s\} > 0$, we get

> C2(s):=0;

$$C2(s) := 0 \qquad (10.175)$$

> eq:=Q(s)=eval(subs(x=0,U(x,s)));

$$eq := 0 = C1(s) + \frac{1}{(1+s)^3(-1+s)} \qquad (10.176)$$

> C1(s):=solve(eq,C1(s));

$$C1(s) := -\frac{1}{-1 - 2s + 2s^3 + s^4} \qquad (10.177)$$

Transformed solution for $U(x, s)$

> U(x,s):=eval(U(x,s));

$$U(x\sim, s) := -\frac{e^{-sx\sim}}{-1 - 2s + 2s^3 + s^4} + \frac{e^{-x\sim}}{(1+s)^3(-1+s)} \qquad (10.178)$$

Solution is the inverse Laplace of $U(x, s)$

> u(x,t):=invlaplace(U(x,s),s,t);

$$u(x\sim, t) := \frac{1}{8} \text{Heaviside}(-t + x\sim) e^{t-x\sim} + \frac{1}{8} e^{-t+x\sim} \text{Heaviside}(t - x\sim)(2t^2 - 4tx\sim$$

$$+ 2x\sim^2 + 2t - 2x\sim + 1) - \frac{1}{8} e^{-x\sim-t}(2t^2 + 2t + 1) \qquad (10.179)$$

ANIMATION

> animate(u(x,t),x=0..12,t=0..5,thickness=3);

The preceding animation command shows the spatial-time wave amplitude distribution $u(x, t)$ on the string. The animation sequence here and in Figure 10.13 shows snapshots of the animation at times $t = 0, 1, 2, 3, 4, 5$.

ANIMATION SEQUENCE

> u(x,0):=subs(t=0,u(x,t)):u(x,1):=subs(t=1,u(x,t)):
> u(x,2):=subs(t=2,u(x,t)):u(x,3):=subs(t=3,u(x,t)):
> u(x,4):=subs(t=4,u(x,t)):u(x,5):=subs(t=5,u(x,t)):
> plot({u(x,0),u(x,1),u(x,2),u(x,3),u(x,4),u(x,5)},x=0..10,thickness=10);

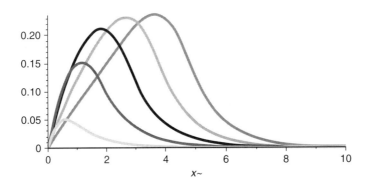

Figure 10.13

Chapter Summary

Laplace transform of a piecewise continuous exponential-order function $f(t)$

$$F(s) = \int_0^\infty f(t) e^{-st} dt$$

Diffusion partial differential equation

$$\frac{\partial}{\partial t} u(x, t) = k \left(\frac{\partial^2}{\partial x^2} u(x, t) \right) + h(x, t)$$

Initial condition

$$u(x, 0) = f(x)$$

Laplace-transformed diffusion equation

$$\frac{\partial^2}{\partial x^2} U(x, s) - \frac{sU(x, s)}{k} = -\frac{f(x)}{k} - \frac{H(x, s)}{k}$$

Wave partial differential equation

$$\frac{\partial^2}{\partial t^2} u(x, t) = c^2 \left(\frac{\partial^2}{\partial x^2} u(x, t) \right) + h(x, t)$$

Initial conditions

$$u(x, 0) = f(x) \quad \text{and} \quad u_t(x, 0) = g(x)$$

Laplace-transformed wave equation

$$\frac{\partial^2}{\partial x^2} U(x, s) - \frac{s^2 U(x, s)}{c^2} = -\frac{sf(x)}{c^2} - \frac{g(x)}{c^2} - \frac{H(x, s)}{c^2}$$

Transformed solution

$$U(x, s) = \int_0^\infty u(x, t) e^{-st} dt$$

Transformed source function

$$H(x, s) = \int_0^\infty h(x, t) e^{-st} dt$$

Solution from inverse transform

$$u(x, t) = \int\limits_{a-i\infty}^{a+i\infty} \frac{U(x, s)\, e^{st}}{2\pi i}\, ds$$

Partitioned transform of solution

$$U(x, s) = U1(x, s) U2(x, s)$$

Convolution integral solution

$$u(x, t) = \int\limits_{0}^{t} u1(x, \tau) u2(x, t - \tau)\, d\tau$$

where $u1(x, t)$ is the inverse transform of $U1(x, s)$ and $u2(x, t)$ is the inverse transform of $U2(x, s)$.

In this chapter, we used the Laplace transform procedure for evaluating solutions to the diffusion and the wave partial differential equations. The spatial domains of interest here were semi-infinite or infinite. We did not consider Laplace transform methods on finite spatial domains because these types of problems generally give rise to having to use the more complicated methods of complex variables and residue theory to evaluate the inverse tranforms. For finite domains, the eigenfunction methods covered in Chapters 3, 4, 6, 7, and 8 are generally much easier to implement.

Exercises

We now consider diffusion and wave equations in a single dimension over semi-infinite intervals with nonhomogeneous boundary conditions. The time variable of interest has the domain $\{t \mid 0 < t < \infty\}$. Because of the possibility of a nonhomogeneous boundary condition, the methods of Chapter 9 may fail here; thus, we use the Laplace transform method to solve these types of problems. The Laplace transforms are taken with respect to the time variable t. We do not consider problems over finite intervals here because the eigenfunction expansion methods of earlier chapters are more convenient to use.

Laplace Transform Methods on the Diffusion Equation

We consider the temperature distribution along a thin rod whose lateral surface is insulated over the semi-infinite interval $I = \{x \mid 0 < x < \infty\}$ for times $\{t \mid 0 < t < \infty\}$. We allow for an external heat source $h(x, t)$ to act on the system. The nonhomogeneous diffusion partial differential equation in rectangular coordinates in a single dimension reads

$$\frac{\partial}{\partial t} u(x, t) = k \left(\frac{\partial^2}{\partial x^2} u(x, t) \right) + h(x, t)$$

with the initial condition $u(x, 0) = f(x)$.

For Exercises 10.1 through 10.20, we seek solutions to the preceding equation for various given initial and boundary conditions. Evaluate the temperature distribution $u(x, t)$ using the Laplace transform method with respect to the variable t. Evaluate the inverse if possible; otherwise, use the convolution method and leave the answer in the form of a convolution integral. When possible, generate the animated solution for $u(x, t)$, and plot the animated sequence for $0 < t < 5$.

10.1. Consider a thin rod whose lateral surface is insulated over the semi-infinite interval $I = \{x \,|\, 0 < x < \infty\}$. The initial temperature distribution in the rod is given as $f(x)$. There is no source term in the system, and the diffusivity has the magnitude $k = 1/4$.

Boundary conditions:

$$u_x(0, t) = 0 \quad \text{and} \quad u(\infty, t) < \infty$$

Initial condition:

$$f(x) = e^{-\frac{x}{2}}$$

10.2. Consider a thin rod whose lateral surface is insulated over the semi-infinite interval $I = \{x \,|\, 0 < x < \infty\}$. The initial temperature distribution in the rod is given as $f(x)$. There is no source term in the system, and the diffusivity has the magnitude $k = 1/4$.

Boundary conditions:

$$u(0, t) = 0 \quad \text{and} \quad u(\infty, t) < \infty$$

Initial condition:

$$f(x) = x e^{-\frac{x}{2}}$$

10.3. Consider a thin rod whose lateral surface is insulated over the semi-infinite interval $I = \{x \,|\, 0 < x < \infty\}$. The initial temperature distribution in the rod is given as $f(x)$. There is no source term in the system, and the diffusivity has the magnitude $k = 1/4$.

Boundary conditions:

$$u(0, t) = 0 \quad \text{and} \quad u(\infty, t) < \infty$$

Initial condition:

$$f(x) = \sin(x)$$

10.4.   Consider a thin cylinder over the semi-infinite interval $I = \{x \mid 0 < x < \infty\}$ containing a fluid that has an initial salt concentration density given as $f(x)$. There is no source term in the system, and the diffusivity has the magnitude $k = 1/4$.

Boundary conditions:

$$u(0, t) = 0 \quad \text{and} \quad u(\infty, t) < \infty$$

Initial condition:

$$f(x) = \text{Dirac}(x - 1)$$

10.5.   Consider a thin rod whose lateral surface is insulated over the semi-infinite interval $I = \{x \mid 0 < x < \infty\}$. The initial temperature distribution in the rod is given as $f(x)$. There is no source term in the system, and the diffusivity has the magnitude $k = 1/4$.

Boundary conditions:

$$u(0, t) = 0 \quad \text{and} \quad u(\infty, t) < \infty$$

Initial condition:

$$f(x) = \sin(x) \, e^{-x}$$

10.6.   Consider a thin rod whose lateral surface is insulated over the semi-infinite interval $I = \{x \mid 0 < x < \infty\}$. The left end of the rod is constrained to a temperature $u(0, t) = q(t)$ given following. The initial temperature distribution in the rod is 0, there is no source term in the system, and the diffusivity has the magnitude $k = 1/4$.

Boundary conditions:

$$u(0, t) = 5 \quad \text{and} \quad u(\infty, t) < \infty$$

Initial condition:

$$f(x) = 0$$

10.7.   Consider a thin rod whose lateral surface is insulated over the semi-infinite interval $I = \{x \mid 0 < x < \infty\}$. The left end of the rod is constrained to the temperature $u(0, t) = q(t)$ given following. The initial temperature distribution in the rod is 0, there is no source term in the system, and the diffusivity has the magnitude $k = 1/4$.

Boundary conditions:

$$u(0, t) = \sin(t) \quad \text{and} \quad u(\infty, t) < \infty$$

Initial condition:

$$f(x) = 0$$

10.8. Consider a thin rod whose lateral surface is insulated over the semi-infinite interval $I = \{x \mid 0 < x < \infty\}$. The left end of the rod is constrained to the temperature $u(0, t) = q(t)$ given following. The initial temperature distribution in the rod is 0, there is no source term in the system, and the diffusivity has the magnitude $k = 1/4$.

Boundary conditions:

$$u(0, t) = e^{-t} \quad \text{and} \quad u(\infty, t) < \infty$$

Initial condition:

$$f(x) = 0$$

10.9. Consider a thin rod whose lateral surface is insulated over the semi-infinite interval $I = \{x \mid 0 < x < \infty\}$. The left end of the rod is constrained to a temperature $u(0, t) = q(t)$ given following. The initial temperature distribution in the rod is 0, there is no source term in the system, and the diffusivity has the magnitude $k = 1/4$.

Boundary conditions:

$$u(0, t) = t\,e^{-t} \quad \text{and} \quad u(\infty, t) < \infty$$

Initial condition:

$$f(x) = 0$$

10.10. Consider a thin rod whose lateral surface is insulated over the semi-infinite interval $I = \{x \mid 0 < x < \infty\}$. The left end of the rod is constrained to the temperature $u(0, t) = q(t)$ given following. The initial temperature distribution in the rod is 0, there is no source term in the system, and the diffusivity has the magnitude $k = 1/4$.

Boundary conditions:

$$u(0, t) = \sin(t)\,e^{-t} \quad \text{and} \quad u(\infty, t) < \infty$$

Initial condition:

$$f(x) = 0$$

10.11. Consider a thin rod whose lateral surface is insulated over the semi-infinite interval $I = \{x \mid 0 < x < \infty\}$. The left end of the rod has the input heat flux $u_x(0, t) = q(t)$ given following. The initial temperature distribution in the rod is 0, there is no source term in the system, and the diffusivity has the magnitude $k = 1/4$.

Boundary conditions:

$$u_x(0, t) = -5 \quad \text{and} \quad u(\infty, t) < \infty$$

Initial condition:

$$f(x) = 0$$

10.12.   Consider a thin rod whose lateral surface is insulated over the semi-infinite interval $I = \{x \mid 0 < x < \infty\}$. The left end of the rod has the input heat flux $u_x(0, t) = q(t)$ given following. The initial temperature distribution in the rod is 0, there is no source term in the system, and the diffusivity has the magnitude $k = 1/4$.

Boundary conditions:

$$u_x(0, t) = -e^{-t} \quad \text{and} \quad u(\infty, t) < \infty$$

Initial condition:

$$f(x) = 0$$

10.13.   Consider a thin rod whose lateral surface is insulated over the semi-infinite interval $I = \{x \mid 0 < x < \infty\}$. The left end of the rod has the input heat flux $u_x(0, t) = q(t)$ given following. The initial temperature distribution in the rod is 0, there is no source term in the system, and the diffusivity has the magnitude $k = 1/4$.

Boundary conditions:

$$u_x(0, t) = -t\,e^{-t} \quad \text{and} \quad u(\infty, t) < \infty$$

Initial condition:

$$f(x) = 0$$

10.14.   Consider a thin rod whose lateral surface is insulated over the semi-infinite interval $I = \{x \mid 0 < x < \infty\}$. The left end of the rod has the input heat flux $u_x(0, t) = q(t)$ given following. The initial temperature distribution in the rod is 0, there is no source term in the system, and the diffusivity has the magnitude $k = 1/4$.

Boundary conditions:

$$u_x(0, t) = -\sin(t) \quad \text{and} \quad u(\infty, t) < \infty$$

Initial condition:

$$f(x) = 0$$

10.15.   Consider a thin rod whose lateral surface is insulated over the semi-infinite interval $I = \{x \mid 0 < x < \infty\}$. The left end of the rod has an input heat flux $u_x(0, t) = q(t)$ given following. The initial temperature distribution in the rod is 0, there is no source term in the system, and the diffusivity has the magnitude $k = 1/4$.

Boundary conditions:

$$u_x(0, t) = -\sin(t)\,e^{-t} \quad \text{and} \quad u(\infty, t) < \infty$$

Initial condition:

$$f(x) = 0$$

10.16. Consider a thin rod whose lateral surface is insulated over the semi-infinite interval $I = \{x \,|\, 0 < x < \infty\}$. The left end of the rod is held at the fixed temperature 0, and the initial temperature distribution in the rod is 0. There is a heat source term $h(x, t)$ (dimension of degrees per second) in the system and the diffusivity has the magnitude $k = 1/4$.

Boundary conditions:

$$u(0, t) = 0 \quad \text{and} \quad u(\infty, t) < \infty$$

Initial condition:

$$f(x) = 0$$

Heat source:

$$h(x, t) = e^{-x}$$

10.17. Consider a thin rod whose lateral surface is insulated over the semi-infinite interval $I = \{x \,|\, 0 < x < \infty\}$. The left end of the rod is held at the fixed temperature 0, and the initial temperature distribution in the rod is 0. There is a heat source term $h(x, t)$ in the system, and the diffusivity has the magnitude $k = 1/4$.

Boundary conditions:

$$u(0, t) = 0 \quad \text{and} \quad u(\infty, t) < \infty$$

Initial condition:

$$f(x) = 0$$

Heat source:

$$h(x, t) = e^{-x} e^{-t}$$

10.18. Consider a thin rod whose lateral surface is insulated over the semi-infinite interval $I = \{x \,|\, 0 < x < \infty\}$. The left end of the rod is held at the fixed temperature 0, and the initial temperature distribution in the rod is 0. There is a heat source term $h(x, t)$ in the system and the diffusivity has the magnitude $k = 1/4$.

Boundary conditions:

$$u(0, t) = 0 \quad \text{and} \quad u(\infty, t) < \infty$$

Initial condition:

$$f(x) = 0$$

Heat source:

$$h(x, t) = \text{Dirac}(x - 1) e^{-t}$$

10.19. Consider a thin rod whose lateral surface is insulated over the semi-infinite interval $I = \{x \mid 0 < x < \infty\}$. The left end of the rod is held at the fixed temperature 0, and the initial temperature distribution in the rod is 0. There is a heat source term $h(x, t)$ in the system, and the diffusivity has the magnitude $k = 1/4$.

Boundary conditions:

$$u(0, t) = 0 \quad \text{and} \quad u(\infty, t) < \infty$$

Initial condition:

$$f(x) = 0$$

Heat source:

$$h(x, t) = e^{-x} \sin(t)$$

10.20. Consider a thin rod whose lateral surface is insulated over the semi-infinite interval $I = \{x \mid 0 < x < \infty\}$. The left end of the rod is held at the fixed temperature 0, and the initial temperature distribution in the rod is 0. There is a heat source term $h(x, t)$ in the system, and the diffusivity has the magnitude $k = 1/4$.

Boundary conditions:

$$u(0, t) = 0 \quad \text{and} \quad u(\infty, t) < \infty$$

Initial condition:

$$f(x) = 0$$

Heat source:

$$h(x, t) = \sin(x) \sin(t)$$

Laplace Transform Methods on the Wave Equation

We now consider transverse wave propagation in taut strings and longitudinal wave propagation in rigid bars over the one-dimensional semi-infinite interval $I = \{x \mid 0 < x < \infty\}$. We allow for an external force to act and we assume no damping in the system. The nonhomogeneous partial differential equation in rectangular coordinates reads

$$\frac{\partial^2}{\partial t^2} u(x, t) = c^2 \left(\frac{\partial^2}{\partial x^2} u(x, t) \right) + h(x, t)$$

with the two initial conditions

$$u(x, 0) = f(x) \quad \text{and} \quad u_t(x, 0) = g(x)$$

For Exercises 10.21 through 10.46, we seek solutions to the preceding equation for various initial and boundary conditions. Evaluate the wave distribution $u(x, t)$ using the Laplace transform method with respect to the variable t. Evaluate the inverse if possible; otherwise, use the convolution method and leave the answer in the form of a convolution integral. When possible, generate the animated solution for $u(x, t)$, and plot the animated sequence for $0 < t < 5$.

10.21. Consider transverse motion over a taut string over the semi-infinite interval $I = \{x \mid 0 < x < \infty\}$. The left end of the string is held secure. The initial displacement distribution $f(x)$ and initial speed distribution $g(x)$ are given following. No external force is acting and $c = 1/2$.

Boundary conditions:

$$u(0, t) = 0 \quad \text{and} \quad u(\infty, t) < \infty$$

Initial conditions:

$$f(x) = e^{-\frac{x}{2}} \quad \text{and} \quad g(x) = 0$$

10.22. Consider transverse motion over a taut string over the semi-infinite interval $I = \{x \mid 0 < x < \infty\}$. The left end of the string is held secure. The initial displacement distribution $f(x)$ and initial speed distribution $g(x)$ are given following. No external force is acting and $c = 1/2$.

Boundary conditions:

$$u(0, t) = 0 \quad \text{and} \quad u(\infty, t) < \infty$$

Initial conditions:

$$f(x) = x e^{-\frac{x}{2}} \quad \text{and} \quad g(x) = 0$$

10.23. Consider transverse motion over a taut string over the semi-infinite interval $I = \{x \mid 0 < x < \infty\}$. The left end of the string is held secure. The initial displacement distribution $f(x)$ and initial speed distribution $g(x)$ are given following. No external force is acting and $c = 1/2$.

Boundary conditions:

$$u(0, t) = 0 \quad \text{and} \quad u(\infty, t) < \infty$$

Initial conditions:

$$f(x) = 0 \quad \text{and} \quad g(x) = e^{-x}$$

10.24.   Consider transverse motion over a taut string over the semi-infinite interval
$I = \{x \mid 0 < x < \infty\}$. The left end of the string is held secure. The initial displacement
distribution $f(x)$ and initial speed distribution $g(x)$ are given following. No external
force is acting and $c = 1/2$.

Boundary conditions:

$$u(0, t) = 0 \quad \text{and} \quad u(\infty, t) < \infty$$

Initial conditions:

$$f(x) = \sin(x)\, e^{-x} \quad \text{and} \quad g(x) = 0$$

10.25.   Consider transverse motion over a taut string over the semi-infinite interval
$I = \{x \mid 0 < x < \infty\}$. The left end of the string is held secure. The initial displacement
distribution $f(x)$ and initial speed distribution $g(x)$ are given following. No external
force is acting and $c = 1/2$.

Boundary conditions:

$$u(0, t) = 0 \quad \text{and} \quad u(\infty, t) < \infty$$

Initial conditions:

$$f(x) = 0 \quad \text{and} \quad g(x) = \text{Dirac}(x - 1)$$

10.26.   Consider transverse motion over a taut string over the semi-infinite interval
$I = \{x \mid 0 < x < \infty\}$. The left end of the string is constrained to move in accordance
with $u(0, t) = q(t)$ given following. The initial displacement distribution $f(x)$ and
initial speed distribution $g(x)$ are given following. No external force is acting and
$c = 1/2$.

Boundary conditions:

$$u(0, t) = e^{-t} \quad \text{and} \quad u(\infty, t) < \infty$$

Initial conditions:

$$f(x) = 0 \quad \text{and} \quad g(x) = 0$$

10.27.   Consider transverse motion over a taut string over the semi-infinite interval
$I = \{x \mid 0 < x < \infty\}$. The left end of the string is constrained to move in accordance
with $u(0, t) = q(t)$ given following. The initial displacement distribution $f(x)$ and
initial speed distribution $g(x)$ are given following. No external force is acting and
$c = 1/2$.

Boundary conditions:

$$u(0, t) = \sin(t)\, e^{-t} \quad \text{and} \quad u(\infty, t) < \infty$$

Initial conditions:

$$f(x) = 0 \quad \text{and} \quad g(x) = 0$$

10.28.   Consider transverse motion over a taut string over the semi-infinite interval $I = \{x \mid 0 < x < \infty\}$. The left end of the string is constrained to move in accordance with $u(0, t) = q(t)$ given following. The initial displacement distribution $f(x)$ and initial speed distribution $g(x)$ are given following. No external force is acting and $c = 1/2$.

Boundary conditions:

$$u(0, t) = t\,e^{-t} \quad \text{and} \quad u(\infty, t) < \infty$$

Initial conditions:

$$f(x) = 0 \quad \text{and} \quad g(x) = 0$$

10.29.   Consider transverse motion over a taut string over the semi-infinite interval $I = \{x \mid 0 < x < \infty\}$. The left end of the string is constrained to move in accordance with $u(0, t) = q(t)$ given following. The initial displacement distribution $f(x)$ and initial speed distribution $g(x)$ are given following. There is no external force acting and $c = 1/2$.

Boundary conditions:

$$u(0, t) = \text{Dirac}(t) \quad \text{and} \quad u(\infty, t) < \infty$$

Initial conditions:

$$f(x) = 0 \quad \text{and} \quad g(x) = 0$$

10.30.   Consider transverse motion over a taut string over the semi-infinite interval $I = \{x \mid 0 < x < \infty\}$. The left end of the string is constrained to move in accordance with $u(0, t) = q(t)$ given following. The initial displacement distribution $f(x)$ and initial speed distribution $g(x)$ are given following. No external force is acting and $c = 1/2$.

Boundary conditions:

$$u(0, t) = -t\cos(t) \quad \text{and} \quad u(\infty, t) < \infty$$

Initial conditions:

$$f(x) = 0 \quad \text{and} \quad g(x) = 0$$

10.31.   Consider longitudinal wave motion in a rigid bar over the semi-infinite interval $I = \{x \mid 0 < x < \infty\}$. The left end of the bar experiences the applied stress force $u_x(0, t) = q(t)$ given following (Young's modulus $E = 1$). The initial displacement distribution $f(x)$ and initial speed distribution $g(x)$ are given following. No external force is acting and $c = 1/2$.

Boundary conditions:

$$u_x(0, t) = -\sin(t) \quad \text{and} \quad u(\infty, t) < \infty$$

Initial conditions:

$$f(x) = 0 \quad \text{and} \quad g(x) = 0$$

10.32. Consider longitudinal wave motion in a rigid bar over the semi-infinite interval $I = \{x \,|\, 0 < x < \infty\}$. The left end of the bar experiences the applied stress force $u_x(0, t) = q(t)$ given following (Young's modulus $E = 1$). The initial displacement distribution $f(x)$ and initial speed distribution $g(x)$ are given following. No external force is acting and $c = 1/2$.

Boundary conditions:

$$u_x(0, t) = -e^{-t} \quad \text{and} \quad u(\infty, t) < \infty$$

Initial conditions:

$$f(x) = 0 \quad \text{and} \quad g(x) = 0$$

10.33. Consider longitudinal wave motion in a rigid bar over the semi-infinite interval $I = \{x \,|\, 0 < x < \infty\}$. The left end of the bar experiences the applied stress force $u_x(0, t) = q(t)$ given following (Young's modulus $E = 1$). The initial displacement distribution $f(x)$ and initial speed distribution $g(x)$ are given following. No external force is acting and $c = 1/2$.

Boundary conditions:

$$u_x(0, t) = -t\,e^{-t} \quad \text{and} \quad u(\infty, t) < \infty$$

Initial conditions:

$$f(x) = 0 \quad \text{and} \quad g(x) = 0$$

10.34. Consider longitudinal wave motion in a rigid bar over the semi-infinite interval $I = \{x \,|\, 0 < x < \infty\}$. The left end of the bar experiences the applied stress force $u_x(0, t) = q(t)$ given following (Young's modulus $E = 1$). The initial displacement distribution $f(x)$ and initial speed distribution $g(x)$ are given following. No external force is acting and $c = 1/2$.

Boundary conditions:

$$u_x(0, t) = -\sin(t)\,e^{-t} \quad \text{and} \quad u(\infty, t) < \infty$$

Initial conditions:

$$f(x) = 0 \quad \text{and} \quad g(x) = 0$$

10.35.   Consider longitudinal wave motion in a rigid bar over the semi-infinite interval $I = \{x \,|\, 0 < x < \infty\}$. The left end of the bar experiences the applied stress force $u_x(0, t) = q(t)$ given following (Young's modulus $E = 1$). The initial displacement distribution $f(x)$ and initial speed distribution $g(x)$ are given following. No external force is acting and $c = 1/2$.

Boundary conditions:

$$u_x(0, t) = -\mathrm{Dirac}(t - 1)\,e^{-t} \quad \text{and} \quad u(\infty, t) < \infty$$

Initial conditions:

$$f(x) = 0 \quad \text{and} \quad g(x) = 0$$

10.36.   Consider transverse wave motion over a taut string over the semi-infinite interval $I = \{x \,|\, 0 < x < \infty\}$. The left end of the string is held secure, and the initial displacement distribution $f(x)$ and initial speed distribution $g(x)$ are given following. There is an external source term $h(x, t)$ (dimension of force per unit mass) acting and $c = 1/2$.

Boundary conditions:

$$u(0, t) = 0 \quad \text{and} \quad u(\infty, t) < \infty$$

Initial conditions:

$$f(x) = 0 \quad \text{and} \quad g(x) = 0$$

External force:

$$h(x, t) = e^{-x} e^{-t}$$

10.37.   Consider transverse wave motion over a taut string over the semi-infinite interval $I = \{x \,|\, 0 < x < \infty\}$. The left end of the string is held secure, and the initial displacement distribution $f(x)$ and initial speed distribution $g(x)$ are given following. There is an external source term $h(x, t)$ (dimension of force per unit mass) acting and $c = 1/2$.

Boundary conditions:

$$u(0, t) = 0 \quad \text{and} \quad u(\infty, t) < \infty$$

Initial conditions:

$$f(x) = 0 \quad \text{and} \quad g(x) = 0$$

External force:

$$h(x, t) = \sin(x)\,e^{-t}$$

10.38. Consider transverse wave motion in a taut string over the semi-infinite interval $I = \{x \mid 0 < x < \infty\}$. The left end of the string is held secure, and the initial displacement distribution $f(x)$ and initial speed distribution $g(x)$ are given following. There is an external source term $h(x, t)$ (dimension of force per unit mass) acting and $c = 1/2$.

Boundary conditions:

$$u(0, t) = 0 \quad \text{and} \quad u(\infty, t) < \infty$$

Initial conditions:

$$f(x) = 0 \quad \text{and} \quad g(x) = 0$$

External force:

$$h(x, t) = x e^{-x} e^{-t}$$

10.39. Consider transverse wave motion in a taut string over the semi-infinite interval $I = \{x \mid 0 < x < \infty\}$. The left end of the string is held secure, and the initial displacement distribution $f(x)$ and initial speed distribution $g(x)$ are given following. There is an external source term $h(x, t)$ (dimension of force per unit mass) acting and $c = 1/2$.

Boundary conditions:

$$u(0, t) = 0 \quad \text{and} \quad u(\infty, t) < \infty$$

Initial conditions:

$$f(x) = 0 \quad \text{and} \quad g(x) = 0$$

External force:

$$h(x, t) = \text{Dirac}(x - 1) e^{-t}$$

10.40. Consider transverse wave motion in a taut string over the semi-infinite interval $I = \{x \mid 0 < x < \infty\}$. The left end of the string is held secure, and the initial displacement distribution $f(x)$ and initial speed distribution $g(x)$ are given following. There is an external source term $h(x, t)$ (dimension of force per unit mass) acting and $c = 1/2$.

Boundary conditions:

$$u(0, t) = 0 \quad \text{and} \quad u(\infty, t) < \infty$$

Initial conditions:

$$f(x) = e^{-x} \quad \text{and} \quad g(x) = 0$$

External force:

$$h(x, t) = e^{-t}$$

10.41. Consider transverse wave motion in a taut string over the semi-infinite interval
$I = \{x \mid 0 < x < \infty\}$. The left end of the string is held secure, and the initial
displacement distribution $f(x)$ and initial speed distribution $g(x)$ are given following.
There is an external source term $h(x, t)$ (dimension of force per unit mass) acting and
$c = 1/2$.

Boundary conditions:

$$u(0, t) = 0 \quad \text{and} \quad u(\infty, t) < \infty$$

Initial conditions:

$$f(x) = 0 \quad \text{and} \quad g(x) = e^{-x}$$

External force:

$$h(x, t) = e^{-t}$$

10.42. Consider transverse wave motion in a taut string over the semi-infinite interval
$I = \{x \mid 0 < x < \infty\}$. The left end of the string is held secure, and the initial
displacement distribution $f(x)$ and initial speed distribution $g(x)$ are given following.
There is an external source term $h(x, t)$ (dimension of force per unit mass) acting and
$c = 1/2$.

Boundary conditions:

$$u(0, t) = 0 \quad \text{and} \quad u(\infty, t) < \infty$$

Initial conditions:

$$f(x) = x e^{-x} \quad \text{and} \quad g(x) = 0$$

External force:

$$h(x, t) = e^{-t}$$

10.43. Consider transverse wave motion in a taut string over the semi-infinite interval
$I = \{x \mid 0 < x < \infty\}$. The left end of the string is held secure, and the initial
displacement distribution $f(x)$ and initial speed distribution $g(x)$ are given following.
There is an external source term $h(x, t)$ (dimension of force per unit mass) acting and
$c = 1/2$.

Boundary conditions:

$$u(0, t) = 0 \quad \text{and} \quad u(\infty, t) < \infty$$

Initial conditions:

$$f(x) = 0 \quad \text{and} \quad g(x) = x e^{-x}$$

External force:

$$h(x, t) = e^{-t}$$

10.44. Consider transverse wave motion in a taut string over the semi-infinite interval $I = \{x \mid 0 < x < \infty\}$. The left end of the string is held secure, and the initial displacement distribution $f(x)$ and initial speed distribution $g(x)$ are given following. There is an external source term $h(x, t)$ (dimension of force per unit mass) acting and $c = 1/2$.

Boundary conditions:

$$u(0, t) = 0 \quad \text{and} \quad u(\infty, t) < \infty$$

Initial conditions:

$$f(x) = e^{-x} \quad \text{and} \quad g(x) = x e^{-x}$$

External force:

$$h(x, t) = e^{-t}$$

10.45. Consider transverse wave motion in a taut string over the semi-infinite interval $I = \{x \mid 0 < x < \infty\}$. The left end of the string is held secure, and the initial displacement distribution $f(x)$ and initial speed distribution $g(x)$ are given following. There is an external source term $h(x, t)$ (dimension of force per unit mass) acting and $c = 1/2$.

Boundary conditions:

$$u(0, t) = 0 \quad \text{and} \quad u(\infty, t) < \infty$$

Initial conditions:

$$f(x) = e^{-x} \quad \text{and} \quad g(x) = x e^{-x}$$

External force:

$$h(x, t) = e^{-x} e^{-t}$$

10.46. Consider transverse wave motion in a taut string over the semi-infinite interval $I = \{x \mid 0 < x < \infty\}$. The left end of the string is held secure, and the initial displacement distribution $f(x)$ and initial speed distribution $g(x)$ are given following. There is an external source term $h(x, t)$ (dimension of force per unit mass) acting and $c = 1/2$.

Boundary conditions:

$$u(0, t) = 0 \quad \text{and} \quad u(\infty, t) < \infty$$

Initial conditions:

$$f(x) = e^{-x} \quad \text{and} \quad g(x) = x e^{-x}$$

External force:

$$h(x, t) = \sin(x) e^{-t}$$

References

Abell, M., Braselton, J., *Differential Equations with Maple V*, Academic Press, 1994.

Abell, M., Braselton, J., *Maple V by Example*, Academic Press, 1994.

Abell, M., Braselton, J., *The Maple V Handbook*, Academic Press, 1994.

Artino, C., Johnson, J., Kolod, J., *Exploring Calculus with Maple*, John Wiley and Sons, 1994.

Bauldry, W., Johnson, J., *Linear Algebra with Maple*, John Wiley and Sons, 1995.

Bauldry, W.C., Fielder, J.R., *Calculus Laboratories with Maple*, Brooks/Cole, 1995.

Burbulla, D.C.M., Dodson, C.T.J., *Self-Tutor for Computer Calculus Using Maple*, Prentice Hall Canada, 1993.

Char, B.W., Geddes, K.O., Gonnet, G.H., Leong, B.L., Monagan, M.B., Watt, S.M., *First Leaves: A Tutorial Introduction to Maple V*, Springer-Verlag, 1992.

Char, B.W., Geddes, K.O., Gonnet, G.H., Leong, B.L., Monagan, M.B., Watt, S.M., *Maple V Language Reference Manual*, Springer-Verlag, 1991.

Char, B.W., Geddes, K.O., Gonnet, G.H., Leong, B.L., Monagan, M.B., Watt, S.M., *Maple V Library Reference Manual,* Springer-Verlag, 1991.

Cheung, C., Harer, J., *A Guide to Multivariable Calculus with Maple V*, John Wiley and Sons, 1994.

Decker, R., *Calculus and Maple V*, Prentice Hall, 1994.

Devitt, J.S., *Calculus with Maple V*, Brooks/Cole, 1993.

Ellis, W., Lodi, E., *Maple for the Calculus Student: A Tutorial*, Brooks/Cole, 1989.

Ellis, W., Lodi, E., Johnson, E., and Schwalbe, D., *The Maple V Flight Manual*, Brooks/Cole, 1992.

Fattahi, A., *Maple V Calculus Labs*, Brooks/Cole, 1995.

Harris, K., Lopez, R., *Discovering Calculus with Maple*, John Wiley and Sons, 1995.

Heck, A., *Introduction to Maple-A Computer Algebra System*, Springer-Verlag, 1993.

Holmes, M.H., Ecker, J.G., Boyce, W.E., Siegmann, W.L., *Exploring Calculus with Maple*, Addison-Wesley, 1993.

Johnson, E., *Linear Algebra with Maple V*, Brooks/Cole, 1993.

Kreyszig, H.E., *Maple Computer Manual for Advanced Engineering Mathematics*, John Wiley and Sons, 1994.

Redfern, D., *The Maple Handbook*, Springer-Verlag, 1995.

Index

A

acceleration, 218
angular frequency
   damped, 222, 246, 417, 440, 472
   multiplicity, 472–473
   natural, 522, 554–555
   undamped, 443, 473, 516
areal density, 274, 472

B

basis vector, 5, 6, 9, 13–14, 32
Bessel function
   first kind, 60
   second kind, 62
boundary condition
   Dirichlet, 78
     mixed, 129
     Neumann, 78
     nonmixed, 73, 74, 77
   periodic, 129
   regular, 74, 77
   Robin, 78

C

characteristic equation, 23
concavity, 162, 163, 218

convergence
   of derivative, 187
   pointwise, 82, 83
   uniform, 82, 83, 187
convolution integral, 568–570
coordinate radius, 190, 244, 365, 437

D

damping coefficient, 217, 414, 460
decay time, 406
   mode, 406
differential equation, 4
   Bessel, 56–63, 65
   Cauchy-Euler, 29–32, 64, 115, 119, 124
   Euler, 9, 24–28, 51–55, 64
   first-order, 19, 166, 371
   first-order homogeneous, 14
   first-order linear nonhomogeneous, 14, 37, 63
   first-order nonhomogeneous, 63
   homogeneous, 13
   linear operator, 10
   normal linear second-order nonhomogeneous, 13
   ordinary, 8–10
   ordinary linear homogeneous, 8